Objective BIOLOGY

Chapter-wise MCQs

for NTA NEET/ AIIMS

• **Corporate Office :** 45, 2nd Floor, Maharishi Dayanand Marg, Corner Market,
Malviya Nagar, New Delhi-110017
Tel. : 011-49842349 / 49842350

By
Sanju Priyadarshi
Raza Akhtar

Typeset by Disha DTP Team
Printed at : Repro Knowledgecast Limited, Thane

DISHA PUBLICATION

For further information about the books from DISHA,
Log on to **www.dishapublication.com** or email to **info@dishapublication.com**

CONTENTS

1 The Living World

1. The living organisms can be unexceptionally distinguished from the non-living things on the basis of their ability for: **Tricky**
   (a) interaction with the environment and progressive evolution
   (b) reproduction
   (c) growth and movement
   (d) responsiveness to touch
2. Which type of organisation is found in only living beings?
   (a) Atomic (b) Molecular
   (c) Mixture (d) Sub-cellular
3. Organisation levels in living beings are:
   (a) Subcellular → cellular → individual → community → population
   (b) Atomic → molecular → subcellular → cellular → tissue → organ → individual
   (c) Individual → population → organ system → tissue → cellular → molecular → atomic
   (d) Atomic → molecular → tissue → individual → ecosystem → community.
4. ICVCN is an abbrevited form of:
   (a) International Code of Virus Nomenclature
   (b) International Code for Virulence Nomenclature
   (c) International Code of Viral Classification and Nomenclature
   (d) International Code of Valuation in Nomenclature
5. Taxonomic hierarchy refers to:
   (a) step-wise arrangement of all categories for classification of plants and animals.
   (b) a group of senior taxonomists who decide the nomenclature of plants and animals.
   (c) a list of botanists or zoologists who have worked on taxonomy of a species or group.
   (d) classification of a species, based on fossil record.
6. Growth, development and functioning of living body is due to:
   (a) Order (b) Homeostasis
   (c) Metabolism (d) Adaptation
7. System of classification used by Linnaeus was:
   (a) Natural system
   (b) Artificial system
   (c) Phylogenetic system
   (d) Asexual system
8. The term "New systematics" was introduced by:
   (a) Bentham and Hooker
   (b) Linnaeus
   (c) Julian Huxley
   (d) A.P. de Condolle
9. *Rattus rattus* scientific name is an example of:
   (a) Autonyms (b) Tautonyms
   (c) Synonyms (d) Homonyms
10. Given below is the botanical name of mango. Mark the option in which the name is correctly written.
    (a) Magnifera Indica (b) Mangifera indica
    (c) *Mangifera Indica* (d) *Mangifera indica*
11. Phylogeny refers to:
    (a) Natural classification
    (b) Evolutionary classification
    (c) Evolutionary history
    (d) Origin of algae
12. The basic unit of classification is:
    (a) Genus (b) Species
    (c) Order (d) All of these
13. Group of organisms that closely resemble each other and freely interbreed in nature constitute a:
    (a) Species (b) Genus
    (c) Family (d) Taxon
14. If an organism is in the same class but not in the same family but it may belong to same:

    (a) Genus (b) Species
    (c) Variety (d) Order
15. The category which includes related families is:
    (a) Class (b) Phylum
    (c) Order (d) Kingdom

16. Match column-I with column-II and select the correct answer using the codes given below.

Column-I	Column-II
A. Growth	I. Production of offspring
B. Reproduction	II. Composed of one or more cells
C. Metabolism	III. Increase in mass and increase in number of individuals
D. Cellular organisation	IV. Sum total of all chemical reactions occuring in body

(a) A – I; B – II; C – III; D – IV
(b) A – III; B – I; C – II; D – IV
(c) A – III; B – I; C – IV; D – II
(d) A – II; B – IV; C – III; D – I

17. Artificial system of classification classifies plants on the basis of:
(a) One or two characters
(b) Phylogenetic trends
(c) Many naturally existing characters
(d) None of the above

18. The most important feature of all living systems is to:
(a) utilise oxygen to generate energy.
(b) replicate the genetic information.
(c) produce gametes.
(d) utilise solar energy for metabolic activities.

19. NBRI is situated at:
(a) Calcutta (b) Bombay
(c) Madras (d) Lucknow

20. Mark the odd one in the following.
(a) Family (b) Class
(c) Taxon (d) Phylum

21. Who was the first to describe, that species is the unit of classification?
(a) John Ray (b) Huxley
(c) Linnaeus (d) Candolle

22. An important criterion for modern day classification is:
(a) Resemblances in morphology
(b) Anatomical and physiological traits
(c) Breeding habits
(d) Presence or absence of notochord

23. Which of the following taxa are not used by botanists?
(a) Kingdom & order
(b) Phylum & family
(c) Suborder & variety
(d) Phylum & sub species

24. Match column-I with column-II and select the correct answer using the codes given below

Column-I	Column-II
A. Herbarium	I. Includes those specimens which can be easily classified on their observable characters.
B. Botanical garden	II. Preserved specimens on sheets become a store house for future use.
C. Museum	III. Generally set up in educational institutes.
D. Zoological Park	IV. Includes those animals which are identified based on their aggregates of characters.
E. Key	V. All animals provided similar conditions to their natural habitat.
	VI. Includes animals of related orders.
	VII. Includes collection of living plants for reference.
	VIII. It identifies animals and plants on the basis of their similarities and dissimilarities.

(a) A – I; B – II; C – VIII; D – V; E – III
(b) A – III; B – I; C – IV; D – II; E – V
(c) A – II; B – VII; C – III; D – V; E – VIII
(d) A – II; B – III; C – VII; D – I; E – IV

25. The scientific naming of plants began with publication of Linnaeus book '________'.
(a) *Genera plantarum* (b) *Systema naturae*
(c) *Species plantarum* (d) *Charaka samhita*

26. Couplet represents the choice made between two opposite options which results in:

(a) rejection of both the option.
(b) acceptance of both the option.
(c) either acceptance or rejection of both the option.
(d) acceptance of only one and rejection of the other.

27. The term "New Systematics" was introduced by:
(a) Bentham and Hooker
(b) Linnaeus
(c) Julian Huxley
(d) A.P. de Candolle

28. Which of the following statements are true?

(i) Genus comprises a group of related species.
(ii) Taxon does not represents a taxonomic group of individual organisms.
(iii) Family comprises a group of related genera.
(iv) Taxonomic category class includes related orders.
(a) (i), (ii), and (iv)
(b) (ii) and (iv)
(c) (i), (iii) and (iv)
(d) (ii), (iii) and (iv)

29. A duplicate of nomenclature type is termed by a taxonomist as:
(a) Syntype (b) Neotype
(c) Paratype (d) Isotype

30. Select the true statement for reproduction.

(i) It is not an all-exclusive defining characteristic of living organisms.
(ii) It is not an all-inclusive defining characteristic of living organisms.
(iii) It is an all-inclusive defining characteristic of plants and fungi only.
(iv) Photoperiod affects reproduction in seasonal plant breeders only.
(v) Photoperiod affects reproduction in seasonal breeders both plants and animals.
(vi) Photoperiod has no role in reproduction.
(a) (ii) and (v) (b) (iii) and (iv)
(c) (iv) and (vi) (d) All of these

31. Which of the following pair of taxa are not present in Linnaean hierarchy?
(a) Genus and variety (b) Family and Phylum
(c) Class and species (d) Species and variety

32. The smallest unit of classification is:
(a) Family (b) Order
(c) Genus (d) Species

33. Morphologically similar but reproductively isolated species are called:
(a) Neontological species
(b) Sibling species
(c) Allopatric species
(d) Morpho-species

34. Which of the following taxonomic category of housefly is incorrectly matched?
(a) Genus – *Musca*
(b) Family – Muscidae
(c) Order – Primata
(d) Class – Insecta

35. Which of the following is less general in characters as compared to genus?
(a) Species (b) Division
(c) Class (d) Family

36. Phenetic classification is based on:

(a) sexual characteristics.
(b) the ancestral lineage of existing organisms.
(c) observable characteristics of existing organisms.
(d) dendograms based on DNA characteristics.

37. "X" being a higher category is the assemblage of families which exhibit a few "Y" characters. The "Z" characters are less in a number as compared to different genera included in a family. Identify "X", "Y", and "Z".
(a) X - Order; Y - Similar; Z - Similar
(b) X - Genus; Y - Similar; Z - Different
(c) X - Species; Y - Different; Z - Similar
(d) X - Class; Y - Different; Z - Different

38. Artificial system of classification was first used by:
(a) Linnaeus
(b) De Candolle
(c) Pliny the Edler
(d) Bentham and Hooker

39. Which one of the taxonomic aids can give comprehensive account of complete compiled information of any one genus or family at a particular time?
(a) Taxonomic key (b) Flora
(c) Herbarium (d) Monograph

40. Classification based on sequencing of DNA and chemical nature of protein is:
(a) Chemotaxonomy
(b) Cytotaxonomy
(c) Adansonian taxonomy
(d) Karyotaxonomy

41. Which of the following is the only taxonomic categories that has a real existence?
(a) Phylum (b) Species
(c) Genus (d) Kingdom

42. Who is the father of biology?
(a) Hippocrates
(b) Theophrastus
(c) Aristotle
(d) Darwin

43. The usage of binomial names, for plant species was accepted by all after the publication of the work by:
(a) Hooker (b) Linnaeus
(c) Bentham (d) Darwin

44. Which of the following subdivisions deals with fossils?
(a) Ecology
(b) Ornithology
(c) Ethology
(d) Palaeontology

45. Biosystematics aims at:
(a) identification and arrangement of organisms on the basis of their cytological characteristics.
(b) the classification of organisms based on broad morphological characters.
(c) delimiting various taxa of organisms and establishing their relationships.
(d) the classification of organisms based on their evolutionary history and establishing their phylogeny on the totality of various parameters from all fields of studies.

46. Need for a proper system of classification arises because
(a) the organisms of the past cannot be studied without it.
(b) classification help in knowing the relationship among the different group of organisms.
(c) it is not possible to study every organism.
(d) All of the above

47. The disadvantage of using common names for species is that:
(a) the names may change.
(b) one name does not apply universally.
(c) one species may have several common name and one common may be applied to two species.
(d) All of the above

48. Which of the following shows the correct example of taxonomic category - Genus?
(a) Potato, tomato and brinjal belong to *Solanum*.
(b) Monkey, gorilla and gibbon are placed in Mammalia.
(c) *Solanum*, *Petunia*, and Datura are placed in Solanacea.
(d) *Mangifera indica*, *Solanum tuberosum*, and *Panthera leo*.

49. On how many criteria living organisms have been classified into five kingdoms?
(a) Two (b) Four (c) Five (d) Three

50. Who developed the 'key' for identification of animals?
(a) Goethe (b) John Ray
(c) Theophrastus (d) George curier

51. Binomial nomenclature means:
(a) One name given by two scientists.
(b) One scientific name consisting of a generic and specific epithet.
(c) Two names, one latinised, other of a person.
(d) Two names of same plant.

52. An animal with same generic, specific and subspecific names is:
(a) Man (b) Gorilla
(c) Rabbit (d) Elephant

Answer KEYs

1	(a)	7	(b)	13	(a)	19	(d)	25	(c)	31	(b)	37	(a)	43	(b)	49	(c)
2	(d)	8	(c)	14	(d)	20	(c)	26	(d)	32	(d)	38	(c)	44	(d)	50	(b)
3	(b)	9	(b)	15	(c)	21	(a)	27	(c)	33	(b)	39	(d)	45	(d)	51	(b)
4	(c)	10	(d)	16	(c)	22	(b)	28	(c)	34	(c)	40	(a)	46	(d)	52	(b)
5	(a)	11	(c)	17	(a)	23	(d)	29	(d)	35	(a)	41	(b)	47	(d)		
6	(c)	12	(b)	18	(b)	24	(c)	30	(a)	36	(c)	42	(c)	48	(a)		

Hints & Solutions

1. (a) 2. (d) 3. (b)
4. (c) ICVCN is the abbreviation of International Code of Viral Classification and Nomenclature. ICNCP stands for International Code of Nomenclature for Cultivated plants.
5. (a) 6. (c)
7. (b) Linnaeus put forward an "Artificial system" of plant classification which was based on sexual characters. It is commonly called as sexual system of plant classification.
8. (c) 9. (b) 10. (d)
11. (c) Phylogeny refers to evolutionary history.
12. (b) The basic unit of classifications is species.
13. (a)
14. (d) Order is a higher taxon and is the assemblage of families having similar characteristics. However, the common characteristic will be fewer than at family or genus level. In mammals, the common orders are primates (monkey, gorilla and human), carnivora, rodentia and cetacea (whale and dolphin).
15. (c) Many related families can be grouped under an order.
16. (c) Growth may be defined as a positive change in size, often over a period of time. Reproduction is the process by which living organisms produce young ones of their own type. Metabolism is defined as the sum total of all the chemical reactions occurring in our body. Cellular organisation refers to the components of a cell and how these individual parts are arranged within the cell.
17. (a)
18. (b) Replication of the genetic information causes transfer of genetic information from one generation to the next generation. It is very important feature of all living organisms.
19. (d)
20. (c) Taxon can be used for any taxonomic rank.
21. (a)
22. (b) Modern day classification is new systematics or biosystematics which includes all the characteristics of organisms gathered from the study of different sections like physiology, ecology, anatomy, biochemistry, cytology.
23. (d) In place of phylum and sub- species, the botanists use division and variety respectively.
24. (c) 25. (c)
26. (d) Couplet represents the choice made between two opposite options which results in acceptance of only one and rejection of the other.
27. (c) The term "New Systematics" was given by Julian Huxley (1940). Characters of plants collected through different branches of science are considered *e.g.*, ecology, physiology, biochemistry, cytology, genetics, etc.
28. (c)
29. (d) Holotype- A specimen on which the original description of the species is based.
Syntype- Any of two or more specimens cited by author when there is no holotype.
Neotype- A specimen is designated when the original types are known to have been destroyed.
Paratype- A specimen other than the holotype available with the author.
Isotype- a duplicate of the holo type.
30. (a)
31. (b) Linnaeus used only 5- taxa in the hierarchy, i.e. Class, Order, Genus, Species and Variety.
32. (d) The smallest unit of classification is species.
33. (b)
34. (c) Housefly or *Musca domestica*, is a fly of the order Diptera. It is the most common of all domestic flies. Primate is a mammal of the order primata.
35. (a) Species is the lowest taxonomic category. Class is a category made of one or more related orders possessing similar correlated characters. Family is composed of one to many related genera. Division comprises of several related classes.

36. **(c)** Phenetic classification is based upon observable characteristics of an organism.

37. **(a)** Order being a higher category is the assemblage of families which exhibit a few similar characters. The similar characters are less in number as compared to different genera included in a family.

38. **(c)** Pliny the Edler introduced first artificial system of classification in his book *Historia Naturalis*.

39. **(d)** **40.** **(a)**

41. **(b)** Species is the lowest or basic taxonomic category which consists of one or more natural populations of individuals that resemble one another more closely than individuals of other species, interbreed freely, have a distinct genetic set up and are reproductively isolated from others *e.g.*, *Panthera leo* (Lion), *Mangifera indica* (Mango), *Solanum tuberosum* (Potato).

42. **(c)**

43. **(b)** Carolus Von Linnaeus in his book *Genera Plantarum* made use of the artificial system of classification. He distinguished between the natural and artificial systems and used the binomial nomenclature system to classify organisms into genus and species.

44. **(d)**

45. **(d)** Biosystematics is the study of diversity of organism and all their comparative and evolutionary relationships.

46. **(d)** Classification helps in understanding diverse varieties of organisms and also gives an idea about the origin and evolution of organisms which are morphologically similar. Classification is needed:

- For convenient study of living organisms.
- For specific identification of various organisms.
- To integrate the idea of life by studying new representative from each group.
- To study about evolutionary relationship by establishing the gradually increasing complexities in group of plants and animals.

47. **(d)**

48. **(a)** Potato, tomato and brinjal are three different species but all belong to the genus *Solanum*. Genus is an assembly of related species which evolved from a common ancestor and have certain common characters. *Solanum* is a polytypic genera with more than one species.

49. **(c)** Five criteria are : (i) complexity of cell, (ii) complexity of organism, (iii) mode of nutrition, (iv) reproduction and (v) phylogenetic relationship.

50. **(b)**

51. **(b)** Binomial nomenclature means the scientific name of any organism which consist of a generic epithet and a specific epithet.

52. **(b)** Trinomial nomenclature of gorilla is *Gorilla gorilla gorilla.*

2 Biological Classification

1. Gram-negative cells stain pink because:
   (a) they have specialised lipids in their cell walls.
   (b) their peptidoglycan layer is thin.
   (c) their peptidoglycan layer is thick.
   (d) they are receptive to antibiotics.
2. The given figure represent a type of virus with two parts labelled as 1 and 2. Identify the virus and the labelled parts.

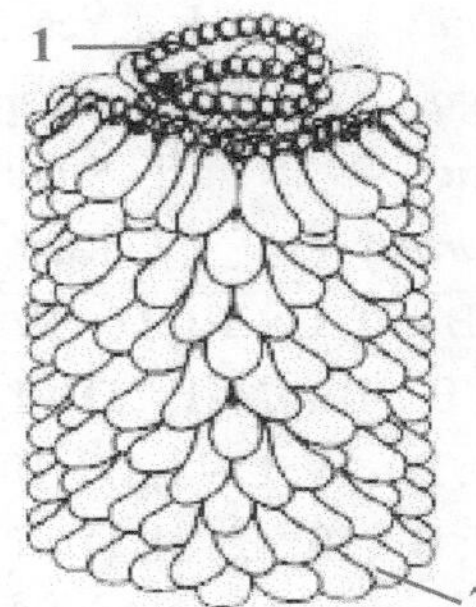

	1	2	Virus
(a)	RNA	Capsid	Tobacco Mosaic Virus
(b)	DNA	Capsid	Tobacco Mosaic Virus
(c)	RNA	Lipid	Tobacco Mosaic Virus
(d)	RNA	Protein	HIV

3. Which of the following regarding protists, in general, is false?
   (a) Protists are always parasitic.
   (b) Protists are single-celled.
   (c) Protists are heterotrophic.
   (d) All of the above
4. An organism having cytoplasm, DNA and RNA but no cell wall is:
   (a) Cyanobacterium (b) Mycoplasma
   (c) Bacterium (d) Virus
5. Most of the antibiotics have no effect on viruses because:
   (a) they kill all bacteria which form host of virus.
   (b) viruses are too small in size for antibiotics to act on them.
   (c) viruses have no metabolism of their own.
   (d) viruses produce a thick covering and endospores.
6. During the evolution of eukaryotes from prokaryotes, which of the following did not occur?
   (a) Infolding of the flexible cell membrane.
   (b) Loss of the cell wall.
   (c) A switch from aerobic to anaerobic metabolism.
   (d) Endosymbiosis of once free-living prokaryotes.
7. Match the following columns.

	Column-I		Column-II
A.	Rod-shaped	I.	*Coccus*
B.	Spherical	II.	*Bacillus*
C.	Spiral-shaped	III.	*Vibrio*
D.	Comma-shaped	IV.	*Spirillum*

   (a) A – III; B – II; C – I; D – IV
   (b) A – IV; B – III; C – II; D – I
   (c) A – II; B – I; C – IV; D – III
   (d) A – I; B – IV; C – III; D – II
8. A tooth scraping yields large numbers of corkscrew shaped bacteria. These bacteria are referred to as:
   (a) Bacilli (b) Cocci
   (c) Spirilli (d) Helici
9. The bacterium *Bacillus thuringiensis* can withstand heating, dryness, and toxic chemicals that would kill most other bacteria. This indicates that it is probably able to form:
   (a) Pseudopodia (b) Endotoxins
   (c) Endospores (d) Pili
10. Which of the following facts differentiates the plant virus from other viruses?
   (a) DNA is the genetic material
   (b) Obligate parasites
   (c) RNA is the genetic material
   (d) Nucleo-protein nature

11. Which of the following does not contain chlorophyll?
(a) Fungi (b) Algae
(c) Bryophyta (d) Pteridophyta

12. Pseudomycelium is a characteristic feature of:
(a) Mushroom (b) *Mucor*
(c) Bread mould (d) Yeast

13. Match the type of protozoans given in column-I with their examples given in column-II and choose the correct option.

Column-I (Type of Protozoans)	Column-II (Examples)
A. Amoeboid protozoans	I. *Paramoecium*
B. Ciliated protozoans	II. *Plasmodium*
C. Flagellated protozoans	III. *Amoeba*
D. Sporozoans	IV. *Trypanosoma*

(a) A – I; B – III; C – IV; D – II
(b) A – III; B – I; C – II; D – IV
(c) A – III; B – I; C – IV; D – II
(d) A – III; B – IV; C – I; D – II

14. Read the following statements and answer the question given below.
(i) They are saprophytic protists.
(ii) Under suitable conditions, they form an aggregation (called plasmodium) which may grow and spread over several feet.
(iii) During unfavourable conditions, the plasmodium differentiates and forms fruiting bodies bearing spores at their tips.
Which of the following class of protists is being described by the above statements?
(a) Euglenoids
(b) Dinoflagellates
(c) Slime moulds
(d) Protozoans

15. Consider the below statements and identify the correct statements regarding classification.
(i) Biological classification is the scientific ordering of organisms in a hierarchical series of groups on the basis of their relationships, i.e. morphological, evolutionary and others.
(ii) Whittaker classified organisms on the basis of autotrophic and heterotrophic mode of nutrition.
(iii) In five kingdom system of classification, living organisms can be divided into prokaryotic and eukaryotic cells on the basis of cell structure.
(a) (i), (ii) and (iii) (b) (i) and (iii)
(c) (ii) and (iii) (d) (i) and (ii)

16. Match column-I with column-II and select the correct answer using the codes given below.

Column-I (Fungus name)	Column-II (Commonly called)
A. *Puccinia*	I. Yeast
B. *Ustilago*	II. Mushroom
C. *Agaricus*	III. Smut fungus
D. *Saccharomyces*	IV. Rust fungus

(a) A – I; B – II; C – III; D – IV
(b) A – II; B – III; C – IV; D – I
(c) A – III; B – IV; C – I; D – II
(d) A – IV; B – III; C – II; D – I

17. Antibiotics are mostly obtained from:
(a) Bacteria (b) Viruses
(c) Angiosperms (d) Fungi

18. Identify the following figures.

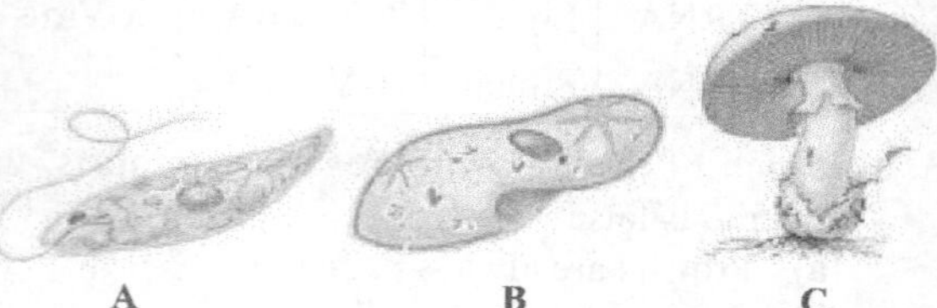

(a) A – *Euglena*, B – *Paramoecium*, C – *Agaricus*
(b) A – *Euglena*, B – *Planaria*, C – *Agaricus*
(c) A – *Planaria*, B – *Paramoecium*, C – *Agaricus*
(d) A – *Euglena*, B – *Paramoecium*, C – *Aspergillus*

19. Bacteria were regarded to be plants because:
(a) some of them are green.
(b) they are present everywhere.
(c) some of them cannot move.
(d) they have a rigid cell wall.

20. Locomotory structures are absent in:
   (a) Sporozoans
   (b) Ciliates
   (c) Zooflagellates
   (d) Rhizopods

21. Which of the following are the characters of Dinoflagellates?
   (i) Planktonic golden yellow algae with soap box like structure.
   (ii) Marine red biflagellated protista.
   (iii) Appear yellow, green, brown, blue and red in colour.
   (iv) Biflagellated organisms with pellicle.
   (v) Saprophytic or parasitic unicellular forms.
   (a) (i), (ii) and (iii) only
   (b) (ii), (iv) and (v) only
   (c) (ii) and (iii) only
   (d) (ii) and (v) only

22. How many kingdoms contain eukaryotes in five kingdom system of classification of R.H Whittaker?
   (a) Four kingdoms (b) One kingdom
   (c) Two kingdoms (d) Three kingdoms

23. Match column-I with column-II and select the correct answer using the codes given below.

Column-I (Kingdom)	Column-II (Class)
A. Plantae	I. Archaebacteria
B. Fungi	II. Euglenoids
C. Protista	III. Phycomycetes
D. Monera	IV. Algae

   (a) A – IV; B – III; C – II; D – I
   (b) A – I; B – II; C – III; D – IV
   (c) A – III; B – IV; C – II; D – I
   (d) A – IV; B – II; C – III; D – I

24. Litmus is obtained from: Tricky
   (a) Lichen (b) Algae
   (c) Fungi (d) Protozoa

25. A bacteriophage is:
   (a) a virus attacking a bacterium.
   (b) a bacterium attacking a virus.
   (c) a stage in the life-cycle of bacterium.
   (d) a virus attacking another virus.

26. Black water fever is caused by:
   (a) *Plasmodium malariae*
   (b) *P. ovule*
   (c) *P. falciparum*
   (d) *P. vivax*

27. Identify the correct statements regarding archaebacteria.
   (i) Archaebacteria differs from other bacteria in having different cell wall structure.
   (ii) Their cell wall is made up of cellulose and contains high amount of unsaturated fatty acid, which is responsible for their survival in extreme conditions.
   (iii) Thermoacidophiles have dual ability to tolerate high temperature as well as high acidity.
   (a) (i) and (ii)
   (b) (ii) and (iii)
   (c) (i) and (iii)
   (d) (i), (ii) and (iii)

28. Read the given statements and answer the question.
   (i) It includes unicellular as well as multicellular fungi.
   (ii) In multicellular forms, hyphae are branched and septate.
   (iii) Conidiophore produces conidia (spores) exogenously in chain.
   (iv) Sexual spores are ascospores produced endogenously in chain.
   (v) Fruiting body is called ascocarp.
   Identify the correct class of fungi which have all the above given characteristics.
   (a) Phycomycetes
   (b) Sac fungi
   (c) Club fungi
   (d) Fungi imperfecti

29. Contractile vacuoles of *Paramoecium* are analogous to: Tricky
   (a) Sweat glands of mammals
   (b) Uriniferous tubules
   (c) Gastrovascular cavity of *Hydra*
   (d) Typhlosole of earthworm

30. Members of Phycomycetes are found:
(i) in aquatic habitats
(ii) on decaying wood
(iii) in moist and damp places
(iv) as obligate parasite on plants
Which of the statements given above are correct?
(a) (i) and (ii)
(b) (ii), (iii) and (iv)
(c) (i), (ii) and (iii)
(d) (i), (ii), (iii) and (iv)

31. The symbiotic association of fungi and algae is called:
(a) Lichen (b) Mycorrhiza
(c) Rhizome (d) Endomycorrhiza

32. Which of the following combinations of characters is true for slime moulds?
(a) Parasitic, plasmodium with true walls, spores dispersed by air currents.
(b) Saprophytic, plasmodium without walls, spores dispersed by water.
(c) Parasitic, plasmodium without walls, spores dispersed by water.
(d) Saprophytic, plasmodium without walls, spores dispersed by air currents.

33. Nutritionally bacteria are:
(a) Autotrophic
(b) Symbiotic
(c) Parasitic
(d) All of the above

34. Which bacteria would function best in hot temperatures (45 – 60°C)?
(a) Psychrophiles
(b) Thermophiles
(c) Mesophiles
(d) All would do equally well

35. If the generation time of a bacterium is 40 minutes and a culture containing 10^7 cells/ml is grown for 4 hours, then calculate its population after that period.

Critical Thinking

(a) 64×10^7 (b) 32×10^7
(c) 6×10^7 (d) 40×10^7

36. Smallest virus is:
(a) Foot and Mouth virus
(b) Alfalfa Mosaic virus
(c) *Pseudomonas*
(d) TMV

37. Yeast is an important source of :
(a) Vitamin C (b) Vitamin B
(c) Vitamin A (d) Vitamin D

38. In manufacturing of bread, it becomes porous due to release of CO_2 by the action of :
(a) Virus (b) Yeast
(c) Bacteria (d) Protozoans

39. Which of the following statements is correct for both blue-green algae and bacteria?
(a) Both show anaerobic respiration.
(b) Both have chlorophyll pigment.
(c) Both are devoid of true nucleus.
(d) None of the above

40. Match column-I with column-II and select the correct answer using the codes given below.

Column-I (Group protista)	Column-II (Example)
A. Chrysophytes	I. *Paramoecium*
B. Dinoflagellates	II. *Euglena*
C. Euglenoids	III. *Gonyaulax*
D. Protozoans	IV. Diatoms

(a) A – I; B – III; C – II; D – IV
(b) A – II; B – IV; C – III; D – I
(c) A – IV; B – II; C – III; D – I
(d) A – IV; B – III; C – II; D – I

41. The phylogenetic system of classification was put forth by:
(a) Carolus Linnaeus
(b) George Bentham and Joseph Dalton Hooker
(c) Theophrastus
(d) Adolf Engler and Karl Prantl

42. *Plasmodium*, the parasite, belongs to class:
(a) Sarcodina (b) Ciliata
(c) Sporozoa (d) Dinophyceae

43. In the five kingdom system of classification, which single kingdom out of the following can include blue-green algae, nitrogen fixing bacteria and methanogenic archaebacteria?
(a) Monera (b) Fungi
(c) Plantae (d) Protista

44. Read the following statements and answer the question.

(i) Some members are saprophytes or parasites while a large number of them are decomposers of litter and help in mineral cycling.

(ii) They reproduce only by asexual spores known as conidia.

(iii) Mycelium is septate and branched.

(iv) *Alternaria, Colletotrichum* and *Trichoderma* are examples of this class.

Which of the following class of fungi is being described by the above statements?

(a) Phycomycetes (b) Deuteromycetes
(c) Basidiomycetes (d) Ascomycetes

45. Match column-I with column-II and select the correct answer using the codes given below.

Column-I (Types of bacteria)	Column-II (Activity)
A. *Streptomyces*	I. Food poisoning
B. *Rhizobium*	II. Source of antibiotics
C. *Nitrosomonas*	III. Nitrogen fixation
D. *Acetobacter*	IV. Nitrification
	V. Vinegar synthesis

(a) A – II; B – III; C – I; D – V
(b) A – II; B – III; C – IV; D – V
(c) A – IV; B – V; C – I; D – III
(d) A – V; B – I; C – III; D – IV

46. Which one of the following pairs is not correctly matched?

(a) *Anabaena* – Cyanobacteria
(b) *Amoeba* – Protozoa
(c) *Gonyaulax* – Dinoflagellates
(d) *Albugo* – Chrysophytes

47. The cell wall of fungi is made up of :

(a) Chitin
(b) Cellulose
(c) Pectin
(d) Suberin

48. The main difference between Gram positive and Gram negative bacteria lies in the composition of :

(a) Cilia
(b) Cell wall
(c) Nucleolus
(d) Cytoplasm

49. LPP-1 is a:

(a) Bacterium
(b) Fungus
(c) Bacteriophage
(d) Cyanophage

50. The conditions which would be favoured by thermoacidophiles are:

(a) Hot and alkaline
(b) Snow and acidic
(c) Hot sulphur spring
(d) Gut of cows

51. Natural system of classification differs from artificial system in: **Tricky**

(a) employing only one floral trait
(b) taking only one vegetative trait
(c) bringing out similarities and dissimilarities
(d) developing evolutionary trends

52. The genetic material in viruses is/are:

(a) Only RNA
(b) Only DNA
(c) RNA and DNA both
(d) RNA or DNA *i.e.* one nucleic acid in a virus

53. Which of the following fungi are edible?

(a) *Agaricus campestris*
(b) *Morchella esculenta*
(c) *Podaxon prodaxis*
(d) All of the above

54. Match the terms given in column-I with their examples given in column-II and choose the correct option.

	Column-I (Terms)	Column-II (Examples)
A.	Ascus	I. *Spirulina*
B.	Basidium	II. *Penicillium*
C.	Protista	III. *Agaricus*
D.	Cyanobacteria	IV. *Euglena*
E.	Animalia	V. Sponges

(a) A – II; B – III; C – IV; D – V; E – I
(b) A – I; B – II; C – III; D – V; E – IV
(c) A – II; B – V; C – III; D – I; E – IV
(d) A – II; B – III; C – IV; D – I; E – V

55. Which protist reproduces both by binary fission and conjugation?

(a) *Amoeba* (b) *Paramoecium*
(c) *Euglena* (d) *Monocystis*

56. Which of the following statements is not correct for methanogens?

(a) They are archaebacteria.
(b) They live in marshy areas.
(c) Methane is their preferred carbon source.
(d) They are present in guts of several ruminant animals (cow, buffaloes) and produce biogas (CH_4) from the dung of these animals.

57. The given figure shows some structures labelled as 1, 2, 3 and 4. Which structure has the protein coat that encloses the nucleic acid?

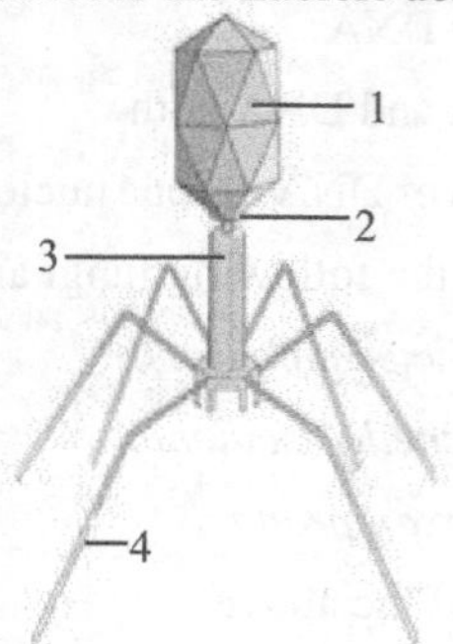

(a) 1 (b) 2 (c) 3 (d) 4

58. Enveloped virus infects/enters the host cell through:

(a) Direct penetration
(b) Injecting nucleic acid
(c) Endocytosis
(d) Endocytosis through chemical stimulators

59. Infective stage of *Entamoeba histolytica* is:

(a) Trophozoite
(b) Pre cyst
(c) Uninucleate cyst
(d) Tetranucleate cyst

60. Cyanobacteria are:

(a) mosses which attack bacteria
(b) bacteria which attack cyanophyceae
(c) autotrophic organisms with phycocyanin
(d) None of the above

61. Whittaker placed prokaryotes and akaryotes in:

(a) Protista (b) Protozoa
(c) Plantae (d) Monera

62. Viruses that infect bacteria, multiply and cause their lysis, are called:

(a) Lysozymes
(b) Lipolytic
(c) Lytic
(d) Lysogenic

63. Archaea differ from eubacteria in which one of the following features?

(a) They have a rigid wall.
(b) Their cell wall lacks peptidoglycan component.
(c) They have 16s rRNA.
(d) They are very ancient.

64. *Leishmania tropica* produces:

(a) Sleeping sickness
(b) Malaria
(c) Dysentery
(d) Oriental sores

65. Common bread mold is:
(a) Yeast (b) *Rhizopus*
(c) Bacteria (d) Virus

66. Which of the following fixes atmospheric N_2?
(a) *Nostoc* (b) Algae
(c) Methanogens (d) None of these

67. Which of the following is not a character of protista?
(a) Protists are prokaryotic.
(b) Some protists have cell walls.
(c) Mode of nutrition is both autotrophic and heterotrophic.
(d) Body organisation is cellular.

68. Which of the following statements is not correct for viruses?
(a) Viruses are obligate parasites.
(b) Viruses can multiply only when they are inside the living cells.
(c) Viruses cannot pass through bacterial filters.
(d) Viruses are made up of protein and DNA or RNA (never both DNA and RNA).

69. Plasmogamy is fusion of :
(a) Two haploid cells including their nuclei
(b) Two haploid cells without nuclear fusion
(c) Sperm and egg
(d) Sperm and two polar nuclei

70. When fungi feed on dead organic matter, they are known as:
(a) Dimorphic
(b) Parasites
(c) Saprophytes
(d) None of these

71. Ergot is caused by:
(a) *Claviceps*
(b) *Penicillium*
(c) *Aspergillus*
(d) *Rhizobium*

72. Koch's postulates are related to:
(a) Cell structure of bacteria
(b) Cell structure of micro-organisms
(c) Laws of diseases
(d) Origin of micro-organisms

Answer KEYs

1	(b)	9	(c)	17	(a)	25	(a)	33	(d)	41	(d)	49	(d)	57	(a)	65	(b)
2	(a)	10	(c)	18	(a)	26	(c)	34	(b)	42	(c)	50	(c)	58	(b)	66	(a)
3	(d)	11	(a)	19	(d)	27	(c)	35	(a)	43	(a)	51	(c)	59	(d)	67	(a)
4	(b)	12	(d)	20	(a)	28	(b)	36	(a)	44	(b)	52	(d)	60	(c)	68	(c)
5	(c)	13	(c)	21	(c)	29	(b)	37	(b)	45	(b)	53	(d)	61	(d)	69	(b)
6	(c)	14	(c)	22	(a)	30	(d)	38	(b)	46	(d)	54	(d)	62	(c)	70	(c)
7	(c)	15	(a)	23	(a)	31	(a)	39	(c)	47	(a)	55	(b)	63	(b)	71	(a)
8	(c)	16	(d)	24	(a)	32	(d)	40	(d)	48	(b)	56	(c)	64	(d)	72	(c)

Hints & Solutions

1. **(b)** Their peptidoglycan layer is thin and does not retain crystal violet stain like Gram-positive bacteria do.
2. **(a)**
3. **(d)** Protists do not have a unifying characteristic other than being eukaryotic; they do not fit into the kingdoms of Plantae, Animalia and Fungi.
4. **(b)** Mycoplasma lacks cell wall but show multiplication like bacteria, so that they are termed as cell wall less bacteria.
5. **(c)** Antibiotics are chemical substance that is produced by an organism which kills another organism. Antibiotics destroy an organism by inhibiting its metabolic process. As viruses have no metabolism of their own, antibiotics have no effect on them.
6. **(c)** The environment was becoming oxygen-rich. There was not a switch from aerobic to anaerobic metabolism but vice versa.
7. **(c)** Coccus-spherical, vibrio-comma shaped, Bacillus-rod shaped, spirillum-spiral shaped.
8. **(c)** Spirilli are corkscrew shaped.
9. **(c)** Endospores are able to survive a wide range of conditions.
10. **(c)** All viruses are characteristically obligate parasites and nucleo-protein in nature. But the nature of genetic material may vary among viruses. In plant viruses, the genetic material is characteristically RNA, except cauliflower mosaic virus, where it is DNA.
11. **(a)**
12. **(d)** Pseudomycelium is characteristic feature of yeast.
13. **(c)** Protozoans are single celled microscopic eukaryotic organisms that are noted for their ability to move independently. Protozoans are classified on the basis of their locomotion - amoeboid (*Amoeba*), ciliated (*Paramoecium*), flagellated (*Trypanosoma*) and sporozoans (*Plasmodium*).
14. **(c)** All the described statements are related to slime moulds. Slime moulds are saprophytic protists. They are simple organism that consists of an acellular mass of creeping jelly-like protoplasm containing nuclei, or a mass of amoeboid cells. When it reaches a certain size, it forms a large number of spore cases.
15. **(a)** The scientific procedure of arranging organisms into groups and sub-groups on the basis of their similarities and dissimilarities and placing the groups in a hierarchy of categories is known as biological classification.

    Whittaker has used five criteria for delimiting the different kingdoms–

    (i) Complexity of cell structure. *i.e.*, prokaryotic aud eukaryotic.

    (ii) Complexity of body structure or structural organisation of unicellular and multicellular.

    (iii) Mode of nutrition, which is divergent in multicellular kingdoms, photoautotrophy in Plantae, absorptive heterotrophy in Fungi and ingestive heterotrophy in Animalia.

    (iv) Methods of reproduction.

    (v) Phylogenetic relationships.
16. **(d)** *Puccinia*, *Ustilago*, *Agaricus* and *Saccharomyces* are commonly called as rust fungi, smut fungi, mushroom and yeast respectively.

17. **(a)** **18.** **(a)** **19.** **(d)** **20.** **(a)**

21. **(c)**
22. **(a)** The kingdoms were named Monera, Protista, fungi, plantae and Animalia. Four kingdoms contain eukaryotes except kingdom Monera, which contains prokaryotes.
23. **(a)** Archaebacteria, Euglenoids, Phycomycetes and algae are classes of monera, protista, fungi and plantae respectively.

24. **(a)** Litmus used as acid-base indicator, is obtained from *Roccella montagnei* and *Laseallia pastulata*.

25. **(a)** **26.** **(c)**

27. **(c)** Bacterial cell wall is made up of protein, non-cellulosic amino acid etc. Archaebacteria are characterised by the absence of peptidoglycan in their wall. Instead, the wall contains proteins and non-cellulosic polysaccharides. Thermoacidophiles have dual ability to tolerate high temperature as well as high acidity. They often live in hot sulphur springs, where the temperature may be as high as 80°C and pH as low as 2, *e.g.*, *Thermoplasma, Thermoproteus.*

28. **(b)** Sac fungi is a common name of ascomycetes which produce spores in a distinctive type of microscopic sporangium called an ascus. They are characterised by well-developed thallus and production of ascospores. This is the largest class of fungi. Examples of sac fungi are yeasts, morels, truffles, and *Penicillium*.

29. **(b)**

30. **(d)** Members of Phycomycetes are found in aquatic habitats and on decaying wood in moist and damp places or as obligate parasites on plants.

31. **(a)** The symbiotic association of fungi and algae is called lichen.

32. **(d)** Myxomycetes are commonly called true slime moulds. These are saprophytic and their vegetative phase is represented by a free living, mass of protoplasm without walls and produce haploid spores, which are dispersed by air currents, rain and mites.

33. **(d)** Bacteria are usually heterotrophic i.e. live either parasitically or saprophytically. Some are symbiotic as well as autotrophic.

34. **(b)**

35. **(a)** Generation time is 40 minutes.

$\therefore$ Number of divisions is $\frac{240}{40} = 6$

Number of bacteria are $= 2^n = 2^6 \times 10^7$

$= 64 \times 10^7$

36. **(a)** **37.** **(b)** **38.** **(b)**

39. **(c)** Cyanobacterial cells (blue-green algal cells) are larger and more elaborate than bacteria. In both the organisms, the cell structure is typically prokaryotic type, one envelope organisation with peptidoglycan wall, naked DNA Blue-green algae have chlorophyll, bacteria is devoid of chlorophyll.

40. **(d)** **41.** **(d)** **42.** **(c)** **43.** **(a)**

44. **(b)** Deuteromycetes is an artificial group without any common relationship. They are commonly known as fungi imperfecti due to absence of perfect sexual stage. They have septate hyphae and reproduce asexually by means of conidia.

45. **(b)**

46. **(d)** *Albugo* is the parasitic fungi which is found on mustard. It is the member of phycomycetes that are found in aquatic habitats and on decaying wood, in moist and damp places.

47. **(a)**

48. **(b)** In Gram-positive bacteria, cell wall is 200–300 Å thick and have 85% mucopeptides and 1–2% lipids while in Gram-negative bacteria cell wall is 100–200 Å thick and 12% mucopeptides are have 10–12% and 80-90% lipids.

49. **(d)** LPP-1 was the first cyanophage to be discovered by Safferman and Morris. It was so named because it can infect three different type of blue green algae which are *Lyngbya, Phormidium* and *Plectonema.*

50. **(c)** Archaebacteria is a primitive group of bacteria.

The three main groups of archaebacteria are methanogens, halophiles and thermoacidophiles. Methanogens are found in the musk of swamps and marshes, the rumen of cattle sewage, sludges and gut of termites.

Halophiles are named so because they usually occur in salt rich substrata like salt pans, salt beds and salt marshes.

Thermoacidophiles have dual ability to tolerate high temperature as well as high acidity. They often live in hot sulphur springs where the temperature may be as high as 80°C and pH as low as 2.

51. **(c)** **52.** **(d)**

53. **(d)** *Agaricus campestris* is a common field mushroom, *Morchella esculenta* having apothecia type of edible ascocarp and *Podaxon prodaxis* is also edible.

54. **(d)** *Ascus* is a sac, typically cylindrical in shape, in which the spores of ascomycetes fungi develop. Basidium is a microscopic club-shaped spore-bearing structure produced by certain fungi. *Euglena*, *Spirulina* and sponges are the examples of protista, cyanobacteria and animalia respectively.

55. **(b)**

56. **(c)** Methanogens are archaebacteria, that live in marshy area and produce methane as a metabolic byproduct in anoxic conditions. Methane is not their preferred carbon source.

57. **(a)** The structure marked as 1, 2, 3 and 4 are respectively head, collar, sheath and tail fibres. In head (structure marked as 1), the protein coat (called the capsid) made of small subunits called capsomeres, protect the nucleic acid.

58. **(b)** Viruses are acellular or non-cellular particles that contain an inner core of genetic material DNA or RNA that is covered by a non-genetic protein 'coat' called a capsid, The protein coat is made up of protein subunits called capsomeres. Together, the capsid and the nucleic acid core form the nucleocapsid. The nucleocapsid may be either enveloped or naked. Plant viruses are generally naked but animal viruses may or may not possess envelop. Plant viruses generally enters into the host through wounds as they cannot penetrate the cuticle of the host cell, so the entire virus come in contact with host cell. The enveloped viruses inject their own nucleic acid inside the host cell.

59. **(d)** **60.** **(c)**

61. **(d)** Whittaker placed prokaryotes and akaryotes in Monera.

62. **(c)** Viruses that get integrated with the bacterial host genome are called lysogenic. Virus which transmit its DNA into bacterial cell and divides within bacterial cell causing breakdown of bacterial cell wall are called lytic virus. Lysozymes are present in the lipolytic enzymes, catalysing breakdown of fats (lipids).

63. **(b)** **64.** **(d)**

65. **(b)** *Rhizopus* (pin mould or black mould) grows on bread and hence also called bread mould.

66. **(a)** **67.** **(a)**

68. **(c)** Virus is a small infectious agent that replicates only inside the living cells of other organisms. Viruses can infect all types of life forms, from animals and plants to microorganisms, including bacteria and archaea. Viruses can pass through bacterial proof filters as they are smaller than bacteria.

69. **(b)** It is the first stage of sexual reproduction in which the cytoplasm of two sex cells fuse with each other. The nuclei of sex cells come close to each other but do not fuse. Thus, the resulting cell becomes binucleate or dikaryon.

70. **(c)** All fungi are heterotrophs in their nutrition. Some depends upon organic matter known as saprophytes.

71. **(a)** *Claviceps purpurea* is a fungus which causes ergotism in rye (*Secale cereale*) and other plants. It also yields a hallucinogenic drug called LSD.

72. **(c)** Robert Koch (1882) working on bacterial diseases, postulated certain fundamental laws of diseases known as Koch's postulates.

3 Plant Kingdom

1. The pigments that gives class-rhodophyceae (red algae) its distinctive red colour are:
   (a) Chlorophyll-*a, b* & phycoerythrin
   (b) Chlorophyll-*a, c* & fucoxanthin
   (c) Chlorophyll-*a, d* & phycoerythrin
   (d) Chlorophyll-*a,c* and phycoeyanin
2. The dominant stage in *Spirogyra* is the sporophytic stage.
   (a) It is correct.
   (b) It is incorrect.
   (c) *Spirogyra* does not display alternation of generation
   (d) Sometimes (a) and (b)
3. What is the similarity between gymnosperms and angiosperms?
   (a) Phloem of both have companian cells.
   (b) Endosperm is formed before fertilisation in both.
   (c) Origin of ovule and seed is similar in both.
   (d) Both have leaves, stem and roots.
4. In the given Life cycle of an angiosperm, 1, 2, 3, 4 and 5 are respectively

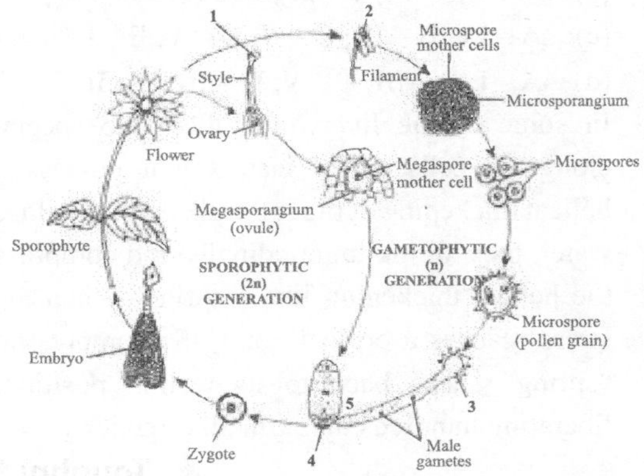

   (a) 1 – Stigma, 2 – Anther, 3 – Male gametophyte, 4 – Egg, 5 – Female gametophyte
   (b) 1 – Stigma, 2 – Anther, 3 – Female gametophyte, 4 – Egg, 5 – Male gametophyte
   (c) 1 – Stigma, 2 – Anther, 3 – Male gametophyte, 4 – Fertilised egg, 5 – Female gametophyte
   (d) 1 – Stigma, 2 – Anther, 3 – Embryo sac, 4 – Egg, 5 – Female gametophyte
5. Which one of the following is the major difference between mosses and ferns?
   (a) Ferns lack alternation of generation while mosses show the same.
   (b) Mosses are facultative aerobes while ferns are obligate aerobes.
   (c) Vascular bundles of ferns show xylem vessels while those of mosses lack it.
   (d) Sporophytes of ferns live much longer as compared to the sporophytes of mosses.
6. Gemma cups are small receptacles that give rise to gemmae (asexual buds). These gemma cups are found in:
   (a) *Sphagnum* and *Marchantia*
   (b) *Marchantia*
   (c) *Funaria*
   (d) Fern
7. Match column-I with column-II and select the correct answer using the codes given below.

	Column-I (Pattern of life cycle in plant)		Column-II (Examples)
A.	Haplontic life cycle	I.	Bryophytes, Pteridophytes, *Ectocarpus*, *Polysiphonia*, Kelps
B.	Diplontic life cycle	II.	Seed bearing plants (Gymnosperm and Angiosperm), *Fucus*
C.	Haplo-diplontic life cycle	III.	Many algae (*Volvox*, *Spirogyra)* and some species of *Chlamydomonas*

   (a) A – III; B – II; C – I
   (b) A – I; B – II; C – III
   (c) A – II; B – I; C – III
   (d) A – III; B – I; C – II

8. Number of cotyledons in *Zea mays, Cycas* and *Pinus* respectively are:
   (a) 1, 2, 4
   (b) 1, 2, many
   (c) 2, 2, many
   (d) 2, 1, 4

9. Mosses and ferns are found in moist and shady places because both:
   (a) require presence of water for fertilisation.
   (b) do not need sunlight for photosynthesis.
   (c) depend for their nutrition on micro-organisms which can survive only at low temperature.
   (d) cannot compete with sun-loving plants.

10. The main plant body in pteridophyte is:
   (a) sporophyte *(2n)* which is differentiated into root, stem and leaf.
   (b) gametophyte *(n)* which is differentiated into root, stem and leaf.
   (c) sporophyte having no root, stem and leaf.
   (d) gametophyte having no root but only stem and leaf.

11. The ploidy level of protonema, apophysis, calyptra and perichaetial leaves in *Funaria* are: **Tricky**
   (a) N, 2N, N and 2N respectively.
   (b) 2N, N, N and 2N respectively.
   (c) 2N, 2N, N and N respectively.
   (d) N, 2N, N and N respectively.

12. Bryophytes can be separated from algae because they:
   (a) are thalloid forms.
   (b) have no conducting tissue.
   (c) possess archegonia with outer layer of sterile cells.
   (d) contain chloroplasts in their cells.

13. The rhizophore is not a true root because it is:

   (a) Endogenous (b) Exogenous
   (c) Mesogenous (d) Endarch

14. Which of the following occurs both in fresh as well as in marine water?
   (a) *Oedogonium*
   (b) *Cladophora*
   (c) *Spirogyra*
   (d) None of the above

15. *Ectocarpus, Dictyota, Laminaria, Sargassum,* and *Fucus* belong to the class:
   (a) Rhodophyceae
   (b) Pheophyceae
   (c) Cyanophyceae
   (d) Chlorophyceae

16. Which of the following is living fossil?
   (a) *Pinus*
   (b) *Ginkgo biloba*
   (c) *Thuja*
   (d) *Deodar*

17. Match column-I with column-II and select the correct answer using the codes given below.

	Column-I		Column-II
A.	Peritrichous flagellation	I.	*Ginkgo*
B.	Living fossil	II.	*Macrocystis*
C.	Rhizophore	III.	*Escherichia coli*
D.	Smallest flowering plant	IV.	*Selaginella*
E.	Largest perennial alga	V.	*Wolffia*

   (a) A – III; B – I; C – IV; D – V; E – II
   (b) A – II; B – I; C – III; D – IV; E – V
   (c) A – IV; B – III; C – II; D – V; E – I
   (d) A – I; B – III; C – V; D – III; E – II

18. In some of the liverworts, there are special elongated cells called elaters that possess a helical thickening of the cell wall. As elaters lose water, they shrink longitudinally and compress the helical thickening like a spring. When the stress reaches a critical point, the compressed "spring" snaps back to its resting position, liberating hundreds of ___ in all directions.

   (a) Moisture particles
   (b) Spores
   (c) Sperm
   (d) Ova

19. Match column-I with column-II and select the correct answer using the codes given below.

	Column-I (features)		Column-II (term)
A.	Presence of tap roots and coralloid roots.	I.	Bryophyte
B.	The synergids and antipodal cells degenerates after fertilisation.	II.	Pteridophytes
C.	The food is stored as floridean starch which is very similar to amylopectin and glycogen in structure.	III.	Red algae
D.	Presence of sporophyte which is not free living but attached to the photosynthetic gametophytes and derives nourishment from it.	IV.	Angiosperms
E.	Members of this group are used for medicinal purposes, as soil binders and frequently grown as ornamentals.	V.	Gymnosperms

(a) A – I; B – II; C – III; D – IV; E – V
(b) A – III; B – V; C – II; D – IV; E – I
(c) A – III; B – I; C – V; D – II ; E – IV
(d) A – V; B – IV; C – III; D – I; E – II

20. Which one of the following is of considerable economic importance?
(a) *Riccia* (b) *Funaria*
(c) *Marchantia* (d) *Sphagnum*

21. In gymnosperm, endosperm is formed by:

(a) fusion between a male gamete and two polar nuclei.
(b) fusion between a male gamete and a polar nuclei.
(c) fusion between egg and male gamete.
(d) germination of megaspore.

22. "Botanical snakes" are:
(a) Algae (b) Fungi
(c) Bryophytes (d) Pteridophytes

23. The plant body constituting a holdfast, stipe and frond are seen in:
(a) *Volvox* (b) *Chara*
(c) *Laminaria* (d) *Chlamydomonas*

24. During monsoon, ground becomes slippery because of :

Tricky

(a) Red algae (b) Blue-green algae
(c) Mosses (d) Liverworts

25. Consider the following statements about brown algae.
(i) *Nereocystis* and *Macrocystis* are the largest kelps.
(ii) Algin is the gelatinous coating outside the cellulosic cell wall in brown algae.
(iii) 'Kombu' is the name given to the food obtained from *Laminaria saccharina.*

Which of the statements given above are correct?
(a) (ii) and (iii) (b) (i) and (ii)
(c) (i) and (iii) (d) (i), (ii) and (iii)

26. Which of the following part of the *Funaria*, sporophyte is involved in the dispersal of spores?
(a) Calyptra (b) Peristome
(c) Operculum (d) Annulus

27. Canada balsam is obtained from:
(a) *Pinus wallichiana*
(b) *Abies balsamea*
(c) *Gnetum gnemon*
(d) *Ephedra vulgaris*

28. When moss spores germinate, they form:
(a) Leafy gametophyte
(b) Capsule
(c) Protonema
(d) Rhizoids

29. Which of the following statements are **correct**?

(i) The commercial product, Agar, obtained from *Gracilaria* and *Gelidium,* as in ice cream and jelly preparation as well as used to grow microbes.
(ii) *Spirogyra* and *Chlorella* are used in sewage disposal ponds.
(iii) Some species of marine algae like *Porphyra*, *Laminaria* and *Sargassum* are used as food.

(a) (ii) and (iii) (b) (i) and (ii)
(c) (i) and (iii) (d) (i), (ii) and (iii)

30. Fern plant is a:
(a) Haploid gametophyte
(b) Diploid gametophyte
(c) Diploid sporophyte
(d) Haploid sporophyte

31. A bryophytic plant has chromosome number 9. Chromosome number in spores and protonema will be: **Tricky**
(a) 9 & 9 (b) 18 & 18
(c) 27 & 18 (d) 9 & 18

32. Heterothallism is a term used for the symbiosis of :
(a) Algae and *Cycas*
(b) Algae and Fungi
(c) Virus and Algae
(d) Fungi and *Pinus*

33. *Cycas* has two cotyledons but is not included under angiosperms because it has:
(a) Circinate ptyxis
(b) Compound leaves
(c) Monocot like stem
(d) Naked seeds

34. Which one of the following is **correct**?
(a) Protonema of moss and prothallus of *Dryopteris* are sporophytic.
(b) Protonema of moss and prothallus of *Dryopteris* are gametophytic.
(c) Moss protonema is sporophytic, *Pteris* prothallus is gametophytic but plant body of *Pteris* and *Funaria* are gametophytic.
(d) Plant body of moss is gametophytic while *Dryopteris* is, both gametophytic and sporophytic.

35. Bryophytes are different from fungi in having:
(a) Land habit
(b) Sterile jacket layers
(c) Multiflagellate gametes
(d) Gametophytic plant body

36. Which of the following statement(s) is/are **correct**?
(a) In angiosperms, each embryo sac has a three-celled egg apparatus – one egg cell and two synergids, three antipodal cells and two polar nuclei.
(b) All seed bearing plants *i.e.,* gymnosperms and angiosperms follow diplontic life patterns of plants.
(c) In gymosperms, roots in some genera have fungal association in the form of mycorrhiza (*Pinus*) while in some others (*Cycas*) small specialised roots called coralloid, are associated with N_2–fixing cyanobacteria.
(d) All of the above

37. The characteristic trait of the plant body of bryophtyes is that:
(a) plant body is more differentiated than that of algae.
(b) plant body is equally differentiated to that of algae.
(c) plant body is less differentiated than that of algae.
(d) plant body is not differentiated.

38. Match column-I with column-II and select the correct answer using the codes given below.

	Column-I		Column-II
A.	*Gelidium*	I.	Peat moss
B.	*Sphagnum*	II.	Agar agar
C.	*Azolla*	III.	Polymorphism
D.	*Mycoplasma*	IV.	Aquatic fern
E.	*Zamia*	V.	Smallest gymnosperm plant

(a) A – II; B – I; C – IV; D – III; E – V
(b) A – I; B – II; C – IV; D – III, E – V
(c) A – II; B – I; C – IV; D – V, E – III
(d) A – II; B – IV; C – I; D – III, E – V

39. The most common type of spore produced during asexual reproduction of algae is:
(a) Aplanospore (b) Endospore
(c) Zoospore (d) Oospore

40. Fruits are not formed in gymnosperms because of :
(a) Absence of pollination
(b) Absence of seed
(c) Absence of fertilisation
(d) Absence of ovary

41. The pyrenoids are made up of:
(a) proteinaceous centre and starchy sheath.
(b) core of protein surrounded by fatty sheath.
(c) core of starch surrounded by sheath of protein.
(d) core of nucleic acid surrounded by protein sheath.

42. Which of the following is known as pond scum?
(a) *Volvox* (b) *Ulothrix*
(c) *Nostoc* (d) *Spirogyra*

43. Match column-I with column-II and select the correct answer using the codes given below.

Column-I		Column-II	
A.	Algae	I.	*Solanum tuberosum*
B.	Fungi	II.	*Equisetum*
C.	Angiosperm	III.	*Cycas*
D.	Pteridophyte	IV.	*Chlamydomonas*
E.	Gymnosperm	V.	*Rhizopus*

(a) A – V; B – IV; C – I; D – II; E – III
(b) A – IV; B – V; C – I; D – II; E – III
(c) A – IV; B – I; C – V; D – II; E – IV
(d) A – IV; B – I; C – V; D – III; E – II

44. Female reproductive part of bryophytes is:
(a) Antheridium (b) Oogonium
(c) Archegonium (d) Sporangium

45. Pollination of pollen grains in *Pinus* takes place at:
(a) One celled stage
(b) Two celled stage
(c) Three celled stage
(d) Four celled stage

46. Which of the following statements are **correct** regarding algae?
(i) Green algae occur in fresh water, brackish water, salt water.
(ii) Habitat of brown algae-fresh water (rare), brackish water, salt water.
(iii) Some red algae are found in fresh water, mostly occur in salt water, some are in brackish water.
(iv) Most of the red algae are multicellular.
(v) Red alga may occur in both well lighted regions close to water-surface and also at great depths in oceans where light penetration is little.
(vi) Cell wall of red algae consists of cellulose + agar.
(a) All of the above
(b) None of the above
(c) (i) and (vi)
(d) (ii), (iii) and (v)

47. Which of the following is/are grouped under phanerograms?
(a) Angiosperms (b) Gymnosperms
(c) Pteridophytes (d) Both (a) and (b)

48. Which of the following statements is/are **correct**?
(i) In *Equisetum,* the female gametophyte is retained on the parent sporophyte.
(ii) In *Ginkgo,* male gametophyte is not independent.
(iii) The sporophyte in *Riccia* is more developed than that in *Polytrichum.*
(iv) Sexual reproduction in *Volvox* is isogamous.
(a) Two (b) Three
(c) Four (d) One

49. Chlorophyll and chloroplast are present in spores of :
(a) *Lycopodium* (b) *Azolla*
(c) *Dryopteris* (d) *Marsillea*

50. *Cycas* and *Ginkgo* are regarded as primitive gymnosperms because:
(a) they are lacking cones.
(b) they need water for fertilisation.
(c) they are living fossils.
(d) they have ciliated sperms.

51. The flagellate and motile spore in algae is known as:
(a) Akinetes (b) Endospore
(c) Zoospore (d) Aplanospore

52. The haploid and thallus like plant body of bryophytes have:
(a) vascular tissues of xylem and phloem.
(b) root, leaf and stem like structures.
(c) complex tissues.
(d) true root, stem and leaves which bear vascular tissues.

53. *Spirogyra* increases its body length by the division of :
(a) every cell of the body.
(b) the apical cell.
(c) the basal cell.
(d) accumulation of food in his body.

54. Which one is a parasitic algae?
(a) *Vaucheria*
(b) *Polysiphonia*
(c) *Cephaleuros*
(d) *Batrachospermum*

55. Match column-I with column-II and select the correct answer using the codes given below.

Column-I	Column-II
A. Amphibian of the plant kingdom	I. *Sphagnum*
B. Specialised structures in liverworts for asexual reproduction	II. Angiosperms
C. Monocotyledons dicotyledons	III. Bryophytes
D. A plant which has capacity to holding water	IV. Gemmae

(a) A – III; B – IV; C – I; D – II
(b) A – III; B – IV; C – II; D – I
(c) A – IV; B – III; C – II; D – I
(d) A – III; B – II; C – IV; D – I

56. Vegetative reproduction in green algae occurs by:
(a) Conidia
(b) Fragmentation
(c) Different types of spores
(d) Both (b) and (c)

57. Agar-Agar is obtained from:
(a) *Gigartina* (b) *Gelidium*
(c) *Gracillaria* (d) All of these

58. Read carefully the following statements about pteridophytes.
(i) They are used for medicinal purposes.
(ii) They are frequently grown as ornaments.
(iii) They are called vascular cryptogams.
(iv) The produce spores rather than seeds.
(v) They are used as soil binders.
Which of the statements given above are **correct**?
(a) (i), (ii) and (v)
(b) (ii), (iv) and (v)
(c) (ii), (iii), (iv) and (v)
(d) All of the above

59. Which of the following structures in *Pinus* are haploid?
(a) Megaspore, integument, root
(b) Endosperm, megaspore, pollen grain
(c) Pollen grain, leaf, root
(d) Megaspore, endosperm, embryo

60. A bryophyte suddenly started reproducing parthenogenetically. The number of chromosomes of the second generation compared to parent plant will be: Tricky
(a) Same (b) One-half
(c) Double (d) Triple

61. A gelatinous coating outside the cellulosic cell wall is observed in the members belonging to brown algae. This gelatinous coating is called as: Tricky
(a) Algin (b) Polyalginate
(c) Glycoalgin (d) Mannitol

62. Sexual reproduction in algae results in the formation of :
(a) Oospore (b) Zoospore
(c) Zygote (d) Zygospore

63. Which of the statements given are **correct**?
(i) *Funaria* possesses unicellular and unbranched rhizoids.
(ii) Mosses along with lichens are the first organisms to colonise rocks.
(iii) Gemmae are asexual buds, which originate from small receptacles called gemma cups.
(iv) *Sphagnum* has magnificant property of retaining water.
(a) (i), (ii), (iii) and (iv)
(b) (i), (iii) and (iv)
(c) (ii), (iii) and (iv)
(d) (i), (ii) and (iii)

64. Asexual reproduction in brown algae happens by the formation of:
(a) multiflagellate gametes and are sickle-shaped.
(b) biflagellate gametes (pear-shaped and have two unequal flagella).
(c) aplanospore (apple-shaped and non-motile).
(d) endospores (round and have one flagella).

65. *Cycas* plant is kept in darkness for quite a long time will have its sago yield:
(a) Increased (b) Decreased
(c) Unaffected (d) None of these

66. If the cells of root in wheat plant have 42 chromosomes, then the no. of chromosome in the cell of pollen grain is: Critical Thinking
(a) 14 (b) 21 (c) 28 (d) 42

67. Only capsule is present in sporophyte of :
(a) *Riccia* (b) *Marchantia*
(c) *Anthoceros* (d) *Funaria*

68. Like angiospermic parasite such as *Cuscuta*, there are some parasitic forms of rhodophyta, which are colourless, heterotrophic and grow on other members of rhodophyta. Select which one is a parasitic form of red algae.
(a) *Gelidium* (b) *Harveyella*
(c) *Choridras* (d) Both (a) and (b)

69. In oogamy, fertilisation involves:
(a) a large non-motile female gamete and a small non-motile male gamete.
(b) a large motile female gamete and a small non-motile male gamete.
(c) a small non-motile female gamete and a large motile male gamete.
(d) a large non-motile female gamete and a small motile male gamete.

70. The leaf gap in the vascular cylinder in ferns, is known as:
(a) Air space
(b) Parenchymatous zone
(c) Collenchymatous zone
(d) Area exclusively of phloem

71. Choose the **incorrect** statement about mosses.
(a) The fusion of antheridia and archegonia produced at the apex of the leafy shoots leads to sexual reproduction.
(b) The sporophtye in mosses is more elaborate than that in liverworts.
(c) Seta and capsule bears spores, which give rise to gametophyte after meiosis.
(d) Sporophyte is differentiated into foot, seta and capsule.

72. Seeds are found in:
(a) Angiosperm (b) Bryophyte
(c) Pteridophyte (d) Algae

73. The walking fern is so named because:
(a) its spores are able to walk.
(b) it is dispersed through the agency of walking animals.
(c) it propagates vegetatively by its leaf tips.
(d) it knows how to walk by itself.

74. Mosses are:
(a) upright and radial in symmetry
(b) leafy
(c) green
(d) All of the above

75. "Chilgoza", a gymnosperm seed, that is eaten as a "dry fruit" is produced by:
(a) *Pinus roxburghii* (b) *Pinus gerardiana*
(c) *Ginkgo biloba* (d) *Cedrus deodara*

76. Pyrenoids in green algal cells are related to:
(a) Starch storage
(b) Protein storage
(c) General metabolism
(d) Enzyme secretion

77. The gametophyte of fern bears:
(a) True roots (b) Antheridia
(c) Archegonia (d) Both (b) and (c)

78. The unique feature of bryophytes that differentiates it from other green plant group is that:
(a) they lack roots.
(b) they lack vascular tissue.
(c) they produce spores.
(d) their sporophytes are attached to the gametophyte.

79. The thallus of *Volvox* is called:
(a) Trichome (b) Coenobium
(c) Coenocyte (d) Parenchymatous

80. In *Funaria*, 20 chromosomes are present in rhizoids, then the number of chromosome in calyptra, theca and foot will be: **Tricky**
(a) 20, 40, 40 respectively
(b) 40, 20, 20 respectively
(c) 20, 40, 20 respectively
(d) 40, 10, 20 respectively

81. A moss plant with mature antheridia and archegonia is exposed to dry and hot air. The egg cell will:
(a) Be fertilised normally
(b) Not be fertilised
(c) Be burned
(d) Come out of the archegonium

82. In *Ulothrix,* meiosis (reduction division) occurs in:
(a) Gamete (b) Zygospore
(c) Zoospore (d) Thallus

83. In moss, the sporophyte:
(a) produces gametes that give rise to the gametophyte.
(b) arises from a spore, produced from the gametophyte.
(c) manufactures food for itself as well as for the gametophyte.
(d) is partially parasitic on the gametophyte.

84. The members of brown algae are found primarily in/on:
(a) Moist rock
(b) Marine habitat
(c) Freshwater habitat
(d) Terrestrial habitat

85. Palmella stage is met within:
(a) *Ulothrix* and *Sphagnum*
(b) *Oedogonium* and *Riccia*
(c) *Chlamydomonas*
(d) *Spirogyra* and *Gnetum*

86. In ferns, the vascular bundles are:
(a) Radial (b) Conjoint
(c) Hadrocentric (d) Leptocentric

87. Brown algae is characterised by the presence of :
(a) Phycocyanin
(b) Phycoerythrin
(c) Fucoxanthin
(d) Carrageenan

88. Sporocarp is a reproductive structure of :
(a) Some algae
(b) Some aquatic ferns having sori
(c) Angiosperms having spores
(d) Bryophytes

89. Which is **incorrect** in respect to bryophytes?
(a) Water is essential for sexual reproduction.
(b) Presence of ciliated sperms.
(c) Presence of antheridium.
(d) Presence of autotrophic independent sporophyte.

90. Mosses are indicator of:
(a) Air pollution
(b) Water pollution
(c) Radioactive pollution
(d) Soil pollution

91. A rigid cell wall is usually seen in members of chlorophyceae. This cell wall is constituted of:
(a) cellulose (outer layer) and algin (inner layer)
(b) chitin (outer layer) and pectose (inner layer)
(c) cellulose (inner layer) and pectose (outer layer)
(d) pectose (inner layer) and peptidoglycan (outer layer)

92. The type of chloroplasts observed in class-Chlorophyceae is:
(a) Discoid and plate-like.
(b) Spiral or ribbon-shaped.
(c) Reticulate and cup-shaped.
(d) All of the above

93. The term 'frond' in fern is used for:
(a) Rachis
(b) Whole plant
(c) Spore bearing leaf
(d) Sexually reproducing structure

94. Apophysis in moss capsule is:
(a) Upper part (b) Middle part
(c) Lower part (d) Fertile part

95. The moss plant is:
(a) sometimes sporophyte and sometimes gametophyte.
(b) predominantly gametophyte with sporophyte attached to it.
(c) only sporophyte
(d) predominantly sporophyte with gametophyte attached to it only.

96. In mosses, vegetative reproduction takes place by:
(a) fragmentation and budding in the secondary protonema.
(b) gemmae and tuber formation
(c) protonema and endospore
(d) gemmae formation

97. Which one of the following is called horsetails?
(a) *Lycopodium* (b) *Marselia*
(c) *Equisetum* (d) *Selaginella*

98. In pteridophytes, phloem is without:
(a) Sieve cells
(b) Sieve tubes
(c) Companion cells
(d) Both (b) and (c)

99. Floridean starch is found in:
(a) Chlorophyceae (b) Rhodophyceae
(c) Myxophyceae (d) Cyanophyceae

Answer KEYs

1	(c)	11	(d)	21	(d)	31	(a)	41	(a)	51	(c)	61	(a)	71	(c)	81	(b)	91	(c)
2	(b)	12	(c)	22	(d)	32	(b)	42	(d)	52	(b)	62	(d)	72	(a)	82	(b)	92	(d)
3	(d)	13	(b)	23	(c)	33	(d)	43	(b)	53	(a)	63	(c)	73	(c)	83	(d)	93	(c)
4	(a)	14	(b)	24	(b)	34	(b)	44	(c)	54	(c)	64	(b)	74	(d)	84	(b)	94	(c)
5	(d)	15	(b)	25	(d)	35	(b)	45	(d)	55	(b)	65	(b)	75	(b)	85	(c)	95	(b)
6	(b)	16	(b)	26	(b)	36	(d)	46	(a)	56	(d)	66	(b)	76	(a)	86	(c)	96	(a)
7	(a)	17	(a)	27	(b)	37	(a)	47	(d)	57	(d)	67	(a)	77	(d)	87	(c)	97	(c)
8	(b)	18	(b)	28	(c)	38	(a)	48	(a)	58	(d)	68	(b)	78	(d)	88	(b)	98	(d)
9	(a)	19	(d)	29	(d)	39	(c)	49	(a)	59	(b)	69	(d)	79	(b)	89	(d)	99	(b)
10	(a)	20	(d)	30	(c)	40	(d)	50	(c)	60	(a)	70	(b)	80	(a)	90	(a)		

Hints & Solutions

1. (c) The pigments that give class-Rhodophyceae (red algae) its distinctive red colour is chlorophyll-*a, d*. These pigments are photosynthetic in nature and are located in the chromatophores. The other pigments present are β-carotene, xanthophylls, phycocyanin (blue in colour) and biliprotein (*r*-phycoerythrin-red in colour).

2. (b) The dominant stage in *Spirogyra* is the gametophytic stage. The sporophyte is a single celled zygote.

3. (d) 4. (a)

5. (d) Mosses and ferns belong to bryophytes and pteridophytes respectively. In bryophytes, the dominant phase in the life cycle is the gametophytic plant body. However, in pteridophytes, the main plant body is a sporophyte, which is differentiated into true root, stem and leaves. These organs possess well – differentiated vascular tissues.

6. (b) Asexual reproduction in *Marchantia* occurs by the formation of gemmae. They are multicellular, green and biconvex lens shaped bodies that are produced in gemma cups. On detaching from the gemma cup, they germinate to produce new plants.

7. (a) 8. (b) 9. (a)

10. (a) The main plant body in pteridophytes is sporophyte ($2n$) which is differentiated into root ($2n$), stem and leaf.

11. (d) 12. (c)

13. (b) The rhizophore is not a true root because it is exogenous.

14. (b)

15. (b) *Ectocarpus, Dictyota, Laminaria, Sargassum*, and *Fucus* belong to the class Phaeophyceae.

16. (b) Presence of *Ginkgo biloba*, a gymnosperm, is confined to the eastern part of China and Japan.

17. (a) Peritrichous flagellation : Flagella all around, e.g. *Escherichia coli*.
Living fossil : Organisms which have undergone very little change since they first evolved, e.g. *Ginkgo*.
Rhizophore : Rhizophores are present in *Selaginella* which is a colourless, leafless, positively geotropic, elongated structure.
Smallest flowering plant : *Wolffia*.
Largest perennial alga : *Macrocystis*.

18. (b) 19. (d)

20. (d) *Sphagnum* is used as fuel. It has the capacity to retain water for long periods and as such used to cover the plant roots during transportation.

21. (d) In gymnosperms, the endosperm is formed in the ovule during germination of the megaspore, it consequently becomes the female prothallus (gametophyte), with a haploid set of chromosomes. The archegonia in gymnosperms develop from certain cells of endosperm.

22. (d)

23. (c) *Laminaria* belong to the class Phaeophyceae. Plant body of *Laminaria* is generally attached to the substratum by a holdfast which has a stalk, the stipe and frond which is the photosynthetic organ.

24. (b) During monsoon, ground becomes slippery because of blue-green algae.

25. (d) Kelps are giant brown algae. *Nereocystis* (20-30 m) and *Macrocystis* (40-50 m) are the largest kelps. Algin is the gelatinous coating outside the cellulosic cell wall in brown algae. Alginic acid is extracted commercially from kelps. Some countries consume certain brown algae as food. 'Kombu' is the name given to the food obtained from *Laminaria saccharina.* It is rich in carbohydrates.

26. (b) Peristome is a ring of teeth around the opening of the capsule in mosses (*Funaria*). It is involved in spore dispersal.

27. (b) Canada balsam is a resin obtained from *Abies balsamea*. The liquid is viscous, yellowish and greenish in colour. The principal use of canada balsam is for mounting microscopic slides and as a cement for optical lenses.

28. (c) When moss spores germinate, they form protonema. It is the first stage.

29. (d) Agar is obtained from *Gracilaria, Gelidium, Ceramium, Chondrus*, etc. It is used in microbiological science to solidify culture media. Green unicellular algae such as *Chlorella* and *Chlamydomonas* are used in sewage disposal ponds. They restore O_2 and remove CO_2 by the process of photosynthesis and make sewage water habitable for many fishes and aerobic bacteria whereas, *Porphyra, Laminaria* and *Sargassum*, are used as food.

30. (c) Fern plant is a diploid sporophyte.

31. (a) Spores are formed after meiosis in diploid cells and protonema by their germination where nuclear divisions are mitotic only.

32. (b) Heterothallism is a condition found in between algae and fungi in which sexual reproduction occurs only between genetically different self-incompatible mating types (strains) of the same species.

33. (d) In *Cycas*, seeds are naked because they are not covered by ovary.

34. (b)

35. (b) Bryophytes are different from fungi in having sterile jacket layers.

36. (d)

37. (a) The characteristic traits of the plant body of bryophytes is that the plant body of bryophytes is more differentiated than that of algae.

Difference between bryophytes and algae–

(i) Tissue differentiation is well-developed in bryophytes while in algae it is found only in higher forms.

(ii) Anisogamous, isogamous, and oogamous types of sexual reproduction occur in algae, while only oogamous type of sexual reproduction is present in bryophytes.

(iii) Sex organs in bryophytes are covered by a sterile jacket while in algae they are not covered.

(iv) Female sex organ in bryophytes is archegonium, while it is oogonium in algae.

(v) In bryophytes, sporophyte is dependent upon gametophyte and in algae sporophyte is independent of gametophyte.

(vi) Embryo is present in bryophytes and absent in algae.

(vii) Sporophyte in bryophytes is differentiated into foot, seta and capsule.

38. (a)

39. (c) Zoospores are the most common type of spores produced during asexual reproduction in algae. These spores are flagellated and thus motile, which gives rise to a new plant on germination.

40. (d) Fruits are not formed in gymnosperms because of absence of ovary.

41. (a) The pyrenoids are small spherical protein bodies surrounded by starch deposition. They are found singly or in numbers embedded in the chloroplast of many algae and bryophytes.

42. (d) *Spirogyra* is a fresh water green algae. It is also known as pond scum, pond silk or water silk.

43. (b)

44. (c) Female sex organ is called archegonium which is flask shaped with a tubular neck and a swollen venter.

45. (d) Pollination in *Pinus* takes place when two prothallial cells, one generative cell and one tube cell are formed. *i.e.* four celled stage.

46. (a) 47. (d)

48. (a) Statement (i) and (ii) are correct. *Riccia* is liverwort in which simplest sporophyte consists of capsule only while *Polytrichum* is moss in which sporophyte consists of foot, seta & capsule. *Volvox* is a fresh water green colonial alga. Reproduction is both sexual and asexual in *Volvox*. Sexual reproduction is of oogamous type.

49. (a) 50. (c)

51. (c) Zoospores are the most common type of spores produced during asexual reproduction in algae. These spores are flagellated and thus motile, which gives rise to a new plant on germination.

52. (b) Bryophytes possess root, leaf and stem like structures. They do not possess true root, leaf and stem structures.

53. (a) *Spirogyra* can reproduce vegetatively as well as sexually. Each cell of the plant body is capable of division. The increase in body length is due to the division of each cell of the body.

54. (c) *Cephaleuros virescens* is a member of class chlorophyceae and causes red rust of tea leaf (*Thea sinensis*).

55. (b) Bryophytes are called as the amphibians of the plant kingdom because these plants can live in soil but are dependent on water for sexual reproduction. Gemma is specialised structure in liverworts for asexual reproduction. Monocotyledons and dicotyledons are two groups of angiosperms. Species of *Sphagnum*, a moss provide peat that have been used as a fuel.

56. (d) Vegetative reproduction in green algae, occurs by fragmentation, cell division, and formation of different type of spores.

57. (d) Agar-agar is a jelly like substance. It is a non-nitrogenous carbohydrate consisting of two polysaccharides namely agarose and agaropectin. It is obtained from several red algae, *e.g., Gracillaria, Gelidium, Gigartina.*

58. (d) Pteridophytes are spore forming, non-seed bearing, non-flowering vascular plants. *Dryopteris*, *Lycopodium* is used in treatment of the disorders of lungs, kidneys and rheumatism. An antihelminthic is obtained from the rhizomes and petioles of the fern. They are used as soil binders. Presence of heterospory (morphologically two levels of spores) is a distinctive feature of pteridophytes.

59. (b) Because endosperm is haploid (*n*) and formed before fertilisation and megaspore mother cell divides reductionally to form a linear tetrad of haploid megaspores and microspore (= pollen grain), is the first stage of the gametophyte (*n*).

60. (a) The number of chromosomes of the second generation will be same because no reduction division takes place.

61. (a) Alginic acid is a phycocolloid, extracted commercially from giant brown algae or kelp. Alginic acid is copolymer of α-1, 4 D-mannuronic acid and α -1, 4 L-glucuronic acid.

62. (d) Zygote secreting a thick wall which acts as perennating spore (zygospore) and helps in perennation, dispersal and multiplication.

63. (c) The rhizoids in *Funaria* arise from the basal region of the stem, which functions as roots and are branched and multicellular. The gemmae are green, multicellular, and biconvex lens shaped bodies produced in gemma cup. Mosses colonise on barren rocks along with lichens and decompose rocks. *Sphagnum* is used as a packing material in the transportation of flower, live plants, tubers, bulbs, seedlings, etc. It is also used in seed-beds and in moss-sticks.

64. (b) Asexual reproduction in brown algae happens by the formation of biflagellate gametes. They are motile zoospores and non-motile neutral spores. The zoospores produced in the zoosporangia are pyriforms have chromatophore, biflagellate, contain contractile vacuoles and an eye spot. Heterodont flagellation is seen, *i.e.,* they possess two unequal flagella, one whip-lash type and the other tinsel-shaped.

65. (b) Sago is stored as starch and in absence of photosynthesis will be used by the plant.

66. (b)

67. (a) Simplest sporophyte in bryophytes, is of *Riccia*, where only capsule is present.

68. (b)

69. (d) Oogamy is the most advanced type of reproduction in which male gamete is motile, smaller and non food storing, called antherozoid while female gamete is food storing, non motile, larger, called oogonium.

70. (b) The parenchymatous zone left behind with main stele after the departure of the leaf trace is called as leaf gap and parenchymatous region left out on the main vascular cylinder due to departure of branch traces is called as branch gap.

71. (c) In mosses, only capsule bears spores, which gives rise to gametophyte after meiosis and the sporophyte in mosses is more elaborate than that in liverworts.

72. (a) 73. (c)

74. (d) Mosses are upright and radial in symmetry, leafy and green. They are highly developed of all the bryophytes.

75. (b)

76. (a) Pyrenoids are the rounded bodies found in the chloroplasts of green algae and are the centres of conversion of glucose to starch and also collection of starch.

77. (d) Gametophyte is a free living, photoautoptrophic thalloid, also called prothallus. Prothallus bears rhizoids, antheridia and archegonia on its under surface.

78. (d) The diploid sporophyte is short lived and dependent upon the gametophyte whereas the haploid gametophyte is long lived, dominant, green and independent.

79. (b) *Volvox* is a hollow ball like colony with a fixed number of cells called coenobium.

80. (a) In *Funaria*, dominant phase of life cycle is haploid or gametophytic, so in rhizoid cell, $n = 20$.
Calyptra develops from the venter-wall of archegonium, so chromosome number is 20.
Theca and foot are the parts of sporophyte hence, chromosome number, $2n = 40$.

81. (b) Fertilisation requires water.

82. (b) Under favourable conditions, meiosis occurs in the zygospore, forming 4-16 meiospores, each of which gives rise to a new filament.

83. (d)

84. (b) The members of brown algae are found primarily in marine habitats except three members namely *Pleurocladia*, *Heribaudiella* and *Bodanella* which are found in fresh water habitat.

85. (c) A colony of considerable size, containing numerous cells embedded in a common mucilaginous mass is formed in *Chlamydomonas*.

86. (c) In the ground tissue of fern, many meristeles are arranged in a ring which form the vascular tissue system. Each meristele has central xylem surrounded by phloem. Thus, each meristele is concentric and amphicribral (hadrocentric).

87. (c) 88. (b)

89. (d) In bryophytes, the diploid sporophyte is dependent upon the gametophyte, and short lived.

90. (a)

91. (c) Class chlorophyceae usually exhibits a two layered rigid cell wall. Inner layer of cell wall is made up of cellulose, while outer layer is made up of pectose.

92. (d) In class Chlorophyceae, the cells possess one or more chloroplasts.
The shape of chloroplasts may be–

Girdle-shaped	–	*Ulothrix*
Spiral	–	*Spirogyra*
Partial reticulate	–	*Oedogonium*
Cup-shaped	–	*Chlamydomonas*
Reticulate	–	*Chlamydomonas reticulata*
Star-shaped	–	*Zygonema*
Disc-shaped	–	*Chara*
Partial band-shaped	–	*Hydrodictyon*

93. (c) In fern, the leaves arise on the node of upper side of rhizome. The leaf in young stage shows circinate venation. The fertile lamina or spore bearing leaf is called frond and the sterile part is called stipes.

94. (c) Apophysis is the basal sterile portion of capsule in continuation with seta. In capsule of *Funaria*, stomata is present only in apophysis.

95. (b) The haploid gametophyte is long lived, dominant, independent and green whereas the diploid sporophyte is dependent upon the gametophyte and short lived.

96. (a) In mosses, vegetative reproduction occurs through fragmentation or through bud in secondary protonema.

97. (c) *Equisetum*, commonly known as 'horsetail', is found with about 25 species all over the world except Australia and New Zealand, and is mostly inhabitant of the cool and temperate regions.

98. (d) 99. (b)

4 Animal Kingdom

1. Which of the following is not a characteristic feature of kingdom animalia?
   (a) Storage of carbohydrates as starch
   (b) Multicellularity
   (c) Obtaining nutrients by ingestion
   (d) Having eukaryotic cells without walls
2. On the basis of organisation, animals are grouped into:
   (a) Metazoa and eumetazoa
   (b) Protozoa and metazoa
   (c) Protozoa and parazoa
   (d) Parazoa and metazoa
3. In coelomates, the problem of diffusion of food from gut to tissues is solved by:
   (a) the presence of coelomic fluid.
   (b) churning the food within the body cavity.
   (c) developing a circulatory system.
   (d) developing gut associated glands.
4. An animal having diploblastic acoelomic condition is:
   (a) *Ascaris* (b) sea anemone
   (c) *Planaria* (d) *Periplaneta*
5. The adults are radially symmetrical but larvae exhibit bilateral symmetry in:
   (a) Mollusca (b) Hemichordata
   (c) Echinodermata (d) Cephalochordata
6. Which of the following statement(s) is/are **correct** regarding phylum aschelminthes?
   (i) The body is circular in cross-section hence known as roundworms.
   (ii) Alimentary canal is complete with a well-developed muscular pharynx.
   (iii) Sexes are separate (dioecious), i.e., males and females are distinct.
   (iv) Nephridia help in osmoregulation and excretion.
   (a) (i) and (ii) (b) (iii) and (iv)
   (c) (i), (ii) and (iii) (d) All of these
7. Which of the following is a correct match of a phylum with its three examples?
   (a) Platyhelminthes–*Planaria, Schistosoma, Enterobius*
   (b) Mollusca – *Loligo, Sepia, Octopus*
   (c) Porifera – *Spongilla, Euplectella, Pennatula*
   (d) Cnidaria – *Bonellia, Physalia, Aurelia*
8. Match column-I with column-II and select the correct answer using the codes given below.

Column-I	Column-II
A. *Millipedes*	I. Roundworm, pseudocoelomate
B. *Ascaris*	II. Diplopoda
C. *Asterias*	III. Softbody animals,mantle
D. *Hirudinaria*	IV. Radial symmetry, water vascular system, tube feet
E. *Pila*	V. Botryoidal tissues, temporary clitellum

   (a) A – II; B – I; C – IV; D – V; E – III
   (b) A – II; B – I; C – IV; D – III; E – V
   (c) A – II; B – IV; C – I; D – V; E – III
   (d) A – II; B – I; C – V; D – IV; E – III
9. Which of the following phylum is being described by the given statements?
   (i) They are bilaterally symmetrical, triploblastic, segmented and coelomate animals.
   (ii) The body consists of head, thorax, abdomen and have jointed appendages.
   (iii) Circulatory system is of open type.
   (iv) Excretion takes place through malphigian tubules.
   (a) Arthropoda (b) Annelida
   (c) Mollusca (d) Echinodermata
10. Which of the following statement(s) (i) to (iv) is/are **incorrect**?
   (i) Circulatory system in arthropods is of closed type.
   (ii) Parapodia in annelids helps in swimming.
   (iii) Phylum mollusca is the second largest animal phylum.
   (iv) Aschelminthes are dioecious.
   (a) (i) only (b) (iii) only
   (c) (i) and (iii) (d) (iii) and (iv)

11. The given figures (1 & 2) shows the germinal layer.

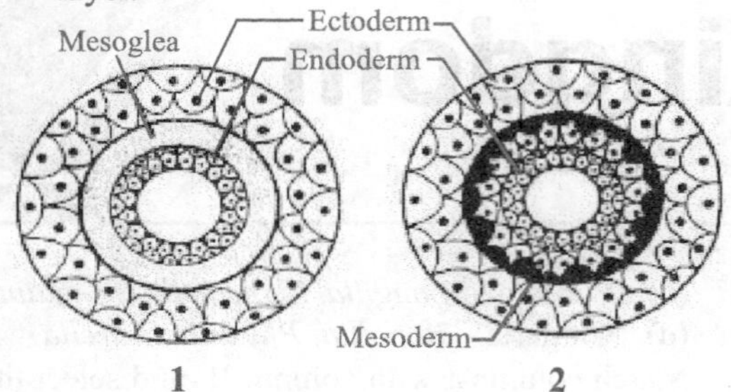

The animals having structures shown in the figures are respectively called:
(a) Diploblastic, triploblastic
(b) Triploblastic, diploblastic
(c) Diploblastic, diploblastic
(d) Triploblastic, triploblastic

12. Cleavage in mammals is:
(a) Holoblastic equal
(b) Holoblastic unequal
(c) Superficial
(d) Discoidal

13. Jaws are absent in:
(a) Protochordata
(b) Protochordata and cyclostomata
(c) *Amphioxus* and *Balanoglossus*
(d) *Herdmania* and *Myxine*

14. Which of the following pairs of animals are similar to each other pertaining to the feature stated against them?
(a) *Pteropus* and *Ornithorhyncus* - Viviparity
(b) Garden lizard and crocodile - Three chambered heart
(c) *Ascaris* and *Ancylostoma* - Metameric segmentation
(d) Sea horse and flying fish - Cold blooded (poikilothermal)

15. Millipede (*Julus*) and Centipede (*Scolopendra*) are both included under:
(a) Arachnida (b) Myriapoda
(c) Scaphopoda (d) Pelecypoda

16. Which of the following statement(s) is/are **correct**?
(i) Organ systems in different group of animals show various patterns of complexities.
(ii) The digestive system in platyhelminthes has only a single opening to the outside of the body that serve as both mouth and anus, and is hence called complete.
(iii) In open type of circulatory system, the blood is pumped out of the heart and the cells and tissues are directly bathed in it.
(iv) In closed circulatory type, the blood is circulated through a series of vessels of varying diameters (arteries, veins and capillaries).
(a) Only (i) (b) Both (ii) and (iii)
(c) (i), (iii) and (iv) (d) All of these

17. The following statement are associated with the occurrence of notochord. Identify the **incorrect** statement. **Tricky**
(a) It is present only in larval tail in ascidians.
(b) It is replaced by a vertebral column in adult frog.
(c) It is absent throughout the life in humans from the very beginning.
(d) It is present throughout life in *Amphioxus*.

18. In which of the following animal, post anal tail is found?
(a) Earthworm (b) Lower invertebrate
(c) Scorpion (d) Snake

19. A rhabditiform larva is formed in the life cycle of:
(a) *Ascaris* (b) Tapeworm
(c) *Hydra* (d) *Leucosolenia*

20. Sense organs of *Aurelia* are:
(a) Tentacle (b) Tentaculocyst
(c) Nematocyst (d) Otolith

21. Book-lungs are found in:
(a) Myriapoda (b) Crustacea
(c) Arachnida (d) Insecta

22. In animals of which class, the number of body segments is definite?
(a) Oligochaeta (b) *Hirudinaria*
(c) Polychaeta (d) None of these

23. Refer the given figures 1, 2, 3 and 4 and identify the option which shows their correct name.

1 2 3 4

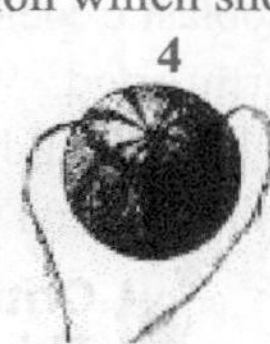

	1	2	3	4
(a)	*Pleurobrachia*	*Cnidoblast*	*Aurelia*	*Adamsia*
(b)	*Aurelia*	*Adamsia*	*Cnidoblast*	*Pleurobrachia*
(c)	*Cnidoblast*	*Pleurobrachia*	*Adamsia*	*Aurelia*
(d)	*Adamsia*	*Aurelia*	*Pleurobrachia*	*Cnidoblast*

24. Which of the following sets of animals give birth to young ones?
(a) *Platypus*, Penguin, Bat, *Hippopotamus*
(b) Shrew, Bat, Cat, Kiwi
(c) Kangaroo, Hedgehog, Dolphin, *Loris*
(d) Lion, Bat, Whale, Ostrich

25. Syncytial epidermis occurs in: 

(a) *Ascaris* (b) *Hydra*
(c) *Taenia* (d) *Leucosolenia*

26. In which of the following classes, locomotory organs are absent?
(a) Rhizopoda (b) Sporozoa
(c) Suctoria (d) Mastigophora

27. Which one of the following animals possesses nerve cells but no nerves?
(a) *Hydra* (b) Tapeworm
(c) Earthworm (d) *Planaria*

28. The given figures 1, 2, 3 and 4 are the examples of first true land vertebrates. They are dominant in mesozoic era and belong to phylum 'X'. Identify 'X' and the animals which have four chambered heart.

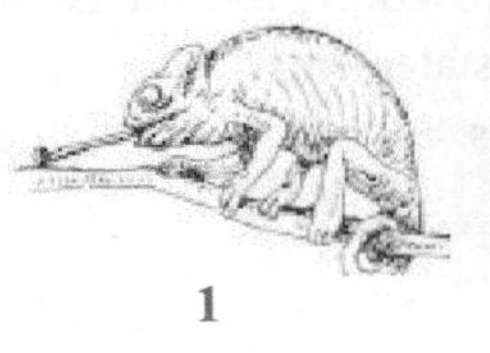
1

2

3

4

(a) X – Reptile; 2 (b) X – Reptile; 1
(c) X – Amphibia; 3 (d) X – Pisces; 4

29. Match column-I with column-II and select the correct answer using the codes given below.

Column-I (Class)	Column-II (Example)
A. Protozoa	I. *Pennatula*
B. Aschelminthes	II. *Beroe*
C. Porifera	III. *Monocystis*
D. Ctenophora	IV. *Wuchereria*
E. Cnidaria	V. *Cliona*

(a) A – III; B – V; C – IV; D – I; E – II
(b) A – IV; B – III; C – V; D – II; E – I
(c) A – III; B – IV; C – V; D – II; E – I
(d) A – II; B – IV; C – V; D – III; E – I

30. Locomotion in Kangaroo is:
(a) Volant (b) Saltatorial
(c) Cursorial (d) Graviportal

31. Match column-I (Characteristic feature) with column-II (Examples) and select the correct answer using the codes given below.

Column-I (Characteristic Features)	Column-II (Examples)
A. Pseudocoelomates	I. *Hydra, Adamsia*
B. Diploblastic	II. *Ctenoplana, Aurelia*
C. Cellular level of organisation	III. *Ascaris, Wuchereria*
D. Radial symmetry	IV. *Sycon, Spongilla*
E. Metamerism	V. *Pheretima, Nereis*

(a) A – V; B – II; C – IV; D – III; E – I
(b) A – III; B – I; C – IV; D – II ; E – V
(c) A – II; B – I; C – III; D – V; E – IV
(d) A – III ; B – II; C – IV; D – I; E – V

32. A student has identified a triploblastic coelomate segmented animal as an arthropod. Which additional character the student should have verified before identifying that animal as an arthropod? **Critical Thinking**

(a) Presence of wings
(b) Presence of antenna
(c) Type of coelom
(d) Type of symmetry

33. Medusa of *Obelia* is:

(a) Carnivorous
(b) Herbivorous
(c) Detritus feeder
(d) Omnivorous

34. Elephantiasis is caused by:

(a) *Trichinella* (b) *Desmoscolex*
(c) *Enterobius* (d) *Wuchereria*

35. Refer the following statement and answer the question.

'Name of "X" is derived from stinging capsules. It exhibits metagenesis containing two body forms in which sessile and cylindrical form is called "Y" and umbrella shaped and free swimming is called "Z". **Toughnut**

Identify X, Y, and Z.

	X	Y	Z
a.	Coelenterate	Polyp	Medusa
b.	Cnidarian	Medusa	Polyp
c.	Ctenophora	Radula	Hypostome
d.	Porifera	Osculum	Radula

36. Identify the characteristic feature marked as 1, 2, 3 and 4 in the given figure.

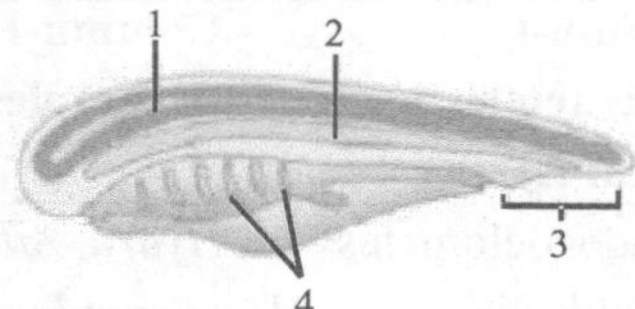

(a) 1-Notochord; 2-Post-anal part; 3-Gill slits; 4-Nerve cord
(b) 1-Nerve cord; 2-Notochord; 3-Post-anal part; 4-Gill slits
(c) 1-Notochord; 2-Nerve cord; 3-Gill slits; 4-Post-anal part
(d) 1-Gill slits; 2-Post-anal part; 3-Nerve cord; 4-Notochord

37. What is the true for mammalia?

(a) *Platypus* is oviparous.
(b) Bats have feather.
(c) Elephant is an ovo-viviparous.
(d) Diaphragm is absent.

38. Which of the following statements (i) to (v) are **incorrect**?

(i) Parapodia are lateral appendages in arthropods used for swimming.
(ii) In molluscs radula is responsible for the process of excretion.
(iii) Aschelminthes are dioecious.
(iv) Echinoderm adults show radial symmetry.
(v) Ctenophorans are diploblastic.

(a) (i) and (ii)
(b) (i) and (iii)
(c) (i), (iv) and (v)
(d) (iii) and (v)

39. Which group of vertebrates comprises the highest number of endangered species?

(a) Mammals (b) Fishes
(c) Reptiles (d) Birds

40. Read the following statements and answer the question.

(i) They are exclusively marine, radially symmetrical, diploblastic organisms with tissue level of organisation.
(ii) Body bears eight external rows of ciliated comb plates, which help in locomotion.
(iii) Digestion is both extracellular and intracellular.
(iv) Reproduction takes place only by sexual means.

Which of the following phylum is being described by above statements?

(a) Platyhelminthes
(b) Arthropoda
(c) Mollusca
(d) Ctenophora

41. Which character is not same in aves and mammals?

(a) Single systemic arch
(b) Metanephric kidney
(c) Seven cervical vertebrae
(d) Homoiotherms

42. In which one of the following the genus name, its two characters and phylum are **incorrectly** matched?

	Genus name	Two characters	Phylum
(a)	*Pila*	(i) Body segmented; Mouth with radula	Mollusca
(b)	*Asterias*	(ii) Spiny skinned; Water vascular system	Echinodermata
(c)	*Sycon*	(iii) Pore bearing; Canal system	Porifera
(d)	*Periplaneta*	(iv) Jointed appendages; Chitinous exoskeleton	Arthropoda

43. Which of the following traits is not shared by both sea anemones and jellyfish?
(a) A medusa as the dominant stage in the life cycle.
(b) Possession of a gastro-vascular cavity.
(c) Sexual reproduction.
(d) Nematocysts present on the tentacles.

44. Which of the following is not the common fundamental feature for animal classification?
(a) Germinal layers
(b) Pathway of water transport
(c) Pattern of organisation of cells
(d) Serial repetition of the segments

45. The given figures 1, 2 and 3 represent types of fluid filled body cavity (coelom) found in many types of animals. Identify the correct option showing the type of coelom. **Tricky**

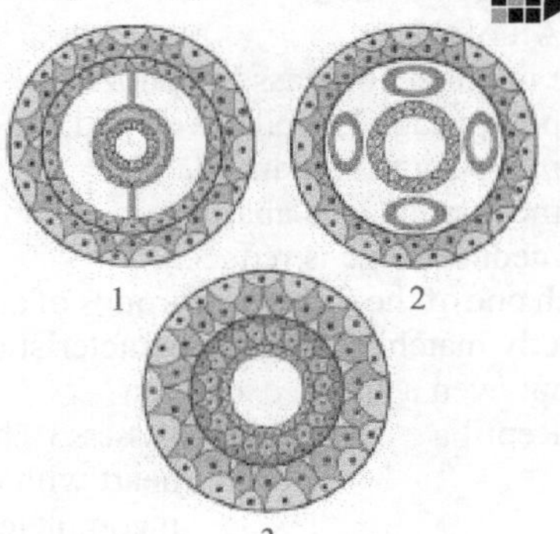

(a) 1 - Pseudocoelomate; 2 - Coelomate; 3 - Acoelomate
(b) 1 - Coelomate; 2 - Pseudocoelomate; 3 - Acoelomate
(c) 1 - Coelomate; 2 - Acoelomate; 3 - Pseudocoelomate
(d) 1 - Coelomate; 2 - Acoelomate; 3 - Eucoelomate

46. Hemichordates have now been placed with the non-chordates, close to echinoderms, because true
(a) notochord is absent.
(b) pharyngeal gill-slits are lacking.
(c) dorsal nerve cord is absent.
(d) heart is lacking.

47. Choanocytes perform:
(a) Nutrition (b) Excretion
(c) Reproduction (d) Secretion of spicules

48. Which of the following is deuterostomate and enterocoelomate invertebrate?
(a) *Pila* (b) *Ascaris*
(c) *Aphrodite* (d) *Asterias*

49. Which of the following characteristics is **correct** for reptilia?
(a) Body covered with dry and cornified skin, scales over the body are epidermal, they do not have external ears.
(b) Body is covered with moist skin and is devoid of scales, the ear is represented by a tympanum, alimentary canal, urinary and reproductive tracts, open into a common cloaca.
(c) Fresh water animals with bony endoskeleton and air-bladder regulate buoyancy.
(d) Marine animals with cartilaginous endoskeleton and body is covered with placoid scales.

50. Which of the following is a living fossil?
(a) *Latimeria* (b) *Lamprey*
(c) Shark (d) Eel

51. In *Hydra*, egestion of undigested food and excretion of nitrogeneous wastes respectively occur through :
(a) mouth and tentacles.
(b) mouth and body wall.
(c) mouth and mouth.
(d) body wall and body wall.

52. "Portuguese man of war" is:
(a) soldier of world war I
(b) portuguese soldier
(c) a sponge
(d) a polymorphic, colonial, colenterata

53. Identify the figures 1, 2 and 3 and choose the correct option.

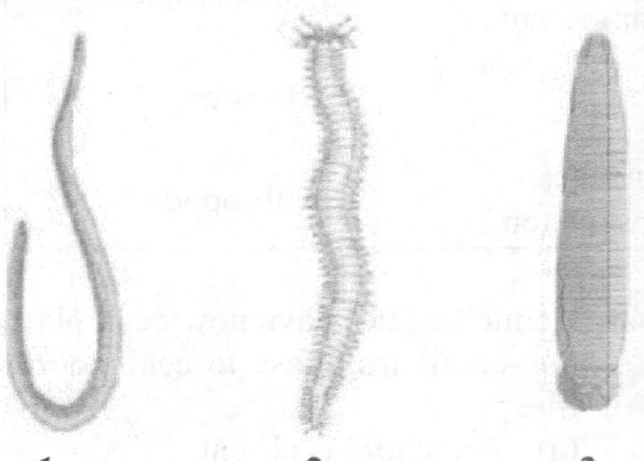

(a) 1 - Male *Ascaris*, 2 - *Hirudinaria* (leech), 3 - *Nereis*
(b) 1 - Female *Ascaris*, 2 - *Nereis*, 3 - *Hirudinaria* (leech)
(c) 1 - Female *Ascaris*, 2 - *Hirudinaria* (leech), 3 - *Nereis*
(d) 1 - Male *Ascaris*, 2 - *Nereis*, 3 - *Hirudinaria* (leech)

54. Which of the following snake is not poisonous?
(a) *Naja-naja* (b) *Python*
(c) *Bungarus* (d) *Hydrophis*

55. Crustacean which show commensalism with some coelenterates is:
(a) *Cypris* (b) *Hippa*
(c) Crab (d) *Eupagurus*

56. Which one of the following statement of animals is **correctly** described with no single exception in it?
(a) In chondrichthyes, notochord is persistent throughout life.
(b) All mammals are viviparous and possess diaphragm for breathing.
(c) All sponges are marine.
(d) All reptiles possess scales, have a three chambered heart and are cold blooded (poikilothermal).

57. Example of living fossil in arthropoda is:
(a) *Cimex* (b) *Triarthrus*
(c) *Peripatus* (d) *Eupagurus*

58. Which of the following characteristic is probably most responsible for the great diversification of insects on land?
(a) Segmentation (b) Antennae
(c) Bilateral symmetry (d) Exoskeleton

59. Which one of the following is a cartilaginous fish?
(a) Silver fish (b) Dog fish
(c) Cray fish (d) Star fish

60. Which of the following statement(s) is/are **correct** for class amphibia?
(i) Body is divisible into head and trunk.
(ii) Respiration is through gills only.
(iii) The heart is two chambered *i.e.* one auricle and one ventricle.
(iv) Fertilisation is internal.
(a) Only (i) (b) Only (iv)
(c) (i), (ii) and (iii) (d) All of these

61. Which of the following endoparasite of man is viviparous?
(a) *Ascaris* (b) *Wuchereria*
(c) *Taenia* (d) *Dracunculus*

62. Refer the following animals and identify those which have a fluid filled body cavity with a complete lining derived from mesoderm? **Tricky**
(i) *Sycon* (ii) Butterfly
(iii) *Nereis* (iv) Sea fan
(v) Scorpion (vi) *Pila*
(a) (i) and (iii) only
(b) (ii) and (iv) only
(c) (ii), (iii), (v) and (vi) only
(d) All of the above

63. *Scoliodon* is called dog fish because of its:
(a) Carnivorous nature
(b) Long mouth
(c) Power of smell
(d) All of these

64. In the members of class Scyphozoa:
(a) polyp stage is well developed.
(b) polyp stage is reduced.
(c) medusa stage is absent.
(d) medusa stage is reduced.

65. Which one of the following groups of animals is correctly matched with its characteristic feature without even a single exception?
(a) Reptilia : possess 3-chambered heart with one incompletely divided ventricle.
(b) Chordata : Possess a mouth provided with an upper and lower jaw.
(c) Chondrichthyes : Possess cartilaginous endoskeleton.
(d) Mammalia : Give birth to young one.

66. Carapace is present on the dorsal surface of:
(a) Elephant (b) Tortoise
(c) Python (d) Alligator

67. Identify the figure with its correct name and phylum.

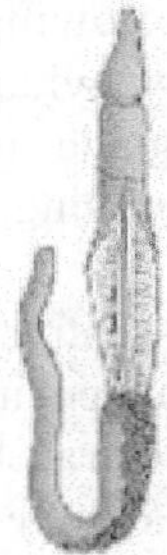

(a) *Cucumaria* – Echinodermata
(b) *Ascidia* – Urochordata
(c) *Balanoglossus* – Hemichordata
(d) *Hirudinaria* – Annelida

68. What distinguishes an insect from a crustacean?
(a) Number of eye
(b) Arrangement of nerve cord
(c) Number of appendages
(d) Presence of wings

69. Which one of the following groups of structures/organs have similar function?
(a) Typhlosole in earthworm, intestinal villi in rat and contractile vacuole in *Amoeba.*
(b) Nephridia in earthworm, Malpighian tubules in cockroach and urinary tubules in rat.
(c) Antennae of cockroach, tympanum of frog and clitellum of earthworm.
(d) Incisors of rat, gizzard (proventriculus) of cockroach and tube feet of starfish.

70. What is common in whale, bat and rat?

(a) Absence of neck
(b) Muscular diaphragm between thorax and abdomen
(c) Extra abdominal testes to avoid high temperature of body
(d) Presence of external ears

71. Pupa of mosquito is called:
(a) Bolus (b) Tumbler
(c) Maggot (d) Wriggler

72. Which of the following statement(s) is/are **correct** regarding class aves?
(i) The forelimbs are modified into wings and the hindlimbs generally have scales and are modified for walking, swimming or clasping the tree branches.
(ii) Heart is completely four-chambered.
(iii) They are warm-blooded (homoiothermous) animals *i.e.*, they are able to maintain a constant body temperature.
(iv) They are oviparous and development is direct.
(a) Both (i) and (iii) (b) Both (i) and (iv)
(c) (i), (ii) and (iii) (d) All of these

73. A student was given a specimen to identify on the basis of the characteristics given below.
(i) They are metamerically segmented.
(ii) They have closed circulatory system.
(iii) They have circular and longitudinal muscles for locomotion.
Identify the specimen.
(a) *Prawn* (b) *Pheretima*
(c) *Wuchereria* (d) *Ctenoplana*

74. Poisonous fangs of a snake are modified:
(a) Mandible (b) Maxillary teeth
(c) Canines (d) Nasals

75. Which one is not a specific trait of birds (Aves)?
(a) Bipedal locomotion (b) Toothless beak
(c) Wings (d) Feathers

76. The free swimming larva of coelenterates is:
(a) Onchosphere (b) Hydrula
(c) Planula (d) Amphiblastula

77. Which of the following is the characteristic feature of fishes?
(a) Tail and venous heart
(b) Venous heart and gills
(c) Epidermal scales and tail
(d) Epidermal scales and gills

78. Which one of the following sets of animals belongs to the same class of a phylum?
(a) *Hydra*, jelly fish, cray fish
(b) Bat, pigeon, whale
(c) Spider, scorpion, centipede
(d) Whale, otter, kangaroo

79. Which of the following class is being described by the given statements?
(i) They are found in a variety of habitats-polar ice-caps, deserts, mountains, forests, grasslands and dark caves.
(ii) Most unique characteristic is the presence of mammary glands by which the young ones are nourished.
(iii) Heart is four-chambered.
(iv) Sexes are separate and fertilisation is internal.
(a) Reptilia (b) Aves
(c) Mammalia (d) Amphibia

80. Amphiblastula is larva of:
(a) *Leucosolenia* (b) *Sycon*
(c) Both (a) and (b) (d) *Clathrina*

81. Which animals become asymmetrical due to torsion in body and shell?
(a) Gastropoda (b) Cephalopoda
(c) Pelecypoda (d) All of these

82. Uricotelism is found in:
(a) fishes and fresh water protozoans.
(b) birds, reptiles and insects.
(c) frogs and toads.
(d) mammals and birds.

83. Trait common amongst earthworm, leech and centipede is:
(a) Absence of legs
(b) Hermaphrodite nature
(c) Ventral nerve cord
(d) Malpighian tubules

84. Which one of the following given character is **incorrect** for phylum chordata?
(a) Presence of notochord
(b) Presence of vertebral column
(c) Paired gill slits
(d) A central nervous system dorsal to digestive canal

85. Which of the following group belongs to deuterostome?
(a) Annelida, Arthropoda, Mollusca
(b) Echinodermata, Hemichordata, Chordata
(c) Annelida, Mollusca, Chordata
(d) Arthropoda, Mollusca, Echinodermata

86. Similarity between fish and tadpole is:
(a) Legs (b) Fins
(c) Lateral line (d) Scales

87. An indigenous fish used in biological control is:
(a) *Catla* (b) *Lebistes*
(c) *Gambusia* (d) *Aplocheilus*

88. Mesogloea of *Hydra* is made up of: **Tricky**
(a) Mucopolysaccharides
(b) Protein
(c) Protein and fat
(d) Reticulate tissue

89. Which of the following feature is not common between *Newt & Hemidactylus*?
(a) Body is divisible into head, neck, trunk and tail.
(b) Head with pair of eyes and tympanic membrane.
(c) Trunk has 2 pair of limbs for locomotion.
(d) Heart is 3-chambered.

90. Flight muscles of bird are attached to:
(a) Clavicle
(b) Coracoid
(c) Keel of sternum
(d) Scapula

91. Which of the following statement(s) is/are **correct** regarding phylum coelenterata?
(i) They are aquatic, mostly marine, sessile or free- swimming, radially symmetrical animals.
(ii) They have a central gastro-vascular cavity with a single opening called hypostome.
(iii) Digestion is extracellular and intracellular.
(iv) Examples are *Sycon, Spongilla* and *Euspongia*.
(a) (i) and (ii) (b) (i) and (iv)
(c) (i), (ii) and (iii) (d) All of these

92. Match column-I (contains the characteristics features) and column-II (contains the function/ location) and select the correct match from the option given below.

	Column-I (Characteristic feature)		Column-II (Function/ Location)
A.	Water canal system	I.	Sponges
B.	Comb plates	II.	Help in swimming
C.	Nephridia	III.	Presentinmollusca
D.	Jointed appendages	IV.	Characteristics of roundworm
E.	Muscular foot	V.	A body part of arthropoda
		VI.	Helps in reproduction
		VII.	Platyhelminthes
		VIII.	Helps in osmoregulation and excretion
		IX.	Eight ciliated external rows present in a body of ctenophora.

(a) A – I; B – IX; C – VIII; D – V; E – III
(b) A – II; B – III; C – VI; D – IX; E – V
(c) A – II; B – V; C – I; D – IV; E – IX
(d) A – III; B – VI; C – IV; D – V; E – I

93. Electric organs occur in:
(a) Sharks (b) Goldfish
(c) Porpoises (d) Rays

94. A common characteristic of all vertebrates without exception is:
(a) the division of body into head, neck, trunk and tail.
(b) body covered with exoskeleton.
(c) the possession of two pairs of functional appendages.
(d) the presence of well- developed skull.

95. A student brought a strange animal at home which he found outside under a rock. It had moist skin, a complete digestive tract, a ventral nerve cord, and had gone through torsion. Identify the phylum of the animal.
(a) Porifera (b) Annelida
(c) Mollusca (d) Echinodermata

96. Ruminants belongs to order:
(a) Proboscida (b) Artiodactyla
(c) Marsupials (d) Edentata

97. Dicondylic skull with 10-pairs of cranial nerves occur in:
(a) Pisces (b) Amphibia
(c) Reptilia (d) Mammalia

98. A poisonous lizard is:
(a) *Heloderma* (b) *Ancistrodon*
(c) *Chamaeleon* (d) *Varanus*

99. Venom of cobra affects:
(a) Respiratory system
(b) Nervous system
(c) Circulatory system
(d) Digestive system

100. Match the phylum given in column-I with their example given in column-II and choose the correct option.

Column-I (Phylum)	Column-II (Examples)
A. Echinodermata	I. *Ascidia, Doliolum*
B. Hemichordata	II. *Asterias, Ophiura*
C. Urochordata	III. *Branchiostoma*
D. Cephalochordata	IV. *Balanoglossus, Saccoglossus*

(a) A – IV; B – II; C – I; D – III
(b) A – II; B – IV; C – I; D – III
(c) A – II; B – IV; C – III; D – I
(d) A – II; B – I; C – IV; D – III

101. Which of the following statement is **incorrect**?
(a) Platyhelminthes has incomplete digestive system.
(b) In coelenterates, the arrangement of cells is more complex.
(c) *Nereis* is monoecious but earthworms and leeches are dioecious.
(d) Simple and compound eyes are present in the animals of those phylum whose over two-thirds of all named species on earth are arthropods.

Answer KEYs

1	(a)	12	(a)	23	(b)	34	(d)	45	(b)	56	(a)	67	(c)	78	(d)	89	(d)	100	(b)
2	(b)	13	(b)	24	(c)	35	(a)	46	(a)	57	(c)	68	(c)	79	(c)	90	(c)	101	(c)
3	(c)	14	(d)	25	(a)	36	(b)	47	(a)	58	(d)	69	(b)	80	(c)	91	(c)		
4	(b)	15	(b)	26	(b)	37	(a)	48	(d)	59	(b)	70	(b)	81	(a)	92	(a)		
5	(c)	16	(c)	27	(a)	38	(a)	49	(a)	60	(a)	71	(b)	82	(b)	93	(d)		
6	(c)	17	(c)	28	(a)	39	(a)	50	(a)	61	(b)	72	(d)	83	(c)	94	(d)		
7	(b)	18	(d)	29	(c)	40	(d)	51	(b)	62	(c)	73	(b)	84	(b)	95	(c)		
8	(a)	19	(a)	30	(b)	41	(c)	52	(d)	63	(c)	74	(b)	85	(b)	96	(b)		
9	(a)	20	(b)	31	(b)	42	(a)	53	(b)	64	(b)	75	(c)	86	(c)	97	(b)		
10	(a)	21	(c)	32	(b)	43	(a)	54	(b)	65	(c)	76	(c)	87	(b)	98	(a)		
11	(a)	22	(b)	33	(a)	44	(b)	55	(d)	66	(b)	77	(b)	88	(a)	99	(b)		

Hints & Solutions

1. (a) Storage of carbohydrates as starch is the characteristic feature of kingdom plantae. In animals, carbohydrates are stored in the form of glycogen while in plants, carbohydrates are stored in the form of sucrose.
2. (b) On the basis of organisation, animal kingdom is classified into two sub-kingdom: protozoa and metazoa. Unicellular animals are placed in protozoa while multicellular organisms are included in metazoa.
3. (c) 4. (b) 5. (c)
6. (c) Nephridia present in phylum- annelida help in osmoregulation and excretion. In aschelminthes, an excretory tube removes body wastes from the body cavity through the excretory pore.
7. (b) *Enterobius* is an example of nematode. *Pennatula,* a colonial coral belongs to coelenterate phylum. *Bonellia* (the green spoon worm) is a marine worm (phylum Echiura) noted for displaying exceptional sexual dimorphism and for the biocidal properties of a pigment in its skin.
8. (a) 9. (a)
10. (a) Circulatory system in arthropods is of open type *i.e.,* blood does not flow in definite vessels. Irregular spaces known as lacunae or sinuses, filled with blood are present.
11. (a) On the basis of the given figure of germinal layers, the animals belongs to 1 and 2 are respectively diploblastic and triploblastic. Diploblastic animals have two germinal layers, outer ectoderm and inner endoderm, *e.g.,* porifera and coelenterate. Triploblastic animals have three germinal layers – outer ectoderm, middle mesoderm and inner endoderm, *e.g.,* platyhelminthes, aschelminthes, annelida, arthropoda, molluscs, echinodermata and chordata.
12. (a)
13. (b) Jaws are absent in protochordata and cyclostomata. Protochordata refers to invertabrate subphylum of phylum chordata. It includes animals such as tunicates. Cyclostomata is a class of group agnatha, where paired appendages, girdles and jaw are absent.
14. (d) Sea horse and flying fish are cold blooded animals. *Ornithorhyncus* is oviparous. Crocodile has four chambered heart. *Ascaris* and *Ancylostoma* are segmented roundworms.
15. (b)
16. (c) The digestive system of platyhelminthes is incomplete. A complete digestive system has two openings, mouth and anus.
17. (c) Notochord is a flexible rod like structure that forms the main support of the body in the lowest chordates. It is not absent in humans throughout their life. Notochord is present in embryonic stage and get changed or replaced by vertebral column in the adult.
18. (d) 19. (a) 20. (b)
21. (c) Book-lungs are found in Arachnida.
22. (b) *Hirudinaria* have definite number of body segments.
23. (b) *Aurelia* is a genus of scyphozoan jellyfish (Cnidaria, Scyphozoa). *Adamsia* is a genus of sea anemones. It is usually found growing on a gastropod shell inhabited by the hermit crab, *Pagurus prideaux*. A cnidocytes (also known as a cnidoblast or nematocyte) is an explosive cell containing one giant secretory organelle that defines the phylum cnidaria (corals, sea anemones, *Hydra*, jellyfish, etc.). *Pleurobranchus* is a genus of sea slugs, specifically side-gill slugs, marine gastropod mollusc.
24. (c) Penguin, kiwi and ostrich all belong to class aves (*i.e.* birds) under phylum chordata and they do not give birth to their young ones. They are oviparous while kangaroo, hedgehog, dolphin and loris, all belong to class mammalia and are viviparous.
25. (a) 26. (b) 27. (a)
28. (a) The given figures (1, 2, 3 and 4) are respectively chameleons, crocodile, turtle and cobra. Reptiles are the first true land vertebrates and dominant in mesozoic era. They are cold blooded vertebrates. Heart in reptiles are usually three chambered but crocodiles is an exception having a four chambered heart.
29. (c) 30. (b)

31. (b) Pseudocoelomate is a group of invertebrates with a three-layered body that has a fluid-filled body cavity called pseudocoelom. E.g., roundworms, rotifers. Diploblastic animals are with two germ layers - ectoderm and endoderm, e.g., cnidaria. In cellular level of organisation, the cells are arranged as loose cell aggregates, e.g., porifera. In radial symmetry, symmetrical arrangement of parts of an organism is around a single main axis, so that the organism can be divided into similar halves by any plane that contains the main axis. The body plans of echinoderms, ctenophores, cnidarians, and many sponges and sea anemones show radial symmetry. Metamerism is the phenomenon of having a linear series of body segments fundamentally similar in structure, e.g., annelida.

32. (b)

33. (a) The medusa is strictly carnivorous. The food includes minute worms, nematodes, insects, crustaceans, etc.

34. (d)

35. (a) Name of coelenterate is derived from stinging capsules. It exhibits metagenesis containing two body forms in which sessile and cylindrical form is called polyp and umbrella shaped and free – swimming is called medusa.

36. (b) The give figure belongs to phylum chordata, which are characterised by the presence of notochord, a dorsal hollow nervechord, paired pharyngeal gill slits and post anal part.

37. (a)

38. (a) Aquatic annelids like *Nereis* possess lateral appendages (called parapodia) which help in swimming. Radula is the rasping organ for feeding which is present in the mouth of molluscs.

39. (a)

40. (d) Ctenophores, also known as comb jellies/ sea gooseberries/sea walnuts, or Venus's girdles, are voracious predators.

41. (c) Except a few, only mammals possess seven cervical (neck) vertebrae.

42. (a) Molluscans are soft bodied animals. Their body is unsegmented with a distinct head, muscular foot and visceral hump. In *Pila,* the buccal cavity contains a rasping organ, the radula with transverse rows of teeth.

43. (a) Medusa and polyp are the stages in the life cycle of cnidarians. Medusa is a sexual free swimming form and polyp is asexual form. The jellyfish possesses both medusa and polyp stage in its life cycle, but the sea anemones do not have the medusa stage and spend the life cycle as polyps only.

44. (b) Pathway of water transport is not the common fundamental feature for animal classification.

45. (b) Pseudocoelomates : Body cavity is not lined by mesoderm continuously, but it is present as scattered pouches in between the ectoderm and endoderm. Aschelminthes (round worms) are pseudocoelomates.
Acoelomates : The animals which do not have coelom *e.g.,* sponges, coelenterates, ctenophorans and flatworms.
Eucoelomates (Coelomates) : The animals which possess true coelom are coelomate True coelom is found in annelids, echinoderms and chordates.

46. (a) Hemichordates have now been placed with the non-chordates because true notochord is absent in them but a buccal diverticulum is present in pre-oral region which is often called stomochord.

47. (a) Digestion in sponges is intracellular like protozoans. It takes place in choanocytes. Choanocytes (also called collar cells), are flagellated cells with a collar of protoplasm at the base of flagellum, numbers of which line the internal chambers of sponges of class porifera.

48. (d)

49. (a) Characteristics given in options (b), (c) and (d), belong to the classes amphibia, osteichthyes and chondrichthyes respectively.

50. (a)

51. (b) In *Hydra,* undigested residues are egested from coelenteron through mouth and body wall.

52. (d)

53. (b) *Ascaris* belongs to nematode phylum which is a non - segmented roundworms / threadworms/pinworms. *Nereis* is a genus of polychaete worms. It possesses setae and parapodia for locomotion. *Hirudinaria,* also known as cattle leech, belongs to annelida phylum. It acts as a parasite on cattle.

54. (b) *Python* is not a poisonous snake captures its prey and directly engulfs it by creating suction pressure inside its mouth.

55. (d)
56. (a) *Platypus* is an oviparous mammal. *Spongilla* and *Euspongia* are fresh water sponges. Crocodiles have four chambered heart. In the members of class chondrichthyes, the notochord is persistent throughout life.
57. (c)
58. (d) Exoskeleton is probably the most responsible feature for the great diversification of insects on land. An exoskeleton is the external skeleton that supports and protects an animal's body, in contrast to the internal skeleton (endoskeleton), for example, human. Examples of exoskeleton animals include insects such as grasshoppers and cockroaches, and crustaceans such as crabs and lobsters.
59. (b) *Elasmobranchii* (dog fish) is one of the two subclasses of cartilaginous fish in the class chondrichthyes, the other being holocephali. Members of elasmobranchii subclass have no swim bladders, five to seven pairs of gills clefts opening individually to the exterior, rigid dorsal fins, and some placoid scales.
60. (a) As the name suggests, amphibians can live in aquatic as well as terrestrial habitats. Respiration is by gills, lungs and through skin. The heart is three chambered (two auricles and one ventricle) and the fertilisation is external.
61. (b)
62. (c) Butterfly, *Nereis*, scorpion and *Pila* are animals which have a fluid filled body cavity with a complete lining derived from mesoderm, Hence, called as coelomate animals.
63. (c) 64. (b)
65. (c) Chondrichthyes are the cartilaginous fish with a flexible skeleton made of cartilage rather than bone.
66. (b)
67. (c) *Balanoglossus* is a deuterostome, and resembles the *Ascidians* or sea squirts. It possesses branchial openings, or "gill slits" and has notochord in the upper part of the body and has no nerve chord.
68. (c) 69. (b) 70. (b) 71. (b)
72. (d) Aves are warm blooded, oviparous, bipedal flying vertebrates with an exoskeleton of feathers. Forelimbs are modified as wings for flying and hind limbs are adapted for walking, perching and swimming.
73. (b) *Pheretima* belongs to annelida phylum. Annelida is a group of segmented worms, and they are found worldwide from the deepest marine sediments to the soils in our city parks and yards.
74. (b) 75. (c) 76. (c)
77. (b) Fishes have two chambered heart (one auricle and one ventricle), which receive only venous blood and pump it to gills for purification.
78. (d) Because they are belong to the class mammalia.
79. (c) Mammals are tetrapod that have hair, a four-chambered heart, a diaphragm, and mammary glands.
80. (c) 81. (a) 82. (b) 83. (c)
84. (b) Vertebral column is present in the vertebrates only. It is not present in all the chordates.
85. (b) Deuterostomes includes echinodermata, hemichordata and chordata. Deuterostomia is a subtaxon of the bilateria branch of the subkingdom eumetazoa, within animalia and are distinguished from protostomes by their deuterostomic embryonic development. In deuterostomes, the first opening (the blastopore) becomes the anus, while in protostomes, it becomes the mouth.
86. (c) Lateral line system in a fish and some aquatic larvae (tadpole) is made up of neuromast organs. It detects vibrations and pressure changes in water.
87. (b) 88. (a) 89. (d) 90. (c)
91. (c) *Sycon*, *Spongilla* and *Euspongia* are examples of phylum porifera. *Physalia*, *Adamsia*, *Pennatula*, *Gorgonia* and *Meandrina* are examples of phylum coelenterata (cnidaria).
92. (a) 93. (d)
94. (d) The sub-phylum vertebrata (or craniata) have a well-developed nervous system that is differentiated into brain and spinal cord. Brain is protected by a brain box called cranium, so they are also called as craniata.
95. (c) The animal found by the boy belongs to mollusca phylum. Mollusca are the second largest phylum after arthropoda and include predominantly marine animals. They are triploblastic, bilaterally symmetrical, schizocoelic and unsegmented protostomes. They have moist skin, a complete digestive tract, a ventral nerve cord, and had gone through torsion.
96. (b) 97. (b) 98. (a) 99. (b) 100. (b)
101. (c) *Nereis* is dioecious but earthworms and leeches are monoecious.

5 Morphology of Flowering Plants

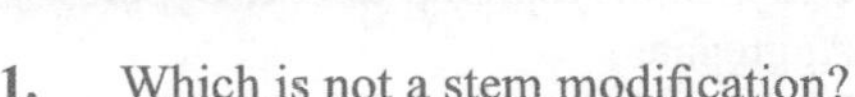

1. Which is not a stem modification?
   (a) Rhizome of ginger
   (b) Corm of *Colocasia*
   (c) Pitcher of *Nepenthes*
   (d) Tuber of potato
2. Branches of stem are:
   (a) exogenous in origin
   (b) endogenous in origin
   (c) partly exogenous and partly endogenous
   (d) superficial
3. The leaves are modified into tendrils, hooks, pitcher and bladder in the following plants respectively:
   (a) Sweet pea, Cat's nail, *Nepenthes*, *Utricularia*
   (b) Sweet pea, Cat's nail, *Utricularia*, *Nepenthes*
   (c) *Nepenthes*, Sweet pea, Cat's nail, *Utricularia*
   (d) *Utricularia*, *Nepenthes*, Cat's nail, Sweet pea
4. Which type of function is performed by the fleshy leaves of onion and garlic?
   (a) Storage (b) Reproduction
   (c) Photosynthesis (d) Protection
5. Fibrous roots develop in maize from: **Tricky**
   (a) Upper nodes (b) Lower nodes
   (c) Upper internodes (d) None of these
6. Match column-I with column-II and select the correct answer using the codes given below.

	Column-I		Column-II
A.	Tubercular storage roots	I.	*Tinospora*
B.	Pneumatophores	II.	*Heritiera*
C.	Haustoria	III.	*Asparagus*
D.	Prop-roots	IV.	*Viscum*
E.	Assimilatory roots	V.	*Pandanus*

   (a) A – II; B – III; C– IV; D – V; E – I
   (b) A – III; B – IV; C–V; D – I; E– II
   (c) A – III; B – I; C– II; D –V; E– IV
   (d) A – III; B – II; C– IV; D –V; E– I
7. Fibrous root system is found in:
   (a) Monocotyledonous plants
   (b) Dicotyledonous plants
   (c) Bryophytes
   (d) Gymnosperms
8. Stilt roots occur in:
   (a) Groundnut (b) Rice
   (c) Sugarcane (d) Wheat
9. Napiform roots are recorded from:
   (a) Radish (b) Carrot
   (c) Beet root (d) Sweet potato
10. Regions of roots from base to root tip are:
   (a) Maturation zone - Cell division zone - Elongation zone
   (b) Maturation zone - Elongation zone - Cell division zone
   (c) Cell division zone - Elongation zone - Maturation zone
   (d) Elongation zone - Cell division zone - Maturation zone
11. Floral formula of Tomato/Tobacco is: **Toughnut**
   (a) $\oplus$ ⚥ $K_{2+2}C_4A_{2+1}G_1$
   (b) $\oplus$ ⚥ $P_2A_3G_1$
   (c) $\oplus$ ⚥ $P_3A_3G_2$
   (d) $\oplus$ ⚥ $K_{(5)}C_{(5)}A_5G_{(2)}$
12. Which one of the following is a fibre yielding plant?
   (a) *Crotalaria juncea*
   (b) *Cicer arietinum*
   (c) *Triticum vulgare*
   (d) *Impatiens balsamina*
13. The anthers in solanaceae are:
   (a) Monothecous, introse
   (b) Dithecous, extrose
   (c) Dithecous, introse
   (d) Monothecous, extrose
14. Name the family having (9) + 1 arrangement of stamens?
   (a) Solanaceae (b) Asteraceae
   (c) Liliaceae (d) Fabaceae

15. Which is correct pair for edible part?
(a) Tomato - Thalamus
(b) Maize - Cotyledons
(c) Guava - Mesocarp
(d) Date palm - Pericarp

16. Which of the following statements are **correct**?
(i) If the stem is jointed with solid nodes and hollow internodes, it is called caudex.
(ii) In *Tridax*, the stem is decumbent.
(iii) Corm is a condensed form of rhizome growing more or less in vertical direction.
(iv) Sucker is an underground modificaton of stem.
(v) Biparous type of cymose branching is seen in *Saraca*.
(a) (i), (iv) and (v) (b) (ii) and (iii)
(c) (ii), (iii) and (v) (d) (iii) and (iv)

17. Match column-I with column-II and select the correct answer using the codes given below.

Column-I	Column-II
A. Entire leaf modified into a spine.	I. *Clematis*
B. Leaf except stipules modified into a tendril.	II. *Citrus*
C. Stipules modified into a tendril.	III. *Euphorbia*
D. First leaf of axillary bud modified into a spine.	IV. *Lathyrus*

(a) A – III; B – IV; C – I; D – II
(b) A – III; B – I; C – IV; D – II
(c) A – II; B – III; C – II; D – IV
(d) A – IV; B – II; C – I; D – III

18. A scar on the seed coat through which the developing seeds are attached to the fruit is called as ________.
(a) Testa (b) Tegmen
(c) Hilum (d) Micropyle

19. A true fruit develops from:
(a) Ovary (b) Thalamus
(c) Petals (d) Receptacle

20. Catkin is a type of:
(a) Flower (b) Inflorescence
(c) Stem (d) All of these

21. Which of the following are schizocarpic fruits?
(a) Siliqua & Legume
(b) Capsule & Berry
(c) Lomentum & Capsule
(d) Carcerulus & Lomentum

22. Which of the following correctly represents the type of fruits given below.

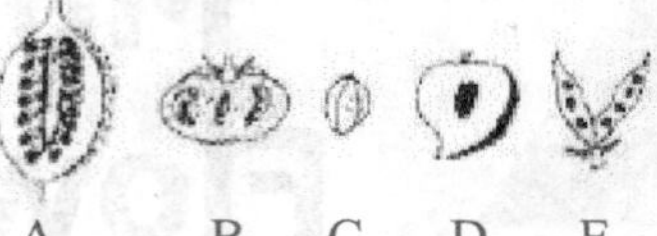

A B C D E

(a) A–Berry; B–Caryopsis; C–Drupe; D–Sorosis; E–Aggregate
(b) B–Berry; C–Caryopsis; D–Drupe; A–Sorosis; E–Aggregate
(c) B–Berry; C–Caryopsis; D–Drupe; E–Legume; A–Aggregate
(d) A–Berry; C–Caryopsis; D–Drupe; B–Sorosis; E–Composite

23. Which one is the correct answer for cohesion of stamens in the given diagram?

(a) Monoadelphous (b) Diadelphous
(c) Polyadelphous (d) Synandrous

24. The perianth is the term used when:
(a) androecium and gynoecium are similar.
(b) androecium and calyx are similar.
(c) corolla and gynoecium are similar.
(d) calyx and corolla are similar.

25. Eye of potato is:
(a) Axillary bud (b) Accessory bud
(c) Adventitious bud (d) Apical bud

26. Flowers are sessile in:
(a) Raceme (b) Spike
(c) Corymb (d) Umbel

27. Sleeping movements commonly occur in plants of family:
(a) Leguminosae (b) Liliaceae
(c) Malvaceae (d) Compositae

28. Fibres are usually obtained from the members of:
(a) Solanaceae (b) Malvaceae
(c) Leguminosae (d) Cruciferae

29. A sterile stamen is known as:
(a) Staminode (b) Anther
(c) Pollen grain (d) Filament

30. Capitulum inflorescence is a modification of:
(a) Raceme (b) Panicle
(c) Spike (d) Cyme

31. An example of false fruit is:
(a) Apple (b) Pomegranate
(c) Grapes (d) Mango

32. Prop roots of Banyan tree are meant for:
(a) Respiration
(b) Absorption of water from soil
(c) Providing support to big tree
(d) All of the above

33. Reserve food is mainly found in:
(a) underground stems.
(b) underground modified tap roots.
(c) underground modified adventitious roots.
(d) All of the above

34. Match column-I with column-II and select the correct answer using the codes given below.

Column-I	Column-II
A. Leguminosae	I. Axile placentation
B. Papilionatae	II. Tetradynamous
C. Solanaceae	III. Marginal placentation
D. Cruciferae	IV. Diadelphous stamens
E. Malvaceae	V. Epipetalous stamens

(a) A – III; B – IV; C – V; D – II; E – I
(b) A – IV; B – III; C – V; D – II; E – I
(c) A – III; B – IV; C – V; D – I; E – II
(d) A– III; B – V; C– IV; D – II; E – I

35. Match column-I with column-II and select the correct answer using the codes given below.

Column-I	Column-II
A. Wheat	I. False fruit
B. Ground nut	II. Caryopsis
C. Grapes	III. Lomentum
D. Mango	IV. Parthenocarpic fruit
E. Apple	V. Drupe

(a) A – III; B – II; C – IV; D – V; E – I
(b) A – II; B – III; C – IV; D – I; E – V
(c) A– II; B – III; C– IV; D –V; E – I
(d) A – II; B – IV; C– III; D – V; E – I

36. The most important function of inflorescence is to help in:
(a) increasing length of plant.
(b) attracting insects for cross pollination.
(c) dispersal of seeds.
(d) release of pollen grains.

37. In water chestnut, the thorns are modification of:
(a) Leaflet (b) Corolla
(c) Leaf (d) Calyx

38. New banana plants develop from:

(a) Rhizome (b) Sucker
(c) Stolon (d) Seed

39. Choose the correct combinations.

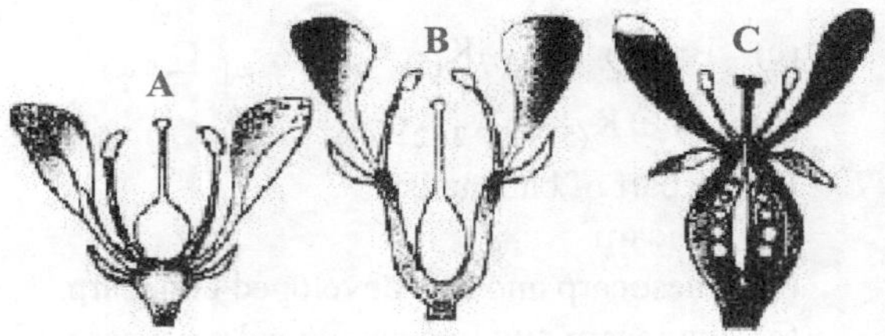

(i) Hypogynous flower
(ii) Perigynous flower
(iii) Epigynous flower
(a) A-(i), B-(ii), C-(iii)
(b) A-(i), B-(iii), C-(ii)
(c) A-(iii), B-(ii), C-(i)
(d) A-(iii), B-(i), C-(ii)

40. Which option is correctly matched with the diagrams?

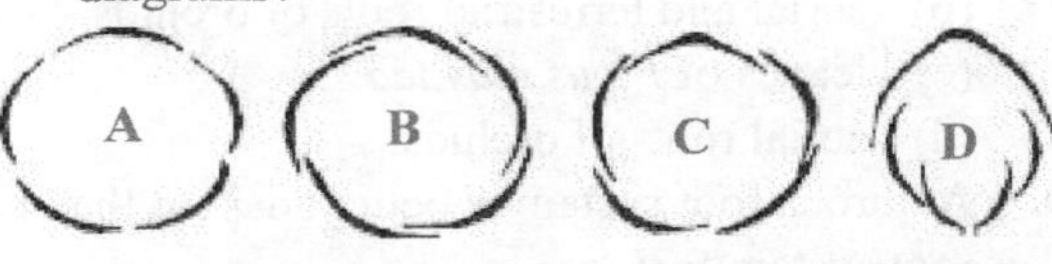

(a) A-Valvate, B-Twisted, C-Imbricate, D-Vexillary
(b) A-Vexillary, B-Valvate, C-Twisted, D-Imbricate
(c) A-Imbricate, B-Vexillary, C-Valvate, D-Twisted
(d) A-Twisted, B-Imbricate, C-Vexillary, D-Valvate

41. The largest leaf belongs to:
(a) *Nerium* (b) Tobacco
(c) *Victoria* (d) *Rafflesia*

42. A modification of leaf is:
(a) Phyllode (b) Phylloclade
(c) Cladode (d) Corm

43. Axis between corolla and androecium is:
(a) Anthophore (b) Gynophore
(c) Gynandrophore (d) Androphore

44. Choose the product that is derived from style and stigma.
(a) Saffron (b) Fenugreek
(c) *Asafoetida* (d) *Psyllium*

45. Opium is obtained from which part of the plant, *Papaver somniferum?*
(a) Inflorescence
(b) Fruit (unripe)
(c) Male part (dried)
(d) Seeds

46. Floral formula of Malvaceae:

(a) $\mathrm{Br}\%\oplus\text{⚥}\ \mathrm{Epi}_{3\text{-}7}K_{(5)}\ \overset{\frown}{C_{(5)}\ A_{(\infty)}}\ \underline{G}_{(2)}$

(b) $\mathrm{Br}\oplus\text{⚥}\ K_{(5)}C_5A_{(\infty)}\underline{G}(5\text{-}\infty)$

(c) $\mathrm{Br}\oplus\text{⚥}\ \mathrm{Epi}_{3\text{-}7}K_{(5)}\ \overset{\frown}{C_{(5)}\ A_{(\infty)}}\ \underline{G}_{(5\text{-}\infty)}$

(d) $\mathrm{Br}\oplus K_{(5)}C_4A_{2+2}\underline{G}_{(2)}$

47. Edible part of banana is:
(a) epicarp.
(b) mesocarp and less developed endocarp.
(c) endocarp and less developed mesocarp.
(d) epicarp & mesocarp.

48. Staminodes commonly occur in:
(a) Liliaceae
(b) Papilionatae
(c) Caesalpinoideae
(d) All of the above

49. Velamen is found in: 

(a) roots of screwpine.
(b) aerial and terrestrial roots of orchids.
(c) leaves of *Ficus elastica*.
(d) aerial roots of orchids.

50. A fibrous root system is better adapted than tap root system for:
(a) Storage of food
(b) Anchorage of plant to soil
(c) Absorption of water and minerals
(d) Transport of water and organic food

51. Buttress roots are found in:
(a) *Sorghum* (b) banyan
(c) *Terminalia* (d) *Pandanus*

52. Which of the following is a modified stem for the protection of plants from browsing animals?
(a) Tendrils (b) Thorns
(c) Rhizome (d) Tuber

53. Select the statements which are not applicable to the family solanaceae?
(i) Epipetalous and syngenesious anthers.
(ii) Bicarpellary and syncarpous ovary.
(iii) Oblique ovary with axile placentation.
(iv) Stamen six, arranged in two whorls.
(v) Bicarpellary, syncarpous and inferior ovary
(a) (ii) and (iii)
(b) (i), (iv) and (v)
(c) (ii), (iv) and (v)
(d) (i) and (iii)

54. A fruit developing from an inflorescence is called:
(a) Etaerio (b) Composite fruit
(c) Achene (d) Berry

55. The inflorescence in cauliflower is:
(a) Compound corymb
(b) Corymb
(c) Umbel
(d) Catkin

56. Water melon is:
(a) Pome (b) Sorosis
(c) Pepo (d) Drupe

57. The most advanced type of inflorescence is:
(a) Corymb (b) Capitulum
(c) Spadix (d) Catkin

58. An edible inflorescence is: Tricky
(a) *Brassica rapa* (b) Mustard
(c) *Raphanus sativus* (d) *Brassica oleracea*

59. Leaves are changed into spines in xerophytic structures called:
(a) Phyllode (b) Stolon
(c) Phylloclade (d) All of these

60. Match column-I with column-II and choose the correct option given below.

	Column-I		Column-II
A.	Apple	I.	Outer portion of receptacle
B.	Coconut	II.	Fleshy thalamus
C.	Jack fruit	III.	Thalamus & pericarp
D.	Guava	IV.	Endosperm
E.	Pineapple	V.	Bract, perianth & seeds

(a) A– II; B – III; C – IV; D – V; E– I
(b) A – V; B – III; C– I; D – IV; E– II
(c) A– II; B – III; C – I; D –V; E– IV
(d) A– II; B – IV; C – V, D – III; E– I

61. Pneumatophores are found in:
(a) the vegetation which is found in marshy and saline lake.
(b) the vegetation which is found in saline soil.
(c) xerophytic condition.
(d) hydrophytic condition.

62. Root hairs are ontogenetically:
(a) Exogenous (b) Endogenous
(c) Both (a) and (b) (d) None of these

63. Outer covering of epiphytic roots is:
(a) Osmophore (b) Rhizophore
(c) Velamen (d) Pneumatophore

64. Stipules are found in:
(a) All dicots
(b) All dicots and some monocots
(c) Some dicots and monocots
(d) All dicots and monocots

65. A biparous cyme ending in uniparous cyme constitutes:
(a) Verticillaster (b) Panicle
(c) Hypanthodium (d) Cyathium

66. When stamens are superior, the flower is:
(a) Hypogynous
(b) Perigynous
(c) Epigynous
(d) Hypogynous

67. Axis of inflorescence is:
(a) Pedicel (b) Peduncle
(c) Petiole (d) All of these

68. The main purpose of phyllotaxy for the leaves is to provide sufficient_______.
(a) soil (b) air
(c) water (d) light

69. Most reduced form of stem is found in:
(a) Bulb (b) Rhizome
(c) Tuber (d) Corm

70. Match column-I with column-II and choose the correct option.

Column-I	Column-II
A. Spike	I. *Bougainvillea*
B. Capitulum	II. *Coleus*
C. Dichasial cyme	III. *Adhatoda*
D. Multiparous cyme	IV. *Zinnia*
E. Verticillaster	V. *Asclepias*

(a) A – III, B – IV, C – I, D – V, E – II
(b) A – III, B – I, C – IV, D – V, E – II
(c) A – II, B – IV, C – I, D – V, E – III
(d) A – IV, B – II, C – V, D – I, E – III

71. In a raceme, the flowers are:
(a) arranged in a basipetal succession
(b) arranged in a acropetal succession
(c) bisexual
(d) monosexual

72. Hypanthodium is:
(a) Thalamus (b) Fruit
(c) Inflorescence (d) Ovary

73. The botanical name of groundnut is:
(a) *Cicer arietinum* (b) *Glycine max*
(c) *Arachis hypogea* (d) *Lens esculentus*

74. Which of the following genera is characterised by the production of geocarpic fruits?
(a) *Cucurbita* (b) *Pisum*
(c) *Glycine* (d) *Arachis*

75. Given inflorescence is a:

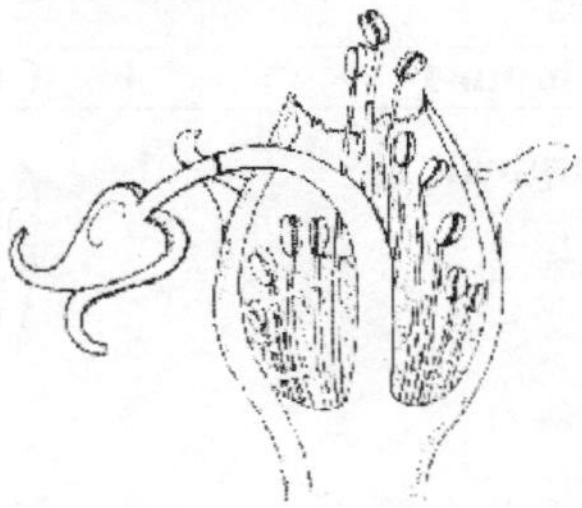

(a) Cyathium
(b) Dichasial cyme
(c) Umbel
(d) Verticillaster

76. Leaves of dicotyledonous plants possess _________ venation, while _________ venation is the characteristic of most monocotyledons.
(a) Reticulate and Parallel
(b) Parallel and Reticulate
(c) Reticulate and Perpendicular
(d) Obliquely and Parallel

77. Prickles of rose are: **Tricky**
(a) Modified leaves
(b) Modified stipules
(c) Exogenous in origin
(d) Endogenous in origin

78. Pentamerous, actinomorphic flowers, bicarpellate ovary with oblique septa and fruit, a capsule of berry, are the characteristic features of:
(a) Brassicaceae
(b) Solanaceae
(c) Liliaceae
(d) Asteraceae

79. The botanical name for 'Shikakai' is:
(a) *Acacia catechu*
(b) *Acacia concinna*
(c) *Acacia senegal*
(d) *Acacia nilotica*

80. Which family yields poor man's protein? **Tricky**
(a) Papilionatae
(b) Cruciferae
(c) Solanaceae
(d) Mimosoidae

81. Why are drupes called stony fruits?
(a) Hard epicarp
(b) Hard mesocarp
(c) Hard endocarp
(d) Both (a) and (b)

82. Which of the following option shows the correct match of type of placentation with their figure and examples?

Column-I	Column-II	Column-III
A. Marginal	I.	p. Sunflower, Marigold
B. Axile	II.	q. *Dianthus*, Primrose
C. Parietal	III.	r. Mustard, *Argemone*
D. Free central	IV.	s. China rose, Tomato, Lemon
E. Basal	V.	t. Pea

(a) A – V, t; B – II, s; C – I, r; D – III, q; E – IV, p
(b) A – I, t; B – II, s; C – III, r; D – IV, p; E – V, q
(c) A – V, p; B – II, s; C – I, q; D – III, r; E – IV, t
(d) A – V, p; B – III, q; C – II, s; D – I, t; E – IV, r

83. Choose the correct option for 1, 2 and 3.

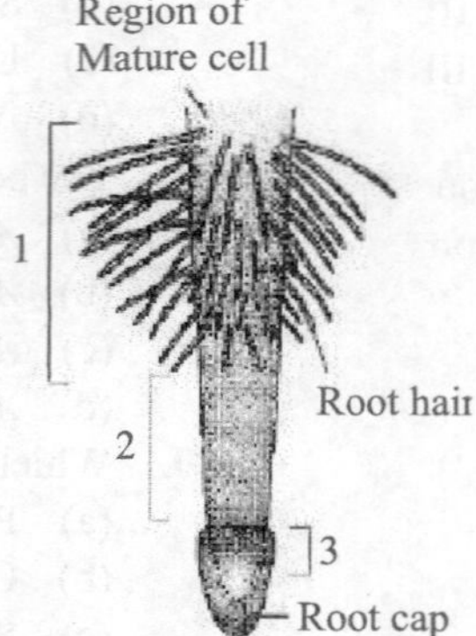

(a) 1 – zone of elongation, 2 – zone of meiosis, 3 – zone of mitosis
(b) 1 – zone of maturation, 2 – zone of meristematic activity, 3 – zone of elongation
(c) 1 – zone of mitosis, 2 – zone of elongation, 3 – zone of root cap
(d) 1 – region of maturation, 2 – region of elongation, 3 – meristematic activity

Answer KEYs

1	(c)	10	(b)	19	(a)	28	(b)	37	(d)	46	(c)	55	(a)	64	(d)	73	(c)	82	(a)
2	(a)	11	(d)	20	(b)	29	(a)	38	(b)	47	(c)	56	(c)	65	(a)	74	(d)	83	(d)
3	(a)	12	(a)	21	(d)	30	(c)	39	(a)	48	(c)	57	(b)	66	(c)	75	(a)		
4	(a)	13	(c)	22	(c)	31	(a)	40	(a)	49	(d)	58	(d)	67	(b)	76	(a)		
5	(b)	14	(d)	23	(b)	32	(c)	41	(c)	50	(b)	59	(c)	68	(d)	77	(c)		
6	(d)	15	(d)	24	(d)	33	(d)	42	(a)	51	(c)	60	(d)	69	(a)	78	(b)		
7	(a)	16	(b)	25	(a)	34	(a)	43	(d)	52	(b)	61	(a)	70	(a)	79	(b)		
8	(c)	17	(a)	26	(b)	35	(c)	44	(a)	53	(b)	62	(a)	71	(b)	80	(a)		
9	(c)	18	(c)	27	(a)	36	(b)	45	(b)	54	(b)	63	(c)	72	(c)	81	(c)		

Hints & Solutions

1. **(c)** Pitcher is insectivorous plant. Leaves of these plants are modified to catch the insects.

2. **(a)** **3.** **(a)**

4. **(a)** In onion and garlic, scale-leaves store food and water and are, therefore, thick and fleshy.

5. **(b)** In maize, fibrous roots or adventitious roots develop from the lower nodes of the stem to provide mechanical support and are called stilt roots e.g. *Pandamus*.

6. **(d)**

7. **(a)** Fibrous root system is found in monocotyledonous plants. In monocotyledons, primary root is short lived and replaced by a large number of roots which originate from the base of the stem and constitute the fibrous root system.

8. **(c)**

9. **(c)** The root is nearly globular or spherical in shape. The basal portion of root is much swollen which suddenly tapers towards the apex giving a top shaped appearance, *e.g.*, turnip and beet root.

10. **(b)** **11.** **(d)** **12.** **(a)** **13.** **(c)**

14. **(d)** In Fabaceae, 10 stamens, diadelphous, 9 fused with each other on anterior side, one free on posterior side (9) + 1.

15. **(d)** **16.** **(b)** **17.** **(a)**

18. **(c)** Hilum is the scar on a seed marking the point of attachment of the developing seed vessel to the fruit.

19. **(a)** A true fruit develops from ovary.

20. **(b)** **21.** **(d)** **22.** **(c)** **23.** **(b)**

24. **(d)** Perianth is the collective name of the non-essential floral organs if there is no distinction between sepals and petals i.e., calyx and corolla.

25. **(a)** **26.** **(b)** **27.** **(a)** **28.** **(b)**

29. **(a)** Staminode is a sterile or abortive stamen, frequently resembling a stamen without its anther.

30. **(c)** **31.** **(a)** **32.** **(c)** **33.** **(d)**

34. **(a)** **35.** **(c)**

36. **(b)** The most important function of inflorescence is to make flowers more conspicuous to pollinating agents (insects/birds) so that chances of cross pollination are high.

37. **(d)** Fruit of water chestnut (*Trapa*) is a one seeded, top shaped drupe, with 2 to 4 spines representing persistent calyx.

38. (b) 39. (a) 40. (a) 41. (c)
42. (a) 43. (d) 44. (a)
45. (b) Opium is a narcotic drug that is obtained from the unripe seed pods of the opium poppy (*Papaver somniferum*), a plant of the family Papaveraceae. Opium is obtained by slightly incising the seed capsules of the poppy after the plant's flower petals have fallen. The slit seed pods exude a milky latex that coagulates and changes colour, turning into a gum like brown mass upon exposure to air. This raw opium may be ground into a powder, sold as lumps, cakes, or bricks, or treated further to obtain such derivatives as morphine, codeine, and heroin. Opium and the drug obtained from it are called opiates.
46. (c)
47. (c) Edible part of banana are endocarp and less developed mesocarp.
48. (c) 49. (d)
50. (b) They are thin thread-like adventitious roots which often develop in groups. Fibrous roots provide better and firm anchorage to the plant, *e.g.*, Grass.
51. (c)
52. (b) Thorn is a stiff, sharp-pointed woody projection on the stem or other part of a plant. Thorns are found in many plants such as *Citrus*, *Bougainvillea*. They protect plants from grazing animals.
53. (b) 54. (b)
55. (a) The inflorescence in cauliflower is compound corymb. In compound corymb, the branching occurs on main axis and the peduncle becomes shorter. The pedicels of lower flowers become longer than the upper flowers due to which all flowers appear to be of same height. E.g., Cauliflower, *Pyrus malus, etc.*
56. (c) 57. (b) 58. (d) 59. (c)
60. (d) 61. (a) 62. (a)
63. (c) In epiphytic roots, the outer covering is made up of spongy tissue, the velamen which absorbs moisture from air.
64. (d) 65. (a) 66. (c) 67. (b)
68. (d) The arrangement of mature leaves on the stem or its branches is called phyllotaxy. These are meant for getting maximum amount of light for photosynthesis.
69. (a) 70. (a) 71. (b) 72. (c)
73. (c) The botanical name of groundnut is *Arachis hypogea.*
74. (d) 75. (a)
76. (a) Leaves of dicotyledonous plants possess reticulate venation while parallel venation is the characteristics of most monocotyledonous. In reticulate venation, the main veins of leaf form numerous irregular branches and as a result a net like arrangement is formed. Reticulate venation is the most common vein formation in leaves. It can be found in the leaves of maple trees, oak trees and rose bushes.

In parallel venation, veins are arranged parallel to each other.
77. (c) Prickles of rose are exogenous in origin.
78. (b)
79. (b) The botanical name for 'Shikakai' is *Acacia concinna.*
80. (a)
81. (c) Drupe is a fleshy, usually one chambered and one seeded fruit developing from monocarpellary syncarpous pistil with pericarp differentiated into mesocarp (fleshy) and the endocarp (stony and hard). Therefore, it is called as stone fruit, e.g., mango, almond, peach, coconut, etc.
82. (a) 83. (d)

6 Anatomy of Flowering Plants

1. Which tissue is considered as most primitive both ontogenetically as well as phylogenetically? **Tricky**
   (a) Parenchyma (b) Collenchyma
   (c) Sclerenchyma (d) Aerenchyma

2. Meristems are not found in:
   (a) *Cycas* stem (b) Pollen of *Pinus*
   (c) Fern leaf (d) Fern rhizome

3. Healing of wound in plants takes place by activity of:
   (a) Intercalary meristem
   (b) Secondary meristem
   (c) Apical meristem
   (d) Lateral meristem

4. The given figure shows apical meristem of root apex with few part marked as 1, 2 and 3. Identify the correct labelling of 1, 2 and 3.

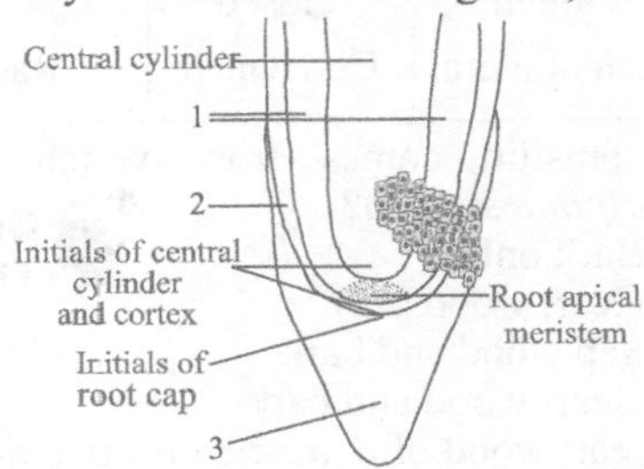

   (a) 1 – Vascular structure, 2 – Protoderm, 3 – Root cap
   (b) 1 – Cortex, 2 – Endodermis, 3 – Root cap
   (c) 1 – Cortex, 2 – Protoderm, 3 – Root cap
   (d) 1 – Tunica, 2 – Protoderm, 3 – Root cap

5. Choose the correct option for 1 and 2.

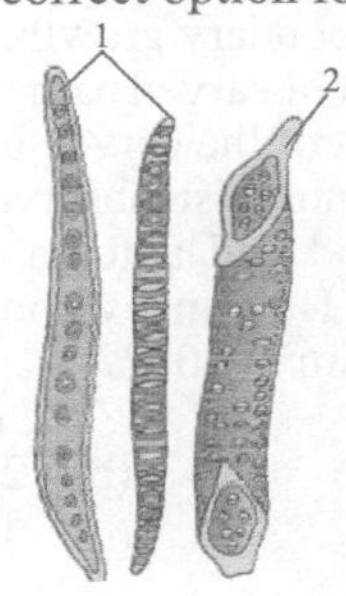

   (a) 1 – Tracheid, 2 – Vessel
   (b) 1 – Vessel, 2 – Tracheid
   (c) 1 – Fibre, 2 – Tracheid
   (d) 1 – Fibre, 2 – Sclereid

6. Monocot leaves grow by: **Tricky**
   (a) Apical meristem
   (b) Lateral meristem
   (c) Intercalary meristem
   (d) Dermatogen

7. In the monocot root, we observe:
   (a) suberised exodermis, polyarch xylem, pith
   (b) exodermis, endarch, tetrarch closed bundles
   (c) conjoint, collateral, open, polyarch vascular bundle
   (d) suberised exodermis, casparian strip, passage cells, cambium

8. Match column-I with column-II and select the correct answer using the codes given below.

Column-I	Column-II
A. Radial vascular bundle	I. *Cucurbita pepo*
B. Collateral vascular bundle	II. *Dracaena*
C. Bicollateral vascular bundle	III. Roots of angiosperms
D. Amphicribral vascular bundle	IV. Sunflower stem
E. Amphivasal vascular bundle	V. fern

   (a) A – III; B – IV; C – I; D – V; E – II
   (b) A – II; B – III; C – I; D – V; E – IV
   (c) A – III; B – IV; C – V; D – I; E – II
   (d) A – III; B – I; C – II; D – IV; E – V

9. In the given columns, column-I contain structures of plant and column-II contain its feature. Select the correct match.

	Column-I		Column-II
A.	Lateral	I.	Fascicular vascular meristem cambium, interfascicular cambium and cork cambium.
B.	Apical meristem	II.	Produces dermal tissue, ground tissues and vascular tissue.
C.	Bast fibres	III.	Generally absent in primary phloem but found in secondary phloem.
D.	Sap wood	IV.	Involved in the conduction of water and minerals from the root to leaf.

(a) A – I; B – II; C – III; D – IV
(b) A – III; B – I; C – II; D – IV
(c) A – I; B – IV; C – III; D – II
(d) A – II; B – IV; C – III; D – I

10. Which one of the following statement is **correct** about bulliform/motor cell?
(a) It is seen in grasses.
(b) It is large-sized, thin-walled colourless, vacuolate cells on the adaxial surface.
(c) It helps in rolling of leaf to minimise water loss when it is flaccid.
(d) All of the above

11. A scion is grafted to a stock. The quality of fruits produced will be determined by the genotype of: Critical Thinking
(a) Stock
(b) Scion
(c) Both stock and scion
(d) Neither stock nor scion

12. The vascular cambium produces xylem on one side and phloem on other because of it: Critical Thinking
(a) interfascicular position
(b) multicellular nature
(c) gravitational force
(d) difference in supply of hormones on two sides

13. Which of the following plant shows multiple epidermis?
(a) *Croton* (b) *Allium*
(c) *Nerium* (d) *Cucurbita*

14. Identify types of vascular bundles in given figures 1, 2 and 3.

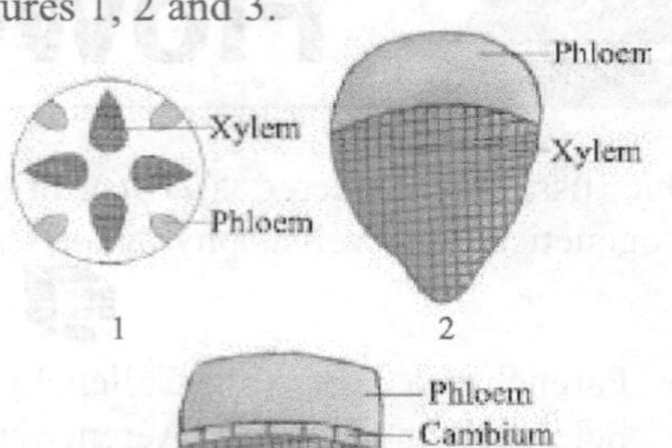

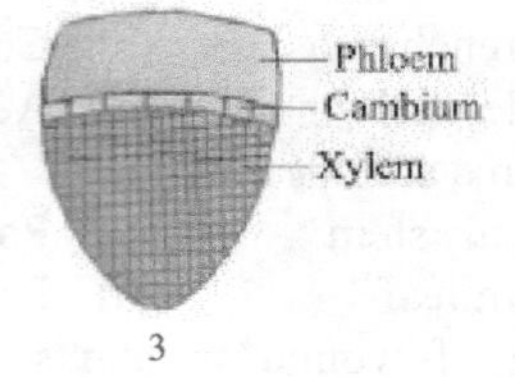

	1	2	3
(a)	Radial	Conjoint closed	Conjoint open
(b)	Conjoint closed	Conjoint open	Radial
(c)	Conjoint open	Conjoint closed	Radial
(d)	Bicollateral	Concentric	Radial

15. Red brasilin comes from which part of *Caesalpinia sappan*? Critical Thinking
(a) Bark only
(b) Heart wood only
(c) Sap wood and bark
(d) Heart wood and bark

16. The heart wood of *Acacia catechu* is rich in:
(a) Gums (b) Resins
(c) Tannins (d) Mucilages

17. A boy standing behind the tree carved his name on the tree, after twenty years: Toughnut
(a) height of the carved name increased.
(b) the name was not there as it gets abolished due to secondary growth.
(c) the height of carved name remain the same.
(d) the height of the carved name decreased.

18. A dendrochronologist observed total 22 rings of spring wood and autumn wood in a dicot stem in 1998. How many annual rings will be observed by him in 2007? Toughnut
(a) 21 (b) 25
(c) 20 (d) 30

19. Bamboo and grasses elongate by the activity of:
(a) Apical meristem
(b) Lateral meristem
(c) Secondary meristem
(d) Intercalary meristem

20. Meristematic tissues are composed of :
(a) mature cell
(b) fully differentiated cell
(c) cells that cannot divide
(d) immature cells with power to divide

21. Which one of the following option shows the correct labelling of the parts marked as 1, 2, 3 and 4 in the given figure of a typical dicot root?

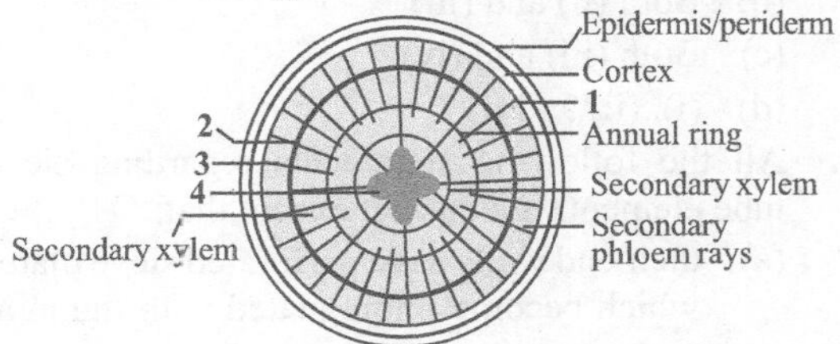

(a) 1 – Primary phloem, 2 – Vascular cambium, 3 – Secondary phloem, 4 – Primary xylem
(b) 1 – Secondary phloem, 2 – Vascular cambium, 3 – Primary phloem, 4 – Primary xylem
(c) 1 – Primary phloem, 2 – Primary xylem, 3 – Secondary phloem, 4– Vascular cambium
(d) 1 – Secondary phloem, 2 – Primary xylem, 3 – Primary phloem, 4 – Vascular cambium

22. Largest number of chloroplast is found in:
(a) Palisade tissue (b) Spongy tissue
(c) Transfusion tissue (d) Bundle sheath cells

23. If a stem is girdled:
(a) root dies first
(b) shoot dies first
(c) Both die together
(d) Neither root nor shoot would die

24. In dicotyledonous stem, fascicular cambium is a meristematic tissue. It is an example of which of the following meristem?
(a) Lateral (b) Secondary
(c) Apical (d) Intercalary

25. Which of the following statement(s) is/are **incorrect**?
(i) Cork cambium is also called phellogen.
(ii) Cork is also called phellem.
(iii) Secondary cortex is also called periderm.
(iv) Cork cambium, cork and secondary cortex are collectively called phelloderm.
(a) (iii) and (iv) (b) (i) and (ii)
(c) (ii) and (iii) (d) (ii) and (iv)

26. Which of the following statement is **correct** about heart wood/duramen?
(i) It does not help in water and mineral conduction.
(ii) It is dark coloured but soft.
(iii) It has tracheary elements filled with tannins, resins, gums, oil, etc.
(iv) It is a peripheral part.
(v) They are sensitive to microbes and insects, hence least durable.
(a) (i) and (iii)
(c) (ii) and (iii)
(b) (iv) and (v)
(d) (iii) and (iv)

27. T.S. of dicot stem is given below, certain parts have been indicated by alphabets. Choose the correct option.

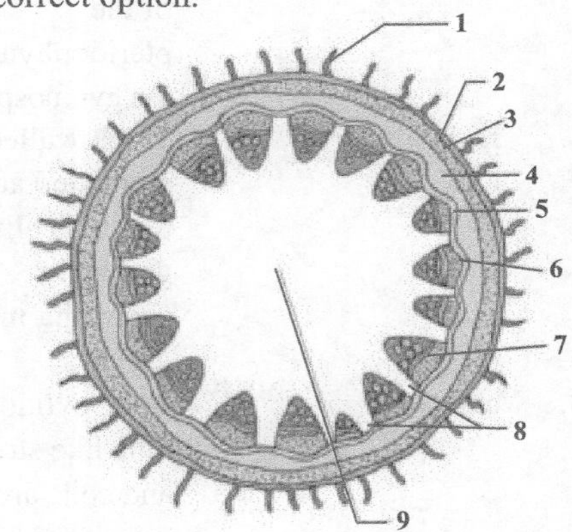

(a) 1 – Epidermis, 2 – Epidermal hairs, 3 – Parenchyma, 4 – Starch Sheath, 5 – Hypodermis (collenchyma), 6 – Vascular bundle, 7 – Bundle cap, 8 – Medulla or pith, 9 – Medullary rays
(b) 1 – Epidermal hairs, 2 – Epidermis, 3 – Hypodermis (collenchyma), 4 – Parenchyma, 5 – Starch sheath, 6 – Bundle cap, 7 – Vascular bundle, 8 – Medullary rays, 9 – Medulla or pith
(c) 1 – Epidermal hairs, 2 – Epidermis, 3 – Hypodermis (collenchyma), 4 – Starch sheath, 5 – Parenchyma, 6 – Vascular bundle, 7 – Bundle cap, 8 – Medulla or pith, 9 – Medullary rays
(d) 1 – Epidermal hairs, 2 – Epidermis, 3 – Parenchyma, 4 – Hypodermis (collenchyma), 5 – Starch sheath, 6 – Vascular bundle, 7 – Bundle cap, 8 – Medulla or pith, 9 – Medullary rays

28. Match the terms given in column-I with their features given in column-II and choose the correct option.

Column-I (Terms)		Column-II (Features)
A. Fibres	I.	Cells are living and thin walled with cellulosic cell wall, store food materials in the form of starch or fat.
B. Sclereids	II.	Main water conductive cells of the pteridophytes and the gymnosperms.
C. Tracheids	III.	Thick walled, elongated and pointed cells, generally occurring in groups.
D. Vessels	IV.	Long cylindrical tube like structure and cells are devoid of protoplasm. Characteristic feature of angiosperms.
E. Xylem parenchyma	V.	Reduced form of sclerenchyma cells with highly thickened lignified cellular walls that form small bundles of durable layers of tissue in most plants.

(a) A – I; B – II; C – III; D – IV; E – V
(b) A – III; B – V; C – II; D – IV; E – I
(c) A – III; B – I; C – V; D – II; E – IV
(d) A – V; B – IV; C – III; D – I; E – II

29. Read the following statements and select the correct one(s) from the options.
(i) In flowering plants, tracheids and vessels are the main water transporting elements.
(ii) The presence of vessels is a characteristic feature of angiosperms.
(iii) Xylem fibres have highly thinned walls and their cell walls are made up of cellulose.
(iv) Xylem parenchyma store food materials in the form of starch or fat and other substances like tannins.
(a) Only (i)
(b) Both (ii) and (iii)
(c) Both (iii) and (iv)
(d) (i), (ii) and (iv)

30. All the following statements regarding sieve tube elements are correct except that:
(a) their end walls have perforated sieve plates which become impregnated with lignin at maturity.
(b) they possess peripheral cytoplasm as well as a large vacuole.
(c) distinct proteinaceous inclusions, the P-proteins are seen evenly distributed throughout the lumen.
(d) long, slender, tube-like structures arranged in longitudinal series.

31. The lacunae in vascular bundles of monocot stems are: **Tricky**
(a) Metaxylem
(b) Mucilagenous canal
(c) Lysigenous water cavity
(d) Large-sized vessel

32. From where the commercial jute fibres are obtained?
(a) Interxylary fibres (b) Xylem fibres
(c) Phloem fibres (d) None of these

33. Abnormal secondary growth is observed in:
(a) *Dracaena* (b) Wheat
(c) Ginger (d) Rice

34. The trees growing in deserts will:
(a) show alternate rings of xylem and sclerenchyma.
(b) show distinct annual rings.
(c) not show distinct annual rings.
(d) have only conjunctive tissue and phloem formed by the activity of cambium.

35. Choose the correct option for T.S. of monocot root.

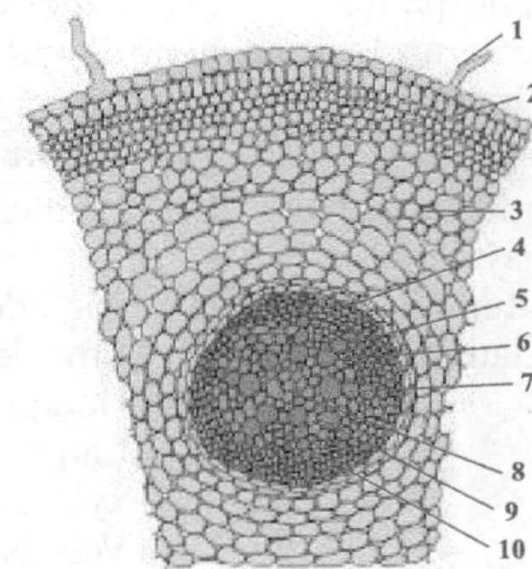

(a) 1 – Root hair, 2 – Epiblema, 3 – Cortex, 4 – Endodermis, 5 – Passage cell, 6 – Pericycle, 7 – Pith, 8 – Phloem, 9 – Metaxylem, 10 – Protoxylem
(b) 1 – Root hair, 2 – Epiblema, 3 – Cortex, 4 – Endodermis, 5 – Passage cell, 6 – Pith, 7 – Pericycle, 8 – Metaxylem, 9 – Phloem, 10 – Protoxylem
(c) 1 – Root hair, 2 – Epiblema, 3 – Cortex, 4 – Endodermis, 5 – Pericycle, 6 – Passage cell, 7 – Phloem, 8 – Pith, 9 – Protoxylem, 10 – Metaxylem
(d) 1 – Root hair, 2 – Cortex, 3 – Epiblema, 4 – Pericycle, 5 – Endodermis, 6 – Pith, 7 – Passage cell, 8 – Phloem, 9 – Protoxylem, 10 – Metaxylem

36. Gaseous exchange between air and internal tissues of old corky stem takes place through:
(a) Pits (b) Stomata
(c) Lenticels (d) Sieve plates

37. In roots of monocotyledon, the histogen present at the apex of the root tip is:
(a) Dermatogen (b) Procambium
(c) Calyptrogen (d) Plerome

38. Active divisions occur in the cells of:
(a) Xylem (b) Phloem
(c) Cambium (d) Collenchyma

39. What happens in plants during vascularisation?

(a) Differentiation of procambium, formation of primary phloem followed by formation of primary xylem.
(b) Differentiation of procambium followed by the formation of primary phloem and xylem simultaneously.
(c) Formation of procambium, primary phloem and xylem simultaneously.
(d) Differentiation of procambium followed by the formation of secondary xylem.

40. In *Cucurbita,* stem vascular bundles are:
(a) Radial (b) Collateral
(c) Concentric (d) Bicollateral

41. In *Bougainvillea,* the primary vascular bundles lie scattered. A cambium ring is formed:
(a) from the pericycle
(b) from intrafascicular cambium
(c) just outside the primary bundles
(d) by inter and intrafascicular cambia

42. When phloem and cambium are present on both sides of xylem, the vascular bundle is called:
(a) Bicollateral (b) Radial
(c) Concentric (d) Collateral

43. Which of the following figure is a type of permanent tissue having many different types of cell?

(a)
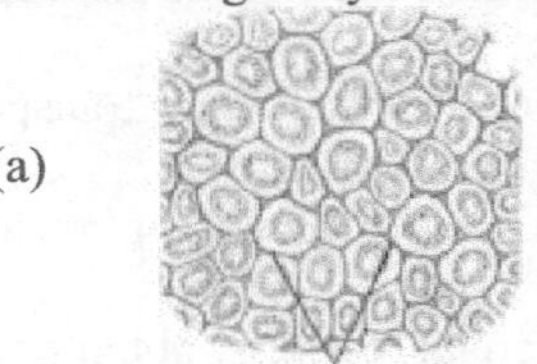

Collenchyma

(b)
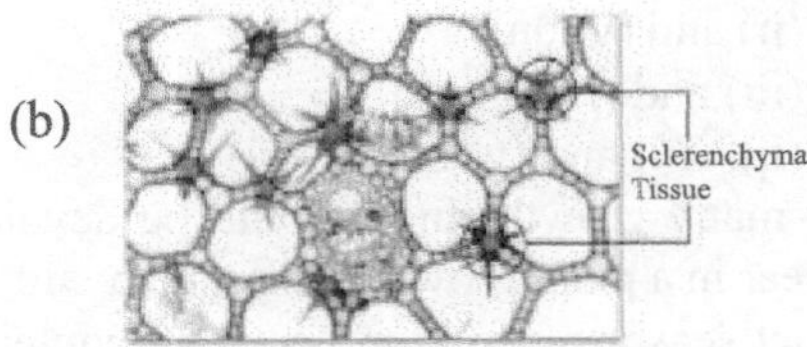

(c)
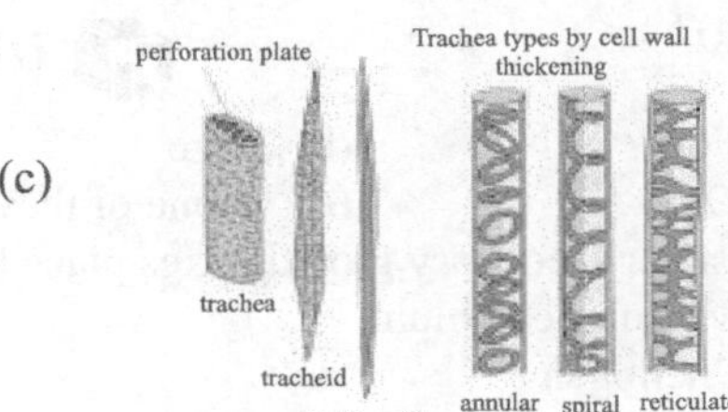

(d)
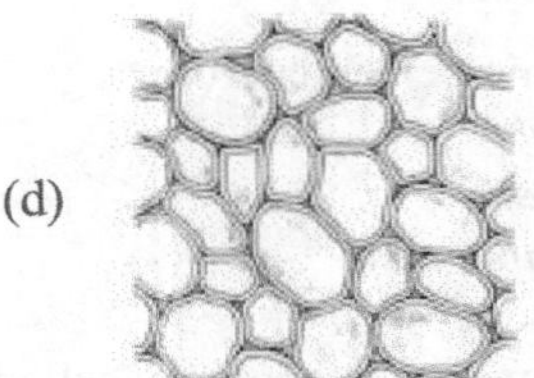
Parenchyma

44. Match column-I with column-II and select the correct answer using the options given below.

	Column-I		Column-II
A.	Xylem vessels	I.	Store food materials
B.	Xylem tracheids	II.	Obliterated lumen
C.	Xylem fibre	III.	Perforated plates
D.	Xylem parenchyma	IV.	Chisel-like ends

(a) A – IV; B – III; C – II; D – I
(b) A – III; B – II; C – I; D – IV
(c) A – II; B – I; C – IV; D – III
(d) A – III; B – IV; C – II; D – I

45. Which of the following statement(s) is/are **correct**?
(i) Uneven thickening of cell wall is characteristic of sclerenchyma.
(ii) Periblem forms cortex of the stem and the root.
(iii) Tracheids are the chief water transporting elements in gymnosperms.
(iv) Companion cell is devoid of nucleus at maturity.
(v) The commercial cork is obtained from *Quercus suber.*
(a) (i) and (iv) only
(b) (ii) and (v) only
(c) (iii) and (iv) only
(d) (ii), (iii) and (v) only

46. How many growth rings should be developed per year in a plant grown in Rajasthan with four distinct seasons (viz, summer, rain, winter and spring)? **Tricky**
(a) Four (b) Two
(c) One (d) None of these

47. Extrastelar secondary growth takes place by:
(a) Vascular cambium
(b) Phellogen
(c) Phellem
(d) Phelloderm

48. The cork of commerce comes from:
(a) *Quercus suber*
(b) *Quercus incana*
(c) *Quercus ithaburensis*
(d) All of the above

49. The pneumatophores have openings called:
(a) Stomata (b) Lenticels
(c) Hydathodes (d) All of these

50. Phloem parenchyma are absent in:
(a) Dicots (b) Monocots
(c) Gymnosperms (d) All of these

51. Match column-I with column-II and choose the correct option.

	Column-I		Column-II
A.	Spring wood or early wood	I.	Lighter in colour
B.	Autumn wood or late wood	II.	High density
		III.	Low density
		IV.	Darker in colour
		V.	Larger number of xylem elements
		VI.	Vessels with wider cavity
		VII.	Lesser number of xylem elements
		VIII.	Vessels with small cavity

Which of the following combination is correct?
(a) A – II; IV; VII; VIII; B – I; III; V; VI
(b) A – I; II; VII; VIII; B – III; IV; V; VI
(c) A – I; III; V; VI; B – II; IV; VII; VIII
(d) A – I; III; VII; VIII; B – II; IV; V; VI

52. In monocots, root cap is formed by:
(a) Dermatogen (b) Calyptrogen
(c) Wound cambium (d) Vascular cambium

53. In woody dicotyledons, the arrangement of vessels is either diffuse porous or ring porous. Based on the these data, which one of the following statements is **correct**?
(a) Ring porous vessels are specialised and are used for conducting more water for a shorter period only, when tyloses occur early in the vessels.
(b) Although diffuse porous vessels are not so specialised as ring porous vessels, they conduct more water at all periods and through new xylem vessels added gradually during development.
(c) Diffuse porous vessels carry more water and also faster because of a greater number of small vessels having greater capillary force.
(d) Ring porous vessels conduct more water as they are formed early during development, when the need for water is great.

54. Periderm is formed from:
(a) Vascular cambium
(b) Phellogen
(c) Fascicular cambium
(d) Interfascicular cambium

55. Annual rings are the bands of:
(a) secondary cortex and cork
(b) all secondary vascular tissues
(c) secondary xylem and xylem rays
(d) secondary phloem and medullary rays

56. Secondary growth in the dicot stems is possible because of the presence of:
(a) Xylem
(b) Phloem
(c) Cambium
(d) Medullary rays

57. Which of the following meristem is responsible for extrastelar secondary growth in dicotyledonous stem?
(a) Phellogen
(b) Intrafascicular cambium
(c) Interfascicular cambium
(d) Intercalary meristem

58. Periblem form:
(a) Endodermis
(b) Cortex
(c) Both (a) and (b)
(d) Epidermis

59. Choose the correct answer for 1 to 6 labelling in the given figure.

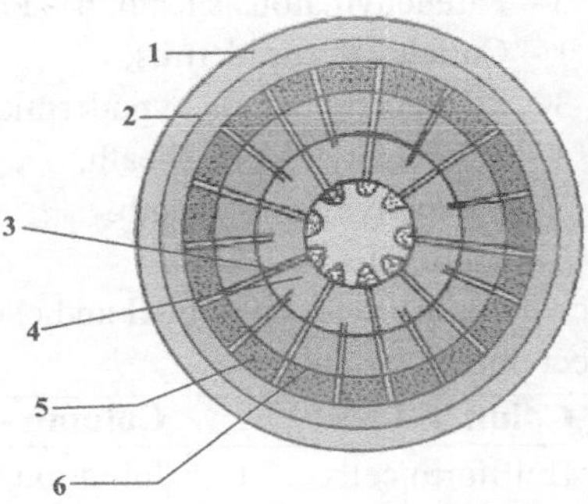

(a) 1–Phellem, 2–Phellogen, 3–Medullary rays, 4–Secondary xylem, 5–Secondary phloem, 6–Cambium ring
(b) 1–Phellem, 2–Phellogen, 3–Medullary rays, 4–Secondary phloem, 5–Secondary xylem, 6–Cambium ring
(c) 1–Phellogen, 2–Phellem, 3–Medullary rays, 4–Secondary xylem, 5–Secondary phloem, 6–Cambium ring
(d) 1–Phellem, 2–Phellogen, 3–Cambium ring, 4–Secondary xylem, 5–Secondary phloem, 6–Medullary rays

60. Identify 1, 2 and 3 in the given figure of shoot apical meristem

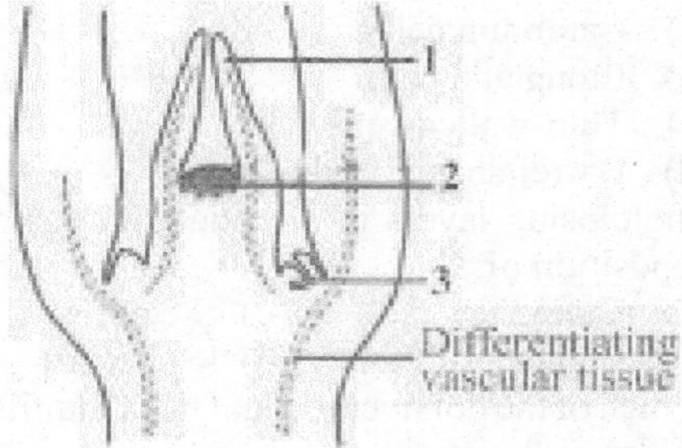

(a) 1 – Leaf primordium, 2 – Shoot apical meristem, 3 – Axillary bud
(b) 1 – Leaf primordium, 2 – Shoot apical meristem, 3 – Apical bud
(c) 1 – Root hair primordium, 2 – Root apical meristem, 3 – Axillary bud
(d) 1 – Root hair primordium, 2 – Root apical meristem, 3 – Terminal bud

61. Select the combination of **correct** statements from the options given below.
(i) Annual rings are formed as a result of seasonal environmental conditions.
(ii) Tracheids/vessels elements are larger during periods when water is abundant.
(iii) Tracheids/vessels elements have thicker wall during periods of water deprivation.
(iv) Wood formed in the previous years is darker than newer wood.
(a) (i) and (ii)
(b) (ii) and (iv)
(c) (i), (ii) and (iii)
(d) (ii), (iii) and (iv)

62. When secondary growth is initiated in dicot stem what will happen first?
(a) The cells of cambium divide periclinally to form xylem mother cells.
(b) Inter fascicular cambium join with intrafascicular cambium.
(c) Parenchymatous cells present between vascular bundle become meristematic.
(d) Pith get obliterated.

63. Protective layer found at the site of abscission is:
(a) Parenchymatous
(b) Collenchymatous
(c) Sclerenchymatous
(d) Suberised

64. Endodermal cells can be distinguished by the presence of:
(a) Cambium cells
(b) Elongated cells
(c) Thin-walled cells
(d) Barrel-shaped cells

65. The closing layers in a lenticel may show the deposition of:
(a) Cutin (b) Pectin
(c) Lignin (d) Suberin

66. Which of the dorsiventral leaf has a multilayered epidermis?
(a) Mulberry (b) Banyan
(c) *Nerium* (d) *Eucalyptus*

67. A meristem may be defined as the group of cells that:
(a) add to the bulk of the plants.
(b) conserve food.
(c) divide continuously to give rise to new cells.
(d) elongate and add to the group of cells.

68. Xylem fibre is:
(a) Bast fibre (b) Wood fibre
(c) Heart wood (d) Libriform fibre

69. The only plant cells without nuclei among following are:
(a) Cambium cells (b) Xylem vessels
(c) Root hairs (d) Companion cells

70. Totipotency is present in:
(a) Meristem (b) Xylem
(c) Phloem (d) Cork

71. Collenchyma is present in:
(a) Herbaceous monocots
(b) Herbaceous dicots
(c) All herbaceous plants
(d) Pteridophytes and monocots

72. In the given figure of phloem tissue, identify the marked part (1, 2 and 3) which help in maintaining the pressure gradient in the sieve tubes.

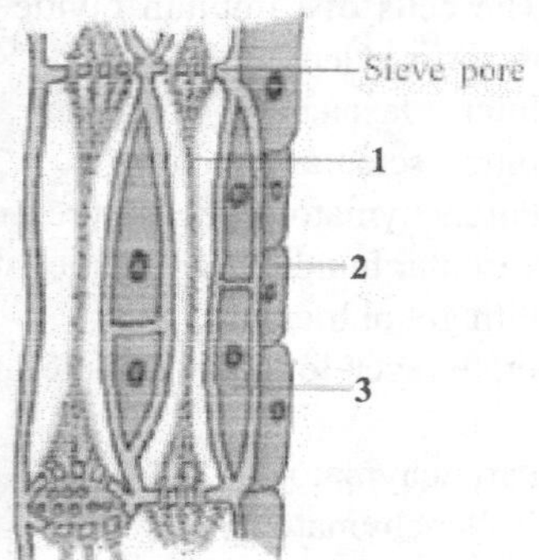

(a) 1 (b) 2
(c) 3 (d) None of the above

73. The given figure shows T.S. of monocot stem. Identify the correct labelling of 1 to 6 marked in the given figure.

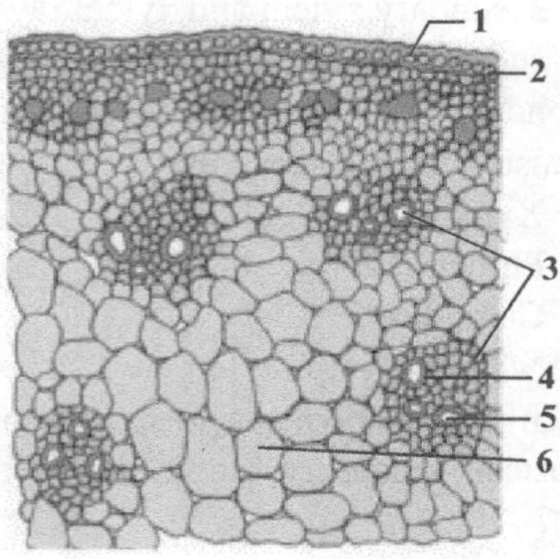

(a) 1 – Epidermis, 2 – Hypodermis, 3 – Vascular bundles, 4 – Phloem, 5 – Xylem, 6 – Ground tissue
(b) 1 – Cuticle, 2 – Epidermis, 3 – Sclerenchymatous sheath, 4 – Sclerenchymatous hypodermis, 5 – Parenchymatous sheath, 6 – Phloem
(c) 1 – Cuticle, 2 – Epidermis, 3 – Sclerenchymatous hypodermis, 4 – Sclerenchymatous sheath, 5 – Parenchymatous sheath, 6 – Phloem
(d) 1 – Cuticle, 2 – Epidermis, 3 – Sclerenchymatous hypodermis, 4 – Sclerenchymatous sheath, 5 – Parenchymatous sheath, 6 – Protoxylem

74. Match column-I with column-II and choose the correct option.

	Column -I		Column -II
A.	Bulliform cells	I.	Initiation of lateral roots
B.	Pericycle	II.	Root
C.	Endarch xylem	III.	Grasses
D.	Exarch xylem	IV.	Dicot leaf
E.	Bundle sheath cells	V.	Stem

(a) A – III; B – V; C – IV; D – I; E – II
(b) A – II; B – V; C – I; D – III; E – IV
(c) A – II; B – IV; C – I; D – III; E – V
(d) A – III; B – I; C – V; D – II; E – IV

75. Which of the following statements are **correct** about phloem?
   (i) Phloem transports food materials, usually from roots to the other parts of the plant.
   (ii) It is composed of sieve tube elements, companion cells, phloem parenchyma and phloem fibres.
   (iii) The companion cells are specialised parenchymatous cells which are closely associated with phloem parenchyma.
   (iv) The first formed primary phloem consists of narrow sieve tubes and referred to as protophloem and the later formed phloem has bigger sieve tubes and referred to as metaphloem.
   (a) Both (i) and (iv)
   (b) Both (ii) and (iii)
   (c) Both (ii) and (iv)
   (d) All of the above

76. Read the following statements and answer the question.
   (i) Cambium is very active and produces a large number of xylary elements having vessels with wider cavities.
   (ii) It is also called early wood.
   (iii) It is lighter in colour and has lower density.
   Which type of wood is described by the above statements?
   (a) Sap wood (b) Heart wood
   (c) Spring wood (d) Autumn wood

77. If there is more than one tunica layer in a stem apex, which among the following is most likely to happen? Critical Thinking
   (a) All the layers will develop into epidermal cells.
   (b) Only the outer layer will develop into epidermal cells.
   (c) All the layers will develop into cortex.
   (d) Inner layer develops into cortex.

78. Which type of meristem is an example of vascular cambium of the root?
   (a) Apical (b) Intercalary
   (c) Secondary (d) Root apical

79. Which combination of tissues acts together to provide the support to the hypocotyl of seedling?

   (a) Xylem and phloem fibres
   (b) Epidermis and parenchyma
   (c) Xylem and parenchyma
   (d) Epidermis and collenchyma

80. P-protein belongs to:
   (a) Xylem parenchyma
   (b) Trichomes
   (c) Sieve tube elements
   (d) Tracheids and vessels

81. The Tunica Corpus Theory was proposed by:
   (a) Nageli (b) Hanstein
   (c) Schmidt (d) Haberlandt

82. Diffuse porous woods are characteristic of plants growing in:
   (a) Tropics
   (b) Alpine region
   (c) Temperate climate
   (d) Cold winter regions

83. T.S. of monocot leaf is given below, certain parts have been marked by alphabets (1 – 7). Which one is the option showing there correct labelling?

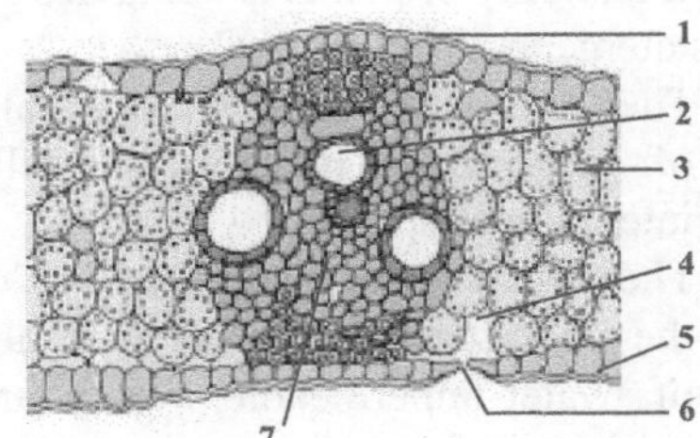

   (a) 1 – Adaxial epidermis, 2 – Xylem, 3 – Mesophyll, 4 – Sub-stomatal cavity, 5 – Abaxial epidermis, 6 – Stoma, 7 – Phloem
   (b) 1 – Adaxial epidermis, 2 – Abaxial epidermis, 3 – Xylem, 4 – Sub-stomatal cavity, 5 – Stoma, 6 – Mesophyll, 7 – Phloem
   (c) 1 – Adaxial epidermis, 2 – Phloem, 3 – Mesophyll, 4 – Sub-stomatal cavity, 5 – Abaxial epidermis, 6 – Xylem, 7 – Stoma
   (d) 1 – Adaxial epidermis, 2 – Xylem, 3 – Stoma, 4 – Sub-stomatal cavity, 5 – Abaxial epidermis, 6 – Phloem, 7 – Mesophyll

84. The given figure shows the T.S of dicot root. Some parts are marked as 1, 2, 3, 4, 5, & 6. Choose the option which shows the correct labelling of marked part.

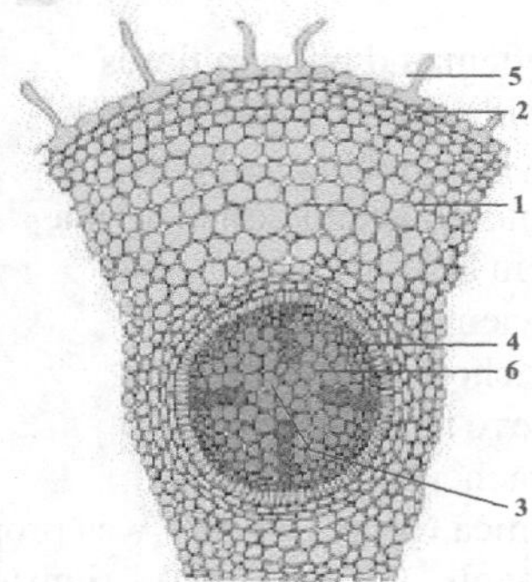

(a) 1 – Epiblema, 2 – Root hair, 3 – Cortex, 4 – Endodermis, 5 – Pith, 6 – Pericycle
(b) 1 – Cortex, 2 – Pith, 3 – Epiblema, 4 – Endodermis, 5 – Root hair, 6 – Pericycle
(c) 1 – Epiblema, 2 – Endodermis, 3 – Cortex, 4 – Root hair, 5 – Pith, 6 – Pericycle
(d) 1 – Cortex, 2 – Epiblema, 3 – Pith, 4 – Endodermis, 5 – Root hair, 6 – Pericycle

85. Which anatomy of plants is being described by the statements given below?
(i) The cortex consists of several layers of thin-walled parenchyma cells with intercellular spaces.
(ii) The tangential as well as radial walls of the endodermal cells have a depostion of water-impermeable, waxy material-suberin, in form of casparian strips.
(iii) Secondary growth takes place.
(iv) Pith is small or inconspicuous.
(a) Dicotyledonous root
(b) Monocotyledonous root
(c) Dicotyledonous stem
(d) Monocotyledonous stem

86. Which of the following statements is **correct**?
(a) Lenticels occur in most woody trees.
(b) Sclerenchymatous cells are usually present in cortex.
(c) The vascular tissue system is divided into three main zones- cortex, pericycle and pith.
(d) The conjoint vascular bundles usually have the xylem located only on the outer side of the phloem.

87. Which one of the following statement is **incorrect**?
(i) Epidermal cell has small amount of cytoplasm and a large vacuole.
(ii) Waxy cuticle layer is absent in roots.
(iii) Root hairs are unicellular, while stem hairs/ trichomes are multicellular.
(iv) Trichomes may be branched or unbranched, soft or stiff and prevent transpiration.
(v) Guard cells are dumbbell shaped in dicots and bean-shaped in monocots (*e.g.*, grass).
(a) Only (i) (b) Only (iv)
(c) Only (iii) (d) Only (v)

88. Maximum growth in root occurs:

(a) at its tip (b) towards light
(c) behind the apex (d) towards apex

89. In *Aristolochia*, the pericycle is:
(a) Parenchymatous
(b) Collenchymatous
(c) Aerenchymatous
(d) Sclerenchymatous

90. The concept envisaging three distinct zones of cells of new tissue builders at the stem and root apices is known as:
(a) Tunica corpus theory
(b) Dermatogen
(c) Meristem theory
(d) Histogen theory

91. Leaf primordium grows into adult lamina by means of:
(a) Marginal meristem
(b) Lateral meristem
(c) First apical and then marginal meristems
(d) Apical meristem

92. The tracheids differ from vessels in having:
(a) Thick wall
(b) Bordered pit
(c) Presence of pitted end wall
(d) Spiral thickening

93. Embryo of a seed is made up of:
(a) Meristematic tissue (b) Parenchyma
(c) Collenchyma (d) Sclerenchyma

94. Non-porous wood is found in:
(a) Dicots (b) Monocots
(c) Gymnosperm (d) Cactus

95. Most distinct annual rings are formed in:
(a) Tropical region (b) Temperate region
(c) Arctic region (d) Equatorial region

96. Which will decay faster if exposed freely to the air?
(a) Heartwood
(b) Sapwood
(c) Wood with lots of fibres
(d) Softwood

97. Diffuse porous woods are characteristic of plants growing in:
(a) Tropics
(b) Alpine region
(c) Temperate climate
(d) Cold winter regions

98. Which one of the followings option is correct for a lenticel?

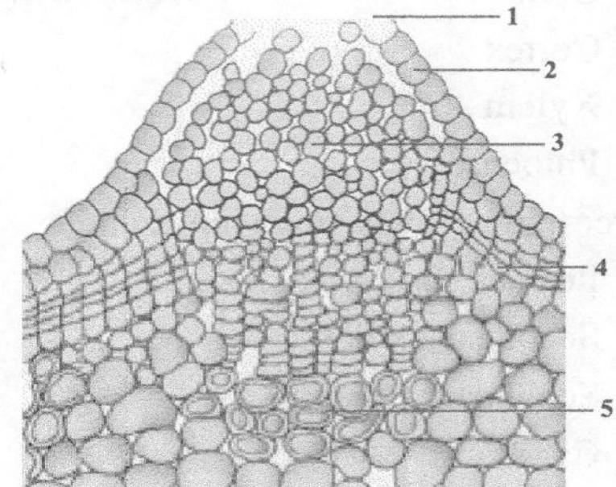

(a) 1 – Pore, 2 – Secondary cortex, 3 – Cork cambium, 4 – Cork, 5 – Complimentary cells
(b) 1 – Pore, 2 – Cork cambium, 3 – Secondary cortex, 4 – Cork, 5 – Complimentary cells
(c) 1 – Pore, 2 – Cork, 3 – Complimentary cells, 4 – Cork cambium, 5 – Secondary cortex
(d) 1 – Pore, 2 – Complimentary cells, 3 – Cork, 4 – Cork cambium, 5 – Secondary cortex

99. Which of the following pair is **incorrectly** matched?
(a) Pith – Large and well developed in monocotyledonous root.
(b) Root hairs – Helps in preventing water loss due to transpiration.
(c) Sieve tube elements – Its functions are controlled by the nucleus of companion cells.
(d) Stomatal apparatus – Consists of stomatal aperture, guard cells and surrounding subsidiary cells.

100. Read the following statements and answer the questions.
(i) It is made up of elongated, tapering cylindrical cells which have dense cytoplasm and nucleus.
(ii) The cell wall is composed of cellulose and has pits through which plasmodesmatal connections exist between the cells.
(iii) It is absent in most of the monocotyledons.

Which part of plant tissue is being described by the above statements?
(a) Sieve tube elements
(b) Companion cells
(c) Phloem parenchyma
(d) Phloem fibres

101. Match column-I with column-II and select the correct option from the codes given below.

	Column-I		Column-II
A.	Stele	I.	Innermost layer of cortex
B.	Endodermis	II.	Suberin
C.	Casparian strips	III.	All the tissues exterior to vascular cambium
D.	Bark	IV.	All the tissues inner to endodermis

(a) A – IV; B – I; C – II; D – III
(b) A – III; B – II; C – I; D – IV
(c) A – I; B – II; C – III; D – IV
(d) A – IV; B – II; C – I; D – III

102. The quiescent centre in root meristem serves as:

(a) region for absorption of water.
(b) reservoir of growth hormones.
(c) reserve for replenishment of damaged cells of the meristem.
(d) site for storage of food which is utilised during maturation.

103. Root cap is not found in:
(a) Lithophytes (b) Xerophytes
(c) Hydrophytes (d) Mesophytes

104. Identify the types of simple tissue indicated by 1, 2, 3 and 4 and their function.

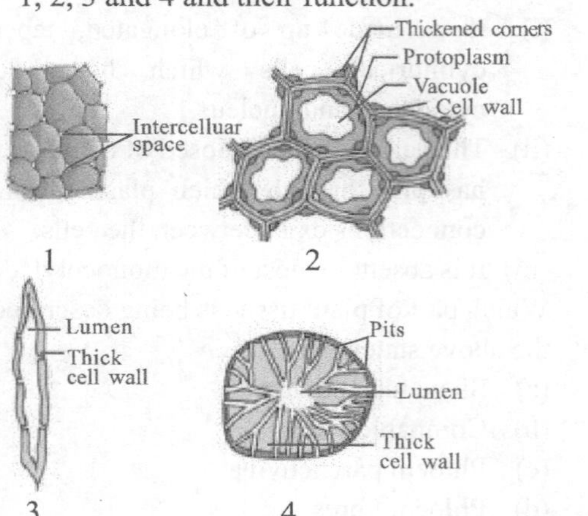

(a) 1 – Parenchyma; Photosynthesis, Storage and Secretion.
(b) 2 – Sclerenchyma, Sclerieds; Transport food material.
(c) 3 – Collenchyma; Provides mechanical support to organs.
(d) 4 – Sclerenchyma Fibres; Provide mechanical support to the growing parts of the plant such as young stem and petiole of a leaf.

105. Lignin is a component of the secondary cell walls of:
(a) Epidermis
(b) Collenchyma
(c) Sclerenchyma
(d) Parenchyma

106. Collenchyma differs from sclerenchyma in:
(a) having sclereids
(b) having thick walls
(c) having inside lumen
(d) retaining protoplasm at maturity

107. "Bast-fibers" obtained from which part of woody stem?
(a) Cork
(b) Cortex
(c) Xylem
(d) Phloem

108. In trees, the growth rings represent:
(a) Primary xylem
(b) Secondary xylem
(c) Secondary phloem
(d) Cambium

Answer KEYs

1	(a)	**12**	(a)	**23**	(a)	**34**	(c)	**45**	(d)	**56**	(c)	**67**	(c)	**78**	(c)	**89**	(d)	**100**	(c)
2	(c)	**13**	(c)	**24**	(a)	**35**	(a)	**46**	(b)	**57**	(a)	**68**	(b)	**79**	(d)	**90**	(d)	**101**	(a)
3	(b)	**14**	(a)	**25**	(a)	**36**	(c)	**47**	(b)	**58**	(c)	**69**	(b)	**80**	(c)	**91**	(c)	**102**	(c)
4	(c)	**15**	(b)	**26**	(a)	**37**	(c)	**48**	(a)	**59**	(a)	**70**	(a)	**81**	(c)	**92**	(c)	**103**	(c)
5	(a)	**16**	(c)	**27**	(b)	**38**	(c)	**49**	(b)	**60**	(a)	**71**	(b)	**82**	(a)	**93**	(a)	**104**	(a)
6	(c)	**17**	(c)	**28**	(b)	**39**	(b)	**50**	(b)	**61**	(c)	**72**	(c)	**83**	(a)	**94**	(c)	**105**	(c)
7	(a)	**18**	(c)	**29**	(d)	**40**	(d)	**51**	(c)	**62**	(c)	**73**	(a)	**84**	(d)	**95**	(b)	**106**	(d)
8	(a)	**19**	(d)	**30**	(a)	**41**	(c)	**52**	(b)	**63**	(d)	**74**	(d)	**85**	(a)	**96**	(b)	**107**	(d)
9	(a)	**20**	(d)	**31**	(c)	**42**	(a)	**53**	(d)	**64**	(d)	**75**	(c)	**86**	(a)	**97**	(a)	**108**	(b)
10	(d)	**21**	(a)	**32**	(c)	**43**	(c)	**54**	(b)	**65**	(d)	**76**	(c)	**87**	(d)	**98**	(d)		
11	(b)	**22**	(a)	**33**	(a)	**44**	(d)	**55**	(c)	**66**	(c)	**77**	(b)	**88**	(c)	**99**	(b)		

Hints & Solutions

1. (a) Parenchyma is the first tissue developed during the course of evolution hence, phylogenetically most primitive also functionally, it is most primitive as it is involved in photosynthesis and storage of food material.
2. (c) 3. (b)
4. (c) In the given figure of root apical meristem, the part marked as 1, 2 and 3 is respectively cortex, protoderm and root cap.

   The apical meristem is a completely undifferentiated meristematic tissue found in the buds and growing tips of roots in plants. Its main function is to begin growth of new cells in young seedlings at the tips of roots and shoots (forming buds, among other things).
5. (a) 6. (c)
7. (a) In the cortex of monocot root, we observe a few layers of cell below the epidermis which have thick walls due to the deposition of a chemical suberin. These layers of suberised cells are called exodermis.

   There are also alternating zone of xylem present in root. The number of xylem patches far exceeds to ten. That is why, they are called polyarch. In monocot root, there is a large sized pith made up of parenchyma tissues.
8. (a)
9. (a) A: Lateral meristem is a meristem that is arranged parallel to the sides of an organ and is responsible for increase in diameter of the organ. Fascicular vascular cambium, interfascicular cambium and cork cambium are the examples of lateral meristem.

   B: Apical meristem is a completely undifferentiated meristematic tissue found in the buds and growing tips of roots in plants. Its main function is to begin growth of new cells in young seedlings at the tips of roots and shoots (forming buds, among other things). During the formation of the primary plant body, specific region of apical meristem produces dermal tissue, ground tissue and vascular cambium.

   C: Bast fibres or phloem fibres are made up of sclerenchymatous cells. These are generally absent in primary phloem but found in secondary phloem.

   D: Sapwood is the peripheral region of the secondary xylem. It is involved in the conduction of water and minerals from root to leaf.
10. (d) All the statements are correct regarding bulliform or motor cells. They are large, bubble-shaped, empty, colourless epidermal cells that occur in groups on the upper surface of the leaves of many grasses. Loss of turgor pressure in these cells causes leaves to "roll up" during water stress.
11. (b) Grafting is the technique of joining parts of two plants so as to form a composite plant. One plant has a strong root system called the stock, and the other plant having better flower or fruit yield is known as scion. When a scion is grafted to a stock, the root system of the plant has the genotype of stock and fruits produced show genotypes of the scion.
12. (a)
13. (c) Epidermis is generally uniseriate. i.e., composed of single layer of epidermal cells. In some cases, it is multilayered also. e.g., *Ficus*, *Nerium*, *Peperomia*, etc.

14. (a) Vascular bundles (VBs) is a strand of conducting vessels in the stem or leaves of a plant, typically with phloem on the outside and xylem on the inside. In the given figures 1, 2 and 3 the types of vascular bundles are respectively radial, conjoint closed and conjoint open. When xylem and phloem are arranged in the radius of the plant in alternative manner then it is called radial vascular bundle. Radial VBs are always closed and present in all types of roots. In conjoint vascular bundle, xylem and phloem are situated at the same radius of vascular bundles. Such VBs are common in stems and leaves.

15. (b) Heart wood is the dark coloured, broad zone of the plant wood. As a stem grows older, its heart wood increases in thickness. The heart wood gets infiltrated with gums, tannins, resins, dyes, etc., e.g. catechu found from *Acacia catechu*, red brasilin from *Caesalpinia sappan* etc.

16. (c) Heart wood is the central xylem of tree-trunks, which is frequently impregnated with tannins, resins, etc. It is considered to be an excretory product.
An important tannin is obtained from the heart wood of *Acacia catechu*.

17. (c)

18. (c) In 1998, 22 rings are observed which means 11 rings of spring wood and 11 of autumn wood i.e., 11 annual rings are there. From 1998 to 2007 in 9 years, 9 more annual rings are formed hence, total number of annual rings in 2007 will be $11 + 9 = 20$.

19. (d) Bamboo and grasses elongate by the activity of intercalary meristems. These meristems are found at the base of leaves of many monocotyledons. Intercalary meristems are mainly responsible for the vertical growth of plant organs.

20. (d) Meristemaic region is the region of active mitotic division in plants, from which permanent tissues are derived. It comprises of immature cells which are packed close to one another without any intercellular space and have the capablity of differentiating to form characteristic tissues of the adult. Meristematic cells are found in roots and shoot tip regions. Depending upon the position meristematic tissues are of three types-intercalary, lateral and apical.

21. (a) In the given figure of a typical dicot root, the marked structure 1, 2, 3 and 4 are respectively primary phloem, vascular cambium, secondary phloem and primary xylem.

22. (a) 23. (a)

24. (a) Lateral meristems are present along the lateral sides of stem and roots. Interstelar cambium ring formed by intrafascicular, interfascicular cambium and cork cambium which are the examples of lateral meristem.

25. (a) Secondary cortex is also called phelloderm. Cork cambium, cork and secondary cortex are collectively called periderm.

26. (a) Statement (i) and (iii) are correct about heartwood/Duramen.
Heartwood is the older, non-living central wood of a tree or woody plant, usually darker and harder than the younger sapwood. The heartwood does not conduct water but it gives mechanical support to the stem.
Sapwood is the peripheral region of the secondary xylem. It is lighter in colour and involved in the conduction of water and minerals from root to leaf.

27. (b)

28. (b) A - Fibres: These are thick walled, elongated and pointed cells, generally occurring in groups, in various parts of the plants.
B - Sclereids: These are thick-walled lignified sclerenchyma cell. Sclereids are variable in shape and are shorter than fibres.
C - Tracheids : These are the more ancient tracheary elements, the main water conductive cells of the pteridophytes and the gymnosperms. They are elongate cells with tapering ends. Their cell walls characteristically bear spiral (pteridiophytes), pitted or bordered-pitted (gymnosperms) thickening. Water moves from cell to cell through the side walls.
D - Vessels: It is a long cylindrical tube like structure made up of many cells called vessels members.
E - Xylem parenchyma: These cells are living and thin walled and their cell walls are made up of cellulose. They store food materials in the form of starch or fats and other substances like tannins.

29. **(d)** Xylem fibres have highly thickened walls and obliterated central lumens whereas xylem parenchyma are thin walled and their cell walls are made up of cellulose.

30. **(a)** The end walls of sieve tube elements are perforated in a sieve like manner to form the sieve plates. A mature sieve element possesses a peripheral cytoplasm and a large vacuole but lacks a nucleus.

31. **(c)** In monocot stem, vascular bundles are scattered. Each bundle is surrounded by a thick walled bundle sheath.

Due to disintegration of protoxylem elements, a lysigneous water cavity is formed.

32. **(c)** **33.** **(a)** **34.** **(c)** **35.** **(a)**

36. **(c)** Lenticel is a pore in the periderm of a woody stem. It is packed with a loose aggregate of cells which is derived from the phelloderm. It acts as an organ of gaseous exchange between air and internal tissues.

37. **(c)** **38.** **(c)** **39.** **(b)** **40.** **(d)**

41. **(c)** In *Bougainvillea*, the primary vascular bundles lie scattered in the ground tissue. A cambium ring arises outside the primary vascular bundles.

This cambium cuts xylem and conjunctive tissue on the inner surface and phloem and conjunctive tissues on the outer surface which arises in centrifugal manner.

42. **(a)**

43. **(c)** Figure (a), (b), (c) and (d) are respectively collenchyma, sclerenchyma, xylem tissues and parenchyma tissues. Collenchyma, sclerenchyma and parenchyma tissues are types of simple permanent tissues. Figure (c) (xylem tissue) is a type of permanent tissue having many different types of cell. Xylem is one of the two types of transport tissue in vascular plants, phloem being the other. The most distinctive xylem cells are the long tracheary elements that transport water. Tracheids and vessel elements are distinguished by their shape; vessel elements are shorter, and are connected together into long tubes that are called vessels.

44. **(d)**

45. **(d)** Statements (ii), (iii) and (v) are correct.

(i) Uneven thickening of the cell wall is not the characteristics of sclerenchyma.

(iv) A mature sieve elements are devoid of nucleus at maturity.

46. **(b)** **47.** **(b)**

48. **(a)** The cork of commerce is obtained from *Quercus suber*. It is a layer of suberised and dead cells, cut off by a phellogen on the outside of the stem or root of a woody plant. The layer is protective and impervious to gases and water vapour.

49. **(b)** Some branches of tap roots grow vertically upward from horizontal secondary roots in the air called pneumatophores.

These pneumatophores have openings called lenticels. They act as organs of gaseous exchange.

50. **(b)** Phloem parenchyma are unspecialised cells found in the phloem. They are absent in the monocots as they have abundant companion cells.

51. **(c)** Spring wood is the part of an annual ring of wood, characterised by large, thin-walled cells, formed during the first part of the growing season. It is lighter in colour with low density. It produces a larger number of xylem elements having vessels with wider cavities.

Autumn wood or late wood is the wood element formed during winter when cambial activity is more. It is darker in colour with high density. Wood elements are lesser in amount and have narrow lumen.

52. **(b)** **53.** **(d)** **54.** **(b)** **55.** **(c)**

56. **(c)** Secondary growth in dicot stems is possible because of the presence of cambium. In dicot stems, the primary permanent tissues make the fundamental parts of the plant.

The further growth in its thickness, is completed by cambial activity, called secondary growth in thickness.

57. (a) 58. (c) 59. (a)

60. (a) In the given figure of shoot apical meristem, the parts marked as 1, 2 and 3 are respectively leaf primordium; shoot apical meristem and axillary bud (present in the axils of leaf and are capable of forming a branch or a flower). The shoot apical meristem is a population of cells located at the tip of the shoot axis. It produces lateral organs, stems, tissues and regenerates itself.

61. (c) In the formation of the heart wood and sap wood, it is not necessarily that the wood formed in previous years is darker than newer wood.

62. (c) The parenchymatous cells present between vascular bundle dedifferentiate and become meristematic and forms interfascicular cambium.

63. (d)

64. (d) Endodermal cells are also called endodermis, which is a single layer of compactly arranged cells. It is characterised by barrel shaped cells.

65. (d) Phellogen cuts off thin-walled phelloderm on inner side and thin wall complementary cell on the outer side at a regular interval. It produces narrow zones of cell called closing layers.
These cells of closing layers show the deposition of suberin i.e., *Magnolia*, *Populus*.

66. (c) 67. (c) 68. (b) 69. (b)

70. (a) 71. (b)

72. (c) In the given fig. of phloem tissue 1, 2 & 3 are respectively sieve tube, phloem parenchyma and campanion cells. Companion cells (3) are specialised parenchymatous cells adjacent to a sieve tube in the phloem of flowering plants. It is believed to regulate the flow of nutrients through the tube.

73. (a) In the given figure of monocot stem, the structure marked as 1, 2, 3, 4, 5 and 6 are respectively epidermis, hypodermis, vascular bundle, phloem, xylem and ground tissue. Monocot stem is characterised by epidermis (2-3 layered), hypodermis and undifferentiated ground tissue stem. Vascular strand is numerous and scattered. Vascular bundles are conjoint, collateral and closed.

74. (d) A - Bulliform cells: They are large, bubble-shaped, empty colourless epidermal cells that occur in groups on the upper surface of the leaves of many grasses.
B - Pericycle: It is a thin layer of plant tissue between the endodermis and the phloem. Initiation of lateral roots and vascular cambium during the secondary growth takes place in these cells.
C - Endarch xylem: When protoxylem lies towards the centre (pith) and metaxylem lies towards the periphery of the organ, it is called endarch xylem. This type of xylem occurs in stems.
D - Exarch xylem: When protoxylem lies towards the periphery and metaxylem lies towards the centre, it is called exarch xylem. This type of xylem occurs in roots.
E - Bundle sheath cells: They are layer of cells in plant leaves and stems that forms a sheath surrounding the vascular bundles. These cells are found in dicot leaf.

75. (c) Phloem transports food materials, usually from leaves to the other parts of the plant. The companion cells are specialised parenchymatous cells which are closely associated with sieve tube elements.

76. (c) The described statements are associated with spring wood or early wood. Spring woods are soft, more porous portion of an annual ring of wood that develops early in the growing season. It consists of large thin-walled xylem cells.

77. (b)

78. (c) In dicot root, there is no primary cambium. Thus, vascular cambium develops later from pericycle and conjunctive parenchyma. It is called as secondary meristem.

79. (d) 80. (c)

81. (c) Tunica corpus theory was proposed by Schmidt. According to this theory, two tissue zones occur at the apex.

82. (a)

83. (a) In the given figure of monocot leaf, the structure marked as 1, 2, 3, 4, 5, 6 and 7 are respectively adaxial epidermis, xylem, mesophyll, sub-stomal cavity, abaxial epidermis, stoma and phloem. Monocot leaf (isobilateral) is somewhat similar to dorsiventral leaf in having epidermis, cuticle, and mesophyll cells. Stomata are present on both the surfaces of epidermis. Mesophyll cells are undifferentiated into palisade and spongy parenchyma. Vascular bundles are conjoint, collateral, closed with phloem towards lower side and xylem towards upper side.

84. (d) In the given figure of T.S of dicot root, the parts marked as 1, 2, 3, 4, 5 and 6 are respectively cortex, epiblema, pith, endodermis, root hairs and pericycle.

Dicot root consists of a single layer of epiblema which bears unicellular root hairs. Endodermis is followed by one or more layers of pericycle. Inner to pericycle lies radially arranged vascular bundles. This arrangement keeps the xylem bundles in direct contact with the outer tissue of the roots which conduct water absorbed by root hairs to the inside. Xylem is exarch.

85. (a) All the statements describe the anatomy of dicotyledonous root. Dicot root consists of a single layer of epiblema which bears unicellular root hairs. Endodermis is followed by one or more layers of pericycle. Inner to pericycle, lies radially arranged vascular bundles. This arrangement keeps the xylem bundles in direct contact with the outer tissue of the roots which conduct water absorbed by root hairs to the inside. Xylem is exarch.

86. (a) Lenticels are raised pores in the stem of a woody plant that allows gas exchange between the atmosphere and the internal tissues. Parenchymatous cells are usually present in cortex. The ground tissue system is divided into three main zones- cortex, pericycle and pith. The conjoint vascular bundles usually have the phloem located only on the outer side of the xylem.

87. (d) Guard cells are specialised cells in the epidermis of leaves, stems and other organs that are used to control gas exchange. The guard cells are produced in pairs with a gap between them that forms a stomatal pore. Guard cells are dumbbell shaped in monocots (e.g., grasses) and bean shaped in dicots.

88. (c)

89. (d) Pericycle is generally made up of sclerenchymatous cells. It is present as a continuous cylinder of sclerenchymatous cells e.g., *Cucurbita, Aristolochia*, etc.

90. (d) 91. (c) 92. (c) 93. (a)

94. (c) 95. (b) 96. (b)

97. (a) In diffuse porous wood, vessels are evenly scattered throughout the xylem and there is a little difference in the size of the vessels formed in different seasons. These woods grow in a tropical regions. e.g., *Salix, Populus*.

98. (d) 99. (b)

100. (c) The described statements are associated with the phloem parenchyma. Phloem parenchyma is living and has thin cell walls. These cells form the packing tissue between all the other types of cells. These cells store compounds such as starch. Phloem parenchyma is absent in monocot because vascular bundle are open that is, they lack cambium, so no secondary growth take place.

101. (a) A - Stele: It is the central core of the stem and root of a vascular plant, consisting of the vascular tissue (xylem and phloem) and associated supporting tissue.

B - Endodermis: It is an inner layer of cells in the cortex of a root and of some stems, surrounding a vascular bundle.

C - Casparian strips: It is a band of cell wall material deposited on the radial and transverse walls of the endodermis, and is chemically different from the rest of the cell wall - the cell wall being made of lignin and without suberin - whereas the casparian strip is made of suberin and sometimes lignin.

D - Bark: Bark is the outermost layers of stems and roots of woody plants. Plants with bark include trees, woody vines, and shrubs.

102. (c) The cells of quiescent centre become active when the previously active initials are damaged. Thus, this centre acts as a reservoir of relatively resistant cells, which constitute a permanent source of active initials.

Therefore, it serves as a reserve for replenishment of damaged cells of the meristems.

103. (c)

104. (a) Figure 1, 2, 3 and 4 are respectively parenchyma, collenchyma, sclerenchyma fibres and sclerenchyma sclereid.

Parenchyma: It is involved in photosynthesis, storage and secretion. Collenchyma: It provides mechanical support to the growing parts of the plant such as young stem and petiole of a leaf. Sclerenchyma fibres: It provides mechanical strength to various plant organs and is specialised to tolerate stress of bending, sheering, compression and pull. Sclerenchyma sclereid: It forms protective covering of testa in many legume seeds.

105. (c) Lignin is the main structural material of vascular plants. It is a component of secondary wall of sclerenchyma, which is deposited during secondary thickening of cell walls.

106. (d) The cells of collenchyma are modified parenchyma cells with cellulose wall thickened at angles where three or more cells are in contact.

Collenchyma resembles parenchyma in being alive and having a moderate amount of protoplasm.

107. (d)

108. (b)

7 Structural Organisation in Animals

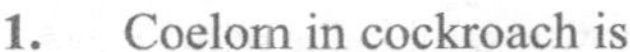

1. Coelom in cockroach is:
   (a) Enterocoel (b) Schizocoel
   (c) Pseudocoel (d) None of these
2. Number of fertilised eggs in ootheca of cockroach is: **Tricky**
   (a) 16 pairs in two rows
   (b) 16 in two rows
   (c) 10 in one rows
   (d) 8 in two rows
3. Read the following statements and answer the question.
   (i) It is made of a single thin layer of flattened cells with irregular boundaries.
   (ii) They are found in the walls of blood vessels and air sacs of lungs.
   (iii) They are involved in forming a diffusion boundary.
   Which of the following tissue is being described by the above statements?
   (a) Squamous epithelium
   (b) Columnar epithelium
   (c) Ciliated epithelium
   (d) Compound epithelium
4. Match the terms given in column-I with their feature given in column-II and choose the correct option.

Column-I (Terms)	Column-II (Features)
A. Exocrine gland	I. They help to stop substances from leaking across a tissue
B. Endocrine gland	II. Hormones are secreted directly into the fluid bathing the gland
C. Tight junctions	III. They perform cementing to keep neighbouring cells together.
D. Adhering junctions	IV. Secretes mucus, saliva, earwax, oil, milk, digestive enzymes and other cell products.

   (a) A – IV; B – II; C – I; D – III
   (b) A – II; B – IV; C – I; D – III
   (c) A – IV; B – II; C – III; D – I
   (d) A – IV; B – I; C – II; D – III
5. Which of the following can be absorbed by hepatic caeca?
   (a) Glucose and amino acid
   (b) Glucose and lipid
   (c) Lipid
   (d) Glucose
6. A student was given a sample of two tissues. He observed the tissues under the microscope and draws their figures as 1 and 2 given below.

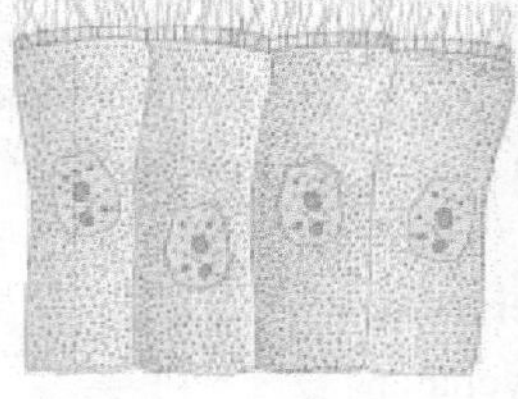

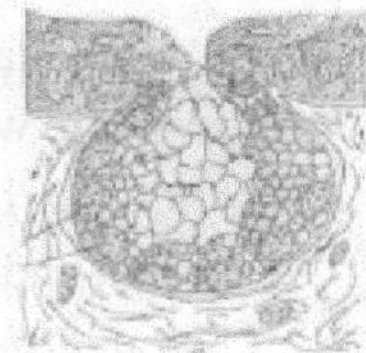

1 2

   Identify the tissues 1 and 2.
   (a) 1: Columnar cells bearing cilia
   2: Unicellular glandular epithelium
   (b) 1: Cuboidal cells bearing cilia
   2: Multicellular glandular epithelium
   (c) 1: Compound cells bearing cilia
   2: Unicellular glandular epithelium
   (d) 1: Columnar cells bearing cilia
   2: Multicellular glandular epithelium
7. Striated or voluntary muscle fibres are found in:

   (a) Lungs (b) Leg muscles
   (c) Gall bladder (d) Blood vessels
8. The wall of heart is thick due to:
   (a) Pericardium (b) Epicardium
   (c) Myocardium (d) Endocardium

9. 'Oxygen debt' is amount of oxygen required for: **Tricky**
   (a) Muscle contraction
   (b) Muscle relaxation
   (c) Muscle recovery
   (d) All of the above

10. Match the description given in column-I with their examples given in column-II and choose the correct option.

Column-I (Description)	Column-II (Example)
A. Aquatic respiratory organ	I. Skin
B. Organ which acts as urogenital duct and opens into the cloaca	II. Ureter
C. A small median chamber that is used to pass faecal matter, urine and sperms to the exterior	III. Cloaca
D. A triangular structure which joins the right atrium and receives blood through vena cava	IV. Sinus venosus

   (a) A – I; B – II; C – III; D – IV
   (b) A – II; B – IV; C – I; D – III
   (c) A – II; B – I; C – III; D – IV
   (d) A – IV; B – II; C – I; D – III

11. The non-excitable, variously shaped and found between neurons are:
   (a) Dendrites (b) Nissl bodies
   (c) Glial cells (d) Schwann cell

12. Consider the diagram and identify 1 to 5.

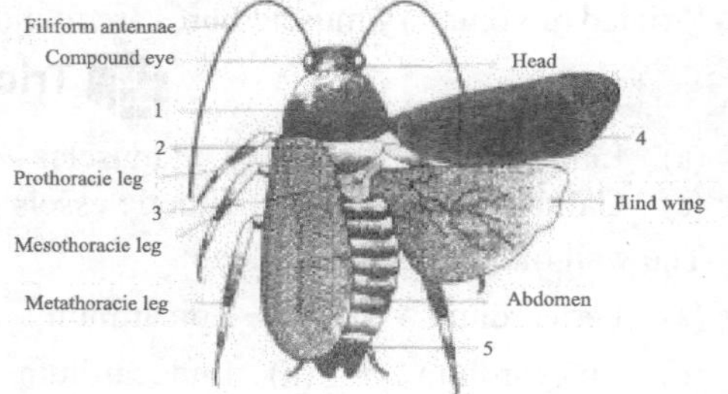

   (a) 1-Pronotum, 2-Mesothorax, 3-Metathorax, 4-Tegmina, 5-Pleura
   (b) 1-Pronotum, 2-Mesothorax, 3-Metathorax, 4-Tegmina, 5-Sterna
   (c) 1-Pronotum, 2-Mesothorax, 3-Metathorax, 4-Tegmina, 5-Anal cerci
   (d) 1-Pronotum, 2-Mesothorax, 3-Metathorax, 4-Tegmina, 5-Anal style

13. Skeletal muscles show resemblance with visceral muscles in which of the aspect?
   (a) Shape of muscle fibres.
   (b) Number of nuclei in muscle fibres.
   (c) Presence of actin and myosin filaments.
   (d) Presence of light and dark bands.

14. Which of the following ion is necessary for the contraction of a muscle and nerve impulse transmission?
   (a) Na^+ (b) K^+
   (c) Ca^{++} (d) None of these

15. Which of the following statement(s) is/are correct about muscle tissue?
   (a) Each muscle is made of many long, cylindrical fibres arranged in parallel arrays.
   (b) Muscle fibres contract (shorten) in response to stimulation, then relax (lengthen) and return to their uncontracted state in a coordinated fashion.
   (c) Muscles play an active role in all movements of the body.
   (d) All of the above

16. RBC are absent in: **Toughnut**
   (a) Cockroach (b) Frog
   (c) Cow (d) Camel

17. Vitamin K is required for:
   (a) Formation of thromboplastin
   (b) Synthesis of prothrombin
   (c) Conversion of fibrinogen to fibrin
   (d) Conversion of prothrombin to thrombin

18. Which among the following is not a characteristic of yellow fibres of connective tissue? **Critical Thinking**
   (a) Presence of elastin
   (b) Fewer in number
   (c) Straight and branched
   (d) Provide toughness and strength

19. Which of the following types of connective tissue is mismatched with its matrix?
   (a) Areolar – Loosely packed matrix of protein fibres
   (b) Bone – Mineralised matrix
   (c) Cartilage – Highly vascular matrix
   (d) Blood – Liquid matrix

20. In the given figure of areolar connective tissue, the different cells and parts have been marked by alphabets (1, 2, 3 & 4). Choose the answer in which these alphabets correctly match with the parts and cells they indicate.

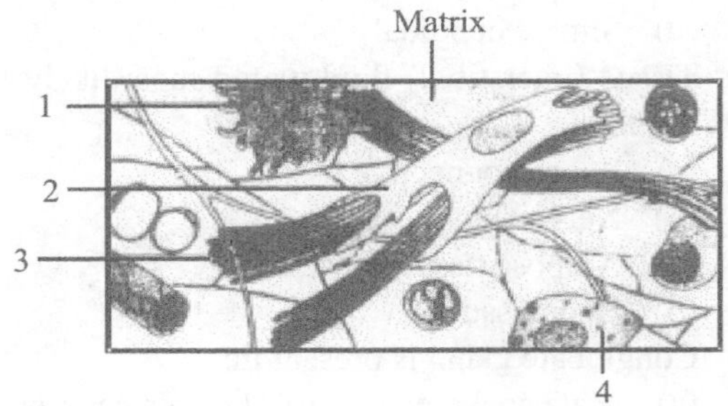

(a) 1-Adipocyte, 2-Collagen fibres, 3-Microfilament, 4-Mast cells
(b) 1-Macrophage, 2-Collagen fibres, 3-Microfilament, 4-Mast cells
(c) 1-Macrophage, 2-Collagen fibres, 3-Microtubule, 4-RBC
(d) 1-Macrophage, 2-Fibroblast, 3-Collagen fibres, 4-Mast cells

21. Which of the following is a purely motor nerve?
(a) Vagus (b) Facial
(c) Abducens (d) Trigeminal

22. Sheath nuclei or Schwann cells and Nodes of Ranvier are found in:
(a) Neurons (b) Osteoblasts
(c) Chondroblasts (d) Gland cells

23. Glisson's capsules are found in:
(a) Kidney of frog
(b) Liver of mammals
(c) Heart of frog
(d) Cerebellum of rabbit

24. Which of the following statement is **not correct** regarding connective tissues?
(a) They are classified into four types.
(b) They are most abundant and widely distributed in the body of animals.
(c) They perform the function of linking and supporting the other tissues.
(d) They include cartilage, bone, adipose tissue and blood.

25. Which of the following statement(s) is/are **correct** regarding cockroaches?
(i) The body of the cockroach is segmented and divisible into three distinct regions – head, thorax and abdomen.
(ii) Blood vascular system is of closed type.
(iii) They are monoecious and both sexes have well developed reproductive organs.
(iv) The development of *P. americana* is paurometabolus, meaning there is development through nymphal stage.
(a) (i) and (ii) (b) (ii) and (iii)
(c) (i) and (iv) (d) All of the above

26. Blood cells are formed in bone marrow. In a process known as:
(a) Haemolysis (b) Erythroblastosis
(c) Haemopoiesis (d) None of these

27. The main difference between bone and cartilage is of:
(a) Mineral salts (b) Blood vessels
(c) Lymph vessels (d) Haversian canals

28. Nerve cells do not possess:
(a) Neurilemma (b) Sarcolemma
(c) Dendrite (d) Axon

29. Which of the following type of tissue is being described by the given statements?
(i) They are named because of their special function of linking and supporting other tissues/organs of the body.
(ii) They include cartilage, bone, adipose tissue and blood.
(iii) They provide strength, elasticity and flexibility to the tissue.
(iv) They also secrete modified polysaccharides, which accumulate between cells and fibres and act as matrix.
(a) Epithelial tissue (b) Connective tissue
(c) Muscle tissue (d) Neural tissue

30. Match column-I (function) with column-II (types of enzymes) and select the correct option.

Column-I	Column-II
A. Squamous	I. Intestinal glands
B. Cuboidal	II. Trachea
C. Columnar	III. Ovary
D. Ciliated	IV. Blood vessels
E. Pseudostratified	V. Bronchioles

(a) A – I; B – II; C – IV; D – III; E – V
(b) A – V; B – IV; C – II; D – I; E – III
(c) A – IV; B – V; C – I; D – II; E – III
(d) A – IV; B – III; C – I; D – V; E – II

31. Krause membrane or Z-line is a myofibril which separates two adjacent:
(a) Sarcomeres (b) H – zones
(c) I – bands (d) A – bands

32. Titin filament connects:
(a) Actin filaments to Z-line
(b) Myosin filaments to Z-line
(c) Actin filaments to myosin filaments
(d) Myosin filaments to M-line

33. The figure below is related with mouth parts of cockroach. Identify A to E.

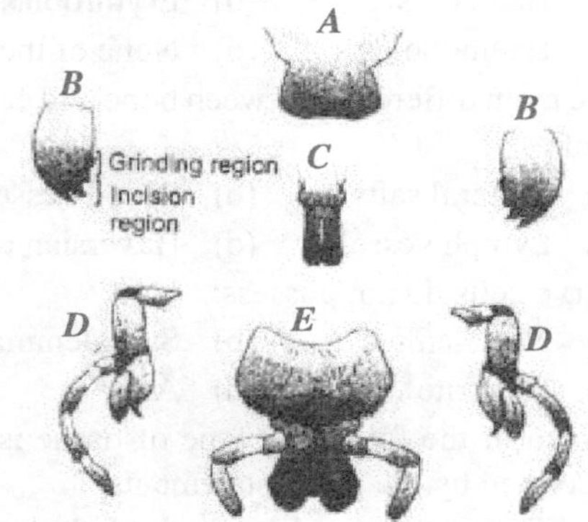

(a) A-Maxilla, B-Hypopharynx, C-Labium, D-Mandible, E-Labrum
(b) A-Mandible, B-Labium, C-Maxilla, D-Labrum, E-Hypopharynx
(c) A- Labrum, B-Mandible, C-Hypopharynx, D-Maxilla, E-Labium
(d) A- Labium, B-Hypopharynx, C-Labrum, D-Maxilla, E-Mandible

34. The efferent process of neuron is known as:
(a) Axon (b) Dendrite
(c) Cyton (d) Neurofibrillae

35. Which of the following statement(s) is/are correct about nervous system of cockroach?

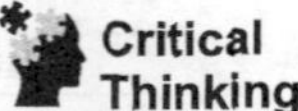

(i) It consists of a series of fused segmentally arranged ganglia joined by paired longitudinal connectives on the ventral side.
(ii) There are six ganglia lie in the thorax, and three in the abdomen.
(iii) The sense organs are antennae, eyes, maxillary pulps, labial pulps and anal cerci etc.
(iv) Each eye consists of about 5000 hexagonal ommatidia.
(a) Both (i) and (iii) (b) Only (ii)
(c) Both (i) and (iv) (d) All of the above

36. The inability to absorb digested nutrients may be due to damage of which type of epithelium? **Tricky**
(a) Ciliated columnar
(b) Simple columnar
(c) Simple squamous
(d) Simple cuboidal

37. Tissue (interstitial) fluid would most likely be found in:
(a) Loose connective tissue
(b) Nervous tissue
(c) Adipose tissue
(d) Bone tissue

38. Conglobate gland is present in:
(a) Male cockraoch (b) Female cockroach
(c) Earthworm (d) *Hydra*

39. Which structure of man is similar to spiracle of cockroach?
(a) Nostril (b) Bronchiole
(c) Lungs (d) Alveoli

40. The most important centre for the formation of lymph is: **Critical Thinking**
(a) Interstitial space
(b) Liver
(c) Pancreas
(d) Kidney

41. The given figure refers to reproductive system of female cockroach. Choose the correct of non showing the correct labellings indicated by A, B and C.

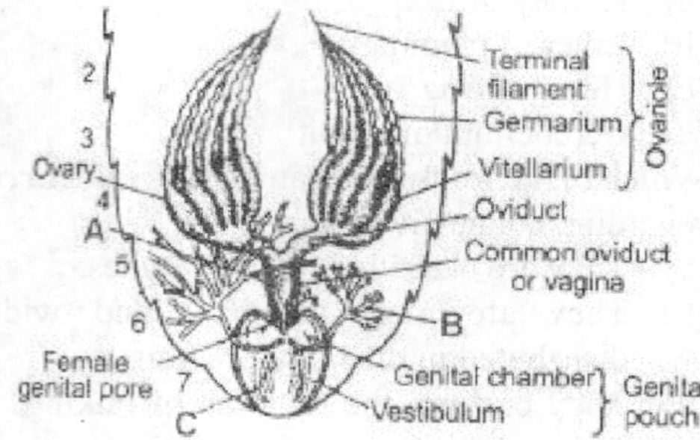

(a) A-Spermatheca, B-Collateral glands, C-Gonapophyses
(b) A-Phallic gland, B-Collateral glands, C-Gonapophyses
(c) A-Spermatheca, B-Seminal vesicle, C-Gonapophyses
(d) A-Spermatheca, B-Collateral glands, C-Tegmina

42. Dried bone possesses (after burn):
(a) Matrix lacunae, canaliculi canals and narrow cavity
(b) Living cells-osteocytes
(c) Only minerals
(d) Only osteocyte and matrix

43. Thousand of year old Mummies are still in their condition as they were before due to non-destruction of: **Toughnut**
(a) Yellow elastin fibres
(b) White elastin fibres
(c) Collagen fibres
(d) Veins

44. Which is secreted during allergic reaction?
(a) Histamine (b) Neutrophils
(c) Basophils (d) Acidophils

45. Which of the following type of muscle tissue is being described on the basis of given statements?
(i) These muscle fibres taper at both ends and do not show striations.
(ii) The wall of internal organs such as the blood vessels, stomach and intestine contain this type of muscle tissue.
(iii) They are 'involuntary' as their function cannot be directly controlled.
(a) Skeletal muscle (b) Smooth muscle
(c) Cardiac muscle (d) All of these

46. An infection would most likely increase phagocytic activity in:
(a) Elastic tissue
(b) Transitional tissue
(c) Adipose tissue
(d) Reticular tissue

47. The lobules containing a central vein and acini containing lumen are respectively found in:

(a) Liver and Pancreas
(b) Spleen and Pancreas
(c) Lungs and Spleen
(d) Lungs and Pancreas

48. Keratinised stratified squamous epithelium is found in:
(a) Trachea
(b) Epidermis
(c) Mouth cavity
(d) Lining of blood vessels

49. The given figure shows open circulatory system of cockroach. Identify 1, 2 and 3 labelled in the figure.

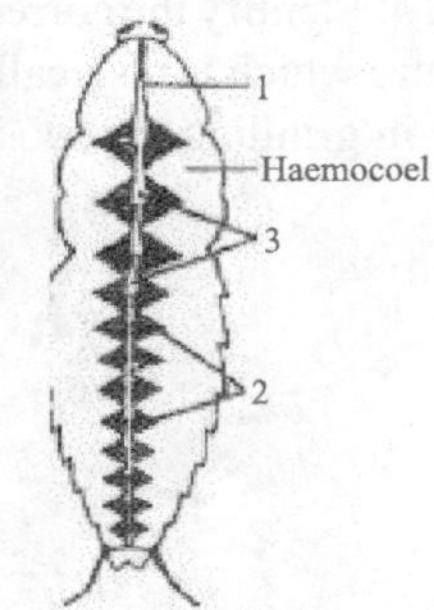

(a) 1-Posterior aorta, 2-Alary muscles, 3-Chambers of heart
(b) 1-Anterior aorta, 2-Ciliary muscles, 3-Chambers of heart
(c) 1-Anterior aorta, 2-Alary muscles, 3-Chambers of heart
(d) 1-Anterior aorta, 2-Circular muscles, 3-Chambers of heart

50. Pacinian corpuscles are abundantly found in: **Tricky**
(a) Skin (b) Blood
(c) Tissue (d) Kidney

51. An exocrine gland in which a portion of the secretory cell is discharged with the secretion is termed:
(a) Apocrine (b) Merocrine
(c) Endocrine (d) Holocrine

52. Glisson's capsule is a delicate connective tissue capsule covering the:
(a) Spleen (b) Liver
(c) Kidney (d) Gall bladder

53. The type of connective tissue that is associated with the umbilical cord is: **Critical Thinking**
(a) Areolar connective tissue
(b) Jelly-like connective tissue
(c) Adipose connective tissue
(d) Reticular connective tissue

54. Match column-I (function) with column-II (Types of enzymes) and select the correct option.

	Column-I		Column-II
A.	Dark band	I.	H-Zone
B.	Krause membrane	II.	A-band
C.	A light line	III.	I-band
D.	Myofibrils	IV.	Sarcostyles

(a) A – III; B – II; C – I; D – IV
(b) A – II; B – III; C – IV; D – I
(c) A – II; B – III; C – I; D – IV
(d) A – II; B – I; C – III; D – IV

55. The figure given below shows the alimentary canal of cockroach with few structures labelled as 1, 2, 3 and 4. Identify the correct identification of the structure which is also called proventriculus and helps in grinding the food particles.

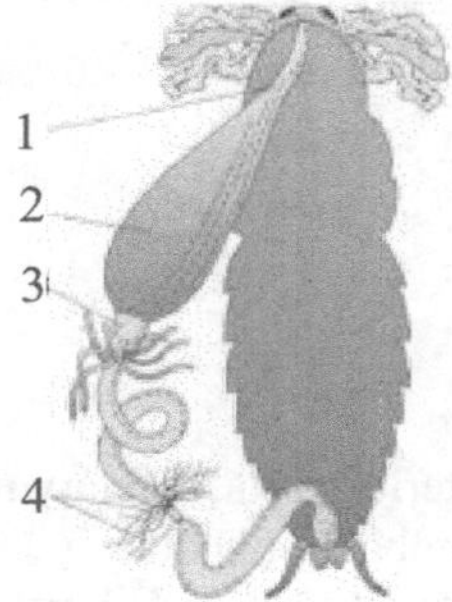

(a) 1 (b) 2 (c) 3 (d) 4

56. Male and female cockroaches can be distinguished externally through: **Tricky**

(a) Anal styles in male
(b) Anal cerci in female
(c) Anal style and antennae in female
(d) Both (b) and (c)

57. In which of the following condition, effect of atomic warfare are the development of buccal ulcers, colon ulcer and lung infection, this is because of:

(a) Leukemia (b) Erythropoiesis
(c) Anaemia (d) Agranulocytosis

58. pH of human blood is:

(a) 7.4 (b) 6.2
(c) 9.0 (d) 7.5 - 8.0

59. Which of the following are required for blood clotting in mammals?

(a) K^+ and vitamin K
(b) Ca^{2+} and vitamin A
(c) Ca^{2+} and vitamin K
(d) K^+ and vitamin E

60. The given diagram represents the reproductive organ of male cockroach. Choose the correct matching of labelling 1 , 2, 3, 4.

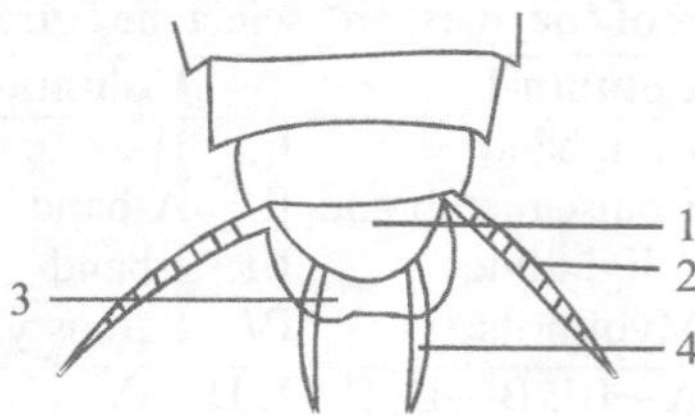

(a) 1 – 8th sternum, 2 – Anal cercus, 3 – 10th tergum, 4 – Anal style
(b) 1 – 10th tergum, 2 – Anal cercus, 3 – Anal style, 4 – 8th sternum
(c) 1 – Anal style, 2 – Anal cercus, 3 – 10th tergum, 4 – 8th sternum
(d) 1 – Anal cercus, 2 – 8th sternum, 3 – 10th tergum, 4 – Anal style.

61. Concave surface of mammalian RBCs is helpful in: **Critical Thinking**

(a) formation of haemoglobin.
(b) increasing surface area.
(c) reducing surface tension.
(d) more space for haemoglobin.

62. Which of the following statement is/are **correct** regarding digestive system of cockroach?

(i) Mouth opens into a short tubular pharynx, leading to a narrow tubular passage called oesophagus.
(ii) Gizzard helps in grinding the food particles.
(iii) A ring of 6-8 blind tubules called hepatic or gastric caecae is present at the junction of foregut and midgut, which secrete digestive juice.
(iv) At the junction of midgut and hindgut is present another ring of 100-150 yellow coloured thin filamentous malphigian tubules which help in removal of excretory products from haemolymph.

(a) Only (i)
(b) Both (ii) and (iii)
(c) Both (i) and (iv)
(d) All of these

63. Match column-I with column-II and choose the correct option.

	Column-I		Column-II
A.	*Periplaneta americana*	I.	Hepatic caeca
B.	A ring of 6-8 blind tubules	II.	Phylum arthropoda
C.	Vascular system	III.	Spiracles
D.	10 pairs of small holes	IV.	Malpighian tubules
E.	Excretion	V.	Open type

(a) A – I; B – II; C – III; D – IV; E – V
(b) A – II; B – I; C – V; D – III; E – IV
(c) A – II; B – I; C – III; D – V; E – IV
(d) A – III; B – IV; C – II; D – V; E – I

64. Which part of cockroach has both exoskeleton and endoskeleton?

(a) Head (b) Thorax
(c) Abdomen (d) All of these

65. In cockroach, the number of ganglia are:
(a) Two pairs thoracic and four pairs abdominal
(b) Three pairs thoracic and six pairs abdominal
(c) Three pairs thoracic and five pairs abdominal
(d) Two pairs thoracic and six pairs abdominal

66. Fat is present in which part of a neuron?
(a) Cyton (b) Node of Ranvier
(c) Dendron (d) Axon

67. Most of the neurons of our body are:
(a) Unipolar (b) Bipolar
(c) Multipolar (d) Pseudounipolar

68. Refer the given figure and identify the correct characteristic feature.

(i) It is a type of loose connective tissue.
(ii) It contains fibroblast, macrophages, collagen fibres and mast cells.
(iii) The cells of this tissue are specialised to store fats.
(iv) The wall of internal organs such as the blood vessels, stomach and intestine contains this type of tissue.
(a) (i) and (ii) (b) (i) and (iii)
(c) (ii) and (iii) (d) (iii) and (iv)

69. Which of the following tissue has power of division and regeneration throughout life?
Tricky
(a) Epithelial tissue
(b) Muscular tissue
(c) Connective tissue
(d) Nervous tissue

70. Epithelial tissues arise from:
(a) Ectoderm (b) Endoderm
(c) Mesoderm (d) All of these

71. A transitional epithelium is found in or where would you look for pseudostratified columnar epithelium:
Critical Thinking
(a) Kidney (b) Ureter
(c) Trachea (d) Blood vessel

72. Which statement regarding anal cerci and anal style in cockroach is correct?
(a) Anal cerci are absent but anal styles are present in male.
(b) Anal cerci are absent and anal styles are present in female.
(c) Anal styles are absent and anal cerci are present in male.
(d) Anal styles are absent and anal cerci are present in female.

73. Which of the following statement(s) is/are **correct** regarding excretory system of cockroach?
(i) Excretion is performed by malphigian tubules.
(ii) They absorb nitrogenous waste products and convert them into uric acid which is excreted out through hindgut. Hence, this insect called ammonetelic.
(iii) In addition, fat body, nephrocytes and uricose glands also help in excretion.
(a) Only (i) (b) Both (ii) and (iii)
(c) Both (i) and (iii) (d) All of the above

74. The function of dense regular connective tissue is:
(a) Elastic recoil
(b) Binding and support
(c) Encapsulation of blood vessels
(d) Articulation

75. Adam's apple represents:
(a) Cartilage of larynx
(b) Thyroid cartilage of trachea
(c) Both (a) and (b)
(d) None of the above

76. In cockroach, the first pair of wings are known as:
(a) Sterna (b) Terga
(c) Tegmina (d) Halteres

77. Bone desolving cells:
(a) Osteoblast (b) Osteocyte
(c) Osteoclast (d) Fibroblast

78. Bipolar nerve cells are present in:
(a) Skin tactile corpuscles
(b) Spinal cord
(c) Retina of eye
(d) All of the above

79. A new born baby and hibernating mammal has the cold resisting device due to:

(a) White fat
(b) Fat rich in reticular tissue
(c) Synapse
(d) Brown fat

80. The vermiform appendix is made up of:
(a) Digestive tissue
(b) Respiratory tissue
(c) Excretory tissue
(d) Lymphatic tissue

81. Identify figures-I and II.

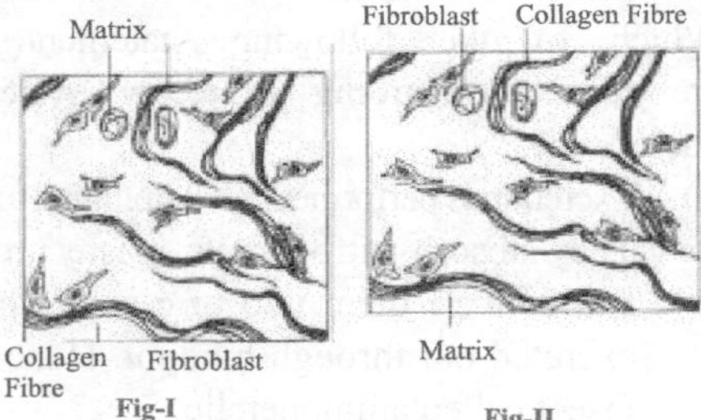

Fig-I Fig-II

	Fig.-I	Fig.-II
(a)	Dense regular connective tissue	Dense irregular connective tissue
(b)	Loose irregular connective tissue	Loose regular connective tissue
(c)	Adipose tissue	Specialised connective tissue
(d)	Connective tissue proper	Areolar tissue

82. Match column-I (function) with column-II (Types of enzymes) and select the correct option.

	Column-I		Column-II
A.	Neuron	I.	Ossein
B.	Bone matrix	II.	Nissl bodies
C.	RBC of man	III.	Antibodies
D.	Lymphocytes	IV.	Non-nucleated

(a) A– IV; B – II; C– III; D – I
(b) A – IV; B – I; C– III; D –II
(c) A – IV; B – II; C– I; D – III
(d) A – II; B – I; C– IV; D – III

83. Read the following statements and mark the right option. **Tricky**
(i) A healthy person has 12 to 16 g of haemoglobin per 100 ml of blood.
(ii) The number of platelets in a normal individual is 1.5 lac to 3.5 lac per mm^3 of blood.
(iii) Eosinophils are involved in allergic reactions.
(a) Only (ii) is correct
(b) (ii) and (iii) are correct
(c) (i) and (iii) are correct
(d) All (i), (ii) and (iii) are correct

84. Cockroach blood does not contain respiratory pigment. It means:
(a) it does not respire.
(b) cockroach respires anaerobically.
(c) oxygen passes to all the tissues through diffusion.
(d) oxygen reaches tissue through tracheoles.

85. The number of RBC in an adult man per mm^3 of blood is:
(a) 5,00,000 (b) 4,50,000
(c) 7,000 (d) 1,00,000

86. A chemical that prevents blood clotting is most useful in the treatment of:
(a) Leukemia
(b) Anaemia
(c) Coronary thrombosis
(d) Haemophilia

87. An example of a holocrine gland is:
(a) A sweat gland
(b) A salivary gland
(c) A pancreatic gland
(d) A sebaceous gland

88. Which of the following is not a specialised type of cell found in connective tissues?
(a) Lymphocyte
(b) Macrophage
(c) Goblet cell
(d) Mast cell

89. Match column-I with column-II and select the correct option.

	Column-I (Epithelial tissue)		Column-II (Location)
A.	Cuboidal	I.	Epidermis of skin
B.	Ciliated	II.	Inner lining of blood vessels
C.	Columnar	III.	Inner surface of gall bladder
D.	Squamous	IV.	Inner lining of fallopian tube
E.	Keratinised squamous	V.	Lining of pancreatic duct

(a) A–V; B–IV; C–II; D–III; E–I
(b) A–III; B–IV; C–V; D–II; E–I
(c) A–V; B–IV; C–III; D–II; E–I
(d) A–III; B–IV; C–V; D–I; E–II

90. Open circulatory system is not of physiological hindrance in cockroach because:
(a) heart is simple but chambered.
(b) blood is colourless.
(c) circulatory and respiratory systems are not connected.
(d) excretion occurs through malpighian tubules.

91. Clot of blood contains:
(a) Prothrombin (b) Thrombin
(c) Fibrinogen (d) Fibrin

92. Which of the following statement about cell junctions is **incorrect**?
(i) All the cells of the epithelium are held together with little intercellular materials.
(ii) In almost all animal tissues, specialised junction provide both structural and functional link between its individual cells.
(iii) Tight junctions help to stop substances from leaking across a tissue.
(iv) Adhering junctions provide cementing to keep neighbouring cells together.
(v) Gap junctions provide cytoplasmic channels between cells for passage of ions, small molecules and sometimes big molecules.
(a) (ii) and (iii) (b) (i) and (ii)
(c) Only (v) (d) None of the above

93. The bone of a mammal contains Haversian canals which are connected by transverse canals known as: **Toughnut**
(a) Semi-circular canals
(b) Volkmann's canals
(c) Inguinal canals
(d) Bidder's canals

94. Cartilage tissues are generally slow to heal injury because:
(a) cartilage is avascular.
(b) cartilage does not undergo mitosis.
(c) the matrix is semisolid.
(d) chondrocytes are surrounded by fluids.

95. The characteristic feature of pernicious anaemia is:
(a) increase in size of RBC.
(b) deficiency of haemoglobin in RBC.
(c) delay in maturation of erythroblasts.
(d) None of the above

96. The intercellular material of the given figure is solid and resists compression. Identify the figure and the label marked as 1 & 2.

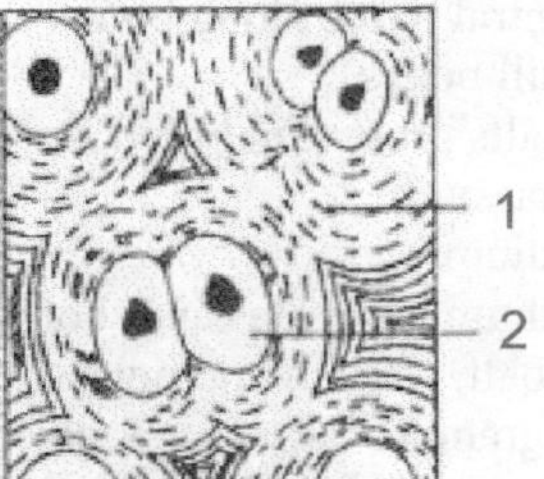

	Figure	1	2
(a)	Cartilage	Collagen	Chondrocyte
(b)	Cartilage	Collagen	Chondroclast
(c)	Bone	Microtubule	Chondroclast
(d)	Bone	Collagen fibres	Osteoblast

97. Which tissue evolved first in animal kingdom?
(a) Connective tissue (b) Skeletal tissue
(c) Epithelial tissue (d) Muscular tissue

98. Which of the following statements are **incorrect**?
(i) Leucocytes disintegrate in the spleen and liver.
(ii) RBC, WBC and blood platelets are produced by bone marrow.
(iii) Neutrophils bring about destruction and detoxification of toxins of protein origin.
(iv) The important function of lymphocytes is to produce antibodies.
(a) (i) and (ii) (b) (i) and (iv)
(c) (i) and (iii) (d) (ii) and (iii)

99. Which of the following statement(s) is/are correct regarding respiratory system of cockroach?
(i) It consists of a network of trachea, that open through 12 pairs of small holes called spiracles present on the lateral side of the body.
(ii) Thin branching tubes carry oxygen from the air to all the parts.
(iii) The opening of the spiracles is regulated by sphincters.
(iv) Exchange of gases take place at the tracheoles by diffusion.
(a) Only (i)
(b) (i), (ii) and (iii)
(c) (ii), (iii) and (iv)
(d) All of the above

100. Red bone marrow are not found in:
(a) Ribs
(b) Vertebrae
(c) Central part of long bones
(d) Skull bones

101. Buffer salts present in blood are:
(a) Potassium
(b) Sodium
(c) Sodium and potassium both
(d) Cobalt, sodium and potassium

102. Nissl's granules are found in cyton of nerve cells. These have affinity for basic dyes. The granules are made up of **Critical Thinking**
(a) Mitochondria (b) Cell metabolites
(c) Fat granules (d) Ribosomes

103. Which of the following possesses nodes of Ranvier?
(a) Medullated nerve fibre
(b) Non-medullated nerve fibre
(c) Muscle fibre
(d) Medullated and non-medullated nerve fibres

104. The medullated or myelinated nerve fibres are found in:
(a) sympathetic nervous system, post ganglionic fibres
(b) invertebrates
(c) parasympathetic nervous system, post ganglionic fibres
(d) CNS

105. The outer covering of non-myelinated neurons in brain is:
(a) Neurilemma
(b) Axolemma
(c) Sarcolemma
(d) Oolemma

106. Which of the following tissue originates exclusively from the ectoderm of the embryo?
(a) Muscular tissue
(b) Epithelial tissue
(c) Nervous tissue
(d) Connective tissue

Answer KEYs

1	(b)	**12**	(c)	**23**	(b)	**34**	(a)	**45**	(b)	**56**	(a)	**67**	(c)	**78**	(c)	**89**	(c)	**100**	(c)
2	(b)	**13**	(c)	**24**	(a)	**35**	(a)	**46**	(d)	**57**	(d)	**68**	(a)	**79**	(d)	**90**	(c)	**101**	(c)
3	(a)	**14**	(c)	**25**	(c)	**36**	(b)	**47**	(a)	**58**	(a)	**69**	(a)	**80**	(d)	**91**	(d)	**102**	(d)
4	(a)	**15**	(d)	**26**	(c)	**37**	(a)	**48**	(b)	**59**	(c)	**70**	(d)	**81**	(a)	**92**	(d)	**103**	(a)
5	(d)	**16**	(a)	**27**	(d)	**38**	(a)	**49**	(c)	**60**	(a)	**71**	(c)	**82**	(d)	**93**	(b)	**104**	(d)
6	(d)	**17**	(b)	**28**	(b)	**39**	(a)	**50**	(a)	**61**	(b)	**72**	(d)	**83**	(d)	**94**	(a)	**105**	(b)
7	(b)	**18**	(d)	**29**	(b)	**40**	(b)	**51**	(a)	**62**	(d)	**73**	(c)	**84**	(d)	**95**	(a)	**106**	(c)
8	(c)	**19**	(c)	**30**	(d)	**41**	(a)	**52**	(b)	**63**	(b)	**74**	(b)	**85**	(a)	**96**	(a)		
9	(c)	**20**	(d)	**31**	(a)	**42**	(c)	**53**	(b)	**64**	(a)	**75**	(a)	**86**	(c)	**97**	(c)		
10	(a)	**21**	(c)	**32**	(b)	**43**	(a)	**54**	(c)	**65**	(b)	**76**	(c)	**87**	(d)	**98**	(c)		
11	(c)	**22**	(a)	**33**	(c)	**44**	(a)	**55**	(c)	**66**	(d)	**77**	(c)	**88**	(c)	**99**	(c)		

Hints & Solutions

1. (b)
2. (b) Ootheca of cockroach contains sixten fertilised egg in two rows.
3. (a) Squamous epithelium is an epithelium whose most superficial layer consists of flat, scale-like cells called squamous epithelial cells.
4. (a) 5. (d)
6. (d) Tissue 1 is columnar cells bearing cilia and tissue 2 is multicellular glandular epithelium. Ciliated columnar epithelial cells possess fine hair-like outgrowths, cilia on their free surfaces. These cilia are capable of rapid, rhythmic, wavelike beatings in a certain direction.
   This movement of the cilia in a certain direction causes the mucus, which is secreted by the goblet cells, to move (flow or stream) in that direction. Ciliated epithelium is usually found in the air passages like the nose. It is also found in the uterus and fallopian tubes of females. Columnar epithelium with goblet cells is called glandular epithelium. Some parts of the glandular epithelium consist of such a large number of goblet cells that there are only a few normal epithelial cells left.
7. (b) In the other choices unstriped muscles are found.
8. (c) Myocardium is the thick middle layer of the heart and is composed of cardiac muscle. Pericardium and endocardium are the linings of epithelium.
9. (c) During active work, the rate of oxygen supply by the lungs falls short according requirement of the muscles. Hence, lactic acid accumulates in the muscles and breathing gradually becomes hard to enhance O_2 intake by the lungs. This is called oxygen debt.
10. (a) 11. (c) 12. (c)
13. (c) Actin and myosin are the structural protein. So they are found in both skeletal and visceral muscle although skeletal muscle is cylindrical and visceral muscles are pear-shaped or spindle-shaped.
14. (c) Ca^{++} is an essential element for the contraction of muscles and transmission of nerve impulse. Release of Ca^{++} ions from sarcoplasmic reticulum trigger muscle contraction process.
15. (d) Muscle tissue is contractile tissue grouped into coordinated systems for greater efficiency. In humans the muscle systems are classified by gross appearance and location of cells. The three types of muscles are striated (or skeletal), cardiac, and smooth (or non-striated).
16. (a) In order to have red blood corpuscle, haemoglobin must be produced and contained in a type of cell which is not present in case of cockroach.
17. (b) 18. (d) 19. (c)
20. (d) The given figure represents loose connective tissue (areolar tissue). Loose connective tissue has cells and fibres which are loosely arranged in a semi-fluid ground substances, for example, areolar tissue present beneath the skin. Often it serves as a support framework for epithelium. It contains fibroblasts (cells that produce and secrete fibres), macrophages and mast cells.
21. (c) Abducens nerve (cranial nerve VI) is one of the pair of nerves that arises from the anterior end of medulla oblongata in the vertebrate brain supplies the posterior rectus muscle of each eyeball. It contains chiefly motor nerve fibres.
22. (a) Schwann cells and Nodes of Ranvier are found in neurons. Myelin sheath covers the axon of most nerve cells. This sheath is formed by the Schwann cell. Node of Ranvier is a depression in the axon of a nerve cell.
23. (b) In mammals, each lobe of the liver is composed of a number of lobules. Each lobule is polygonal in outline, and is separated from one another by a thin layer of connective tissue, called Glisson's capsule.

24. (a) Connective tissues are classified into three types loose connective tissue, dense connective tissue and specialised connective tissue.

25. (c) Blood vascular system of cockroach is an open type. They are dioecious and both sexes have well developed reproductive organs.

26. (c) **27. (d)** **28. (b)**

29. (b) Connective tissue is a group of tissues in the body that maintains the form of the body and its organs and provides cohesion and internal support. The connective tissues include several types of fibrous tissue that vary only in their density and cellularity, as well as the more specialised and recognisable variants-bone, ligaments, tendons, cartilage, and adipose (fat) tissue.

30. (d) **31. (a)** **32. (b)** **33. (c)**

34. (a) The afferent process of neuron is dendrite and the efferent process is axon.

35. (a) There are three ganglia lie in the thorax, and six in the abdomen. Each eye consists of about 2000 hexagonal ommatidia.

36. (b) The GI tract is lined with simple columnar epithelium, which allows a maximum number of cells to contact food particles.

37. (a) Tissue (interstitial) fluid fills the space between fibres and cells of connective tissue. Loose connective tissue has the most space for fluid to accumulate.

38. (a) **39. (a)** **40. (b)** **41. (a)**

42. (c) **43. (a)** **44. (a)**

45. (b) The smooth muscle fibres are involuntary, taper at both ends and do not show striation. The wall of internal organs such as the blood vessels, stomach and intestine contains this type of muscle tissue.

46. (d) Reticular tissue within lymphatic organs contains large numbers of phagocytic cells, which engulf invading pathogens.

47. (a) **48. (b)** **49. (c)** **50. (a)**

51. (a) Based on development and mode of secretion, mammary glands and certain sweat glands are classified as apocrine glands.

52. (b) Each liver lobe is formed of hexagonal lobules surrounded by a connective tissue sheath called Glisson's capsule.

53. (b) **54. (c)**

55. (c) In the given figure part marked as 1, 2, 3 and 4 are oesophagus, crop, gizzard and malpighian tubules. Gizzard is a short muscular organ which is used for grinding food.

56. (a)

57. (d) Agranulocytes like lymphocytes (20-30%) and monocytes (2-8%) are responsible for antibody formation and phagocytosis respectively.

58. (a) **59. (c)** **60. (a)** **61. (b)**

62. (d) Digestive system of cockroach is divided into - foregut, midgut and hindgut. The mouth opens into a short tubular pharynx which further opens into a narrow tubular structure called oesophagus. The oesophagus opens into a sac-like structure called crop where food is stored. The crop is followed by gizzard and proventriculus. The gizzard helps in grinding the food particles. The entire foregut is lined by cuticle. At the junction of foregut and midgut, a ring of 6 - 8 blind tubules are present. These are called gastric or hepatic cecae which secrete digestive juice. The hindgut is broader than the midgut. The hindgut is differentiated into ileum, colon and rectum. The rectum opens through anus. Malpighian tubules are the excretory organs which absorb nitrogenous wastes and convert them into uric acid. Uric acid is excreted out through the hindgut.

63. (b) **64. (a)**

65. (b) The total number of ganglia in ventral nerve cord of cockroach is nine pairs, *i.e.*, three pairs thoracic and six pairs abdominal.

66. **(d)** **67.** **(c)**

68. **(a)** The given figure shows the image of areolar tissue, a type of loose connective tissue. Loose connective tissue (which includes areolar tissue, reticular tissue, and adipose tissue) is the most common type of connective tissue which holds organs in place and attaches epithelial tissue to other underlying tissues. It also surrounds the blood vessels and nerves. Areolar tissues are present beneath the skin and serve as a support framework for epithelium. It contains fibroblast, macrophages, collagen fibres and mast cells.

69. **(a)** Epithelial tissue has great regeneration power because when epithelia are injured, they regenerate more rapidly than other tissues and thus facilitate rapid healing of wounds.

70. **(d)** Following embryonic layers give rise to–

Ectoderm	–	Epidermis
Mesoderm	–	Mesothelium
Endoderm	–	Epithelium of alimentary canal

71. **(c)** In pseudostratified epithelium the cells have one basement membrane but their heights are different.

72. **(d)** In cockroaches, males bear a pair of short, thread like anal styles which are absent in females. In both sexes, the 10th segment bears a pair of jointed filamentous structures called anal cerci.

73. **(c)** Cockroaches absorb nitrogenous waste products and convert them into uric acid which is excreted out-through hindgut. Hence they called as uricotelic.

74. **(b)** Dense regular connective tissue forms tendons and ligaments, as well as the capsules surrounding various organs.

75. **(a)** **76.** **(c)** **77.** **(c)** **78.** **(c)**

79. **(d)** **80.** **(d)**

81. **(a)** The figure I & II represent dense regular and dense irregular connective tissue respectively. In the dense regular connective tissues, the collagen fibres are present in rows between many parallel bundles of fibres. Tendons and ligaments are example of this tissue. Dense irregular connective tissue has fibroblasts and many fibres (mostly collagen) that are oriented differently. This tissue is present in the skin.

82. **(d)** **83.** **(d)**

84. **(d)** All body tissue receive oxygen directly through tracheoles.

85. **(a)** $5x10^6$ RBC are found per cubic mm of blood in male (man).

86. **(c)** Coronary thrombosis is formation of thrombus (clot) in the coronary artery. Any chemical which is anticoagulant will prevent the formation of thrombus.

87. **(d)** Sebaceous glands are holocrine glands because they secrete by discharging entire cells full of product (sebum).

88. **(c)** Goblet cells are found in lining of respiratory tract and the GI tract, where they are needed to secrete their lubricating and protective mucus. Macrophages and lymphocytes are both found in connective tissue, where they aid the immune response.

89. **(c)** **90.** **(c)**

91. **(d)** Fibrin is the clot which is formed when fibrinogen is acted upon by thrombin in the presence of calcium. If forms a Fibrous mesh that impedes the flow of blood.

92. **(d)**

93. **(b)** Semi-circular canal is found in the internal ear. Inguinal canal connects scrotum with the abdominal cavity in the male mammal. Bidder's canal is found in the male frog.

94. **(a)** Lacking a capillary blood supply, cartilage tissue heals poorly.

95. **(a)** Perniuous anaemia is a chronic illnass, caused by impaired absorption of vitamin B_{12} needed to make enough healthy red blood cells. When there is folic acid and vit. B_{12} deficiency, the pro-erythroblasts of the bone marrow do not mature into normoblasts, instead, they give rise to immature cells called megaloblasts which lose their nuclei and appear as large RBC, called macrocytes. In the peripheral circulation, therefore, we find RBC bigger than normal.

96. (a) The given figure represents the image of cartilage. Cartilage is an important structural component of the body. It is a firm tissue but is softer and much more flexible than bone. It is a connective tissue found in joints between bones e.g. the elbows, knees and ankles; ends of the ribs; between the vertebrae in the spine; ears and nose; bronchial tubes or airways. Cartilage is made up of specialised cells called chondrocytes (2). These chondrocytes produce large amounts of extracellular matrix composed of collagen (1) fibres, proteoglycan, and elastin fibres. There are no blood vessels in cartilage to supply the chondrocytes with nutrients.

97. (c) **98. (c)**

99. (c) In cokroach, respiratory system consists of a network of trachea, that open through 10 pairs of small holes called spiracles present on the lateral side of the body.

100. (c) The yellow bone marrow is composed of fatty tissue. It stores fat and produces blood corpuscles only in emergency situations. It is found in the central part of long bones.

101. (c) Cobalt does not constitute the buffer system of blood.

102. (d) **103. (a)** **104. (d)**

105. (b) **106. (c)**

8 Cell: The Unit of Life

1. Identify the statement that does not hold true for a eukaryotic cell? **Tricky**
(a) Presence of 80S type of ribosome in the mitochondria.
(b) Presence of 80S type of ribosome in the cytoplasm.
(c) Presence of mitochondria containing circular DNA.
(d) Presence of membrane bound organelles.

2. Living beings are made up of cells. This statement belongs to:
(a) Lamarck
(b) Von Helmont
(c) Hugo de Vries
(d) Schleiden and Schwann

3. Largest unicellular eukaryotic plant is:
(a) Egg of *Rhinodon*
(b) *Acetabularia*
(c) Nerve cells of squid
(d) RBC of *Amphiuma*

4. The acidic condition within the lysosome is maintained by:
(a) digestive enzymes synthesised on RER.
(b) pumping Cl^- ion out of lysosome.
(c) pumping protons (H^+) into the lysosome.
(d) All of the above

5. Mitochondria supply most of the necessary biological energy by:
(a) Breaking down of sugar
(b) Oxidising substrates of TCA cycle
(c) Reducing NADP
(d) Breaking down of protein

6. The following diagram shows some of the missing structures in a plant cell (1 – 5). Choose the correct option.

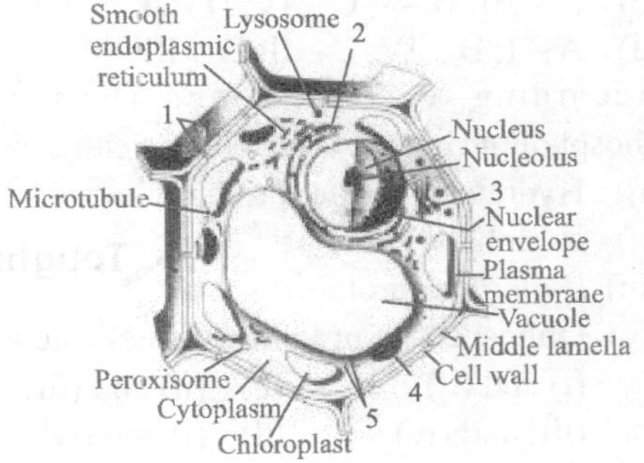

(a) 1 – Plasmodesmata, 2 – Rough endoplasmic reticulum, 3 – Golgi apparatus, 4 – Mitochondrion, 5 – Ribosomes
(b) 1 – Desmosome, 2 – Rough endoplasmic reticulum, 3 – Golgi apparatus, 4 – Mitochondrion, 5 – Ribosomes
(c) 1 – Plasmodesmata, 2 – Smooth endoplasmic reticulum, 3 – Golgi apparatus, 4 – Mitochondrion, 5 – Ribosomes
(d) 1 – Tight junction, 2 – Rough endoplasmic reticulum, 3 – Golgi apparatus, 4 – Mitochondrion, 5 – Ribosomes

7. Identify the **correct** statements in comparison to the cell doctrine and choose the option.
(i) Cells are the basic units of structure in the body of living organisms.
(ii) The bodies of all living beings are made up of cells and their products.
(iii) Genetic information is stored and expressed inside the cells.
(iv) Cells are the basic units of function in living organism that is, the activities of an organism are the sum total of the activities of its cells.
(a) (ii) and (iii) (b) (i) and (ii)
(c) Only (i) (d) All of the above

8. Prokaryotic cell does not have:
(a) Nucleolus
(b) Membrane bound organelles
(c) Centrioles
(d) All of the above

9. Which one of the following cell organelles is found only in plants?
(a) Golgi complex (b) Mitochondria
(c) Plastids (d) Ribosomes

10. In which of the following organisms, plastids are not found? **Tricky**
(a) Blue green algae (b) Bacteria
(c) Fungi (d) All of these

11. Function of centriole is:
 (a) Formation of spindle fibres
 (b) Formation of nucleolus
 (c) Initiation of cell division
 (d) Formation of cell plate

12. Order the following terms in a way that depicts how vesicles are transported to membranes, are likely to move through the cell. **Critical Thinking**
 (i) ER
 (ii) Nuclear envelope
 (iii) Golgi apparatus
 (iv) Vesicle
 (v) Plasma membrane.
 (a) (i), (ii), (iii), (iv), (v)
 (b) (i), (iv), (ii), (iv), (iii), (iv), (v)
 (c) (ii), (iii), (i), (iv), (v)
 (d) (ii), (i), (iv), (iii), (iv), (v)

13. Sodium and potassium pumps are common in many cells. Which of the following are necessary for the pumps to work? **Tricky**
 (a) ATP-driven pumping proteins.
 (b) A signal to activate the pumps.
 (c) A concentration gradient to work against.
 (d) All of the above

14. Peroxisomes, in plant cells, are involved in:
 (a) Photooxidation
 (b) Photorespiration
 (c) Photophosphorylation
 (d) Photolysis of water

15. Arrange the following steps in a **correct** sequence as per Gram's staining technique.

 (i) Treatment with 0.5% iodine solution.
 (ii) Washing with water.
 (iii) Treatment with absolute alcohol/acetone.
 (iv) Staining with weak alkaline solution of crystal violet.
 The correct sequence is:
 (a) (ii), (i), (iv), (iii) (b) (iv), (i), (ii), (iii)
 (c) (iii), (ii), (i), (iv) (d) (iv), (iii), (ii), (i)

16. Identify the components labelled as 1, 2, 3, 4 and 5 in the diagram given below from the list (i) to (viii).

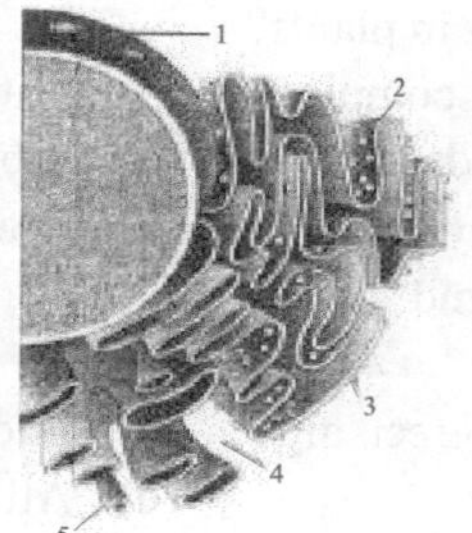

 (i) Cristae of mitochondria
 (ii) Inner membrane of mitochondria
 (iii) Cytoplasm
 (iv) Smooth endoplasmic reticulum
 (v) Rough endoplasmic reticulum
 (vi) Mitochondrial matrix
 (vii) Ribosome
 (viii) Nucleus

	1	2	3	4	5
(a)	(viii)	(v)	(vii)	(iii)	(iv)
(b)	(i)	(iv)	(vii)	(vi)	(iii)
(c)	(vi)	(v)	(iv)	(vii)	(i)
(d)	(v)	(i)	(iii)	(ii)	(iv)

17. The cells are held together by a Ca-pectate layer called.
 (a) primary cell wall (b) secondary cell wall
 (c) middle lamella (d) tertiary cell wall

18. Cristae are found in:
 (a) Surface of grana
 (b) Surface of plasma membrane
 (c) Wall of mitochondria
 (d) Nuclear membrane

19. Match column-I (scientists) with column-II (discovery) and select the correct option.

Column-I		Column-II	
A.	Leeuwenhoek	I.	First saw and described a living cell.
B.	Robert Brown	II.	Presence of cell wall is unique to plant cells.
C.	Schleiden	III.	Discovered the nucleus.
D.	Schwann	IV.	All plants are composed of different kind of cells.

 (a) A – I; B – III; C – IV; D – II
 (b) A – I; B – III; C – II; D – IV
 (c) A – III; B – I; C – IV; D – II
 (d) A – I; B – IV; C – II; D – III

20. According to Danielli and Davson, the phospholipid bilayer is covered on either side by: **Toughnut**
 (i) Hydrated globular proteins
 (ii) α-globular proteins
 (iii) β-globular proteins
 (iv) Only spectrin protein with sialic acid
 (a) (i) and (ii) (b) (ii) and (iii)
 (c) (iii) and (iv) (d) (i) and (iv)

21. Three morphological forms of golgi complex are:

(a) Lamellae, tubules and vesicles
(b) Cisternae, tubules and vesicles
(c) Cisternae, tubules and lamellae
(d) Granum, thylakoids and vesicles

22. Channel proteins allow ions that would not normally pass through the cell membrane to go through the channel. What properties of the proteins are responsible for this? **Tricky**

(a) The channels are often composed of charged or polar R groups.
(b) The channels are often composed of hydrophobic R groups.
(c) Both (a) and (b)
(d) None of the above

23. Which of the following statement(s) depict the fundamental features of cell theory?

(i) All cells share similar chemistry and physiology.
(ii) All living organisms are composed of cells and their products.
(iii) Each cell is made up of a small mass of protoplasm containing a nucleus inside and a plasma membrane with or without a cell wall outside.
(iv) Activities of an organism are the sum total of activities and interaction of its constituents cells.

(a) Only (i) is correct
(b) (ii) and (iii) are correct
(c) (ii), (iii) and (iv) are correct
(d) All are correct

24. Which of the following organelles are double membrane-bound?

(a) Nucleus
(b) Chloroplast
(c) Mitochondria
(d) All of the above

25. Which of the following substances is most likely to permeate a membrane by diffusing directly through the lipid bilayer? **Tricky**

(a) K^+ (b) Glucose
(c) Cl^- (d) O_2

26. Match column-I with column-II and choose the correct option.

Column-I	Column-II
A. Tonoplast	I. Contain digestive enzyme
B. Contractile vacuole	II. Store metabolic gases
C. Food vacuole	III. Excretion
D. Air vacuole	IV. Transport of ions in plants

(a) A – IV; B – III; C – I; D – II
(b) A – II; B – III; C – IV; D – I
(c) A – IV; B – II; C – III; D – I
(d) A – I; B – III; C – II; D – IV

27. Match column-I with column-II and select the correct option.

Column-I	Column-II
A. Golgi apparatus	I. Storage
B. Mitochondria	II. Photosynthesis
C. Vacuoles	III. Transport
D. Grana	IV. Secretion
	V. Respiration

(a) A – IV; B – V; C – I: D – II
(b) A – I; B – II; C – IV: D – III
(c) A – IV; B – I; C – II: D – III
(d) A – I; B – II; C – III: D – IV

28. Match the components 1, 2, 3, 4 and 5 in the diagram (cell membrane) below from the list (i) to (vii).

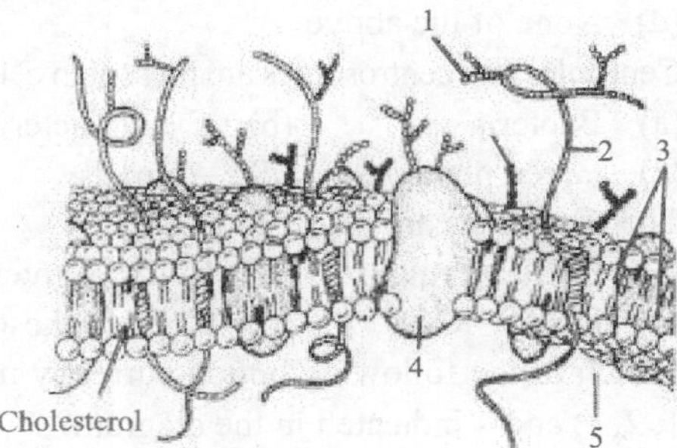

(i) Sugar
(ii) Protein
(iii) Lipid bilayer
(iv) Integral protein
(v) Cytoplasm
(vi) Cell wall
(vii) External protein

(a) 1 – (i), 2 – (ii), 3 – (iii), 4 – (iv), 5 – (v)
(b) 1 – (ii), 2 – (i), 3 – (iii), 4 – (iv), 5 – (v)
(c) 1 – (i), 2 – (ii), 3 – (iii), 4 – (iv), 5 – (vi)
(d) 1 – (i), 2 – (ii), 3 – (iii), 4 – (vii), 5 – (v)

29. Plasmodesmata connections help in:
(a) Cytoplasmic streaming.
(b) Synchronous mitotic divisions.
(c) Locomotion of unicellular organisms.
(d) Movement of substances between cells.

30. Correct sequence of layers of bacterial cell envelope is: **Tricky**
(a) cell membrane → glycocalyx → cell wall
(b) glycocalyx → cell wall → cell membrane
(c) cell wall → glycocalyx → cell membrane
(d) glycocalyx → cell membrane → cell wall

31. Both chloroplasts and mitochondria:
(a) have multiple membranes.
(b) have highly structured innermost membranes.
(c) are found only in eukaryotic cells.
(d) All of the above

32. Which of the face of golgi complex is associated with ER?
(a) Forming face *i.e.,* Trans-face
(b) Maturing face *i.e.,* Trans-face
(c) Both forming and maturing face
(d) Forming face or Cis-face

33. Schleiden and Schwann proposed:
(a) phenomenon of Brownian movement
(b) cell theory or cell doctrine
(c) protoplasm as a physical basis of life
(d) None of the above

34. Centrioles and centrosomes are present in cells of:
(a) Bacteria (b) Cyanobacteria
(c) Green plants (d) Animals

35. The ribosomes are made up of:
(a) DNA + Protein (b) RNA + Protein
(c) DNA + RNA (d) None of these

36. Which of the following option correctly match 1, 2, 3, and 4 indicated in the diagram?

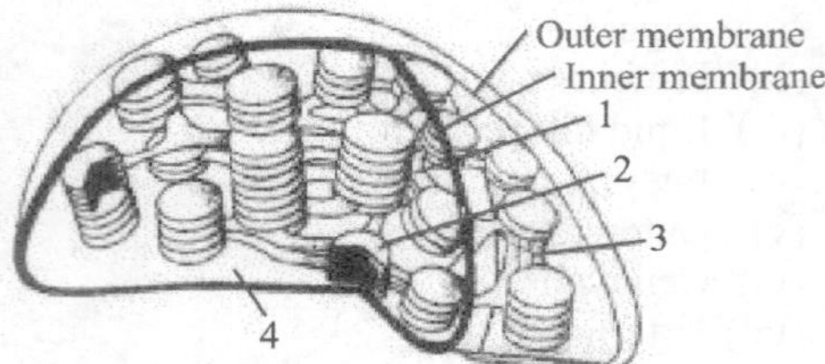

(a) 1–Thylakoid, 2–Stromal lamella, 3–Stroma, 4 – Granum
(b) 1 – Granum, 2 – Thylakoid, 3 – Stromal lamella, 4 – Stroma
(c) 1 – Thylakoid, 2 – Granum, 3 – Stromal lamella, 4 – Stroma
(d) 1 – Granum, 2 – Thylakoid, 3 – Stroma, 4 – Stromal lamella

37. In which of the following cell types would you see more cellular detail using an electron microscope as compared to a light microscope? **Critical Thinking**
(a) Animal (b) Plant
(c) Bacteria (d) All of the above

38. The principal protein of cilia and flagella is:
(a) Tubulin (b) Nexin
(c) Myosin (d) Albumin

39. Ion carriers are located in:
(a) Nucleus (b) Cell wall
(c) Cellular space (d) Plasma membrane

40. The four basic shapes of bacteria are:
(a) amoeboid, elongated, biconcave and coccus.
(b) elongated, bacillus, coccus and amoeboid.
(c) bacillus, coccus, vibrio and spirillum.
(d) bacillus, amoeboid, elongated and coccus.

41. Which of the following organelles is directly connected to the outer membrane of the nucleus in a eukaryotic cell?
(a) Mitochondrion
(b) Lysosome
(c) Golgi apparatus
(d) Endoplasmic reticulum

42. Match column-I and column-II and select the correct option.

Column-I	Column-II
A. Bacteria without walls	I. Lysosome
B. Small circular DNA	II. Mycoplasma cells
C. Flattened sacs in a chloroplast	III. Thylakoid
D. A vesicle in which hydrolytic enzymes are stored	IV. Plasmid

(a) A – III; B – IV; C – II; D – I
(b) A – II; B – IV; C – III; D – I
(c) A – I; B – II; C – III; D – IV
(d) A – IV; B – III; C – I; D – II

43. Which one of the following pairs is not **incorrectly** matched?
(a) Cristae – The tubular structure formed by the folding of the inner membrane of the mitochondrion.
(b) Plasmodesmata – The membrane surrounding the vacuole in plants.
(c) Grana – Membrane bound discs in chloroplasts that contain chlorophylls and carotenoids.
(d) Middle lamella – Layer between adjacent cells walls in plants derived from cell plate.

44. Which of the following statements are **correct**?
(a) Na^+/K^+ pump is an example of active transport.
(b) In plant cells, lipid like steroidal hormones are synthesised in SER.
(c) In plant cells, the vacuoles can occupy up to 10% of the volume of the cell.
(d) Chlorophyll and leucoplast are responsible for trapping light energy essential for photosynthesis.

45. The simplest way to distinguish a prokaryotic from a eukaryotic cell, is to check for:
(a) Plasma membrane (b) Nucleus
(c) DNA (d) Proteins

46. The polynucleated condition appeared by free nuclear division without cytokinesis is:
(a) Syncytium
(b) Coenocytic
(c) Found in amphibian oocyte
(d) All of these

47. Which of the following limits the movement of molecules when carrier-mediated facilitated diffusion is involved?
(a) Concentration gradient
(b) Availability of carrier molecules
(c) Temperature
(d) All of the above

48. Grana and stroma lamellae are the parts of:
(a) Mitochondria
(b) Chloroplast
(c) Endoplasmic reticulum
(d) Vacuoles

49. Cell wall in higher plants is made up of:
(a) Cellulose + lignin
(b) Cellulose + pectin
(c) Cellulose + suberin
(d) Cellulose + lipid

50. Identify the **correct** statements and choose the option.
(i) The smallest of all cells are nerve cells.
(ii) Bacteria are 3-5 μm in length.
(iii) The Ostrich egg is the largest cell.
(iv) Mycoplasma is the smallest cell (0.3 μm in length).
(a) (i), (ii), (iii) and (iv)
(b) Only (ii)
(c) Only (i)
(d) (ii), (iii) and (iv)

51. The following diagram represents a structure of chromosome. Identify the structures of marked as 1, 2 and 3.

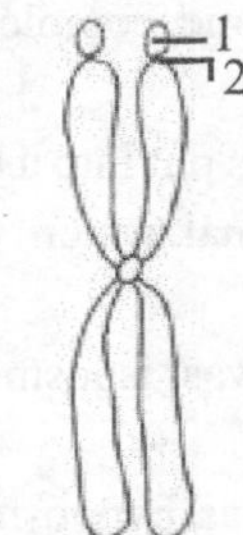

Chromosome - 3

(a) 1 – Satellite, 2 – Primary constriction, 3 – Acrocentric
(b) 1 – Satellite, 2 – Secondary constriction, 3 – Metacentric
(c) 1 – Satellite, 2 – Centromere, 3 – Telocentric
(d) 1 – Satellite, 2 – Centromere, 3 – Submetacentric

52. Which one of the following combination is mismatched?
(a) Glycocalyx – May be capsule or slime layer
(b) Pili – Reproduction
(c) Cell wall – Protective, determines shape, prevents from bursting
(d) Flagella, pili and fimbriae – Surface structures of bacterial cell

53. Match column-I (cell organelle) with column-II (membrane) and select the correct option from the codes given below.

Column-I	Column-II
A. Mitochondria	I. Without membrane
B. Lysosomes	II. Single membrane
C. Ribosomes	III. Double membrane

(a) A – I; B – II; C – III
(b) A – III; B – I; C – II
(c) A – III; B – II; C – I
(d) A – II; B – III; C – I

54. In prokaryotes, chromatophores are:

(a) specialised granules responsible for colouration of cells.
(b) structures responsible for organising the shape of the organism.
(c) inclusion bodies lying free inside the cells for carrying out various metabolic activities.
(d) internal membrane system which becomes extensive and complex in photosynthetic bacteria.

55. An animal cell is put into a bathing solution and it gains additional water. Choose the **correct** statements.
(a) The cell was isoosmotic to the bathing solution.
(b) The cell was hypoosmotic to the bathing solution.
(c) The cell was hyperosmotic to the bathing solution.
(d) The bathing solution was hyperosmotic to the cell.

56. Select the statements which are related to Schwann.
(i) He reported that cells have a thin outer layer which is today known as plasma membrane.
(ii) Cell wall is a unique character of the plant cell.
(iii) Body of plants and animals are composed of cells and products of cells.
(a) Only (i) (b) Only (iii)
(c) (i) and (iii) (d) All of the above

57. Which of the following is related to glycosylation of protein?
(a) Endoplasmic reticulum
(b) Peroxisome
(c) Lysosome
(d) Mitochondria

58. Which of the following statements is/are **correct?**
(i) The endomembrane system includes plasma membrane, ER, Golgi complex, lysosomes and vacuoles.
(ii) ER helps in the transport of substances, synthesis of proteins, lipoproteins and glycogen.
(iii) Ribosomes are involved in protein synthesis.
(iv) Mitochondria help in oxidative phosphorylation and generation of ATP.
(a) (ii), (iii) and (iv) (b) (i) only
(c) (ii) only (d) (iii) only

59. Choose the **correct** statement from the options given below.

Critical Thinking

(i) Separation from extracellular medium allows the cells to maintain its chemical pool, orderliness of structure and reactions in contrast to disorderly distribution and randomly interacting molecules in the extra-cellular medium.
(ii) Cells are unable to recognise one another due to the presence of specific chemicals on their surface.
(iii) Cells of plant tissue are often connected with one another through cytoplasmic bridges called plasmodesmata.
(iv) Different cells of an organism communicate as well as exchange materials with one another.
(a) (ii) and (iii)
(b) (iii) and (iv)
(c) (i), (iii) and (iv)
(d) (i), (ii), (iii) and (iv)

60. Which of the following enzymes are found in matrix of peroxisomes?

Tricky

(a) NADH cytochrome reductase
(b) Glyoxidases and malate dehydrogenase
(c) Acid phosphatases and isocitric lysase
(d) Catalases and oxidases

61. Which of the following layer is present nearest plasma membrane in plant cell?
(a) Secondary wall (b) Middle lamella
(c) Primary wall (d) Tonoplast

62. Which of the following is present in both prokaryotes and eukaryotes?
(a) Lysosome (b) Vesicles
(c) Chloroplast (d) Plasma membrane

63. Which structure is present in chromosomes?
(a) Nucleus (b) Centromere
(c) Centrosome (d) Golgi body

64. Select the **incorrect** statement about prokaryotic ribosomes.
(a) 50S and 30S subunits unite to form 70S ribosomes.
(b) Polysome/polyribosome consists of many ribosomes only.
(c) Ribosome is the site of protein synthesis.
(d) Polysome indicate the synthesis of identical polypeptide in multiple copies.

65. Consider the following statements and choose the **correct** statement.
(i) The endomembrane system includes mitochondria, chloroplast and peroxisomes.
(ii) Smooth endoplasmic reticulum is the major site for synthesis of lipid.
(iii) Rough endoplasmic reticulum is actively involved in protein synthesis.
(iv) Mitochondrial matrix possesses single circular DNA, a few RNA and 80S ribosomes.
(a) (i) and (iii) (b) (ii) and (iv)
(c) (iii) and (iv) (d) (ii) and (iii)

66. Which of the following pair are correctly matched ?
(i) Amyloplasts – Store proteins
(ii) Mitochondrion – 'Power house' of the cell
(iii) Stroma – Chlorophyll pigment
(iv) Axoneme – 9 + 2 array
(a) (i) and (iii) only
(b) (ii), (iii) and (iv) only
(c) (iii) and (iv) only
(d) (ii) and (iv) only

67. In a prokaryotic cell, the correct option is:
(i) enveloped genetic material is present
(ii) ribosomes are absent
(iii) an organised nucleus is absent.
(a) Only (i) (b) Only (ii)
(c) Only (iii) (d) All of the above

68. Which of the substance is a complicated mixture of glycerol + phellonic acids deposited on cork cells to make it impermeable to water?

(a) Silica (b) Cutin
(c) Lignin (d) Suberin

69. In nucleoplasm, a conspicuous body of spherical shape attached to a particular chromosome on a definite position is called.
(a) plasmid (b) karyolymph
(c) nucleolus (d) nuclear reticulum

70. F_1 particles / oxysome/ elementary particles are present in:
(a) Endoplasmic reticulum
(b) Chloroplast
(c) Mitochondria
(d) Golgi complex

71. Which one is the correct option for the labelled diagrams 1, 2, 3 and 4?

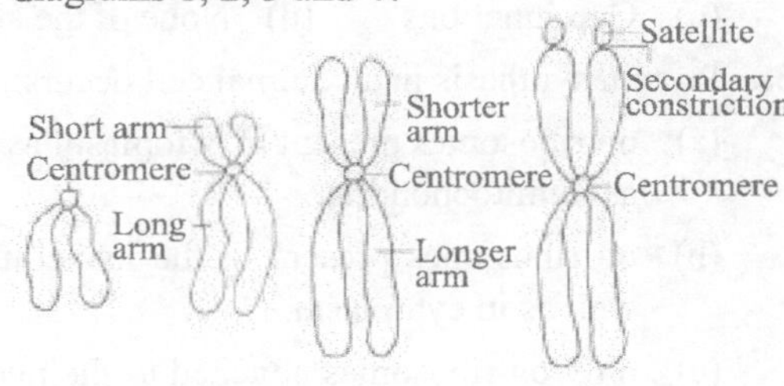

(a) 1 – Telocentric chromosome,
2 – Acrocentric chromosome,
3 – Submetacentric chromosome,
4 – Metacentric chromosome
(b) 1 – Acrocentric chromosome,
2 – Telocentric chromosome,
3 – Metacentric chromosome,
4 – Submetacentric chromosome
(c) 1 – Submetacentric chromosome,
2 – Metacentric chromosome,
3 – Telocentric chromosome,
4 – Acrocentric chromosome
(d) 1 – Metacentric chromosome,
2 – Submetacentric chromosome,
3– Acrocentric chromosome,
4 – Telocentric chromosome.

72. Choose the **incorrect** match.
(a) Nucleus — RNA
(b) Lysosome — Protein synthesis
(c) Mitochondria — Respiration
(d) Cytoskeleton — Microtubules

73. Which one of the following cellular parts is correctly matched with their characters?

(a) Centrioles – Sites for active RNA synthesis.
(b) Lysosomes – Optimally active at a pH of about 8.5
(c) Thylakoids – Flattened membranous sacs forming the grana of chloroplasts.
(d) Ribosomes – Those on chloroplasts are larger (80S) while those in the cytoplasm are smaller (70S).

74. Which of the following lacks cell wall?

(a) Gametes (b) *Amoeba*
(c) *Mycoplasma* (d) All of these

75. Which of the following types of junctions are responsible for communication between cells?

(a) Tight junctions (b) Desmosomes
(c) Gap junctions (d) None of the above

76. Protein synthesis in an animal cell occurs:

(a) on ribosomes present in cytoplasm as well as in mitochondria.
(b) on ribosomes present in the nucleolus as well as in cytoplasm.
(c) only on ribosomes attached to the nuclear envelope and endoplasmic reticulum.
(d) only on the ribosomes present in cytosol.

77. When mitochondria are actively engaged in phosphorylation and electron transport, it is:

(a) in active state
(b) in condensed state
(c) having enlarged intermembrane space
(d) All of these

78. Gas vacuoles are not reported in:

(a) Cyanobacteria
(b) Purple/green photosynthetic bacteria
(c) *Amoeba*
(d) Prokaryotes

79. Which organelle is surrounded by a double phospholipid bilayer with many large pores?

(a) Nuclear envelope
(b) Plasma membrane
(c) Golgi apparatus
(d) Mitochondrion

80. Similarity between plant and animal flagella microtubules is:

Tricky

(a) 9 + 3 (b) 9 + 2
(c) 9 + 6 (d) 9 + 1

81. Which of the following is not the function of cell wall?

(i) Provides shape to the cell.
(ii) Protects the cell from mechanical damage and infection.
(iii) Helps in cell to cell interaction.
(iv) Provides barrier to undesirable macromolecules.

(a) Only (i)
(b) Only (iv)
(c) Only (ii), (iii) and (iv)
(d) None of the above

82. Match column-I with column-II and select the correct option.

Column-I	Column-II
A. RER	I. Intracellular and extracellular digestion
B. Cell wall	II. Provide structural support to the cell
C. Flagella	III. Protein synthesis and secretion
D. Lysosomes	IV Responsible for cell movement

(a) A – III; B – II; C – IV; D – I
(b) A – II; B – III; C – IV; D – I
(c) A – I; B – III; C – II; D – IV
(d) A – IV; B – II; C – III; D – I

83. Which of the following contributes to differences in the two sides of the cell membrane?

(a) Differences in peripheral proteins.
(b) Different domains expressed on the ends of integral proteins.
(c) Differences in phospholipid types.
(d) All of the above

84. The function of centrosome is:
(a) Inhibition of cell division
(b) Initiates cell division
(c) To increase protein synthesis
(d) Initiate cytokinesis

85. Cell wall of bacteria is made up of:
(a) Murein (b) Protein
(c) Polysaccharide (d) Lipid

86. You have found a mass of cells in the sediment surrounding a thermal vent in the ocean floor. The salinity in the area is quite high. Upon microscopic examination of the cells you find no evidence of membrane-bound organelles. How would you classify this cell?
(a) As a eukaryotic cell
(b) As a prokaryotic cell
(c) As a member of domain Archaebacteria
(d) Both (b) and (c)

87. Match column-I with column-II and select the correct answer using the codes given below.

Column-I (Chromosome)	Column-II (Position of centromere)
A. Metacentric	I. At the tip
B. Submetacentric	II. Almost near the tip
C. Acrocentric	III. At the middle
D. Telocentric	IV. Slightly away from the middle

(a) A– III ; B–IV; C–II; D–I
(b) A–IV; B–III; C–II; D–I
(c) A–I; B–II; C–III; D–IV
(d) A–IV; B–III; C–I; D–II

88. Read the following statements and identify the correct options given.

Critical Thinking

(i) Sap vacuoles-contain digestive enzymes with the help of which nutrients are digested.
(ii) Contractile vacuoles – take part in osmoregulation and excretion
(iii) Food vacuoles – store and concentrate mineral salts as well as nutrients
(iv) Air vacuoles – store metabolic gases and help in buoyancy of cells.

(a) (i) and (ii) (b) (i) and (iii)
(c) (ii) and (iv) (d) (i) and (iv)

89. Which of the following statements regarding cilia is **incorrect** ?
(a) Cilia contain an outer of nine doublet microtubules surrounding two single microtubules.
(b) The organised beating of cilia is controlled by fluxes Ca^{2+} across the membrane.
(c) Cilia are hair-like cellular appendages.
(d) Microtubules of cilia are composed of tubulin.

90. A single unit membrane organelle is:
(a) Ribosomes (b) Mitochondria
(c) Chloroplast (d) Lysosomes

91. In which of the following cell organelles would you expect to find the biochemical reactions that harness energy from the breakdown of sugar molecules to synthesise large amounts of ATP?
(a) Lysosome
(b) Vesicles
(c) Chloroplast
(d) Mitochondrion

92. Folding of inner membrane of mitochondria are called.
(a) cristae
(b) grana
(c) calcium oxalate crystals
(d) sacs

93. Which type of set of chromosomes has one genome?
(a) Haploid (b) Diploid
(c) Triploid (d) Polyploid

94. Study the following statements on cilium or flagellum and choose the correct option.
(i) Cilium / flagellum contains an outer ring of nine doublet microtubules surrounding two singlet microtubules.
(ii) Cilia are smaller which work like oars, causing the movement of either the cells or surrounding fluid.
(iii) Flagella are comparatively longer and responsible for cell movement.
(iv) Cilium and flagellum are covered with plasma membrane.
(a) (i) and (ii) (b) (i), (ii), (iii) and (iv)
(c) (i) and (iv) (d) (ii) and (iii)

95. Given below are some characters of a cell organelle. Identify the **correct** organelle which shows all the characters described above.
(i) It is a membrane bound space found in the cytoplasm.
(ii) It is bound by a single membrane called tonoplast.
(iii) It contains water, sap, excretory products and other materials, not useful to the cell.
(iv) It has higher concentration of sap than the cytoplasm.
(a) Golgi apparatus
(b) Lysosomes
(c) Endoplasmic reticulum
(d) Vacuoles

96. Which of the following terms is **incorrectly** matched with its feature?
(a) Osmosis – Movement of water by diffusion.
(b) Nucleoplasm – Site of active synthesis of ribosomal RNA.
(c) Mesosome – Infolding of cell membrane and characteristics of eukaryotes.
(d) Pili – Elongated tubular surface structures (made of special protein) of bacteria.

97. The strength and rigidity of a cell wall is due to the substance known as.
(a) suberin (b) cellulose
(c) lignin (d) pectin

98. Semi autonomous cell organelles of cell are:
(a) Nucleus and chloroplast
(b) Chloroplast and mitochondria
(c) Vacuoles and golgi complex
(d) Ribosome and lysosome

99. Part of chromosome which joins with spindle fibres is:
(a) Chromatid (b) Chromonema
(c) Chromomere (d) Centromere

100. What would happen if lysosomes get ruptured inside the cells in which they are present?
(a) Cells will swell
(b) Cells will shrink
(c) Cells will die
(d) Nothing would happen

Answer KEYs

1	(a)	11	(a)	21	(b)	31	(d)	41	(d)	51	(b)	61	(a)	71	(a)	81	(d)	91	(d)
2	(d)	12	(d)	22	(a)	32	(d)	42	(b)	52	(b)	62	(d)	72	(b)	82	(a)	92	(a)
3	(b)	13	(d)	23	(d)	33	(b)	43	(b)	53	(c)	63	(b)	73	(c)	83	(d)	93	(a)
4	(c)	14	(b)	24	(d)	34	(d)	44	(a)	54	(d)	64	(b)	74	(d)	84	(d)	94	(b)
5	(b)	15	(b)	25	(d)	35	(b)	45	(b)	55	(c)	65	(d)	75	(c)	85	(a)	95	(d)
6	(a)	16	(a)	26	(a)	36	(b)	46	(b)	56	(d)	66	(d)	76	(a)	86	(d)	96	(c)
7	(d)	17	(c)	27	(a)	37	(d)	47	(d)	57	(a)	67	(c)	77	(d)	87	(a)	97	(c)
8	(d)	18	(c)	28	(a)	38	(a)	48	(b)	58	(a)	68	(d)	78	(c)	88	(c)	98	(b)
9	(c)	19	(a)	29	(d)	39	(d)	49	(b)	59	(c)	69	(c)	79	(a)	89	(b)	99	(d)
10	(d)	20	(a)	30	(b)	40	(c)	50	(d)	60	(d)	70	(c)	80	(b)	90	(d)	100	(c)

Hints & Solutions

1. **(a)** In eukaryotic cell, ribosomes are of 80S type. However, 70S ribosomes, occur in plastids and mitochondria.

2. **(d)** 3. **(b)**

4. **(c)** The lysosome is filled with 40 types of acid hydrolases (digestive enzymes). The acidic pH 5 (4.6 to 5) is due to the action of ATP proton pump in the membrane of the lysosome. All the enzymes do not occur in the same lysosome. There digestive enzymes are synthesised on RER and packed into lysosomes.

5. **(b)** 6. **(a)** 7. **(d)** 8. **(d)**

9. **(c)** Plants are autotrophs and synthesise their food in the process of photosynthesis with the help of chloroplast (plastid).

10. **(d)** Plastids are the characteristic feature of photosynthetic eukaryotes. Blue green algae and bacteria are prokaryotes and fungi are saprophytic eukaryotes.

11. **(a)** During cell division, spindle fibres are attached on centriole.

12. **(d)** The typical flow of vesicles within the cell is from nuclear envelope to rough ER, from rough ER to Golgi apparatus, and from Golgi apparatus to plasma membrane. Membrane movement between nuclear envelope, Golgi apparatus, and ER is via vesicles.

13. **(d)** Temperature, molecule size, molecule charge, and concentration gradients all affect the rate at which diffusion takes place.

14. **(b)**

15. **(b)** Steps of Gram's staining technique.
    (i) Staining with weak alkaline solution of crystal violet.
    (ii) Treatment with 0.5% iodine solution.
    (iii) Washing with water.
    (iv) Treatment with absolute alcohol/acetone.

16. (a) In the given diagram of endoplasmic reticulum, the components marked as 1, 2, 3, 4 and 5 are respectively nucleus (viii), rough endoplasmic reticulum (v), ribosomes (vii), cytoplasm (iii) and smooth endoplasmic reticulum (iv). Endoplasmic reticulum is a network of membranous tubules within the cytoplasm of a eukaryotic cell, continuous with the nuclear membrane. It usually has ribosomes attached and is involved in protein and lipid synthesis.

17. (c) The cells are held together by a Ca-pectate layer called middle lamella.

18. (c) Cristae are found in wall of mitochondria.

19. (a) Leeuwenhoek – He is commonly known as "the Father of Microbiology", and considered to be the first microbiologist. First saw and described a living cell.

Robert Brown was a Scottish botanist and palaeobotanist who made important contributions nucleus and cytoplasmic streaming.

Schleiden was a German botanist and co-founder of the cell theory, along with Theodor Schwann and Rudolf Virchow. He discovered that all plants are composed of different kind of cells.

Schwann - He discovered that presence of cell wall is unique to plant cells.

20. (a) Danielli and Davson says that biomembrane contains four molecular layers, two of phospholipid and two of proteins. Phospholipid form a double bilayer. The phospholipid bilayer is covered on either side by a layer of hydrated globular or α-protein molecules.

21. (b)

22. (a) The charged or polar lining of the channel proteins allows passage of polar and charged molecules.

23. (d)

24. (d) The nucleus, the mitochondria, and the chloroplasts are the only double membrane-bound organelles.

25. (d) Usually only small, nonpolar molecules can move through the lipid bilayer of the membrane by simple diffusion. Because a molecule of oxygen has a balanced electron distribution, it is nonpolar.

26. (a) Tonoplast is a membrane separating a vacuole from the surrounding cytoplasm in a plant cell. It controls ionic movement in and around the cell. Contractile vacuole is a sub-cellular structure found in unicellular algae. It absorbs excess water and wastes from a microorganism's cell and excretes them into the environment. Food vaculole is a membrane-unclosed cell vacuole with a digestive function. It contains material taken up in by the process of phagocytosis. Air vacuole is a small cavity or space in the tissues. It stores metabolic gases.

27. (a) Golgi apparatus is a cellular organelle which is involved in the modifying, sorting and packaging of proteins for secretion. Mitochondria are known as the power house of the cell as they are the site of chemical reactions that transfer energy from organic compounds in ATP. Vacuole is a membrane bound organelle and filled with water containing inorganic and organic molecules including enzymes in solution. Vacuoles might store food or any variety of nutrients, a cell might need to survive. Grana are a stacked membranous structure within the chloroplasts of plants and green algae that contains the chlorophyll and is the site of the light reactions of photosynthesis.

28. (a)

29. (d) Plasmodesmata connections help in movement of substances between cells.

30. (b) Bacterial cell envelope consists of three components glycocalyx, cell wall and cell membrane.

Glycocalyx - It is the outermost mucilage layer of the cell envelope.

Cell Wall - It is rigid solid covering, which provides shape and structural support to the cell. Cell wall lies between plasma membrane and glycocalyx.

Plasma/cell membrane - It is selectively permeable covering of the cytoplasm that forms the innermost component of cell envelope.

31. **(d)** **32.** **(d)**

33. **(b)** Cell theory was proposed in 1838 by Matthias Schleiden and by Theodor Schwann. The cell theory says that:
All organisms are composed of one or more cells. All cells come from pre-existing cells. All vital functions of an organism occur within cells. Cells contain the hereditary information necessary for regulating cell functions and for transmitting information to the next generation of cells.

34. **(d)** **35.** **(b)** **36.** **(b)**

37. **(d)** Electron microscopes have much greater magnification power than light microscopes. This feature permits us to see more detail for whatever cell type we are viewing.

38. **(a)**

39. **(d)** Surface of plasma membrane has ion carriers which participates in exchange of ions.

40. **(c)** Prokaryotic cells may vary greatly in shape and size. The four basic shapes of bacteria are bacillus (rod-like), coccus (spherical), vibrio (comma shaped) and spirillum (spiral).

41. **(d)** This connection allows material produced in the nucleus to directly enter the ER via the nuclear pores.

42. **(b)** Mycoplasma cells are bacteria without walls.
Plasmid is a genetic structure in a cell that can replicate independently of the chromosomes. It is typically a small circular DNA strand in the cytoplasm of a bacterium or protozoan are sac-like membranes that make up grana in a chloroplast.
Lysosomes are a single membrane bound vesicular structure. It contains many hydrolytic enzymes and commonly called as suicidal bag of the cells. It is involved in intracellular and extracellular digestion.

43. **(b)** Plasmodesmata are cytoplasmic bridges between two adjacent cells which connect the neighbouring cells.

44. **(a)** In animal cells, lipid like steroidal hormones are synthesised in SER. In plant cells, the vacuoles can occupy up to 90% of the volume of the cell. Chlorophyll and carotenoid pigments are responsible for trapping light energy essential for photosynthesis.

45. **(b)** All cells have plasma membranes, DNA, and proteins. There are single-celled representatives from both prokaryotes and eukaryotes. The most reliable way to tell the difference between cell types is the presence of a nucleus, which is found only, in eukaryotes.

46. **(b)** Syncytium- If a multinucleated condition arises due to fusion of cells, it is called syncytium, e.g., body of slime moulds, striated muscles etc.

47. **(d)** Anything that affects the rate of diffusion will affect carrier-mediated facilitated diffusion. In addition to those, carrier-mediated facilitated diffusion also relies on the availability of carrier molecules.

48. **(b)** Lamellae of chloroplast are known as thylakoids.

49. **(b)**

50. **(d)** Muscle and nerve cells are comparatively very large. The nerve cells are the longest cells of human body, which may reach a length of upto 90 cm.

51. **(b)** The figure shows the structure of metacentric (3) chromosomes. The parts marked as (1) and (2) are respectively are satellite and secondary constriction. Metacentric chromosomes have the centromere in the centre, due to which both sections are of equal length.

52. (b) Pili are organelles of adhesion allowing bacteria to colonise and resist flushing.

53. (c) Mitochondria are double membrane cell organelles with the outer membrane and inner membrane. Lysosomes are a single membrane bound vesicular structure. It contains many hydrolytic enzymes and commonly called as suicidal bag of the cells. Ribosome is a granular structure, without membrane. It consists of RNA and proteins and is the protein builders or the protein synthesisers of the cell.

54. (d) In some prokaryotes like cyanobacteria, there are membranous extensions into the cytoplasm called chromatophores which contain pigments.

55. (c) Because the cell absorbs water, we know that its osmotic potential was more negative than the osmotic potential of the environment.

56. (d) All the given statements are related to Theodor Schwann. Schwann was a German physiologist. His contributions to biology include the development of cell theory, the discovery of Schwann cells in the peripheral nervous system, the discovery and study of pepsin, the discovery of the organic nature of yeast, and the invention of the term metabolism.

57. (a) Almost all the proteins synthesised by ER bounded ribosomes are converted into glycoproteins and this glycosylation occurs in the rough ER.

58. (a) Endomembrane system is a grouping of some membrane organelles which function in close coordination with one another, *viz.*, endoplasmic reticulum, golgi complex, lysosomes and vacuoles.

59. (c) The ability to distinguish different neighbouring cells is important for organisms's function. Glycolipids are lipids with attached carbohydrates, which acts as recognition sites during cell-cell interaction, as well as sites of attachment in a tissue.

Glycoproteins are often integral membrane proteins and are also important for cell recognition.

60. (d)

61. (a) Secondary wall situated near the plasma membrane after the formation of primary wall.

62. (d) All cells have a plasma membrane. The organelle such as lysosome, chloroplast, vacuoles, etc. are present only in eukaryotes.

63. (b)

64. (b) Prokaryotes have 70S ribosomes, each consisting of a small (30S) and a large (50S) subunit. Polyribosomes (or polysomes) are a cluster of ribosomes, bound to a mRNA molecule in translation during the process of protein synthesis. It was first discovered and characterised by Jonathan Warner, Paul Knopf, and Alex Rich in 1963. Polysomes indicate the synthesis of identical polypeptide in multiple copies.

65. (d) Statements (ii) and (iii) are correct. The endomembrane system include endoplasmic reticulum (ER), golgi complex, lysosomes and vacuoles. Since the functions of the mitochondria, chloroplast and peroxisomes are not coordinated with the above components, these are not considered as part of the endomembrane system. The mitochondrial matrix possesses single circular DNA molecule, a few RNA molecule, ribosomes (70S) and the components required for the synthesis of proteins.

66. (d) Amyloplasts store carbohydrates (starch), *e.g.*, potato whereas proteins are stored by aleuroplasts. Chlorophyll pigments are present in the thylakoid.

67. (c) In prokaryotes, a nucleus is absent but nucleoid is found which is equivalent to a single chromosome or prochromosome.

68. (d) **69.** (c)

70. (c) Oxysomes (Elementary particles or inner membranes sub-units or F_1 particles); these are responsible for respiratory chain phosphorylation.

71. (a)

72. (b) Protein synthesis takes place in ribosomes, which are attached to surface of endoplasmic reticulum by ribophorin-I and ribophorin-II. About 50 hydrolytic enzymes are found in the lysosome. They include proteases, nucleases, glycosidases, lipases, phospholipases, phosphatases and sulphatases. All lysosomal enzymes are acid hydrolases and optimally active at pH-5.0.

73. (c) Centrioles serve as basal bodies for cilia and flagella. They are concerned with spindle formation during cell division. Lysosomes are membrane-bound organelles containing many hydrolytic enzymes, which are optimally active at an acidic pH (near pH 5). Thylakoids are membrane-bound organelles found within chloroplasts. The thylakoid membrane, forms many flattened, fluid-filled tubules that enclose a single convoluted compartment. These tubules tend to stack on top of each other to form a structure called a granum.

74. (d)

75. (c) Gap junctions are involved in chemical and electrical signaling between cells.

76. (a) Protein synthesis is the process by which amino acids are linearly arranged into proteins through the involvement of ribosomal RNA, transfer RNA, messenger RNA, and various enzymes. In animal cells, protein synthesis occurs on ribosomes present in cytoplasm as well as in mitochondria.

77. (d) When ATP concentration is low or the respiratory chain is inhibited, the mitochondria are seen in an inactive or orthodox state, while reverse happens in condensed state in which intermembrane space is also increased.

78. (c) Gas vacuoles (Pseudovacuoles) occur in some prokaryotes and are without tonoplast. It store metabolic gases and take part in buoyancy regulation. These vacuoles protect cell from radiations.

79. (a) The pores in the nuclear membrane permit large molecules such as proteins and RNA to enter or exit this organelle.

80. (b) 9 + 2 similarity occurs between plant and animal flagella microtubules.

81. (d) All the given function is related to cell wall. Cell wall is a very tough, flexible and fairly rigid layer that surrounds some types of cells. It surrounds the cell membrane and provides these cells with structural support and protection. It also acts as a filtering mechanism.

82. (a) RER (rough endoplasmic reticulum) bears ribosomes on their surface. RER is involved in protein synthesis and secretion. Cell wall is a characteristic feature of plant cell, bacteria, fungi, algae and some archaea. The major function of the cell wall is to provide rigidity, tensile strength, structural support, protection against mechanical stress and infection. Flagella are a slender thread-like structure, especially a microscopic whip-like appendage which enables many protozoa, bacteria, spermatozoa, etc. to swim. It is responsible for cell movement. Lysosome is a single membrane bound vesicular structure and contains many hydrolytic enzymes. It is commonly called as suicidal bag of the cells and involved in intracellular and extracellular digestion.

83. (d) The cell membrane is asymmetric and has different properties, and functions of the cytoplasmic side versus the extracellular side. These properties arise from differences in the constituents of the membrane.

84. (d) Centrosome is responsible for initiation of cytokinesis or splitting of mother cell into two.

85. (a)

86. (d) Based on the absence of membrane bound organelles, it can be either prokaryote or archaebacteria. The presence in extreme conditions also approve this. There are several characteristics that suggest this is a prokaryote. It survives in high salinity and high temperature, but the sure indication is that it contains no membrane-bound organelles.

87. (a) 88. (c) 89. (b) 90. (d)

91. (d) The mitochondrion is the powerhouse of the cell. It is the organelles where the bulk of the reactions that manufacture ATP, take place.

92. (a) Cristae are finger like process projecting inwards in mitochondria.

93. (a) One genome is haploid type of set of chromosomes.

94. (b) All the given statements regarding cilia and flagella are correct.

95. (d) The given characters are related to cell organelle - vacuole. Vacuole is a membrane bound organelle, found in both animal and plant cells but are much larger in plant cells. Vacuoles are essentially enclosed compartments which are filled with water containing inorganic and organic molecules including enzymes in solution, though in certain cases they may contain solids which have been engulfed. Vacuoles might store food or any variety of nutrients a cell might need to survive. They can even store waste products so the rest of the cell is protected from contamination.

96. (c) Mesosome is special structure, formed by the extensions of plasma membrane. It helps in the formation of cell wall, DNA replication and distribution to daughter cells. They also help in respiration, secretion processes, to increase the surface area of the plasma membrane and enzymatic content. It is the characteristic of prokaryotes.

97. (c) 98. (b) 99. (d)

100. (c) Cell wall die due to the release of enzyme of lysosome (autolysis).

9 Biomolecules

1. Phospholipids are:
(a) Hydrophilic (b) Amphibolic
(c) Hydrophobic (d) Amphipathic

2. Deionised water is obtained by passing hard water through: Tricky
(a) Zeolite
(b) Cation exchanger
(c) Anion exchanger
(d) Both cation and anion exchanger

3. Chargaff's rule is applicable to:
(a) Single stranded DNA
(b) Double stranded DNA
(c) Single stranded RNA
(d) Single stranded DNA and RNA

4. The ground susbtance of connective tissue is basically composed of:
(a) Lipids
(b) Phospholipids
(c) Monosaccharides
(d) Mucopolysaccharides

5. Match the following nutritional vitamin deficiencies in column-I with the causes/ deficiencies in column-II and choose the correct option.

	Column-I		Column-II
A.	Kwashiorkar	I.	Iron
B.	General anaemia	II.	Menadione
C.	Dermatitis	III.	Protein
D.	Marasmus	IV.	Pyridoxine
E.	Bleeding	V.	Biotin

(a) A – I; B – V; C – II; D – III; E – IV
(b) A – V; B – II; C – III; D – IV; E – I
(c) A – V; B – III; C – IV; D – I; E – II
(d) A – III; B – IV; C – I; D – V; E – II

6. Identify the graph labelling as per the statements given and choose the correct option. Tricky

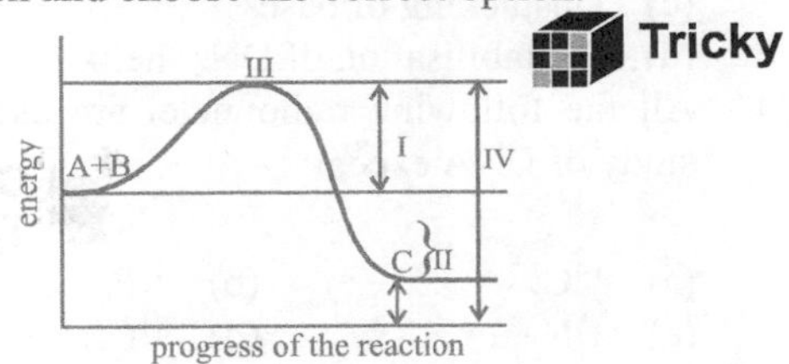

(i) Segment representing the energy of activation
(ii) Segment representing the amount of free energy released by the reaction
(iii) Transition state
(iv) Segment would be the same regardless of whether the reaction were uncatalysed or catalysed.

	(i)	(ii)	(iii)	(iv)
(a)	I	III	II	IV
(b)	I	II	III	II
(c)	I	III	II	IV
(d)	I	II	IV	III

7. Allosteric enzymes have allosteric sites for: Critical Thinking
(a) Inhibition only
(b) Activation only
(c) Reduction in activation energy
(d) Both activation and inhibition

8. Which of the following compounds listed below is not a monoamine? Tricky
(a) Histamine (b) Serotonin
(c) Spermine (d) Dopamine

9. The cellular pool is formed by:
(a) inorganic molecules
(b) organic molecules
(c) all the molecules in dissolved or suspended state
(d) water

10. Milk protein is:
(a) Lactogen (b) Myosin
(c) Casein (d) Pepsin

11. Enzymes, vitamins and hormones can be classified into a single category of biological chemicals, because all of these:
(a) enhance oxidative metabolism.
(b) are conjugated proteins.
(c) are exclusively synthesised in the body of a living organism.
(d) help in regulating metabolism.

12. All the following statements are true with regard to coconut oil, except:
(a) it is composed of medium chain triglycerides.
(b) it contains mainly saturated fatty acids.
(c) it has a higher iodine number than human fat.
(d) it contains cholesterol.

13. The base pairs of DNA are correctly shown as:
(a) A ≡ T and C = G
(b) A = T and C = G
(c) A = T and C ≡ G
(d) A ≡ T and C ≡ G

14. Ammonia is very toxic to brain cells. This is immediately trapped in brain by one of the following mechanisms.

(a) Glutamine synthetase reaction
(b) Glutaminase reaction
(c) Urea synthesis cycle
(d) Glutamate dehydrogenase reaction

15. Which one of the given graph shows the effect of pH on the velocity of a typical enzymatic reaction (V)?

Critical Thinking

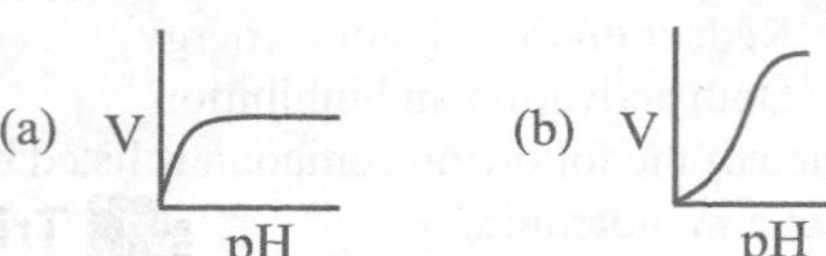

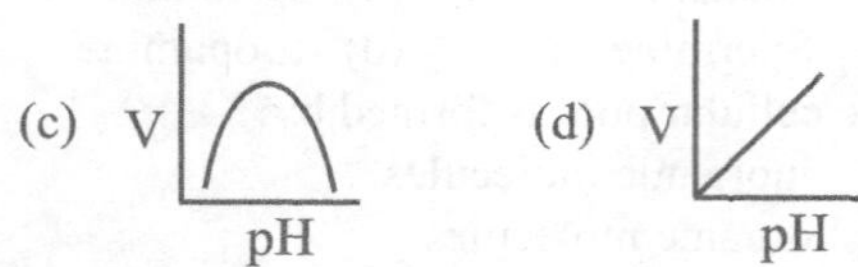

16. Which one of the following pairs of nitrogenous bases of nucleic acids, is wrongly matched with the category mentioned against it ?
(a) Adenine, thymine – Purines
(b) Thymine, uracil – Pyrimidines
(c) Uracil, cytosine – Pyrimidines
(d) Guanine, adenine – Purines

17. Glycogen is stored in:
(a) Liver and muscles
(b) Liver only
(c) Muscles only
(d) Pancreas

18. One of the following proteins does not have a quaternary structure.
(a) Haemoglobin
(b) Myoglobin
(c) Lactate dehydrogenase
(d) Immunoglobin G

19. Iron porphyrin occurs in:
(a) Cytochrome (b) Phytochrome
(c) Anthocyanin (d) Both (b) and (c)

20. The high energy bonds of ATP are present between:

Tricky

(a) C–C (b) C–O
(c) C–N (d) O–P

21. Match Column-I with Column-II and select the correct answer using the codes given below.

Column-I	Column-II
A. Collagen	I. Glucose transport
B. Trypsin	II. Binding with some chemical like for small taste and hormones
C. Insulin	III. Homones
D. Antibody	IV. Enzyme
E. Receptor	V. Intercellular ground substance
F. GLUT-IV	VI. Fight infectious agents

(a) A – V; B – IV;C – III; D – VI; E – II; F – I
(b) A – II; B – III; C – IV; D – V; E – VI; F – I
(c) A – VI; B – II; C – I; D – V; E – VI; F – III
(d) A – I; B – IV; C – III; D – VI; E – II; F – V

22. Which of the following is **incorrectly** matched?
(a) Ribozyme – Proteinaceous in nature
(b) Apoenzyme – The protein part of enzyme
(c) Co-enzyme – Loosely attached organic co-factor of holoenzyme
(d) Co-factors – Non-protein part of holoenzyme

23. Denaturation of DNA is accompanied by the following changes except:

Tricky

(a) Increase in viscosity
(b) Hyperchromicity
(c) Unstacking of bases
(d) Destabilisation of DNA helix

24. All the following radionuclei are used in the study of DNA except:

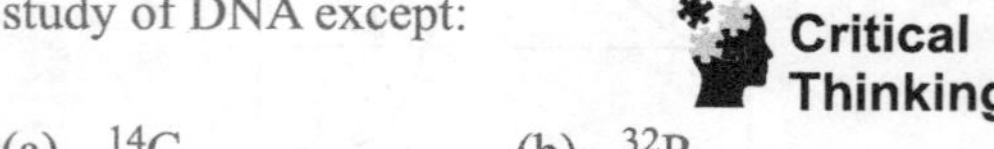

(a) ^{14}C (b) ^{32}P
(c) ^{15}N (d) ^{51}Cr

25. The adjoining graph shows change in conc. of substrate on enzyme activity. Identify 1, 2 and 3.

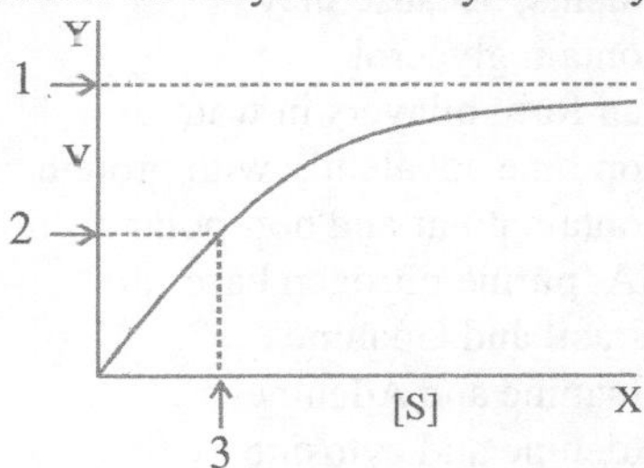

	1	2	3
(a)	Ki	K_m	V_{max}
(b)	$\frac{V_{max}}{2}$	K_m	K_i
(c)	V_{max}	K_m	$\frac{V_{max}}{2}$
(d)	K_m	V_{max}	$\frac{V_{max}}{2}$

26. If base order in one chain of DNA is "ATCGA" then how many no. of H-bond found in DNA duplex? **Toughnut**

(a) 20 (b) 12 (c) 10 (d) 11

27. Raffinose is a:
(a) Trisaccharide
(b) Disaccharide
(c) Polysaccharide
(d) Monosaccharide

28. Sugar and amino acids are:
(a) Primary metabolites
(b) Secondary metabolites
(c) Feed stock
(d) Inoculum

29. Long chain molecules of fatty acids are formed by:
(a) polymerisation of 2-carbon compounds
(b) decomposition of fats
(c) polymerisation of glycogen
(d) conversion of glycogen

30. Keratin is:
(a) Tripeptide
(b) Chromoprotein
(c) Fibrous protein
(d) Conjugate protein

31. Match the column-I (Component) with column-II (% of the total cellular mass) and identify the correct options.

	Column-I (Component)		Column-II (% of the total cellular mass)
A.	Water	I.	3
B.	Proteins	II.	70-90
C.	Carbohydrates	III.	2
D.	Lipids	IV.	5-7
E.	Nucleic acids	V.	10-15

(a) A – V; B – II; C – III; D – IV; E – I
(b) A – II; B – V; C – I; D – III; E – IV
(c) A – III; B – I; C – IV; D – V; E – II
(d) A – V; B – IV; C – III; D – II; E – I

32. Identify the following molecule.

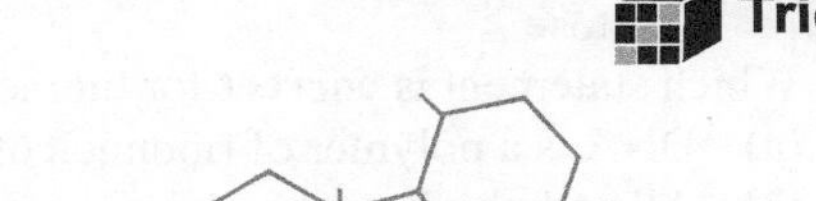

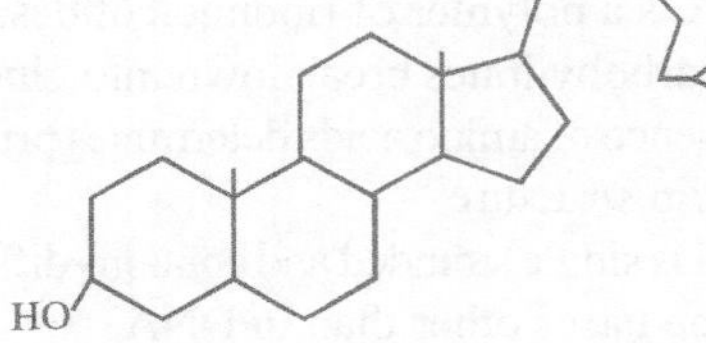

(a) Phospholipid (b) Lecithin
(c) Cholesterol (d) Oleic acid

33. Study the given statements and select the option.
(i) Nitrogen bases are open chain hydrocarbons.
(ii) A nucleoside differ from nucleotide in not having phosphate.
(iii) One turn of B-DNA has ten base pairs.
(iv) Length of one turn of DNA is 3.4 Å.
(a) (i) and (ii) are correct.
(b) (i) and (iv) are correct.
(c) (ii) and (iii) are incorrect.
(d) (i) and (iv) are incorrect.

34. According to Watson and Crick how many base pairs are there in each turn of the helix of DNA?
(a) 10 (b) 11 (c) 12 (d) 13

35. Phospholipid can be neutralised by:
(a) Glycerol (b) Fatty acids
(c) Choline (d) Aspartic acid

36. Which of the following is a non-reducing sugar?

(a) Maltose (b) Trehalose
(c) Lactose (d) Isomaltose

37. Thymine is:
(a) 5-Methyl uracil
(b) 4-Methyl uracil
(c) 3-Methyl uracil
(d) 1-Methyl uracil

38. Sucrose is made up of glucose and fructose, which are reducing sugars; but sucrose is a non-reducing sugar, because the linkage involves:

(a) 1st carbon of glucose and 2nd carbon of fructose
(b) 2nd carbon of glucose and 1st carbon of fructose
(c) 1st carbon of glucose and 4th carbon of fructose
(d) 4th carbon of glucose and 1st carbon of fructose

39. Which statement is **correct** for biomolecules?
(a) DNA is a polymer of ribonucleotides.
(b) All carbohydrates breakdown into glucose.
(c) Sequence of amino acids determines primary protein structure.
(d) RNA is single stranded and contains different purine bases other than in DNA.

40. Ligase enzyme is used for:
(a) Denaturation of DNA
(b) Splitting DNA into small bits
(c) Joining bits of DNA
(d) Digestion of lipids

41. Which of the following is an essential fatty acid?
(a) Linoleic acid
(b) Linolenic acid
(c) Arachidonic acid
(d) All of these

42. Which of the following discharide gives two molecules of glucose on hydrolysis?
(a) Maltose
(b) Lactose
(c) Both (a) and (b)
(d) Sucrose

43. In protoplasm, the amount of organic compounds, in decreasing order, are: Critical Thinking
(a) Protein, Lipid, Carbohydrate, Nucleic acid
(b) Water, Protein, Lipid, Nucleic acid
(c) Carbohydrate, Protein, Lipid, Nucleic acid
(d) Protein, Carbohydrate, Nucleic acid, Lipid

44. Phospholipids are important cell membrane constituents, because they:
(a) contain glycerol
(b) can form bilayers in water
(c) combine covalently with protein
(d) contain polar and non-polar portions

45. In DNA, purine nitrogen bases are:
(a) Uracil and Guanine
(b) Guanine and Adenine
(c) Adenine and cytosine
(d) Cytosine and Thymine

46. tRNA attaches amino acid at its:
(a) 3' end (b) 5' end
(c) Loop (d) Anticodon

47. The most abundant structural polysaccharide is cellulose and the second largest structural polysaccharide is: Tricky
(a) Chitin
(b) Mannose
(c) Glycogen
(d) Hyaluronic acid

48. What is denoted by 1 and 2 in the given graph?

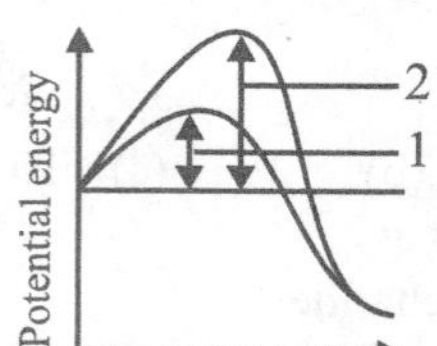

	1	2
(a)	Activation energy without enzyme	Activation energy with enzyme
(b)	Activation energy with enzyme	Activation energy without enzyme
(c)	Substrate concentration with enzyme	Substrate concentration without enzyme
(d)	Substrate concentration without enzyme	Substrate concentration with enzyme

49. Allosteric modulation is due to the inhibition action of enzyme by:
(a) Competitive inhibition
(b) Substrate concentration
(c) Products of reaction
(d) Enzyme concentration

50. All the following statements are true with regard to coconut oil, except:
(a) it is composed of medium chain triglycerides.
(b) it contains mainly saturated fatty acids.
(c) has a higher iodine number than human fat.
(d) contains cholesterol.

51. The set of three elements included in micronutrients is:
(a) Na, Cu, Mg (b) Fe, Zn, Cu
(c) Fe, K, Ca (d) Mn, Ni, P

52. The structure of linolenic acid can be described as: Tricky
(a) 16 carbon, one double bond
(b) 18 carbon, one double bond
(c) 18 carbon, two double bonds
(d) 18 carbon, three double bonds

53. Which of the following amino acids will give rise to an inhibitory neurotransmitter? Tricky
(a) Histidine
(b) Glutamic acid
(c) Ornithine
(d) Tyrosine

54. Pectin is:
(a) Monosaccharides
(b) Homopolysaccharides
(c) Mucopolysaccharides
(d) Heteropolysaccharides

55. Insulin activates:
(a) Lipolysis (b) Ketogenesis
(c) Glycogenolysis (d) Glycolysis

56. Maltose is composed of:
(a) Glucose & Fructose
(b) Glucose & Glycogen
(c) Two molecules of Glucose
(d) Glycogen & Fructose

57. Which of the following is a reducing sugar?
(a) Sucrose
(b) Galactose
(c) Gluconic acid
(d) β-methyl galactoside

58. Bond present between two nucleosides of a polynucleotide is:
(a) Covalent bond
(b) Hydrogen bond
(c) Phosphodiester bond
(d) High energy phosphate bond

59. Which one of the following vegetable oils has a very low content of PUFA (polyunsaturated fatty acid)?

(a) Safflower oil
(b) Groundnut oil
(c) Coconut oil
(d) Mustard oil

60. Which is not a micronutrient?
(a) Zn (b) Ca (c) Cu (d) Mo

61. Fats in the body are formed when: Critical Thinking
(a) glycogen is formed from glucose.
(b) sugar level becomes stable in blood.
(c) extra glycogen storage in liver and muscles is stopped.
(d) All of the above

62. Bond between phosphate and sugar in a nucleotide is:
(a) H-bond
(b) Covalent bond
(c) Phosphodiester bond
(d) Sulphide bond

63. Which of the following is an amino acid, which is metabolised by the brain? Tricky
(a) Alanine (b) Histidine
(c) Glycine (d) Glutamic acid

64. Proteins are conducted in the body in the form of:
(a) Amino acids
(b) Natural proteins
(c) Enzymes
(d) Nucleic acids

65. The differences in D and L forms of sugars are based on: Tricky
(a) spatial arrangement of carbon atom to which the functional group is attached.
(b) configuration of H and OH groups around the penultimate carbon.
(c) if the plane of polarised light is rotated to left (levo), it is L form; and if shifted to right (dextro), it is D form.
(d) special arrangement of H and OH groups at the last carbon atom.

66. Which is not polysaccharide?
(a) Sucrose
(b) Starch
(c) Glycogen
(d) Cellulose

67. Which of the following hormone is a derivative of amino acid?
(a) Oestrogen
(b) Epinephrine
(c) Progestrone
(d) Prostaglandin

68. Cholesterol is a crucial molecule in animals as: **Tricky**
(a) it is necessary for survival.
(b) it is the energy source.
(c) it is helpful in hydrolysis of glycogen.
(d) it is a source of many vertebrate hormones and other steroids.

69. Inulin is a:
(a) Lipid (b) Protein
(c) Polysaccharide (d) Human insulin

70. Glycolysis cannot occur in the absence of:
(a) Ca (b) Mg (c) Mn (d) Co

71. Which of the following is not a wax? **Tricky**
(a) Tripalmitin
(b) Myricyl palmitate
(c) Myricyl cerotate
(d) Lanolin

72. The number of essential amino acids in a man is: **Tricky**
(a) 9 (b) 8 (c) 10 (d) 14

73. How many pair of nucleotides are present in one helix of β- DNA? **Critical Thinking**
(a) 10 (b) 12 (c) 5 (d) 6

74. All the following are pentoses, except:
(a) Arabinose
(b) Xylose
(c) Xylulose
(d) Mannose

75. DNA polymerase enzyme is required for the synthesis of:
(a) DNA from DNA
(b) DNA from RNA
(c) Both (a) and (b)
(d) DNA from nucleosides

76. There is no net synthesis of glucose from fatty acids, because of the irreversible nature of the reaction. **Critical Thinking**
(a) Pyruvate to oxaloacetate
(b) Pyruvate to acetyl CoA
(c) Phosphoenol pyruvate to puruvate
(d) Oxaloacetate to phosphoenol pyruvate

77. Which of the following is a micronutrient?
(a) Copper
(b) Calcium
(c) Phosphorus
(d) Magnesium

78. Pentoses and hexoses are the most commonly:
(a) Disaccharides
(b) Monosaccharides
(c) Oilgosaccharides
(d) Polysaccharides

79. Specificity of protein is due to:
(a) Types of amino acid
(b) Sequence of amino acid
(c) Number of amino acid
(d) Quantity of amino acid

80. In an experiment it was found that adenine constitutes 31% and guanine 19%. The quantity of cytosine in this DNA is likely to be: **Critical Thinking**
(a) 19% (b) 40%
(c) 31% (d) 50%

81. Which of the following mineral is essential for the normal excitability of cardiac muscles?
(a) Calcium ions
(b) Potassium ions
(c) Magnesium ions
(d) All of these

Answer KEYs

1	(d)	10	(c)	19	(a)	28	(a)	37	(a)	46	(a)	55	(d)	64	(a)	73	(a)		
2	(d)	11	(d)	20	(d)	29	(a)	38	(a)	47	(a)	56	(c)	65	(b)	74	(d)		
3	(b)	12	(d)	21	(a)	30	(c)	39	(c)	48	(b)	57	(b)	66	(a)	75	(c)		
4	(d)	13	(c)	22	(a)	31	(b)	40	(c)	49	(c)	58	(c)	67	(b)	76	(b)		
5	(d)	14	(a)	23	(a)	32	(c)	41	(d)	50	(d)	59	(c)	68	(d)	77	(a)		
6	(b)	15	(c)	24	(d)	33	(d)	42	(a)	51	(b)	60	(b)	69	(c)	78	(b)		
7	(d)	16	(a)	25	(c)	34	(a)	43	(a)	52	(d)	61	(c)	70	(b)	79	(b)		
8	(c)	17	(a)	26	(b)	35	(c)	44	(d)	53	(b)	62	(b)	71	(a)	80	(a)		
9	(c)	18	(b)	27	(a)	36	(b)	45	(b)	54	(b)	63	(d)	72	(a)	81	(d)		

Hints & Solutions

1. **(d)** Phospholipid (phosphoglyceride) is a bipolar molecule. Its two long fatty acid (tails) represent the hydrophobic end, while the phosphate containing end is hydrophilic end.
   Hydrophilic represents the polar groups and hydrophobic represents the non-polar groups. Thus they are called amphipathic in nature.

2. **(d)** Deionised or demineralised water is obtained by passing hard water through both cation and anion exchangers one after the other.

3. **(b)**

4. **(d)** The intercellular substance (extracellular matrix) in connective tissues is usually abundant. That is why, the cells are widely spaced. It consists of fibres and a large amount of an amorphous ground substance. This ground substance is mainly composed of acidic mucopolysaccharides bound with proteins to form hyaluronic acid.

5. **(d)** 6. **(b)** 7. **(d)** 8. **(c)**

9. **(c)**

10. **(c)** Milk protein is casein.

11. **(d)**

12. **(d)** Cholesterol is an animal sterol and therefore will not be seen in vegetable oils.

13. **(c)**

14. **(a)** Glutamine synthetase catalyses the reaction, glutamic acid + ammonia $\rightarrow$ glutamine; this traps the ammonia in brain. Glutaminase catalyses the reverse reaction, glutamine $\rightarrow$ glutamic acid + ammonia. Urea cycle is taking place in liver. Glutamate dehydrogenase removes ammonia from glutamic acid, and this is taking place in liver.

15. **(c)** Parabolic curve in option (c) correctly represents the effect of pH on the velocity of a typical enzymatic reaction (V). Enzyme shows its highest activity at a particular pH called the optimum pH. Activity declines both below and above the optimum value.

16. (a) Adenine is purine while thymine is pyrimidine.

17. (a)

18. (b) Myoglobin has only the one sub-unit so it does not have quaternary structure. Haemoglobin has 2 alpha and 2 beta chains; LDH has 4 subunits; IgG has 2H and 2L chains; CPK has 2 subunits.

19. (a) Cytochrome is a conjugated protein, having apoprotein and a prosthetic group (haem).

20. (d) ATP is a universal energy carrier of living cells. It is a nucleotide. It consists of adenine and ribose with three phosphate groups attached to it. The hydrolysis of high energy terminal phosphate bond (O–P) releases energy (30.5 kJ $mole^{-1}$).

21. (a)

22. (a) Ribozymes (ribonucleic acid enzymes) are also called catalytic RNA. They are RNA molecules capable of catalysing specific biochemical reactions, similar to the act of protein enzymes. Ribozyme, discovered in 1982, demonstrated that RNA can be both genetic material (like DNA) and a biological catalyst (like protein enzymes). Examples of ribozymes include the hammerhead ribozyme, the VS ribozyme, Leadzyme and the hairpin ribozyme.

23. (a) Viscosity decreases when DNA is denatured. The hyperchromacity is dye to unstacking of bases and strand separation. Unwinding occurs due to disruption of hydrogen bonds.

24. (d) Chromium is not used to label nucleic acids since it is not an atom found in nucleic acids and nucleoprotein. It is used for the study of RBC kinetics. ^{35}S is tagged to methionine which is then incorporated to newly synthesised proteins. ^{32}P and ^{15}N are incorporated into newly synthesising DNA strands.

25. (c) 26. (b)

27. (a) Raffinose is a trisaccharide. It is made up of three molecules of glucose, *i.e.*, glucose, fructose and galactose.

28. (a) 29. (a)

30. (c) Keratin is a fibrous and structural protein. It contains a large amount of sulphur. Feathers are made up of keratin.

31. (b) 32. (c)

33. (d) Nitrogen bases are heterocyclic compounds. Length of one turn of DNA is 34Å.

34. (a) 35. (c)

36. (b) Trehalose is a non-reducing sugar. It is made up to 2 molecules of glucose.

37. (a) Thymine is 5-Methyl uracil.

38. (a) As the 1st carbon of glucose and 2nd carbon of fructose are combined, the anomeric carbon atoms of both are not available for the reduction property.

39. (c)

40. (c) Ligase enzyme is used for joining bits of DNA.

41. (d) A essential fatty acid is a polyunsaturated acid which can not be synthesised in the animal body. Therefore it must be supplied with food.

Linoleic acid, linolenic acid and arachidonic acids are some examples of essential fatty acids.

42. (a) Maltose gives two molecules of glucose on hydrolysis.

43. (a)

44. (d) Phospholipids are composed of phosphate group and one or more fatty acids. They have hydrophilic (polar) phosphate group and long hydrophobic (non-polar) hydrocarbon 'tails'. The phospholipids readily form membrane like structure in water.

45. (b) In DNA purine nitrogen bases are Guanine and Adenine.

46. (a) Transfer RNA (t-RNA) or soluble RNA (s-RNA) or acceptor RNA is the smallest molecular RNA.

Its 3' end carries activated amino acid and contains trinucleotide with the base sequence of CCA (i.e., cytidylic acid-cytidylic acid-adenylic acid).

47. (a) Chitin is a polysaccharide, and principal component of exoskeleton of insects and crustaceans. It is a polymer of N-acetyl glucosamine. Thus it the second largest structural polysaccharide.

48. (b) In the given graph, 1 and 2 are respectively activation energy with enzyme and activation energy without enzyme.

49. (c) 50. (d) 51. (b)

52. (d) In linolenic acid double bonds are between carbon atoms 9-10, 12-13 and 15-16.

53. (b) Glutamic acid on decarboxylation gives rise to gamma amino butyric acid (GAMA), an inihibitory neurotransmitter. Tyrosine will produce dopamine and epinephrine, which are excitatory neurotransmitters.

54. (b) Pectin are abundant in fruits, particularly in the citrus fruits like oranges and lemons. They are homopolysaccharides present in the cell wall and in the intracellular substance. They contain arabinose, galactose and galacturonic acid.

55. (d) 56. (c)

57. (b) Glucose, fructose, mannose and galactose are hexose monosaccharides. The monosaccharides have free aldehyde or ketone group which can reduce Cu^{++} to Cu. Therefore these are called reducing sugars.

58. (c) 59. (c) 60. (b)

61. (c) Fats in the body are formed when extra glycogen storage in liver and muscles is stopped.

62. (b) Covalent bond exists between phosphate and sugar in a nucleotide.

63. (d) Glutamic acid is an amino acid formed in hydrolysis of proteins. It is the only amino acid, which is metabolised by the brain.

64. (a) Proteins are conducted in the body in the form of amino acids.

65. (b) Depending on the shifting of plane of polarised light, the sugars are denoted as 'L' or 'D' forms. The D and L forms are mirror images based on the penultimate carbon atom.

66. (a)

67. (b) Epinephrine is secreted by medulla of adrenal gland alongwith closely related noradrenaline.

It is used in medicine as a heart stimulant and to constricist blood vessels. Chemically, epinephrine is a derivative of amino acid (aminohydroxy phenyl-propionic acid).

68. (d) Cholesterol is considered as a crucial molecule in animals as it is the source of many vertebrate hormones and other steroids.

69. (c) Inulin is a polysaccharide food reserve of some higher plants, particularly in the family Compositae e.g., dahlia root tubers. It is polymer of fructose.

70. (b) In glycolysis Mg^{++} ion is a co-factor of enzymes. e.g., hexokinase, phosphofructokinase.

71. (a) Waxes are esters formed between a long chain alcohol and saturated fatty acids. Tripalmitin is a triglyceride.

72. (a) Essential amino acids in human are 9. They are histidine, threonine lysine, isoleucine, leucine, methionine, phenylalanine, tryptophan and valine.

They must be ingested in food for the survival, as they are not synthesised in the body.

73. (a)

74. (d) Mannose is a hexose.

75. (c)

76. (b) The reaction sequence, glucose → pyruvate → acetyl CoA → fatty acid is possible because, the reaction pyruvate → acetyl CoA is catalysed by pyruvate dehydrogenase. The reaction is absolutely irreversible. Fat is degraded to acetyl CoA; this 2 carbon unit enters into the TCA cycle, where 2 carbon atoms are oxidised to 2 molecules of carbon dioxide; hence no net conversion of fat to carbohydrate is possible. However, glycerol part of neutral fat and propionyl CoA (3 carbon unit) from odd chain fatty acids can generate glucose; but quantitatively these are minimal sources.

77. (a) Micronutrients. These are boron (B), chlorine (Cl), molybdenum (Mo), sodium (Na), manganese (Mn), copper (Cu) and Zinc (Zn).

78. (b) Pentoses and hexoses are the example of monosaccharides.

79. (b) 80. (a)

81. (d) Proper balance of calcium, potassium, and magnesium ions is essential for the normal excitability of muscles, especially cardiac muscles.

10 Cell Cycle and Cell Division

1. A cell is bound to divide, if it has entered:
   (a) G_2 - phase (b) G_1 - phase
   (c) Prophase (d) S - phase
2. In plants, active mitosis occurs in:
   (a) Cambium (b) Leaf tip
   (c) Root base (d) Mid vein
3. In meiosis, actual haploidy in terms of DNA comes in:
   (a) Metaphase-I (b) Anaphase-II
   (c) Anaphase-I (d) Interkinesis
4. Match the terms (given in column-I) with their explanation (given in column-II) and choose the correct combination from the options given below. **Critical Thinking**

Column-I (Terms)	Column-II (Explanation)
A. Terminalization	I. Pairing of homologous chromosomes.
B. Synapsis	II. Point of attachment between homologous chromosomes.
C. Chiasmata	III. Nuclear protein complex that helps in adherence of sister chromatids and then homologous chromosomes.
D. Synaptonemal complex	IV. Shifting of chiasmata outwards complex towards the ends of a bivalent.

   (a) A – IV; B – I; C – II; D – III
   (b) A – II; B – III; C – IV; D – I
   (c) A – II; B – IV; C – III; D – I
   (d) A – IV; B – I; C – III; D – II
5. If the number of bivalents are 8 in metaphase - I, what shall be the number of chromosomes in daughter cells after meiosis - I and meiosis - II respectively? **Tricky**
   (a) 8 and 4 (b) 4 and 4
   (c) 8 and 8 (d) 16 and 8
6. Programmed cell death (apoptosis):
   (a) occurs in cells that have been deprived of essential nutrients.
   (b) occurs only in cells that have damaged DNA.
   (c) is a natural process during development.
   (d) is signaled by the initiated mitosis.
7. Number of chromosomes in a zygote may be indicated by: **Tricky**
   (a) 4x (b) x (c) 3x (d) 2x
8. At which stage of cell cycle division the homologous chromosome separate, while sister chromatid remain associated at their centromere:
   (a) Metaphase-I of meiosis
   (b) Anaphase-I of meiosis
   (c) Metaphase of mitosis
   (d) Anaphase of mitosis
9. Identify the given figures (1, 2 & 3) showing meiotic phases and select the correct option.

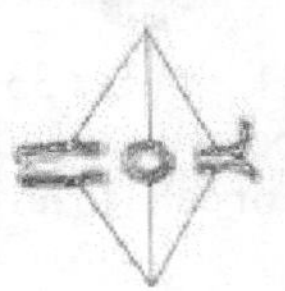
1

2

3

   (a) 1-Metaphase 2-Anaphase 3-Telophase
   (b) 1-Metaphase-I 2-Anaphase-I 3-Telophase-I
   (c) 1-Metaphase-II 2-Anaphase-II 3-Telophase-II
   (d) 1-Anaphase-I 2-Metaphase-I 3-Telophase-I
10. During which stages (or prophase I substages) of meiosis do you expect to find the bivalents and DNA replication respectively? **Critical Thinking**
   (a) Pachytene and interphase (between two meiotic divisions).
   (b) Pachytene and interphase (just prior to prophase I).
   (c) Pachytene and S phase (of interphase just prior to prophase I).
   (d) Zygotene and S phase (of interphase prior to prophase I).

11. Which of the following is correct for the mitotic spindle?
    (a) It is composed of actin and myosin microfilaments.
    (b) It is composed of kinetochores at the metaphase plate.
    (c) It is composed of microtubules, which help to separate the chromosomes to opposite poles of the cell.
    (d) It originates only at the centrioles in the centrosomes.
12. Which of the following is not a function of mitotic cell division in animals?
    (a) Asexual reproduction
    (b) Growth
    (c) Repair of damaged organs
    (d) Production of gametes
13. In G_1-phase of cell division:
    (a) RNA and proteins are synthesised for cell growth and subsequent DNA replication.
    (b) DNA and proteins are synthesised.
    (c) Formation of RNA and protein continues for spindle formation and cell growth.
    (d) Cell undergoes duplication.
14. If there are 22 chromosomes in the nucleus of a toad skin cell, a toad egg would contain ______ chromosomes. **Tricky**
    (a) 22 (b) 44 (c) 11 (d) 33
15. Cell division is a normal process in organisms, but sudden and rapid mitosis of cells in an organ may result into:
    (a) New organ (b) Gastrula
    (c) Zygote (d) Cancer
16. A plant has number of chromosome groups arranged at equatorial plane of metaphase-I whose 2n = 50; the number of chromosomes visible will be:
    (a) 100 (b) 25 (c) 50 (d) 75
17. A drug called colchicine is shown to interfere specially with spindle microtubules formation during mitosis. This would result in:
    (a) dehydration of chromosomes
    (b) despiralisation of chromosomes
    (c) arrest of chromosome movements
    (d) arrest of centrioles movements
18. Chromosomes in salivary gland cells of *Drosophila* larvae are:
    (a) Polytene (b) Lampbrush type
    (c) β-type (d) Normal type
19. Crossing over during meiosis occurs at four strand stage. This is proved by the observation that:
    (a) rarely all four gametes resulting from a meiotic division could be recombinants.
    (b) usually only two gametes resulting from meiotic division could be recombinants.
    (c) chaismata are seen only at the four strand stage.
    (d) occasionally none of the four gametes resulting from meiosis could be a recombinant.
20. Meiosis and mitosis differ from each other because in meiosis:
    (a) The four nuclei formed are not similar to parental ones.
    (b) Homologous chromosomes pair are exchange parts.
    (c) Number of chromosomes gets halved.
    (d) All of the above
21. Non-Sister chromatids:
    (a) cross over during prophase I of meiosis.
    (b) separate during the first meiotic division.
    (c) are produced during S phase between cell divisions.
    (d) cross over during prophase II of meiosis.
22. In some organisms, such as certain fungi and algae, cells undergo mitosis repeatedly without subsequently undergoing cytokinesis. What would result from this? **Critical Thinking**
    (a) A decrease in chromosome number.
    (b) Inability to duplicate DNA.
    (c) Division of the organism into many cells, most lacking nuclei.
    (d) Large cells containing many nuclei.
23. In the diagram, a cell undergoing meiosis

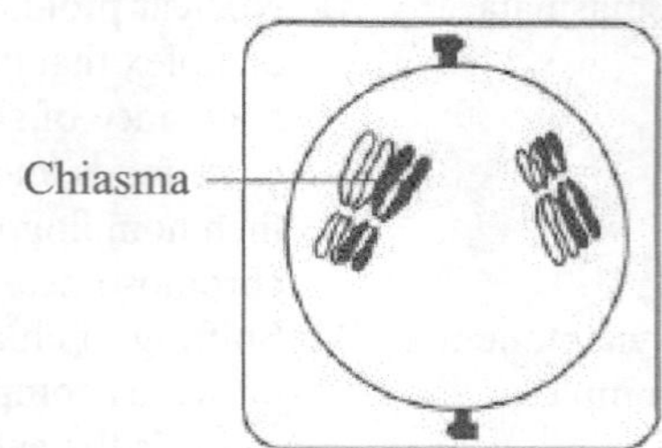

Which one of the diagrams below shows the next stage in the process?

(a) (b)
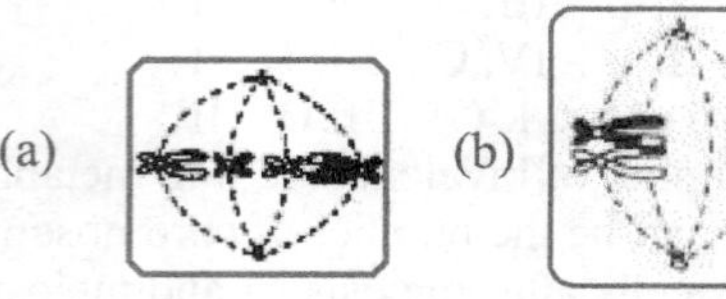

(c)

(d)

24. Match the column-I and column-II and choose the correct option. Critical Thinking

	Column-I		Column-II
A.	Chromosomes in matching pairs are called	I.	Metaphase plate
B.	In mitosis, chromosome is copied and	II.	Homologous
C.	Precursor for cell wall formation that represents the middle lamella between the walls of two adjacent cells is called	III.	Cell plate
D.	The plane of alignment of chromosomes at metaphase is called	IV.	Cell divides to give two daughter cells.

(a) A – II; B – IV; C – III; D – I
(b) A – IV; B – III; C – I; D – II
(c) A – I; B – IV; C – II; D – III
(d) A – III; B – II; C – IV; D – I

25. At what phase of meiosis there are two cells, each with sister chromatids aligned at the spindle equator?
(a) Anaphase II (b) Metaphase II
(c) Metaphase I (d) Anaphase I

26. Who stated that new cells develop from pre-existing cells?
(a) Remak
(b) Virchow
(c) Prevost and Dumas
(d) Strasburger

27. One chromosome of a homologous pair carries the genes J and K. The other chromosome of the pair carries the genes j and k at corresponding loci. Crossing over results in exchange of chromosome segments and production of gametes with new combinations of genes. A recombinant type gamete resulting from this crossover might contain Critical Thinking
(a) Genes J and K (b) Genes j and K
(c) Genes J and j (d) Genes j and k

28. Chiasmata are formed during meiosis:
(a) before metaphase I
(b) after metaphase I
(c) during prophase II
(d) during metaphase II

29. Replication of centriole occurs during:
(a) Early anaphase
(b) Mid metaphase
(c) Late metaphase
(d) Interphase

30. Recombinant nodules are found during:

(a) Anaphase (b) Prophase
(c) Telophase (d) Metaphase

31. Prophase is longer in:
(a) Mitosis (b) Meiosis
(c) Equal in both (d) Amitosis

32. In telophase of mitosis, the mitotic spindle breaks down and nuclear membranes form. This is essentially the opposite of what happens in: Tricky
(a) Prophase (b) Interphase
(c) Metaphase (d) S phase

33. Centromere is required for:
(a) Movement of chromosomes towards poles
(b) Cytoplasmic cleavage
(c) Crossing over
(d) Transcription

34. Most cells divide if they receive the proper signal at a checkpoint in the ____ phase of the cell cycle. Tricky
(a) M (b) G_1 (c) S (d) G_2

35. Which of the following does not lead to genetic variability?
(a) random fertilisation
(b) crossing over during meiosis.
(c) division of chromosomes during anaphase of mitosis.
(d) orientation of chromosomes during metaphase I of meiosis.

36. During pachytene stage of meiosis, the chromosomes appear:
(a) Single stranded
(b) Four stranded
(c) Six stranded
(d) Eight stranded

37. Which of the following statements related to G_0 stage of the cell cycle are correct?
(i). It is a quiescent stage.
(ii). In this phase, cell cycle is stopped.
(iii). G_0 cells do not grow or proliferate but metabolically active.
(iv). G_0 cells can divide in response to some stimulus.
(a) (i) and (ii) (b) (i), (ii) and (iii)
(c) Only (i) and (iv) (d) All of the above

38. Given below is a schematic break-up of the phases / stages of cell cycle with few parts labelled as 1, 2, 3 & 4.

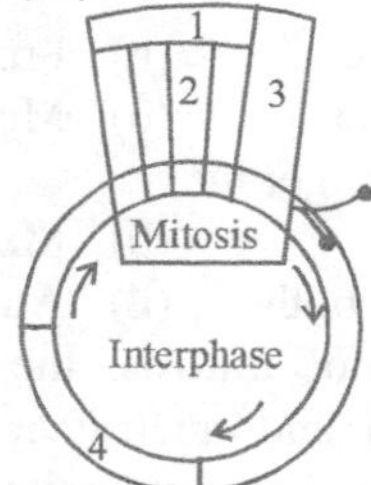

Which one of the following is the correct indication of the stage/phase in the cell cycle?
(a) 3 - Karyokinesis (b) 4 - Synthetic phase
(c) 1 - Cytokinesis (d) 2 - Metaphase

39. Pick out the correct sequence of two main events in mitosis.
(a) Karyokinesis is followed by cytokinesis.
(b) Cytokinesis is followed by karyokinesis.
(c) Karyokines is followed by separation of the daughter cells.
(d) Cytokinesis is followed by separation of the daughter cells.

40. Cell division in blue green algae is similar to that in:
(a) Bacteria (b) Brown algae
(c) Green algae (d) Higher plants

41. Strasburger is famous for:
(a) discovery of cell division
(b) proposing the term nucleoplasm
(c) proposing the term cytoplasm
(d) All of the above

42. When synapsis is complete all along the chromosome, the cell is said to have entered a stage called:
(a) Zygotene (b) Pachytene
(c) Diplotene (d) Diakinesis

43. In pachytene stage of meiosis the chromosomes appear:
(a) Single stranded (b) Double stranded
(c) Three stranded (d) Four stranded

44. Choose the event of cell cycle which shows the importance of synapsis and the formation of chiasmata.

Critical Thinking

(a) An increase in the variation of progeny occurs.
(b) Reciprocal exchange of chromosomal sections occurs.
(c) The DNA on homologous chromosomes mix.
(d) All of the above

45. Meiosis which occur at the time of spore formation is called.
(a) Zygotic meiosis (b) Haplontic meiosis
(c) Terminal meiosis (d) Intermediate meiosis

46. The cellular structure which disappear during mitosis is:
(a) Plasma membrane
(b) Nuclear membrane
(c) Mitochondria
(d) Nuclear membrane and nucleolus

47. Mature nerve cells cannot undergo cell division. These cells are probably considered in:
(a) G_2-phase (b) S-phase
(c) Mitosis (d) G_0-phase

48. After meiosis - I, the two chromatids of a chromosome are:
(a) Genetically similar
(b) Genetically different
(c) There occurs only one chromatid in each chromosome
(d) None of the above

49. Fully differentiated cells which do not divide are supposed to be in:
(a) G_1 - phase (b) G_2 - phase
(c) S - phase (d) G_0 - phase

50. A zoologist examined an intestine cell from a crayfish and counted 200 chromosomes, each consisting of 2 chromatids, at prophase I of mitosis. What would he expect to see in each of the four cells at telophase II of meiosis if he looked in the crayfish ovary?

Critical Thinking

(a) 50 chromosomes, each consisting of 2 chromatids
(b) 50 chromosomes, each consisting of 1 chromatid
(c) 100 chromosomes, each consisting of 2 chromatids
(d) 100 chromosomes, each consisting of 1 chromatid

51. The diagram shows the nuclear contents of a cell. Tricky

Which one of the following option is correct?

	Number of pairs of homologous chromosomes	Number of chromatids	Number of centromeres
(a)	3	6	12
(b)	3	12	6
(c)	6	6	12
(d)	6	12	6

52. Which of the following pair is correctly matched?
(a) Anaphase I – Homologous chromosomes are separated.
(b) Metaphase I – Pairing of maternal and paternal chromosomes.
(c) Interphase – A nuclear envelop encloses each haploid set of chromosomes.
(d) Prophase I – Non-homologous chromosomes are separated.

53. Meiosis in AaBb will produce gametes:
(a) AB, aB, Ab, ab (b) AB, ab
(c) Aa, bb (d) Aa, Bb

54. Which one of the following list contain the correct order of meiotic events? Critical Thinking
(a) Separation of sister chromatids, recombination, formation of the synaptonemal complex, separation of homologous chromosomes.
(b) Separation of homologous chromosomes, formation of the synaptonemal complex, recombination, separation of sister chromatids.
(c) Formation of synaptonemal complex, recombination, separation of sister chromatids, separation of homologous chromosomes.
(d) Formation of the synaptonemal complex, recombination, separation of homologous chromosomes, separation of sister chromatids.

55. "Bouquet-stage" occur in which sub stages of prophase I:
(a) Leptotene (b) Zygotene
(c) Pachytene (d) Diplotene

56. Which of the following human cells unlikely to see dividing? Tricky
(a) Nerve cell
(b) Skin cell
(c) Cancer cell
(d) Cell from an embryo

57. Meiosis takes place in:
(a) Apical meristem
(b) Intercalary meristem
(c) Reproductive cells
(d) Vegetative cells

58. Which division maintains genetic similarity?
(a) Mitosis (b) Meiosis
(c) Amitosis (d) Reduction division

59. Match the description given in column-I with their stages given in column-II and identify the correct answer.

Column-I	Column-II
A. Initiation of the assembly of mitotic spindle	I. Anaphase
B. Proteins are synthesized in preparation for mitosis while cell growth continues.	II. Prophase
C. Spindle fibres attach to kinetochores of chromosomes.	III. Interphase
D. Movement of chromatids towards opposite poles	IV. Metaphase

(a) A – II; B – III; C – IV; D – I
(b) A – III; B – II; C – I; D – IV
(c) A – I; B – III; C – II; D – IV
(d) A – IV; B – III; C – I; D – II

60. When dividing cells are observed under a light microscope, chromosomes become visible in:
(a) Interphase (b) S-phase
(c) Prophase (d) G_1-phase

61. Which cell division is found during cleavage?
(a) Amitosis (b) Mitosis
(c) Closed mitosis (d) Meiosis

62. If a grasshopper cell contained 200 units of DNA during G_2, what would be the quantity of DNA in one of the grasshopper cells at the end of telophase II of meiosis? **Toughnut**
(a) 50 units
(b) 100 units
(c) Between 50 and 100 units
(d) 200 units

63. If the n=16 in plant cell then what is possible in metaphase - I of meiosis? **Toughnut**
(a) 32 bivalents (b) 16 tetravalents
(c) 16 bivalents (d) 32 bivalents

64. During cell division, sometimes there will be failure of separation of sister chromatids. This event is called. **Tricky**
(a) interference
(b) complementation
(c) coincidence
(d) non-disjunction

65. During mitosis, after the separation of centromeres, the chromatids move towards opposite poles of the spindle. Name the term used for these chromatids.
(a) Daughter chromosomes
(b) Kinetochores
(c) Half spindles
(d) Centrosomes

66. Colchicine prevents the mitosis of cells at which of the following stage?
(a) Anaphase (b) Metaphase
(c) Prophase (d) Interphase

67. Number of chromosomes in human karyotype is:
(a) 48 (b) 44 (c) 46 (d) 23

68. Four different steps that occur during meiosis are given in the following list.
(i). Complete separation of chromatids.
(ii). Pairing of homologous chromosomes.
(iii). Lining up of paired chromosomes on equator.
(iv). Crossing over between chromatids.
These steps would occur in the order:
(a) (ii), (iii), (iv), (i)
(b) (iii), (ii), (iv), (i)
(c) (ii), (iv), (iii), (i)
(d) (iii), (i), (ii), (iv)

69. Each chromosome composed of one chromatid in:
(a) Anaphase - I
(b) Anaphase - II
(c) Metaphase - I
(d) Metaphase - II

70. In mitosis nucleolus and nuclear membrane disappear at:
(a) Interphase (b) Prophase
(c) Metaphate (d) Telophase

71. What is the number of DNA strand in chromosome at G_2 stage of cell cycle? **Tricky**
(a) One (b) Two (c) Four (d) Eight

72. Homologous chromosomes show maximum attraction during:
(a) Leptotene (b) Zygotene
(c) Pachytene (d) Diplotene

73. Which of the following events not occurs in Anaphase -I?
(a) Segregation of homologous chromosomes
(b) Contraction in spindle
(c) Poleward movement of chromosomes
(d) Division of centromere

74. Repulsion of homologous chromosomes takes place in:
(a) Zygotene (b) Leptotene
(c) Diakinesis (d) Diplotene

75. How many reduction divisions are necessary for the formation of 200 grains of wheat? **Critical Thinking**
(a) 250 (b) 150 (c) 200 (d) 360

76. Constancy of the chromosome number in sexually producing generation is brought by the process of:
(a) Meiosis (b) Mitosis
(c) Amitosis (d) None of these

77. Phragmoplast is the precursor of:
(a) Chloroplast
(b) Chromoplast
(c) Cell plate
(d) Leucoplast

78. In mitosis, duplication of chromosomes occurs during:
(a) Interphase (b) Prophase
(c) Metaphase (d) Telophase

Answer KEYs

1	(d)	9	(b)	17	(c)	25	(b)	33	(a)	41	(d)	49	(d)	57	(c)	65	(a)	73	(d)
2	(a)	10	(d)	18	(a)	26	(b)	34	(b)	42	(b)	50	(d)	58	(a)	66	(b)	74	(d)
3	(b)	11	(c)	19	(a)	27	(b)	35	(c)	43	(d)	51	(b)	59	(a)	67	(c)	75	(a)
4	(a)	12	(d)	20	(d)	28	(a)	36	(b)	44	(d)	52	(a)	60	(c)	68	(c)	76	(b)
5	(c)	13	(a)	21	(a)	29	(d)	37	(d)	45	(d)	53	(a)	61	(c)	69	(b)	77	(c)
6	(c)	14	(c)	22	(d)	30	(b)	38	(b)	46	(d)	54	(d)	62	(a)	70	(b)	78	(a)
7	(d)	15	(d)	23	(b)	31	(b)	39	(a)	47	(d)	55	(a)	63	(c)	71	(b)		
8	(b)	16	(c)	24	(a)	32	(a)	40	(a)	48	(b)	56	(a)	64	(d)	72	(c)		

Hints & Solutions

1. **(d)** A cell is bound to divide, if it has entered S - phase..
2. **(b)** 3. **(a)**
4. **(a)** Terminalization which occurs in diakinesis stage of meiosis I, is the disappearance of chiasmata of sliding towards the tips of the chromosomes due to tight condensation. Synapsis is the pairing of two homologous chromosomes. It allows matching-up of homologous pairs prior to their segregation, and possible chromosomal crossover between them. The process of attachment of homologous chromosomes leads to the formation of bivalents. Chiasmata is a point at which paired chromosomes remain in contact during the first metaphase of meiosis, and at which crossing over and exchange of genetic material occur between the strands. Synaptonemal complex is formed at the region of synapsis. It is a protein structure that forms between homologous chromosomes (two pairs of sister chromatids) and is thought to mediate chromosome pairing, synapsis, and recombination.
5. **(c)**
6. **(c)** Programmed cell death occurs during the development of many organisms (for instance, tadpoles lose their tails to become adult frogs). One of the stimuli for programmed cell death. Necrosis (cell death that is not programmed) occurs when cells have been deprived of cell cycle, in which cells reproduce, and is not a step in programmed cell death.
7. **(d)** 8. **(b)**
9. **(b)** The given figures (1, 2 and 3) showing meiotic phases are respectively metaphase I, anaphase I and telophase I. Metaphase I: Spindle apparatus appear and the chromosomes are arranged on equatorial plate, with the centromeres towards the pole. Spindle fibres become attached to the centromeres of the two homologous chromosomes. Anaphase I: The homologous chromosomes break apart while sister chromatids remains associated at their centromere. At the end of anaphase I, two groups of chromosomes are produced at two poles, having half the number of parental chromosomes. Telophase I: The homologous chromosome pairs complete their migration to the two poles as a result of spindle action. As a result a haploid set of chromosomes is present at each pole, with each chromosome still having two chromatids. A nuclear envelope reforms around each chromosome set, the spindle disappears, and cytokinesis follows.

10. (d)
11. (c) The mitotic spindle is composed of microtubules, not actin and myosin filaments. The spindle originates from the centrosome, which may or may not have centrioles.
12. (d) Gametes are produced by meiosis.
13. (a) In G_1 phase, cell synthesises RNAs, proteins and other biomolecules for cell growth and subsequent replication of DNA. While in G_2-phase, synthesis of RNAs, protein and other biomolecules for spindle formation takes place.
14. (c) Gametes are haploid, somatic cells are diploid.
15. (d)
16. (c) The number of chromosome will be same (50), but each chromosome will have be 2 chromatids.
17. (c) 18. (a) 19. (a)
20. (d) Four nuclei formed are not similar to parent ones because they are haploid in nature. Homologous chromosome pairs have exchange parts because in crossing over exchange of chromatid arms takes place and number of chromosomes gets halved.
21. (a) Crossing over and genetic recombination occur during prophase I.
22. (d)
23. (b)
24. (a)
    A. Chromosomes in matching pairs are called homologous.
    B. In mitosis, each chromosome is copied and cell divides to give two daughter cells.
    C. Precursor for cell wall formation that represents the middle lamella between the walls of two adjacent cells is called cell plate.
    D. The plane of alignment of the chromosomes at metaphase is referred to as the metaphase plate.
25. (b) 26. (b)
27. (b) Crossing over does not alter the traits that are coded for, it only affects the variant that is coded for.
28. (a)
29. (d) During the S-period, the centrioles separate and undergo duplication which produces two pairs of centrioles still contained within the radiating masses of microtubules.
30. (b) Recombination nodules are multicomponent proteinaceous ellipsoids found in association with the synaptonemal complex (sc) during prophase I of meiosis.
31. (b)
32. (a) In telophase nuclear membrane and nucleolus reappear. Chromosomes become less compact and the chromonemata coils unwind. Thus it is just reverse of prophase.
33. (a) Centromere is required for movement of chromosomes towards poles.
34. (b) If at G_1 cells are given the signal to divide, they are unlikely to be stopped at subsequent checkpoints.
35. (c) The result of cell division is two cells, genetically identical to each other and to the parental cell. During anaphase centromere splits from the middle, and two chromatids get separated.
36. (b)
37. (d) All the statements regarding G_0 stage of the cell are correct. G_0 stage or resting phase is a period in the cell cycle in which cells exist in a quiescent state. G_0 phase is viewed as either an extended G_1 phase, where the cell is neither dividing nor preparing to divide, or a distinct quiescent stage that occurs outside of the cell cycle. Some (like nerve, cells and heart muscle cells), become quiescent when they reach maturity (*i.e.*, when they are terminally differentiated) but continue to perform their main functions for the rest of the organism's life.
38. (b) In schematic break-up of the phases/stages of cell cycle, 4-synthetic phase is the correct indication of the stage/phase in the cell cycle.
39. (a) There are two main events in mitosis: karyokinesis or duplication of the nucleus, followed by division of the cytoplasm called cytokinesis. This results in the separation of the daughter cells.
40. (a) 41. (d)
42. (b) When synapsis is complete all along the chromosome, the cell is said to have entered a stage called pachytene.

43. (d) During the pachytene, chromosomes get crossed with each other and forms a four stranded or tetrad appearance.

44. (d) Synapsis is the pairing of homologous chromosomes during meiosis. While autosomes undergo synapsis during meiosis, sex chromosomes often remain unpaired. A consequence of recombinant synapsis is to increase genetic variability within both the offsprings and population.

45. (d) 46. (d)

47. (d) Some cells in the adult animals do not show division (e.g., heart cells, and many other cells divide only occasionally i.e., when there is need to replace cells that have been lost due to injury or cell death). These cells do not divide further and exit G_1-phase to enter an inactive stage called quiescent stage (G_0) of the cell cycle. Cells in this stage remains metabolically active but no longer proliferate.

48. (b) After meiosis - I, the two chromatids of a chromosome are genetically different.

49. (d) G_C represents a stage in G_1 in which cells are supposed to be withdrawn from division.

50. (d) Meiosis reduces the chromosome count from diploid to haploid and halves the amount of genetic material.

51. (b)

52. (a) Pair of anaphase I with their feature is correctly matched.

Metaphase I: Spindle apparatus appear and the chromosomes are arranged on equatorial plate, with the centromeres towards the pole. Spindle fibres become attached to the centromeres of the two homologous chromosomes.

Interphase: It is the time during which cell is preparing for division by undergoing both cell growth and DNA replication.

Prophase I: It is the lengthy phase when compared with mitotic phase. It is subdivided into 5 sub-phases: leptotene, zygotene, pachytene, diplotene and diakinesis.

53. (a)

54. (d) The correct order of meiotic events which occur during meiosis is: Formation of synaptonemal complex, recombination, separation of homologous chromosomes, separation of sister chromatids.

55. (a)

56. (a) Once past the first few years of life, nerve cells do not divide.

57. (c) Meiosis takes place in reproductive cells.

58. (a) Mitosis division maintains genetic similarity.

59. (a) A – II, B – III, C – IV, D – I

Anaphase	–	Movements of chromatids towards opposite poles.
Prophase	–	Initiation of assembly of mitotic spindle.
Interphase	–	Proteins are synthesized in preparation for mitosis while cell growth continues.
Metaphase	–	Spindle fibres attach to the kinetochores of chromosomes.

60. (c) The S and G_2-phases of interphase are followed by prophase. Prophase is identified by the initiation of condensation of chromosomal material. The chromosomal material during the process of chromatin condensation become untangled. Centriole, now begins to move towards opposite poles of the cell. Therefore, when dividing cells are examined under a light microscope, in prophase only the chromosomes become visible.

61. (c) During cleavage cell divides mitotically without taking any gap, called closed mitosis

62. (a) During meiosis I DNA content get reduced by half and after meiosis II, further it reduced by half.

63. (c) 64. (d)

65. (a) During mitosis chromosomes that results from the separation of sister chromatids during cell division are called daughter chromosomes. During anaphase of mitosis, paired chromosomes (sister chromatids) separates to form daughter chromosomes. Each daughter chromosome migrates to centromere, towards the opposite ends of the cell. At the end of cell division, two distinct daughter cells are formed from a single cell.

66. (b) 67. (c) 68. (c)

69. (b) Each chromosome composed of one chromatid in Anaphase - II.

70. (b)

71. (b) The duplication of DNA takes place in 'S' stage of interphase after that cell reaches in G_2 phase. That is why the strand of DNA in G_2 phase will be two.

72. (c) 73. (d)

74. (d) In meiosis, crossing over takes place in pachytene, after that chromosome get repulsive in diplotene.

75. (a) Fortheformationof 200grainsofwheat-200 pollen grains and 200 megaspores are required. Each - microspore mother cell produces 4 microspores (pollen grains) as a result of reduction division. Hence for the formation of 200 pollen grains, = 50 PMC are needed and each will undergo one reduction division. Each megaspore mother cell will produce 4 megaspores in which one become functional and other three degenerate. Thus each MMC produces one megaspore. So 200 MMC are required to produce 200 megaspores. Total number of reduction divisions to produce 200 grains is 200 + 50 = 250.

76. (b) 77. (c) 78. (a)

11 Transport in Plants

1. Which of the following is used to determine the rate of transpiration in plants?
(a) Porometer/hygrometer
(b) Potometer
(c) Auxanometer
(d) Tensiometer/Barometer

2. Which is produced during water stress that brings stomatal closure? Critical Thinking
(a) Ethylene (b) Abscisic acid
(c) Ferulic acid (d) Coumarin

3. Water absorption by root hairs occurs until:
(a) concentration of water in the cell sap is higher.
(b) salt concentration in soil is higher.
(c) they are separated from soil by a selectively permeable membrane.
(d) water potential is lower.

4. A bottle filled with previously moistened mustard seeds and water was screw capped tightly and kept in a corner. It blew up suddenly after about half an hour. The phenomenon involved is:
(a) Diffusion (b) Imbibition
(c) Osmosis (d) D.P.D.

5. What point can be cited for 10% sodium chloride solution and 10% sugar solution present? Critical Thinking
(a) Both have equal OP.
(b) The concentration of sodium chloride solution will be less than concentration of sugar solution.
(c) The OP of sugar solution will be higher than OP of sodium chloride solution.
(d) DPD of sodium chloride solution will be higher than DPD of sugar solution.

6. A plant cell placed in pure water will:
(a) expand until the osmotic potential or solute potential reaches that of water.
(b) becomes more turgid until the pressure potential of cell reaches its osmotic potential.
(c) become more turgid until the osmotic potential reaches that of pure water.
(d) becomes less turgid until the osmotic potential reaches that of pure water.

7. Water potential of actively absorbing cells is:
(a) Always +ve (b) Always –ve
(c) Always 0 (d) Always >1

8. Analyse the figure and choose the correct statement regarding this.

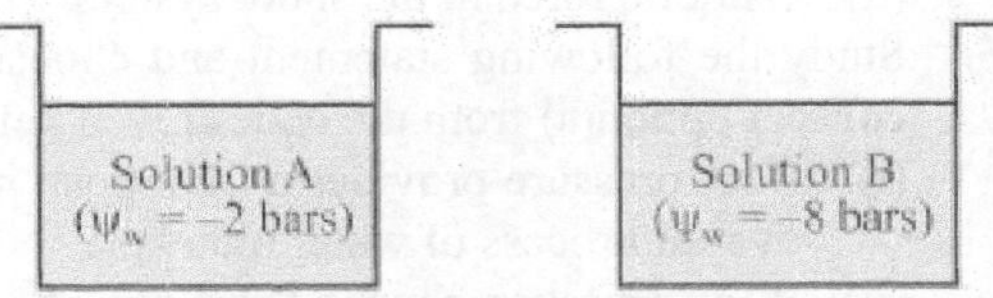

(a) Kinetic energy (K.E.) of H_2O in A solution > K.E. of water in B solution.
(b) K.E. of water in B solution > K.E. of water in A solution.
(c) K.E. of water in A solution = K.E. of H_2O in B solution.
(d) Water potential has nothing to do with K.E. of water in a solution.

9. Refer the experiment given below: Tricky

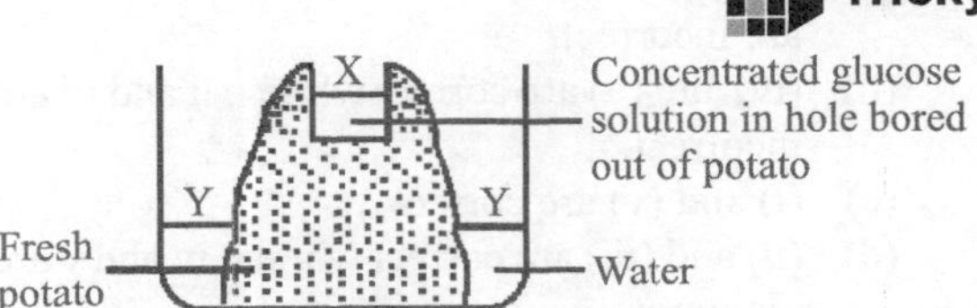

After a few days, which of the following will have occurred?
(a) A rise in level X and a drop in level Y.
(b) A drop in level X and a drop in level Y.
(c) A rise in level X and a rise in level Y.
(d) A drop in level X and a rise in level Y.

10. Stomata of a plant open due to:
(a) influx of calcium ions
(b) influx of potassium ions
(c) efflux of potassium ions
(d) influx of hydrogen ions

11. Potometer works on the principle of:
(a) osmotic pressure.
(b) amount of water absorbed equals the amount transpired.
(c) root pressure
(d) potential difference between the tip of the tube and that of the plant.

12. The metal ion involved in the stomatal regulation is:
(a) Iron (b) Magnesium
(c) Zinc (d) Potassium

13. Movement of H_2O through cell wall is called:
(a) Apoplast (b) Symplast
(c) Tonoplast (d) None of these

14. Passive absorption of water by the root system, is the result of:
(a) Forces created in the cells of the root
(b) Increased respiratory activity in root cells
(c) Tension on the cell sap due to transpiration
(d) Osmotic force in the shoot system

15. Study the following statement and choose the **correct** option(s) from the codes given below.
(i) Root pressure provides a light push in the overall process of water transport.
(ii) Root pressure causes the flow of water faster through xylem than it can be lost by transportation.
(iii) In symplast pathway, water move exclusively through the cell wall and intercellular spaces.
(iv) Guttation is a cause of transpiration pull.
(v) Most plants fulfill their water requirement by transpiration pull.
(a) (i), (ii) and (iii) are correct while iv and v are incorrect.
(b) (iv) and (v) are correct while i, ii and iii are incorrect.
(c) (i) and (v) are correct
(d) (ii) and (iii) are correct while i, iv and v are incorrect.

16. Which of the following statements is/are **incorrect**?
(i) In CAM plants, stomata open during dark and remain closed during the day.
(ii) Role of Na^+ in stomatal opening is now universally accepted.
(iii) The water potential of root cells is higher than the water potential of soil.
(iv) Capillarity theory is the most accepted theory of water movement through plants.
(v) The walls of xylem vessels made up of lignocellulose have strong affinity for water molecules.
(a) (ii), (iii) and (v) only
(b) (ii), (iii) and (iv) only
(c) (i), (ii) and (iii) only
(d) (ii) and (iii) only

17. Water will move from the root hairs through cortex if the water potentials are:
(a) Root hairs = 0; Cortex = 0; Xylem = 0
(b) Root hairs = 0; Cortex = – 1; Xylem = –2
(c) Root hairs = –2; Cortex = –1; Xylem = 0
(d) Root hairs = 0; Cortex = + 1; Xylem = + 2

18. Which of the atmospheric factor act as anti-transpirant? **Tricky**
(a) SO_2 (b) CO
(c) CO_2 (d) All pollutant gases

19. Adhesion is caused by:
(a) formation of hydrogen bond between water molecules.
(b) transpiration pull.
(c) higher surface tension.
(d) attraction of water molecule to polar surface.

20. When stomata open, the pH of guard cells:
(a) Increases (b) Decreases
(c) Remains same (d) Both (a) and (b)

21. Influx of K^+ ions of subsidiary cells into guard cells and efflux of H^+ ions from guard cells lead to:
(a) Closing of stomata
(b) Opening of stomata
(c) Exosmosis
(d) Plasmolysis

22. Which of the following will be zero when the cell becomes fully turgid?
(a) Turgor pressure
(b) Wall pressure
(c) Suction pressure
(d) Osmotic pressure

23. If the pressure potential is +0.16 megapascals (MPa) and the osmotic potential is –0.24 megapascals, then the water potential would be: **Tricky**
(a) + 0.4 MPa (b) +0.08 MPa
(c) – 0.08 MPa (d) + 0.16 MPa

24. Water is lost in liquid state in some plants through hydathodes. These hydathodes:
(a) always remain open.
(b) do not show any specificity in opening and closing.
(c) remain closed at night.
(d) remain closed during day.

25. Lenticels and hydathodes are small pores with one of the following common attribute:

(a) They allow exchange of gases.
(b) Their opening and closing is not regulated.
(c) They always remain closed.
(d) They are found on the same part of plants.

26. When stomata close, which of the following events does not occur? Toughnut

(a) Guard cells become flaccid.
(b) Sugar is converted to starch.
(c) O.P. of the guard cells decreases.
(d) Accumulation of O_2 takes place.

27. Based on Munch's pressure-flow hypothesis, which of the following conditions would increase the rate of translocation?

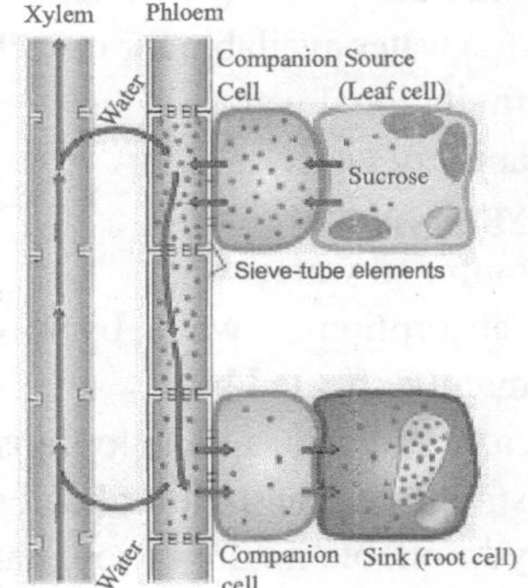

(a) An increase in the humidity in the outside air.
(b) A decrease in phloem unloading at the sink.
(c) An increase in sucrose production at the source.
(d) A decrease in photosynthesis.

28. Match the following columns.

Column-I	Column-II
A. Semipermeable membrane	I. Dixon and Jolly
B. Transpiration Pull	II. Manometer
C. Pulsation theory	III. Holard
D. Root pressure	IV. J. C. Bose
E. Whole wate content in soil	V. Osmosis
F. Cohesion and Adhesion theory	

(a) A – II; B – I; C – IV; D – V; E – III; F – I
(b) A – V; B – I; C – IV; D – II; E – III; F – I
(c) A – V; B – IV; C – I; D – II; E – III; F – I
(d) A – I; B – I; C – II; D – IV; E – III; F – V

29. What happens when concentration of solutes decreases in guard cells? Toughnut

(a) Water potential decreases
(b) Water potential increases
(c) Osmotic pressure increases
(d) None of these

30. Munch hypothesis is based on:

(a) translocation of food due to turgor pressure (TP) gradient.
(b) translocation of food due to imbibition force.
(c) translocation of food due to TP gradient and imbibition force.
(d) None of these

31. The translocation sugars in angiosperms occurs in the form of:

(a) Glucose (b) Sucrose
(c) Fructose (d) Lactose

32. Transport of different types of solute substances takes place by:

(a) Bulk flow system
(b) Combined response
(c) Facilitated diffusion
(d) Pressured transport

33. Excessive loss of water causes wilting of leaves, it can be prevented by: Critical Thinking

(a) keeping the plant in bright light.
(b) spraying the plant with alcohol.
(c) applying vaseline on the leaf surface.
(d) adding high amounts of fertilisers to the soil.

34. The spray of Phenyl Mercuric Acetate (PMA) causes: Tricky

(a) Decrease in transpiration
(b) Increase in transpiration
(c) Increase in absorption
(d) Increase in guttation

35. Mesophyll tissue of leaf is at water potential of –10.0 bars while epidermal cells of leaf have water potential of –8.0 bars and the air outside has water potential of –5.0 bars. Water will move from: Critical Thinking

(a) air → epidermis → mesophyll
(b) mesophyll → air → epidermis
(c) mesophyll → epidermis → air
(d) water will not move.

36. The lowest water potential is found in:
(a) Leaves (b) Xylem
(c) Root hair (d) Phloem

37. Water in the adjacent xylem moves into the phloem by the process of:
(a) Facilitated diffusion
(b) Acute transport
(c) Simple diffusion
(d) Osmosis

38. The graph shows the relationship between ψ_w (water potential), ψ_s (solute potential) and ψ_p (pressure potential) for a plant cell placed in pure water. Toughnut

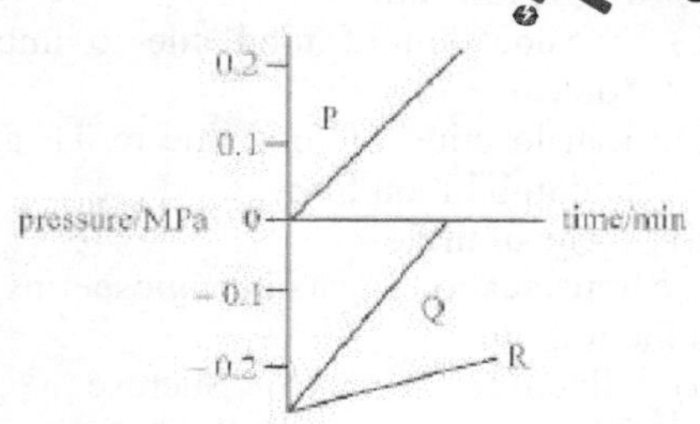

What are the correct labels for the graph?

	P	Q	R
(a)	ψ_w	ψ_p	ψ_s
(b)	ψ_w	ψ_s	ψ_p
(c)	ψ_p	ψ_w	ψ_s
(d)	ψ_p	ψ_s	ψ_w

39. Which one of the following graphs most closely represents the relationship between the rate of transpiration of a mesophytic leaf and the atmospheric humidity? Critical Thinking

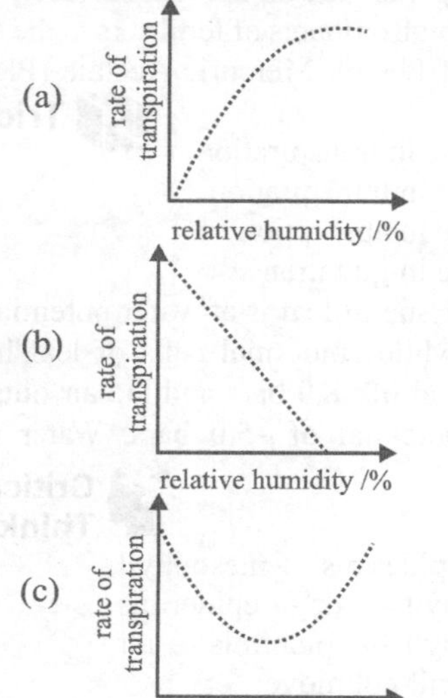

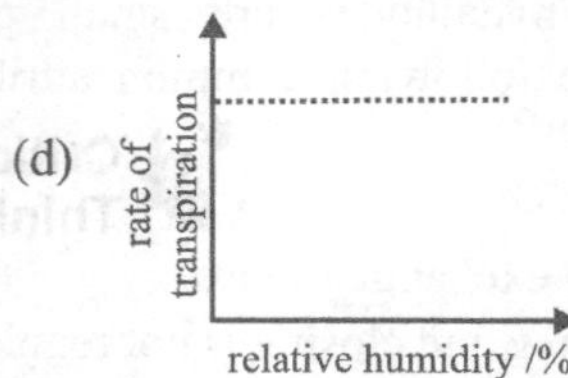

40. Which of the following theories of ascent of sap was proposed by an eminent Indian scientist J.C. Bose?
(a) Pulsation theory
(b) Relay pump theory
(c) Transpiration pull theory
(d) Root pressure theory

41. Oozing out of water drops from injured edges or tips is called:
(a) Oozation (b) Bleeding
(c) Guttation (d) Transpiration

42. In soil, the water available for root absorption is:
(a) Gravitational water
(b) Capillary water
(c) Hygroscopic water
(d) Combined water

43. Active absorption of water by roots from the soil is mainly affected by:
(a) Tension in cell sap due to transpiration
(b) Osmotic concentration of cell sap
(c) Hydrophobic nature of root hair
(d) Typical tissue organisation

44. Read the following statement and choose the **correct** one from the codes given below.
(i) The apoplastic movement of water takes place exclusively through intercellular spaces and cell wall without crossing any membrane.
(ii) Symplastic movement occurs from cell to cell through plasmodesmata, i.e., adjacent cells are connected through plasmodesmata.
(iii) Permeability of a membrane depends on its composition and chemical nature of the solute.
(iv) Solutes present in a cell increase the free energy of the water or water potential.
(a) (i), (ii) and (iii) (b) (i), (ii) and (iv)
(c) (ii) and (iv) (d) (i) and (iv)

45. Match column-I with column-II and select the correct answer using the codes given below.

Column-I	Column-II
A. Transpiration	I. Uses energy to pump molecules against a concentration gradient.
B. Guttation	II. Active absorption of water.
C. Exudation	III. Loss of water vapour from plant parts.
D. Active transport	IV. Loss of liquid water from leaves.
	V. Loss of water from injured plant parts.

(a) A–I; B–II; C–III; D–V
(b) A–II; B–I; C–IV; D–III
(c) A–III; B–IV; C–V; D–I
(d) A–IV; B–V; C–II; D–III

46. According to Steward's starch hydrolysis theory, which one of the following is the principal reason for the opening of stomata during daytime?
(a) Influx of K^+ ions into guard cells under the influence of ABA hormone.
(b) Conversion of sugar into starch in guard cells
(c) Efflux of K^+ ions from guard cells under the influence of ABA hormone.
(d) Photosynthetic utilisation of CO_2 in guard cells.

47. What will happen if stem of plant is cut under a state of tension in xylem sap? **Tricky**
(a) The xylem sap will flow out.
(b) Xylem sap will accumulate at cut surface.
(c) The cut surface will form air bubbles, when placed in water.
(d) Air will be pulled into the xylem.

48. According to the transpiration-cohesion theory, the upward pull of water is transferred to other water molecules by cohesion, which is caused by? **Critical Thinking**
(a) Hydrogen bond
(b) Turgor pressure
(c) Hydrophilic cell walls
(d) Osmosis

49. A group of students are studying the transport of certain type of molecules in a cell. They observe that the transport slows down when the cells are treated with poison, a chemical, which inhibits energy production. But when studying molecules, which are in normal conditions, it is estimated that molecules are probably transported by the process of? **Critical Thinking**
(a) osmosis
(b) active transport
(c) facilitated diffusion
(d) simple diffusion

50. Examples of bulk flow by a positive hydrostatic pressure gradient and a negative hydrostatic pressure gradient are: **Critical Thinking**
(a) suction through straw and swelling of wood, respectively.
(b) imbibition and a garden hose.
(c) garden hose and suction through a straw, respectively.
(d) swelling of wood and imbibition, respectively.

51. Which of the following is **incorrect** for the carrier molecules involved in facilitated diffusion? **Toughnut**
(a) They increase the speed of transport across a membrane.
(b) They concentrate solute molecules on one side of the membrane.
(c) They have specific binding sites for molecules they transport.
(d) They may undergo conformational change upon binding of solutes.

52. Carbohydrates are commonly found as starch in plant storage organs. Which of the following five properties of starch (i-v) make it useful as a storage material? **Critical Thinking**
(i) Easily translocated
(ii) Chemically non-reactive
(iii) Easily digested by animals
(iv) Osmotically inactive
(v) Synthesised during photosynthesis
(a) (ii) and (iii) (b) (ii) and (iv)
(c) (i), (iii) and (v) (d) (i) and (v)

53. Aquaporins are:
(a) cytoplasmic connections between cortex cells.
(b) openings in the lower epidermis of leaves through which water vapour escapes.
(c) openings into root hairs through which water enters.
(d) water specific channels in membranes that may regulate the rate of osmosis.

54. At the endodermis, water movement through the apoplast pathway is obstructed by which labelling?

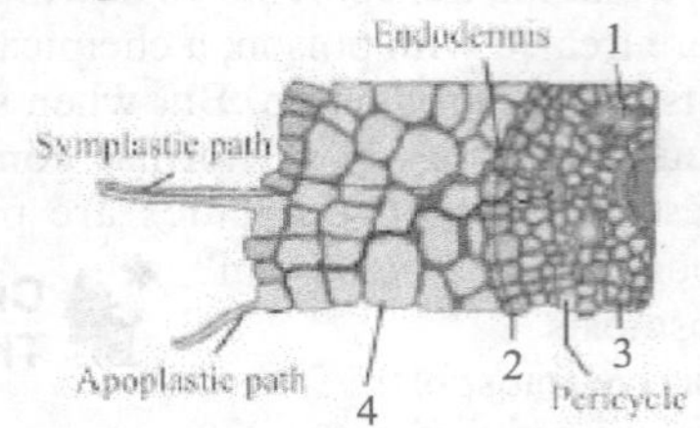

(a) 1 (b) 2
(c) 3 (d) 4

55. Read the following statements.
(i) Most minerals must enter the cytoplasm and epidermal cells of roots by active absorption.
(ii) Ions are absorbed by both passive and active transport.
(iii) Active absorption does not require energy.
(iv) Active uptake of ions is responsible for osmosis.

Choose the **correct** option in reference to the statements given above.
(a) (i) and (ii) are correct and (iii) and (iv) are incorrect
(b) (i), (ii) and iv are correct and (iii) is incorrect
(c) (i), (ii), (iii) and (iv)
(d) (i), (iii) and (ii) are correct (iv) is incorrect

56. Passive absorption of water by root system is the result of:
(a) tension on the sap due to transpiration.
(b) osmotic force in the shoot system.
(c) force created in the cells of the root.
(d) increased respiratory activity in root cells.

57. What mechanism explains the movement of sucrose from source to sink? Critical Thinking
(a) Evaporation of water and active transport of sucrose from the sink.
(b) Osmotic movement of water into the sucrose-loaded sieve-tube cells creating a higher hydrostatic pressure in the source than in the sink.
(c) Tension created by the differences in hydrostatic pressure in the source and sink.
(d) Active transport of sucrose through the sieve-tube members driven by proton pumps.

58. Starch is insoluble in water, yet it is accumulated in large quantities in potato tuber because:

Critical Thinking

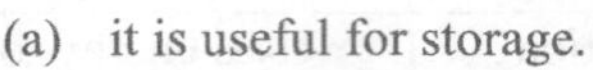
(a) it is useful for storage.
(b) starch is synthesised in tubers.
(c) tubers respire slowly.
(d) it is translocated in the form of sugar from leaves.

59. The translocation of organic solutes in sieve tube members is supported by: Tricky
(a) P-proteins.
(b) mass flow involving a carrier and ATP.
(c) cytoplasmic streaming.
(d) it is translocated in the form of sugar from leaves

60. What will be the effect of accumulation of K^+ ions in guard cells? Tricky
(a) Loss of turgidity
(b) Exosmosis
(c) Water potential increases
(d) Water potential decreases

61. Root pressure is higher when:
(a) Transpiration is very low and absorption is high.
(b) Transpiration is very high and absorption is high.
(c) Transpiration is low and absorption is low.
(d) Transpiration is high and absorption is low.

62. Opening of stomata is affected by:
(a) Light
(b) K^+ ions
(c) CO_2
(d) All of the above

63. Ganong's potometer is used to measure:
(a) Photosynthesis
(b) Respiration
(c) Transpiration
(d) Guttation

64. In terrestrial habitats, temperature and rainfall conditions are influenced by:
(a) Water transformations
(b) Transpiration
(c) Thermoperiodism
(d) Translocation

65. Leaves which appear wilted in the day time recover at night because: Critical Thinking

(a) Light is essential for photosynthesis.
(b) The stomata close down, temperature decrease, transpiration is reduced and the plant is able to absorb more water from the soil.
(c) Respiration and translocation of organic substance both increase.
(d) The plant is sleeping because of dark conditions.

66. Which of the following shows the correct explanation of the given figures?

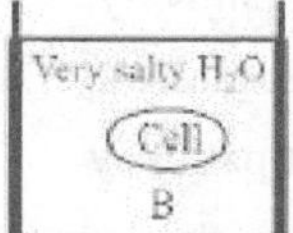

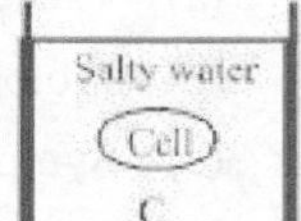

(a) Cell "A" will lose H_2O, Cell "B" will gain H_2O, Cell "C" neither gain nor loses H_2O.
(b) Cell "A" neither gain nor loses H_2O, Cell "B" will gain H_2O, Cell "C" will lose H_2O.
(c) Cell "A" will gain, Cell "B" neither gain nor loses H_2O, Cell "C" lose H_2O.
(d) Cell "A" will gain H_2O, Cell "B" will lose H_2O, Cell "C" neither gain nor loses H_2O.

67. Read the following statements regarding porins and select the **correct** option given below.

(i) Porins are transport proteins.
(ii) Channel proteins are usually gated and are a type of transport protein.
(iii) Carrier protein binds the particular solute to be transported.
(iv) Particular solute is delivered to the other side of the membrane by carrier proteins.

(a) (i), (ii) and (iii) (b) (i), (iii) and (iv)
(c) (i), (ii), (iii) and (iv) (d) (i) and (iv)

68. The osmotic potential and pressure potential of three cells (A, B, C) located in different parts of an actively transpiring plant are given below.

	Cell Osmotic Potential (MPa)	Pressure Potential (MPa)
A	– 0.87	0.44
B	– 0.92	0.34
C	– 0.68	0.27

Which of the cells (A, B, C) have the highest water potential?

(a) A (b) B
(c) C (d) All of these

69. Consider the following statements and choose the **correct** answer from the options given below.

(i) A dry a live seed still contains water.
(ii) A mature maize plant absorbs about 3L water per day.
(iii) A mustard plant take up water equal to its weight in about five hours.
(iv) Water is not considered as the limiting factor for plant growth and productivity.

(a) (i), (ii), (iii) and (iv)
(b) (iv) and (ii)
(c) (i), (ii) and (iii)
(d) Only (iv)

70. The potential energy of water is referred as:

(a) Water potential
(b) Osmotic potential
(c) Gravity potential
(d) Pressure potential.

71. In a fully turgid cell:

(a) ψ_s will be negative and ψ_p will be positive.
(b) ψ_p will be negative and ψ_s will be positive.
(c) Both ψ_p and ψ_s will be positive.
(d) Both ψ_s and ψ_p are negative.

72. Addition of solute in the cell develops:

(a) Turgor pressure
(b) Osmotic pressure
(c) Diffusion pressure
(d) Water potential

73. In rainy season, the doors get swell due to:

(a) Imbibition (b) Absorption
(c) Diffusion (d) Endosmosis

74. A negative effect of transpiration is:

(a) development of water stress.
(b) increase in mineral absorption.
(c) maintenance of leaf temperature.
(d) causing cooling.

75. Ringing experiment is related with:
 (a) Ascent of sap
 (b) Translocation of food
 (c) Both (a) and (b)
 (d) Mineral nutrition

76. Which of the following theory gives the latest explanation for the closure of stomata?
 (a) Munch theory
 (b) ABA theory
 (c) Starch glucose theory
 (d) Active K^+ transport theory

77. Root cap has no role in water absorption because:
 (a) it has no direct connection with the vascular system.
 (b) it has no cells containing chloroplasts.
 (c) it has no root hairs.
 (d) it has loosely arranged cells.

78. Passage cells are thin walled cells found in:
 (a) Endodermis of roots facilitating rapid transport of water from cortex to pericycle.
 (b) Phloem elements that serve as entry point for substances for transport to other plant parts.
 (c) Testa of seeds to enable emergence of growing embryonic axis during seed germination.
 (d) Central region of style through which the pollen tube grows towards the ovary.

79. Two cells A and B are contiguous. Cell A has osmotic pressure 10 atm, turgor pressure 7 atm and diffusion pressure deficit 3 atm. Cell B has osmotic pressure 8 atm, turgor pressure 3 atm and diffusion pressure deficit 5 atm. The result will be: Tricky
 (a) no movement of water.
 (b) equilibrium between the two.
 (c) movement of water from cell A to B.
 (d) movement of water from cell B to A.

80. Which among the following has highest water potential?
 (a) 1 m salt solution
 (b) 1 m glucose solution
 (c) Distilled water
 (d) None of the above

81. Which one of the following is a driving force for the process of passive absorption of water in roots?
 (a) The increase in imbibitional pressure in root cells
 (b) Root pressure
 (c) Activity of aquaporions
 (d) Transpiration in leaves

82. Which one of the pathways of water movement involves cell wall and intercellular spaces?
 (a) Vacuolar pathway
 (b) Protoplast pathway
 (c) Symplast pathway
 (d) Apoplast pathway

83. Stomatal opening is affected by: Tricky
 (a) nitrogen concentration, carbon dioxide concentration and light.
 (b) carbon dioxide concentration, temperature and light.
 (c) nitrogen concentration, light and temperature.
 (d) carbon dioxide concentration, nitrogen concentration and temperature.

84. Sunken stomata are usually found in:
 (a) Phanerograms
 (b) Insectivorous plants
 (c) C_3 plants
 (d) CAM plants

85. In both transpiration and evaporation, water is lost in the form of vapour yet they differ, because:

 (a) Both transpiration and evaporation are similar but the rate of water loss differs.
 (b) Frequency of water loss is different in both of them.
 (c) Transpiration is a physical process and evaporation is a physiological process.
 (d) Transpiration is a physiological process and evaporation is a physical process.

86. Identify the correct relationship with reference to water potential of a plant cell?

(a) $\psi_w = \psi_m - (\psi_s + \psi_p)$

(b) $\psi_w = \psi_m - \psi_s - \psi_p$

(c) $\psi_w = \psi_m + \psi_s + \psi_p$

(d) $\psi_w = \psi_m + (\psi_s - \psi_p)$

87. Read the following statements and choose the **correct** option given below:

(i) Surface/margin of leaves is a major area for transpiration.

(ii) Little amount of water is lost through stem, this is referred to cauline transpiration.

(iii) Transpiration is comparatively a slow process then evaporation.

(iv) Transpiration driven ascent of sap does not depend on cohesion, adhesion and surface tension properties of water

(a) (i), (ii), (iii) and (iv)

(b) (i), (iii) and (ii)

(c) (i), (ii) and (iii)

(d) (ii), (iii) and (iv)

88. Match the terms of column-I with the appropriate terms of column-II.

Column-I	Column-II
A. K^+ pump theory	I. Semi-permeable membrane
B. Plasmolysis	II. Hydathodes
C. Imbibition	III. Stomatal movement
D. Guttation	IV. Exosmosis
E. Osmosis	V. Hydrophilic colloids

(a) A – III; B – IV; C – V; D – II; E – I

(b) A – I; B – II; C – III; D – IV; E – V

(c) A – II; B – III; C – V; D – I; E – IV

(d) A – IV; B – V; C – II; D – III; E – I

89. Match the following columns.

Column-I	Column-II
A. Relay Pump Theory	I. Stocking
B. Transpiration Cohesion Theory	II. Sir J. C. Bose
C. Mass Flow Theory	III. Godlewski
D. Pulsation Theory	IV. Dixon and Jolly
	V. Ernst Munch

(a) A–III; B–II; C–V; D–I

(b) A–II; B–I; C–V; D–III

(c) A–III; B–IV; C–V; D–II

(d) A–IV; B–III; C–I; D–II

90. Identify the **correct** statements from the following.

(i) Accumulation of K^+ ions in the guard cells does not require energy.

(ii) A high pH favours stomatal opening.

(iii) Movement of chloride ion into guard cells as in response to the electric differential created by K^+ ions.

(iv) With entry of several K^+ ions and chloride ions, water potential of guard cell increases.

The correct combination is:

(a) (i) and (iii) (b) (i) and (ii)

(c) (ii) and (iii) (d) (iii) and (iv)

91. Transpiration ratio is the ratio of moles of H_2O transpired/moles of CO_2 fixed. This ratio is measure of: **Toughnut**

(a) The efficiency of guard cells on stomatal movement.

(b) Effectiveness of stomata is maximising photosynthesis while minimising water loss.

(c) Distinguishing a xerophyte from a glycophyte.

(d) Stomatal pore size of the leaves.

Answer KEYs

1.	(b)	**11.**	(b)	**21.**	(b)	**31.**	(b)	**41.**	(b)	**51.**	(b)	**61.**	(a)	**71.**	(a)	**81.**	(d)	**91.**	(b)
2.	(b)	**12.**	(d)	**22.**	(c)	**32.**	(a)	**42.**	(b)	**52.**	(b)	**62.**	(d)	**72.**	(b)	**82.**	(d)		
3.	(d)	**13.**	(a)	**23.**	(c)	**33.**	(c)	**43.**	(b)	**53.**	(d)	**63.**	(c)	**73.**	(a)	**83.**	(b)		
4.	(b)	**14.**	(c)	**24.**	(a)	**34.**	(a)	**44.**	(a)	**54.**	(b)	**64.**	(b)	**74.**	(a)	**84.**	(d)		
5.	(d)	**15.**	(c)	**25.**	(b)	**35.**	(a)	**45.**	(c)	**55.**	(b)	**65.**	(b)	**75.**	(c)	**85.**	(d)		
6.	(b)	**16.**	(b)	**26.**	(d)	**36.**	(c)	**46.**	(d)	**56.**	(a)	**66.**	(d)	**76.**	(d)	**86.**	(c)		
7.	(b)	**17.**	(b)	**27.**	(c)	**37.**	(d)	**47.**	(d)	**57.**	(b)	**67.**	(c)	**77.**	(c)	**87.**	(b)		
8.	(a)	**18.**	(c)	**28.**	(b)	**38.**	(c)	**48.**	(a)	**58.**	(d)	**68.**	(c)	**78.**	(a)	**88.**	(a)		
9.	(a)	**19.**	(d)	**29.**	(b)	**39.**	(b)	**49.**	(b)	**59.**	(a)	**69.**	(c)	**79.**	(c)	**89.**	(c)		
10.	(b)	**20.**	(a)	**30.**	(a)	**40.**	(a)	**50.**	(c)	**60.**	(d)	**70.**	(a)	**80.**	(c)	**90.**	(c)		

Hints & Solutions

1. **(b)** **2.** **(b)** **3.** **(d)** **4.** **(b)**
5. **(d)** **6.** **(b)** **7.** **(b)** **8.** **(a)**
9. **(a)** **10.** **(b)** **11.** **(b)** **12.** **(d)**
13. **(a)** **14.** **(c)**

15. **(c)** Root pressure, a manifestation of active water absorption is developed in xylem sap of roots of same plants. It maintains optimum metabolic activity and reduce transpiration and provide a light push in overall process of water transport because root pressure cannot transport water up to the whole length of plant. Movement of water is shown through xylem. In symplastic movements, movement of water occurs from cell to cell through their protoplasm, which are connected by a bridge called plasmodesmata.

16. **(b)** **17.** **(b)** **18.** **(c)**

19. **(d)** Movement of water inside the roots from soil to xylem and then in most of the plant parts takes place by transpiration forces, which provides both energy and necessary pull. Cohesion force is responsible to join the water molecule with one another in water column. While, force between tracheary wall and water molecule produces surface tension which account high capillarity through tracheary elements, which is called as adhesion force. These forces help to ensure the continuity of water column in xylem.

20. **(a)** **21.** **(b)**

22. **(c)** In a turgid cell, SP = O.P. – T.P. When O.P. = T.P. in turgid cell, then S.P. will be zero.

23. **(c)** The water potential is the sum of the osmotic potential (usually negative) and the pressure potential (usually positive), so W.P. = – 0.24 + 0.16 = –0.08 mPa.

24. **(a)** **25.** **(b)** **26.** **(d)** **27.** **(c)**
28. **(b)** **29.** **(b)** **30.** **(a)** **31.** **(b)**

32. **(a)** Transport of substances over longer distances through vascular tissue is termed as translocation and this translocation of different substances either inorganic minerals or organic substance (like sugar) occurs through a mass or bulk flow system. This mass flow of substance is unidirectional in case of water, while it is multidirectional in case of organic solute and minerals.

33. **(c)** **34.** **(a)**

35. **(a)** Water potential is highest in air & lowest in mesophyll.

36. **(c)**

37. **(d)** Water in adjacent xylem moves into the phloem by the process of osmosis (diffusion of a solvent through a semipermeable membrane from a less concentrated to a more concentrated solution until both the solutions are of the same concentration). As osmotic pressure builds the phloem sap, will move to areas of lower pressure.

38. **(c)** **39.** **(b)** **40.** **(a)** **41.** **(b)**

42. **(b)** **43.** **(b)**

44. **(a)** Addition of solutes in a system of cell decreases the energy of water. Pure water has the maximum diffusion pressure. Water potential or chemical potential of pure water is the difference in the free energy per unit molal volume of water in a system in reference to pure water at normal temperature and pressure.

45. **(c)** **46.** **(d)**

47. **(d)** If the stem of plant is cut under a state of tension in xylem sap, the air will be pulled into the xylem and the transport of water (xylem sap) remains in continuity. However, it can be discontinued with the introduction of air bubble in the xylem. Copeland (1902) believed that air bubbles enter into the xylem and break the tensile strength or cohesion force between the water molecule.

48. **(a)** Water is a polar molecule and forms hydrogen bonds between the positively charged hydrogen atoms and negatively charged oxygen atom. Hydrogen bonds made water molecules stick together, a process known as cohesion. When water molecules form hydrogen bonds with other molecules, such as carbohydrates, it is called adhesion. The hydrogen bonds have tension between them. So, water molecules stick together and move together.

49. **(b)** In active transport, the movement of solutes occur against the concentration gradient or chemical potential gradient with the expenditure of energy.

50. **(c)** A characteristic of mass flow is that the substances, whether in a solution or in a suspension, are swept along at the same pair, as in flowing river. Bulk flow can be achieved either through a positive hydrostatic pressure gradient (e.g., a garden hose) or a negative hydrostatic pressure gradient (e.g., suction through a straw).

51. **(b)** **52.** **(b)** **53.** **(d)** **54.** **(b)**

55. **(b)** **56.** **(a)** **57.** **(b)** **58.** **(d)**

59. **(a)** **60.** **(d)**

61. **(a)** Stephentiales (1727) coined the term root pressure, Root pressure is developed when rate of water absorption is more than the rate of transpiration.

62. **(d)**

63. **(c)** Ganong's potometer is used for measuring the rate of transpiration.

64. **(b)** **65.** **(b)** **66.** **(d)**

67. **(c)** Cell membranes possess certain pores, which are called porin. These are present in the outer membrane of mitochondria, plastids and in some bacteria. These large protein pores provide passage for small sized proteins. These are also called transport proteins, which can be divided into two types; carrier proteins and channel proteins. Carrier proteins bind to the particular substances, while channel proteins allow movement of an appropriate sized solute to be transported and this in turn is delivered to the outer side of the membrane by carrier proteins.

68. **(c)** Water potentials of the given cells are

$\psi = \psi_s + \psi_p$

$A = -0.87 + 0.44 = -0.43$

$B = -0.92 + 0.34 = -0.58$

$C = -0.68 + 0.27 = -0.41$

As water moves from greater water potential to less, root hair, root cortical and leaf mesophyll cells are *C, A, B* respectively.

69. **(c)** A dry alive seed contains around 10-15% water of its weight. Water is a major component of life which is generally absorbed by plants in variable quantity according to their metabolic requirement. A mature maize plant absorbs about 3 litres of water/day, while a mustard plant can absorb the water equal to its weight in only 5 hrs. Hence, water is considered the limiting factor for plant life, growth and development.

70. **(a)** **71.** **(a)** **72.** **(b)**

73. **(a)** Due to adsorption of water molecules into wooden furniture, it gets swelled.

74. **(a)**

75. **(c)** Ascent of sap takes place through xylem and translocation of food takes place through phloem. Both are proved by ringing experiment.

76. **(d)** **77.** **(c)** **78.** **(a)** **79.** **(c)**

80. **(c)** Distilled water has the highest water potential. The value of water potential is always negative or less than zero. Water potential of a solution is determined by using pure water as the standard of reference, which has zero water potential at normal temperature and pressure. The presence of solute particles reduces free energy of the water. Hence, it decreases the water potential in negative value. So, water potential of a solution is always less than zero.

81. **(d)** **82.** **(d)** **83.** **(b)** **84.** **(d)**

85. **(d)** Transpiration occurs in plants, only from living cells. Evaporation occurs from any free surface. *i.e.*, both from living and non-living surfaces.

86. **(c)**

87. **(b)** An account of 90% total transpiration occurs through leaves, i.e. foliar. Remaining 10% takes place through stem, flower and fruits etc. Cauline transpiration is the loss of water from stem. Ascent of sap in conducting tissues of plant is affected by cohesion, adhesion and properties of water.

88. **(a)** **89.** **(c)** **90.** **(c)** **91.** **(b)**

12 Mineral Nutrition

1. Sulphur is component of:
   (a) Protein (b) DNA
   (c) RNA (d) Both (b) and (c)
2. Which of the following element is very essential for uptake and utilisation of Ca^{2+} and membrane function?
   Critical Thinking
   (a) Phosphorus (b) Molybdenum
   (c) Manganese (d) Boron
3. Various fungi are known to accumulate considerable quantities of divalent metals. It includes:
   (a) Cd (b) Zn
   (c) Pb (d) All of these
4. Chlorosis is produced in leaves due to deficiency of Fe, Mg, Mn, N or S. Of these essential elements, those that are exclusive constituents of the chlorophyll molecule are:
   (a) Fe and S (b) N and S
   (c) Mg and S (d) Mg and N
5. If the size of fruits diminishes in plants, which mineral ion should be added to soil?
   (a) Copper (b) Calcium
   (c) Boron (d) Chlorine
6. Which of the following contains copper?
   (a) Quinone (b) Plastoquinone
   (c) Plastocyanin (d) None of these
7. All of the following statements concerning the Actinomycetous filamentous soil bacterium *Frankia* are correct except, the *Frankia*:
   Critical Thinking
   (a) can induce root nodules on many plant species.
   (b) cannot fix nitrogen in the free-living state.
   (c) like *Rhizobium*, it usually infects, its host plant through root hair deformation and stimulates cell proliferation in the host's cortex.
   (d) forms specialised vesicles in which the nitrogenase is protected from oxygen by a chemical barrier involving triterpene hopanoids.
8. The mode of catching insects in *Drosera* plants is by means of:
   (a) sensitive glandular hairs which secrete a sweet, viscous, shining substance.
   (b) specially sensitive trigger hairs.
   (c) leaves which are modified into pitcher.
   (d) leaf segments modified into bladder.
9. The component of nitrogenase and nitrate reductase is:
   (a) N
   (b) Mo
   (c) Co
   (d) No specific component
10. In the choices below, which is not a correct pairing of a macronutrient and the major functions it performs in the life of a plant?
    Critical Thinking
    (a) Potassium in enzyme activation, water balance, ion balance.
    (b) Calcium in activity of membranes and cytoskeleton, secondary messenger.
    (c) Sulphur in proteins and coenzymes.
    (d) Iron in active sites of many redox enzymes and electron carriers.
11. "Mottled chlorosis" on the leaves occurs due to the deficiency of:
    (a) Nitrogen (b) Potassium
    (c) Sulphur (d) Phosphorus

12. An essential element which:

(a) is found in plant ash.
(b) is available in the soil.
(c) improves health of the plant.
(d) is irreplaceable and indispensable for growth of plants.

13. Which of the following is essential mineral element and is not a constituent of any enzyme but stimulate the activity of many enzymes?
(a) Zn (b) Mg
(c) Mn (d) K

14. Molybdenum causes:
(a) Mottling (b) Reclamation
(c) Wilting (d) Chlorosis

15. Ratio of C : O in plant materials is:
:

(a) 45 : 45 (b) 10 : 20
(c) 25 : 50 (d) 50 : 25

16. Which one of the following is not an essential element for plants?
(a) Potassium (b) Iron
(c) Iodine (d) Zinc

17. Which of the following is/are not an essential micronutrient?
(a) Boron
(b) Nickel and cadmium
(c) Molybdenum
(d) Zinc

18. Appearance of brown spots surrounded by chlorotic veins is a prominent toxicity symptom of :

Critical Thinking

(a) Mn (b) Mo
(c) Mg (d) Zn

19. Which of the following is a flowering plant with nodules containing filamentous nitrogen-fixing microorganism?
(a) *Cicer arietinum*
(b) *Casuarina equisetifolia*
(c) *Crotalaria juncea*
(d) *Cycas revoluta*

20. Which of the following pigment is essential for nitrogen fixation by leguminous plants?
(a) Anthocyanin (b) Phycocyanin
(c) Phycoerythrin (d) Leghaemoglobin

21. Which one of these do, plants require for the formation of ATP?
(a) N and Ca (b) N and P
(c) N and Cu (d) K

22. Which element is required comparatively, in least quantity for the growth of plant?
(a) Zn (b) N
(c) P (d) Ca

23. Which element is not considered as macronutrient?
(a) Mg (b) Ca
(c) Mn (d) P

24. Which one of the following is an amide involved in nitrogen assimilation by plants?
(a) Glutamate (b) Alanine
(c) Asparagine (d) Serine

25. The most abundant element present in the plants is:
(a) Iron (b) Carbon
(c) Nitrogen (d) Manganese

26. Which of the following does NPK denote?

(a) Nitrogen, Potassium, Kinetin
(b) Nitrogen, Protein, Kinetin
(c) Nitrogen, Protein, Potassium
(d) Nitrogen, Phosphorus, Potassium

27. Insectivorous plants live in a soil that is usually deficient in:
(a) Nitrate (b) Chloride
(c) Potassium (d) Magnesium

28. Where does most nutrient uptake occur in roots?
(a) At the root tip, where root tissue first encounters soil away from the zone of nutrient depletion.
(b) At the casparian strip, where ions must enter the symplast prior to entering xylem cells.
(c) In the symplastic and apoplastic pathways.
(d) In root hairs and in the zone of maturation.

29. Gray spots of Oat are caused by deficiency of:

(a) Cu (b) Zn
(c) Mn (d) Fe

30. Deficiency of iron causes:
(a) interveinal chlorosis first on young leaves.
(b) decrease in protein synthesis.
(c) reduced leaves and stunted growth.
(d) bending of stem tip.

31. Which of the following will not make minerals more available to plants?

(a) Increasing the rainfall in a wet forested area.
(b) Raising the pH of a very acidic soil.
(c) Tillering a packed-down or water-logged soil.
(d) Introducing fungi that can form mycorrhiza into a soil that lacks them.

32. Phosphorous and nitrogen ions generally get depleted in soil because they usually occur as:

(a) Neutral ions
(b) Negatively charged ions
(c) Positively charged ions
(d) Both positively and negatively charged but in disproportionate mixture.

33. The following reaction points out the process of

$$\alpha\text{-ketoglutaric acid} + NH_4^+ + NADPH \xrightarrow[\text{dehydrogenase}]{\text{Glutamate}} \text{Glutamate} + H_2O + NADP$$

(a) Nitrification
(b) Transamination
(c) Amination
(d) Reductive amination

34. Function of leghaemoglobin (a red pigment) in root nodules of leguminous plants is to regulate:

(a) O_2 supply in cells
(b) CO_2 supply in cells
(c) Production of phenolic compounds
(d) Mo supply in cells

35. Essential macroelements are:
(a) absorbed from soil.
(b) manufactured during photosynthesis.
(c) produced by enzymes.
(d) produced by growth hormones.

36. Generally, plants absorb N_2 in the form of:
(a) NO_2^- (b) NO_3^-
(c) $N \equiv N$ (d) HNO_2

37. Which of the following characteristics defines an element as essential for a particular species?

(a) It has to be added as fertiliser to achieve maximum seed production.
(b) If it is missing, a plant cannot grow or reproduce normally.
(c) If it is present in high concentration, plant growth increases.
(d) If it is absent, other nutrients may be substituted for it.

38. Hydroponics is a technique in which plants are grown in:
(a) Green house
(b) Water saturated sand
(c) Balanced nutrient solution
(d) Purified distilled water

39. The plant ash is an indication of:
(a) Mineral salts absorbed by the plant
(b) Organic matter of the plant
(c) Both the mineral salts and organic matter
(d) None of the above

40. Which element is related with Khaira disease of paddy & auxin synthesis?
(a) Fe (b) Zn
(c) B (d) Cu

41. Plants absorb mineral salts from the soil solution through: Critical Thinking
(a) a semipermeable membrane into the cytoplasm.
(b) perforations at the apex of root hair cells.
(c) the cell wall which is semipermeable.
(d) None of the above

42. Active uptake of minerals depends upon:
(a) Active water absorption
(b) Transpiration
(c) Photorespiration
(d) Dephosphorylation

43. The diagram shows the development of root nodule in soyabean.

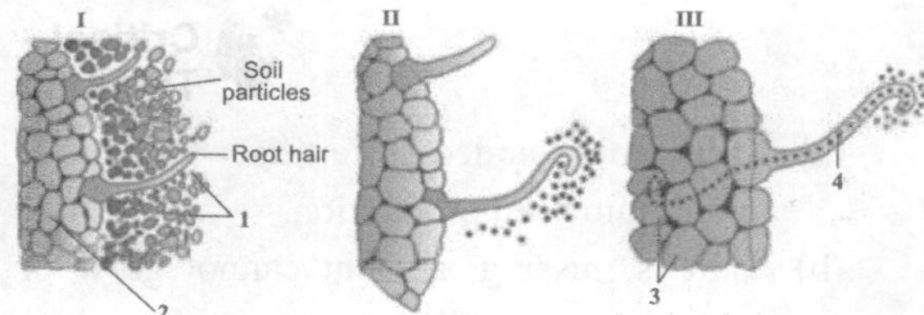

Identify 1, 2, 3 & 4.
(a) 1-*Rhizobium* bacteria; 2-Cortex cell; 3-Outer cortex; 4-Infection thread
(b) 1-*Rhizobium* bacteria; 2-Cortex cell; 3-Inner cortex and pericycle cells; 4-Infection thread
(c) 1-*Rhizobium* bacteria; 2-Endodermal cell; 3-Inner endodermis; 4-Infection thread
(d) 1-*Nitrosomonas* bacteria; 2-Cortex cell; 3-Inner cortex and pericyle cells; 4-Infection thread

44. Refer the figure given below and select the option which gives correct words for A, B, C and D.

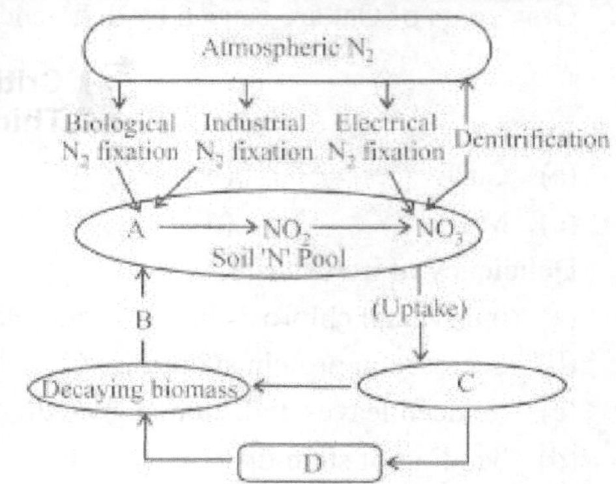

	A	B	C	D
(a)	K	Ammonification	Animal biomass	Plant biomass
(b)	NH_3	Ammonification	Plant biomass	Animal biomass
(c)	CO_2	Denitrification	Animal biomass	Plant biomass
(d)	CHO	Nitrification	Plant biomass	Animal biomass

45. Which of the following is a free living aerobic non-photosynthetic nitrogen-fixer?
(a) *Rhizobium* (b) *Azotobacter*
(c) *Azospirillum* (d) *Nostoc*

46. Which of the following statements are **incorrect** in accordance with hydroponics?
(i) It determines the mineral nutrients essential for plants.
(ii) The hydroponics involves the culture of plants in a soil with defined mineral solution.
(iii) Hydroponics requires purified water with non-defined mineral nutrient salts.
(iv) In the hydroponics technique, plants are grown in sandy soil with nutrient solution.
(v) By this method, essential elements required for the growth of plants can be identified and their deficiency symptoms can also be discovered.
(a) Only (iv) (b) Only (v)
(c) Only (iii) (d) None of the above

47. *Nif* genes occur in:
(a) *Rhizobium* (b) *Aspergillus*
(c) *Penicillium* (d) *Streptococcus*

48. Match the following and choose the correct combination from the options given.

	Column–I		Column–II
A.	Sulphur	I.	Chlorophyll
B.	Zinc	II.	Nitrogenase
C.	Magnesium	III.	Methionine
D.	Molybdenum	IV.	Auxin

(a) A – I; B – II; C – III; D – IV
(b) A – III; B – IV; C – I; D – II
(c) A – III; B – I; C – II; D – IV
(d) A – II; B – IV; C – I; D – III

49. Match the following and choose the correct combination from the options given.

	Column–I		Column–II
A.	Photolysis of water	I.	Zinc
B.	Diazotrophy	II.	Copper
C.	Cytochrome 'c' oxidase	III.	Manganese
D.	Biosynthesis of IAA	IV.	Molybdenum
		V.	Boron

(a) A – III, B – IV, C – II, D – I
(b) A – V, B – II, C – III, D – IV
(c) A – III, B – II, C – I, D – IV
(d) A – IV, B – I, C – III, D – II

50. Which of the following gene clusters in bacteria is responsible for nitrogen fixation?

(a) *nod, nif, fix* (b) *nod, nif, nfx*
(c) *nod, nix, nfx* (d) *ndx, nif, fix*

51. Mycorrhiza is a symbiotic relationship between roots of higher plants and:
(a) Virus (b) Fungi
(c) Bacteria (d) Blue green algae

52. Choose a **correct** statement in accordance with essential mineral elements of plants.
(a) Minerals present in the soil cannot enter plants.
(b) Gold is the only element which cannot be accumulated by plants.
(c) Plants growing near nuclear test sites take up radioactive strontium.
(d) Minerals present in very low concentration cannot be detected and hence, they remain undiscovered.

53. Consider the functions given below and identify the concerned element.
(i) Activator of catalase.
(ii) Important constituent of cytochrome.
(iii) Important constituent of proteins involved in ETS.
(iv) Essential for chlorophyll synthesis.
(a) Mo (b) Fe
(c) Cu (d) Ca

54. Match Column-I with Column - II and select the correct answer using the codes given below.

	Column-I		Column-II
A.	Potassium	I.	Constituent of ferredoxin
B.	Sulphur	II.	Involved in stomatal movement
C.	Molybdenum	III.	Needed in the synthesis of auxin
D.	Zinc	IV.	Component of nitrogenase

(a) A – II; B – I; C – IV; D – III
(b) A – I; B – II; C – III; D – IV
(c) A – IV; B – III; C – II; D – I
(d) A – I; B – III; C – IV; D – II

55. The enzyme responsible for the reduction of molecular nitrogen to the level of ammonia in leguminous root nodule is:
(a) Nitrogenase (b) Nitrite reductase
(c) Nitrate reductase (d) Hydrogenase

56. Which of the following is a bacterium involved in denitrification?
(a) *Nitrococcus* (b) *Azotobacter*
(c) *Pseudomonas* (d) *Nitrosomonas*

57. Which is the most abundant gas in our atmosphere that cannot be utilised by plants directly in its atmospheric form and is, therefore, captured by certain bacteria that live symbiotically in the nodules of roots?

Critical Thinking

(a) Oxygen (b) Nitrogen
(c) Neon (d) Hydrogen

58. Enzyme nitrogenase contains element:
(a) Potassium (b) Zinc
(c) Molybdenum (d) Manganese

59. Legume's roots have swellings called nodules that:

Critical Thinking

(a) produce antibiotics that protect the plant from soil bacteria.
(b) provide a steady supply of sugar to the host plant.
(c) increases the surface area for water uptake.
(d) contain nitrogen-fixing bacteria.

60. The ability of the Venus Fly trap to capture insects is due to:
(a) chemical stimulation by the prey.
(b) a passive process requiring no special ability on the part of the plant.
(c) specialised "muscle-like" cells.
(d) rapid turgor pressure changes.

61. Nitrogen gas is reduced to ammonia by:
(a) *Rhizobium* (b) Nitrogenase
(c) Nitrification (d) Denitrification

62. *Azotobacter* and *Beijerinckia* are the examples of:
(a) symbiotic nitrogen-fixers
(b) non-symbiotic nitrogen-fixers
(c) ammonifying bacteria
(d) disease causing bacteria

63. Non-biological nitrogen fixation is:
(a) Electrochemical and photochemical
(b) Rhizobial
(c) Cyanobacterial
(d) None of the above

64. Consider the statements and choose the **correct** options.
(i) The deficiency of any element can cause multiple symptoms.
(ii) Same symptoms may be caused by the deficiency of one or several different elements.
(iii) The concentration of the essential element below which plant growth is retarded is termed as critical concentration.
(iv) Chlorosis is the loss of chlorophyll due to deficiency of N, K, Mg, Fe, S, Mn, Zn, Mo.
(v) Different plants respond differently to the deficiency of the same element.
(a) All are the above
(b) (i) and (iv)
(c) (i) and (iii)
(d) (iii) and (iv)

65. An element must be considered essential when:
(i) the element is necessary for supporting normal growth and reproduction of the plants.
(ii) the deficiency of that particular element cannot be met by supplying some other element.
(iii) the element is directly involved in the metabolism of the plants.
Choose the **correct** statements.
(a) (i) and (iii)
(b) Only (ii)
(c) (ii) and (iii)
(d) (i), (ii) and (iii)

66. Which of the following statement represent hydroponics?
(i) is useful in areas having thin; infertile and dry soils.
(ii) can regulate pH optimum for a particular crop.
(iii) increases labour cost.
(iv) increases problem of weeding.
(a) (i), (ii) and (iii)
(b) (i) and (ii)
(c) (ii) and (iv)
(d) (i) and (iii)

67. Match column-I with column-II and select the correct answer using the codes given below.

Column-I (Element)	Column-II (Function)	Column-III (Deficiency Symptom)
A. Calcium	I. Required for ionic-balance.	i. Grey blot on leaves.
B. Boron	II. Essential for constitution of nucleic acid	ii. Fruit-yield decreases.
C. Phosphorus	III. Required for absorption of calcium.	iii. Red blots on leaves.
D. Chlorine	IV. Required to activate respiratory enzyme.	iv. Fruit-size diminishes.
E. Manganese	V. Required for synthesis of bipolar spindle	v. Young root tip begin to die.

(a) A – I-iv; B – II-v; C – III-iii; D – IV-i; E – V-ii
(b) A – V-v; B – IV-iv; C – III-i; D – II-iii; E – I-ii
(c) A – IV-iii; B – I-iv; C – V-v; D – III-ii; E – II-i
(d) A – V-v; B – III-iv; C – II-iii; D – I-ii; E – IV-i

68. Which of the following statements are **correct**?
(i) The deficiency of any element can cause multiple symptoms.
(ii) Same symptoms may be caused by the deficiency of one or several different elements.
(iii) The concentration of the essential element below which plant growth is retarded is termed as critical concentration.
(iv) Chlorosis is the loss of chlorophyll due to deficiency of N, K, Mg, Fe, S, Mn, Zn Mo.
(v) Different plants respond differently to the deficiency of the same element.

(a) (iii) and (iv) (b) (i) and (iv)
(c) (i) and (iii) (d) All of the above

69. Which of the following statement are **incorrect**?
(i) The morphological changes are indicative of certain element deficiencies and are called deficiency symptoms.
(ii) The part of plants that show the deficiency symptoms depend on the mobility of the element in the plant.
(iii) Deficiency symptoms appear first in the young tissues whenever the element are relatively mobile.
(iv) The deficiency symptoms of Cl, Zn, N, O, are visible first in the senescent leaves.

(a) (iii) and (iv) (b) (i) and (iii)
(c) (i) and (iv) (d) (ii) and (iv)

70. Which of the following statements about nitrification is correct?
(a) *Nitrobacter* oxidises nitrate to nitrite.
(b) *Nitrosomonas* and *Nitrosococcus* convert ammonium ions to nitrite.
(c) Nitrification reactions are energy consuming reactions.
(d) Heterotrophic plants are more directly dependent on the nitrifying bacteria for usable nitrogen than autotrophic plants.

71. Match the items given in column-I with their examples given in column-II and choose the correct answer.

Column-I	Column-II
A. Free living aerobic nitrogen fixers	I. *Anabaena* and *Nostoc*
B. Anaerobic nitrogen fixers	II. *Pseudomonas* and *Thiobacillus*
C. Nitrogen fixing cyanobacteria	III. *Nitrosomonas* and *Nitrococcus*
D. Denitrifying bacteria	IV. *Azotobacter* and *Beijernickia*
E. Nitrifying bacteria	V. *Rhodospirillum*

(a) A – IV; B – V; C – I; D – II; E – III
(b) A – V; B – IV; C – I; D – III; E – II
(c) A – IV; B – V; C – II; D – III; E – I
(d) A – IV; B – III; C – I; D – II; E – V

72. Find the incorrectly matched pair.
(a) *Rhizobium* → *Alfalfa*
(b) *Frankia* → *Alnus*
(c) *Rhodospirillum* → Aerobic
(d) *Bacillus* → Free -living

73. Leguminous plants are able to fix atmospheric nitrogen through the process of symbiotic nitrogen fixation. Which one of the following statements is **incorrect** during this process of nitrogen fixation?
(a) Leghaemoglobin scavanges oxygen and is pinkish in colour.
(b) Nitrogenase is insensitive to oxygen.
(c) Nodules act as sites for nitrogen fixation.
(d) The enzyme nitrogenase catalyses the conversion of atmospheric N_2 to NH_3.

74. Which of the following statement(s) is/are **correct**?
(i) Conversion of organic nitrogen to NH_4^+ by soil microbes is called ammonification.
(ii) Ammonia is first oxidised to nitrite by *Nitrosomonas* and *Nitrosococcus*.
(iii) The nitrite is further oxidised to nitrate with the help of the bacterium *Thiobacillus*.
(iv) In leaves, nitrate is reduced to form ammonia that finally forms the $-NH_2$ group of amino acids.
(v) *Nitrosomonas*, *Nitrosococcus* and *Nitrobacter* are chemoautotrophs.
(a) (i), (ii) and (iii)
(b) (ii), (iii), (iv) and (v)
(c) (i), (ii) and (iv)
(d) (i), (ii), (iv) and (v)

75. Plant obtains nitrogen from soil in the form of:
(a) Nitrite
(b) Nitrate
(c) Ammonia
(d) Hydroxylamine

Answer KEYs

1	(a)	9	(b)	17	(b)	25	(b)	33	(d)	41	(a)	49	(a)	57	(b)	65	(d)	73	(b)
2	(d)	10	(d)	18	(a)	26	(d)	34	(a)	42	(d)	50	(a)	58	(c)	66	(b)	74	(d)
3	(d)	11	(b)	19	(b)	27	(a)	35	(a)	43	(b)	51	(b)	59	(d)	67	(d)	75	(b)
4	(d)	12	(d)	20	(d)	28	(d)	36	(b)	44	(b)	52	(c)	60	(d)	68	(d)		
5	(c)	13	(d)	21	(b)	29	(c)	37	(b)	45	(c)	53	(b)	61	(b)	69	(a)		
6	(c)	14	(d)	22	(a)	30	(a)	38	(c)	46	(c)	54	(a)	62	(b)	70	(b)		
7	(b)	15	(a)	23	(c)	31	(a)	39	(a)	47	(a)	55	(a)	63	(a)	71	(a)		
8	(a)	16	(c)	24	(c)	32	(d)	40	(b)	48	(b)	56	(c)	64	(a)	72	(c)		

Hints & Solutions

1. (a) 2. (d) 3. (d) 4. (d)
5. (c) 6. (c) 7. (b) 8. (a)
9. (b) 10. (d) 11. (b) 12. (d)
13. (d) According to Webster (1953), potassium is an activator of more than 40 enzymes connected with phosphorylation, photosynthesis, starch synthesis, respiration, synthesis of chlorophyll, DNA and proteins.
14. (d) 15. (a) 16. (c) 17. (b)
18. (a) 19. (b) 20. (d) 21. (b)
22. (a) 23. (c) 24. (c) 25. (b)
26. (d) 27. (a) 28. (d) 29. (c)
30. (a) The characteristic symptom of iron deficiency is the interveinal chlorosis. The symptom may be general or localised to a single leaf or a single branch because of the limited mobility of iron in the tissues. The deficiency symptom first appears in young leaves.
31. (a) As in heavy rainfall, minerals are also drained along with water, hence it will not make minerals more available to plants.
32. (d)
33. (d) The organic acid α-ketoglutaric acid plays an important role in the synthesis of amino acid. The ammonia formed by nitrogen assimilation (i.e., reduction of nitrates), reacts with α-ketoglutaric acid to form an amino acid, i.e., glutamic acid. In this process, α-ketoglutaric acid comes from Krebs cycle and hydrogen is donated by the coenzyme NADH or NADPH. The reaction takes place in the presence of enzyme glutamic dehydrogenase.
34. (a) 35. (a) 36. (b) 37. (b)
38. (c)
39. (a) Plant ash contains only inorganic substances or minerals.
40. (b) 41. (a)
42. (d) Active uptake of minerals depends upon dephosphorylation.
43. (b) 44. (b) 45. (c)
46. (c)
47. (a) *Nif* gene is present in *Rhizobium*. It is responsible for the synthesis of enzyme nitrogenase which can fix atmospheric nitrogen. *Rhizobium leguminosarum* bacteria is found symbiotically associated in the root nodules of leguminous plant.
48. (b) 49. (a) 50. (a) 51. (b)
52. (c)
53. (b) Iron (Fe) is involved in the transfer of electrons like ferredoxin and cytochromes. During electron transfer, it is reversibly oxidised from Fe^{2+} to Fe^{3+}. It activates catalase enzymes and is essential for the formation of chlorophyll.
54. (a) 55. (a) 56. (c)
57. (b) Despite the fact that the atmosphere is 79% nitrogen gas, plants cannot use the element in that form and frequently suffer from nitrogen deficiency.
58. (c) 59. (d) 60. (d)
61. (b) Nitrogenase catalyses the reduction of nitrogen gas to ammonia. This is an energy-expensive process.
62. (b) 63. (a) 64. (a)
65. (d) The criteria for essentiality of an element are given below
    (i) The element must be absolutely necessary for supporting normal growth and reproduction. In the absence of the element, the plants do not complete their life cycle or set the seeds.
    (ii) The requirement of the element must be specific and not replaceable by any another element. In other words, deficiency of any one element cannot be met by supplying some other element.
    (iii) The element must be directly involved in the metabolism of the plant.
66. (b) 67. (d) 68. (d)

69. (a) The deficiency symptoms tend to appear first in the young tissues whenever the elements are relatively immobile and are not transported out the mature organs, for example, elements like sulphur and calcium are a part of the structural component of the cell and hence are not easily released. The deficiency symptoms of nitrogen, potassium and magnesium are visible first in the senescent leaves.

70. (b) Nitrification is the conversion of ammonia into nitrite and nitrates by soil bacteria (nitrifying bacteria). Heterotrophic plants are less dependent on nitrogen obtained from nitrification since they receive some nitrite and nitrate through their parasitic or carnivorous nutritional modes.

71. (a)

72. (c) *Rhodospirillum* is a free-living, anaerobic photoautotrophic nitrogen-fixing bacteria.

73. (b)

74. (d) The nitrite is further oxidised to nitrate with the help of the bacterium *Nitrobacter*.

75. (b)

13 Photosynthesis in Higher Plants

1. Bacteriochlorophyll differs from chlorophyll '*a*' in having: **Toughnut**
   (a) One pyrrole nucleus with one hydrogen
   (b) One pyrrole nucleus with two hydrogen
   (c) One pyrrole nucleus with three hydrogen
   (d) One pyrrole nucleus with four hydrogen
2. Which of the following pigment is yellow in colour?
   (a) Chlorophyll '*a*' (b) Chlorophyll '*b*'
   (c) Carotene (d) Xanthophyll
3. C_4 plants are adapted to:
   (a) Hot and dry climate
   (b) Temperate climate
   (c) Cold and dry climate
   (d) Hot and humid climate
4. Enzymes necessary for oxidative phosphorylation occur in:
   (a) Chloroplast (b) Ribosomes
   (c) Mitochondria (d) ER
5. Important requirement of phosphorylation is:
   (a) Release of electrons
   (b) Release of protons
   (c) Transfer of electrons
   (d) Activation of oxygen
6. Photosynthetically active radiation is represented by the range of wavelength of:
   (a) 340-450 nm (b) 400-700 nm
   (c) 500-600 nm (d) 400-950 nm
7. Plants give off oxygen because: **Tricky**
   (a) oxygen results from the incorporation of carbon dioxide into sugars.
   (b) plants do not respire since they photosynthesise.
   (c) water is the initial proton donor, leaving oxygen as a photosynthetic by-product.
   (d) All of the above
8. Study the figure showing graph of light intensity on the rate of photosynthesis. **Critical Thinking**

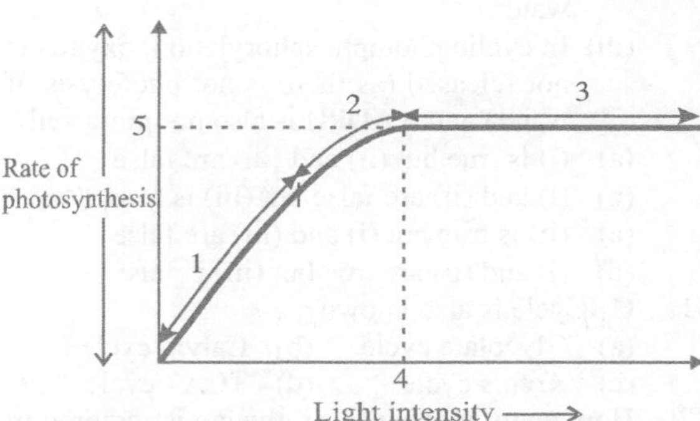

Column-I		Column-II
A. Limiting factor in region (1)	I.	Some factor other than light intensity is becoming the limiting factor
B. (2) represents to	II.	Light is no longer
C. (3) represents to	III.	Light intensity
D. (4) represents to	IV.	Maximum rate of photosynthesis
E. (5) represents to	V.	Saturation point for light intensity

   (a) A – I; B – II; C – III; D – IV; E – V
   (b) A – III; B – I; C – II; D – V; E – IV
   (c) A – IV; B – II; C – V; D – III; E – I
   (d) A – V; B – IV; C – III; D – II; E – I
9. Match column-I with column- II and select the correct answer using the codes given below.

Column-I	Column-II
A. Grana of chloroplast	I. Krebs cycle
B. Stroma of chloroplast	II. Light reaction
C. Cytoplasm	III. Dark reaction
D. Mitochondrial matrix	IV. Glycolysis

   (a) A – IV; B – III; C – II; D – I
   (b) A – I; B – II; C – IV; D – III
   (c) A – IV; B – I; C – III; D – II
   (d) A – II; B – III; C – IV; D – I

10. Consider the following statements. Choose the **correct** option.

(i) The portion of spectrum between 500 nm and 800 nm is also referred to as photosynthetically active radiation (PAR).

(ii) Magnesium, calcium and chloride ions play prominent roles in the photolysis of water.

(iii) In cyclic photophosphorylation, oxygen is not released (as there is not photolysis of water) and NADPH is also not produced.

(a) (i) is true but (ii) and (iii) are false
(b) (i) and (ii) are false but (iii) is true
(c) (ii) is true but (i) and (iii) are false
(d) (i) and (ii) are true but (iii) is false

11. C_2-Cycle is also known as:

(a) Glycolate cycle (b) Calvin cycle
(c) Kreb's cycle (d) TCA - cycle

12. How many molecules of glycine is required to release one CO_2 molecule in photorespiration?

(a) One (b) Two
(c) Three (d) Four

13. Which one of the following have high CO_2 compensation point?

(a) C_2 plants (b) C_3 plants
(c) C_4 plants (d) Alpine herbs

14. Cyclic and non-cyclic electron flow is used in plants to: **Toughnut**

(a) Meet the ATP demands of the Calvin Benson cycle.
(b) Avoid producing excess $NADPH+H^+$.
(c) Balance ATP and $NADPH+H^+$ ratios in the chloroplast.
(d) All of the above

15. Organism that does not evolve O_2 during photosynthesis is:

(a) *Funaria* (b) *Anabaena*
(c) *Rhodospirillum* (d) *Nostoc*

16. Which of the following best represents the components that are necessary for photosynthesis to take place? **Critical Thinking**

(a) Mitochondria, accessory pigments, visible light, water and carbon dioxide.
(b) Chloroplasts, accessory pigments, visible light, water and carbon dioxide.
(c) Mitochondria, chlorophyll, visible light, water and oxygen.
(d) Chloroplasts, chlorophyll, visible light, water and carbon dioxide.

17. What is/are the function(s) of accessory pigments?

(a) They enable a wider range of wavelength of incoming light to be utilised for photosynthesis.
(b) They absorb light and transfer the energy to reaction centre.
(c) They protect reaction centre from photo-oxidation.
(d) All of the above

18. Photosynthesis cannot continue for long if during light reaction, only cyclic photophosphorylation takes place. This is because:

(a) photosystem I stops getting excited at a wavelength of light beyond 680 nm.
(b) only ATP is formed, $NADPH + H^+$ is not formed.
(c) there is no evolution of O_2.
(d) there is unidirectional cyclic movement of the electrons.

19. The diagram below shows the movement of substances into in and out of a chloroplast with labelling 1, 2, 3 & 4. What do labels 1 to 4 represent?

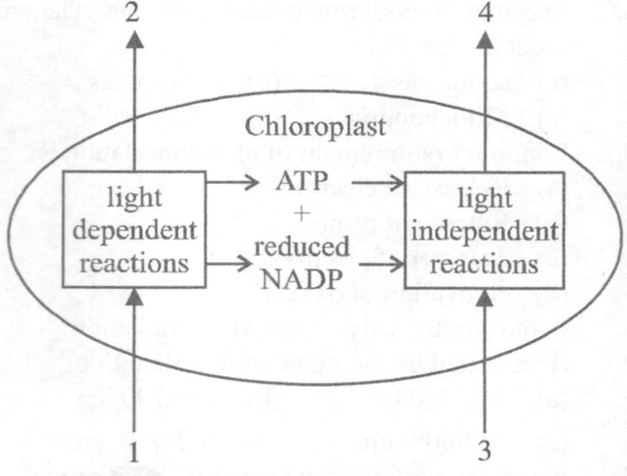

	1	2	3	4
(a)	CO_2	ATP	H_2O	Starch
(b)	CO_2	H_2O	Sugar	O_2
(c)	H_2O	O_2	CO_2	Sugar
(d)	Sugar	H_2O	ATP	O_2

20. The specific characteristic of C_4-plants is:

(a) Bulliform cells
(b) Isobilateral leaf
(c) Kranz anatomy
(d) Parallel veins configuration

21. Substrate for photorespiration is:
(a) Serine (b) Glycolate
(c) Indole acetic acid (d) Malic acid

22. Which is the evidence to show that O_2 released in photosynthesis comes from water?
(a) Isotopic O_2 supplied as H_2O appears in the O_2 released in photosynthesis.
(b) Isolated chloroplast in water releases O_2 if supplied potassium ferricyanide or some other reducing agent.
(c) Photosynthetic bacteria use H_2S and CO_2 to make carbohydrates.
(d) All of the above

23. The diagram below shows ATP synthesis through chemiosmosis.

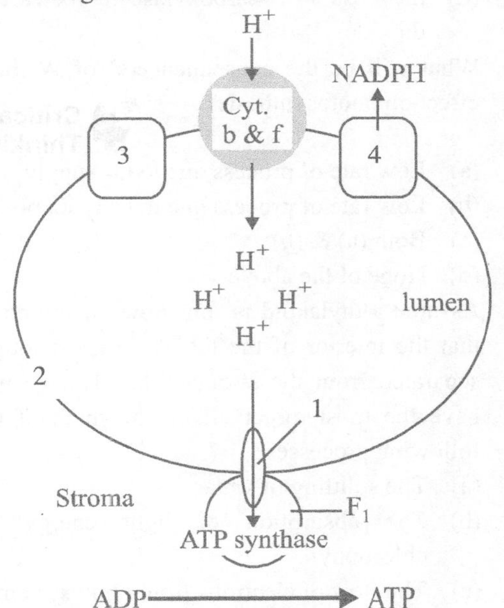

Which option shows the correct labelling of 1, 2, 3 and 4 in the diagram?
(a) 1 - F_1, 2 - Thylakoid membrane, 3 - Photosystem-I, 4 - Photosystem-II
(b) 1 - F_0, 2 - Thylakoid membrane, 3 - Photosystem-I, 4 - Photosystem-II
(c) 1 - F_1, 2 - Thylakoid membrane, 3 - Photosystem-II, 4 - Photosystem-I
(d) 1 - F_0, 2 - Thylakoid membrane, 3 - Photosystem-II, 4 - Photosystem-I

24. Which of the following statements are **correct**? **Tricky**
(i) Photorespiration is favoured by high O_2, low CO_2, rise in temperature, high light intensity.
(ii) Chloroplasts of cells of bundle sheath of C_4 plants are granal, have PEPCo PEPCase but no RuBisCo.
(iii) Chloroplasts of mesophyll cells of C_4 plant are granal, have PEPCo PEPCase but no RuBisCo.
(iv) Maize and Sorghum are C^4 plants.
(a) All are correct except (ii)
(b) (i) and (iv)
(c) (ii) and (iii)
(d) Only (iv)

25. Tropical plants like sugarcane show high efficiency of CO_2 fixation because of:
(a) Calvin cycle
(b) Hatch - Slack cycle
(c) Cyclic photophosphorylation
(d) TCA Cycle

26. Contribution of Ingen-Housz in elucidation of process of photosynthesis is that: **Tricky**
(a) only green parts of plants exposed to light can convert foul air (CO_2) into pure air (O_2).
(b) green plants convert light energy into chemical energy.
(c) plants have the capacity to purify foul air.
(d) sunlight is the ultimate source of energy for plants and animals.

27. The main site for dark reaction of photosynthesis is:
(a) Stroma (b) Grana
(c) Intergrana (d) Mitochondria

28. The scientist, who proved that bacteria use H_2S gas and CO_2 to synthesise carbohydrate, is:
(a) Van Niel
(b) Ruben
(c) Jean Senebier
(d) Julius Robert Mayer

29. Accumulation of food in assimilatory cells results in:
(a) Increase in the rate of photosynthesis
(b) Decrease in the rate of photosynthesis
(c) No effect
(d) May increase or decrease.

30. Maximum photosynthesis takes place in:
(a) Red light (b) Blue light
(c) Bright light (d) Green light

31. The form of pigment which inhibits germination is:
(a) P_{760} (b) P_{730}
(c) P_{650} (d) All of these

32. Which fractions of the visible spectrum of solar radiations are primarily absorbed by carotenoids of the higher plants?
(a) Violet and blue (b) Blue and green
(c) Green and red (d) Red and violet

33. NADP is converted into $NADPH_2$ in:
(a) Photosystem-I
(b) Non-cyclic photophorylation
(c) Calvin cycle
(d) Photosystem - II

34. The experiment material used by Van Neil, to prove that O_2 comes out from water was studied on:
(a) *Chlorella pyrenoidosa*
(b) *Cladophora*
(c) Purple & green sulphur bacteria
(d) Blue green algae

35. The graph below shows the result of an experiment in which isotopes of oxygen were used to distinguish between oxygen absorbed by plants and oxygen given out.
A mixture of the oxygen istopes $^{16}O_2$ and $^{18}O_2$ was supplied to a suspension of the unicellular alga *Chlorella* which had previously been exposed to $^{16}O_2$ only. During the following hour changes in the concentration of these gases in the suspension were measured.

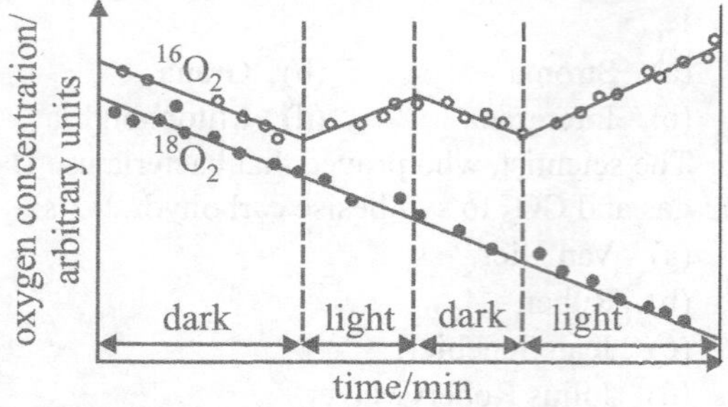

The $^{18}O_2$ concentration fell in light because it was:

Critical Thinking

(a) undergoing radioactive decay to from $^{16}O_2$.
(b) absorbed in respiration but was not being produced in photosynthesis.
(c) absorbed in respiration slower than it was being produced in photosynthesis.
(d) absorbed as $C^{18}O_2$ in photosynthesis.

36. The type of CO_2 fixation seen in many succulent plant species is:
(a) C_2-pathway (b) C_3-pathway
(c) C_4-pathway (d) CAM-pathway.

37. CAM plants differ from C_4 plants in that:
(a) carbon dioxide is stored as malic acid.
(b) photosynthesis can occur at night in these plants.
(c) their stomata close during periods that favour photorespiration.
(d) they use PEP carboxylase to fix carbon dioxide.

38. What will be the consequence(s) of Warburg effect on photosynthesis?

Critical Thinking

(a) Low rate of process due to O_2 supply
(b) Low rate of process due to CO_2 supply
(c) Both (a) & (b)
(d) None of the above

39. Assume a thylakoid is somehow punctured so that the interior of the thylakoid is no longer separated from the stroma. This damage will have the most direct effect on which of the following processes?
(a) The splitting of water.
(b) The absorption of light energy by chlorophyll.
(c) The flow of electrons from photosystem II to photosystem I.
(d) The synthesis of ATP.

40. CAM-plants avoid photorespiration by:
(a) performing Calvin cycle at night.
(b) fixing CO_2 into organic acid in night and releasing CO_2 in day.
(c) using PEP carboxylate to fix CO_2 to RuDP.
(d) keeping the stomata closed during the day.

41. Stomata of CAM plants:
(a) never open
(b) are always open
(c) open during the day and close at night
(d) open during the night and close during the day

42. Half leaf experiment proves that:

(a) light is essential for photosynthesis.

(b) CO_2 is essential for photosynthesis.

(c) O_2 releases during photosynthesis.

(d) chlorophyll is essential for photosynthesis.

43. Number of ATP molecules required for regeneration phase of RuBP during synthesis of 1 glucose molecule:

(a) 6 (b) 12

(c) 18 (d) 30

44. Formation of ATP in photosynthesis and respiration is an oxidation process which utilises the energy from:

(a) Cytochromes

(b) Ferredoxin

(c) Electrons

(d) Carbon dioxide

45. The rate of photosynthesis of a freshwater plant is measured using five spectral colours. Which sequence of colours would give an increasing photosynthetic response?

smallest → largest response

(a) Blue Green Yellow Orange Red

(b) Green Yellow Orange Red Blue

(c) Red Orange Yellow Green Blue

(d) Yellow Green Orange Blue Red

46. Which one of the following pigment is not present in the chloroplast?

(a) Carotene (b) Xanthophyll

(c) Chlorophyll 'b' (d) Anthocyanin

47. The wavelength of light absorbed by Pr form of phytochrome is:

(a) 680 nm (b) 720 nm

(c) 620 nm (d) 640 nm

48. To form one molecule of glyceraldehyde phosphate in Calvin cycle: **Critical Thinking**

(a) 9 ATP and 36 NADPH are required

(b) 6 ATP and 6 NADPH are required

(c) 3 ATP and 3 NADPH are required

(d) 9 ATP and 6 NADPH are required

49. Given below is the pathway of light reaction. Identify the given blanks indicated by 1, 2, 3 and 4.

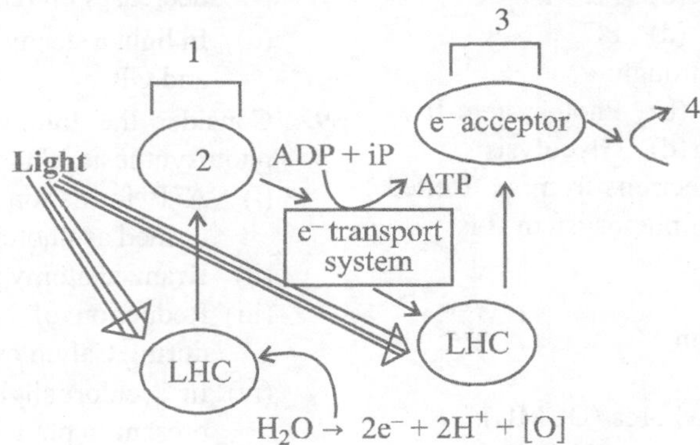

	1	2	3	4
(a)	P700	H^+ acceptor	P680	$NADP^+$
(b)	Photosystem I	e^- acceptor	Photosystem II	$NADPH_2^+$ ATP
(c)	Photosystem II	H^+ acceptor	P700	NADPH
(d)	Photosystem II	e^- acceptor	Photosystem I	$NADPH^+$ H^+

50. Two groups of isolated thylakoids are placed in an acidic bathing solution so that H^+ diffuses into the thylakoids. They are then transferred to a basic bathing solution, and one group is placed in the light, while the other group is kept in the dark.

Select the choice given below that describes what you expect each group of thylakoids to produce.

	In Light	In Dark
(a)	ATP only	Nothing
(b)	ATP, O_2	ATP only
(c)	ATP, O_2, Glucose	ATP, O_2
(d)	ATP, O_2	O_2

51. Final product of HSK cycle is formed in:
(a) Mesophyll cells (b) Bundle sheath cells
(c) Epidermal cells (d) Guard cells

52. The first step in photosynthesis is:
(a) Joining of three carbon atoms to form glucose.
(b) Formation of ATP.
(c) Ionisation of water.
(d) Excitement of an electron of chlorophyll by a photon of light.

53. During non-cyclic photophosphorylation, in which of the following, $4e^-$ produced through photolysis will enter? Toughnut
(a) PQ (b) PS – I
(c) PS – II (d) PC

54. NADPH is generated through:
(a) Photosystem-I (b) Photosystem-II
(c) Anerobic respiration (d) Glycolysis

55. The first acceptor of electrons from an excited chlorophyll molecule of photosystem II is:
(a) Quinone
(b) Cytochrome
(c) Iron-sulphur protein
(d) Ferredoxin

56. Dichlorophenyl dimethyl urea (DCMU): Tricky
(a) Inhibits PS-I
(b) Inhibits PS-II
(c) Destroy chloroplast
(d) Inhibits oxidative phosphorylation

57. Which of the following statements concerning the Calvin-Benson cycle are **incorrect**?
(a) Light energy is not required for the cycle to proceed.
(b) Carbon dioxide is assimilated into sugars.
(c) Ribulose 1, 5-bisphosphate (RuBP) is regenerated.
(d) None of the above

58. The graph below refers to an experiment involving species of alga. The relative concentrations of GP and RuBP present in the cells were monitored when the plants were in light and then in darkness Toughnut

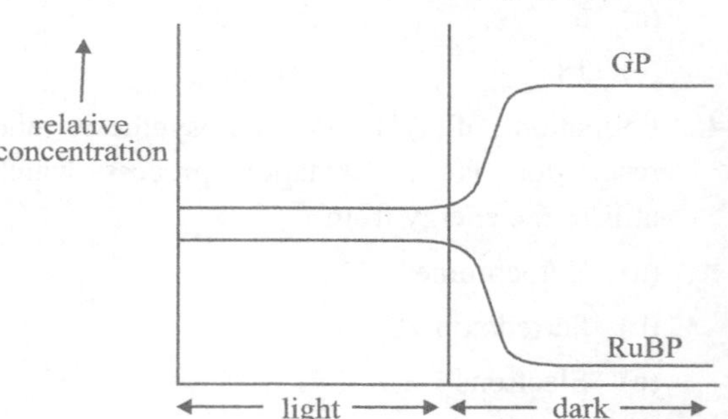

Which of the following conclusions cannot be drawn from these results?
(a) In darkness the relative concentration of GP increases.
(b) During the experiment RuBP may be converted into GP.
(c) The relative concentration of RuBP decreases on removal of CO_2.
(d) In light a steady state exists between RuBP and GP.

59. Consider the following statements regarding photosynthesis. Choose the **correct** option.
(i) ATP formation during photosynthesis is termed as photophosphorylation.
(ii) Kranz anatomy pertains to leaf.
(iii) Reduction of NADP to NADPH occurs during Calvin cycle.
(iv) In a chlorophyll molecule magnesium is present in phytol tail.
(a) (i) and (ii) (b) (iii) and (iv)
(c) (i) and (iii) (d) (i) and (iv)

60. The loss of water in C_4 plants compared to C_3 plants for the same amount of CO_2 fixed is:

(a) Half (b) One-third
(c) One-fourth (d) Double

61. Which of the following is likely to be the first substance that a green plant makes in photosynthesis?
(a) A simple sugar (b) Starch
(c) Fats (d) Proteins

62. Assimilatory power refers to:
(a) Generation of ATP and $NADPH_2$
(b) Reduction of CO_2
(c) Splitting of water
(d) Disintegration of plastids

63. Which of the following is used as an experimental material for making rapid advances in tracing path of carbon in Calvin cycle & other photosynthetic research?

(a) *Chlorella and Scenedesmus*
(b) *Chlamydomonas and Chlorella*
(c) *Hydrilla*
(d) Spinach

64. Of the total incident solar radiation the proportion of PAR is: Tricky
(a) About 70% (b) About 60%
(c) Less than 50% (d) More than 80%

65. Cyclic photophosphorylation results in the formation of:
(a) ATP and NADPH
(b) ATP, NADPH and O_2
(c) ATP
(d) NADPH

66. Pigment acting as a reaction centre during photosynthesis is:
(a) Carotene (b) Phytochrome
(c) P_{700} (d) Cytochrome

67. The diagram below shows the Calvin cycle. Identify 1, 2 and 3

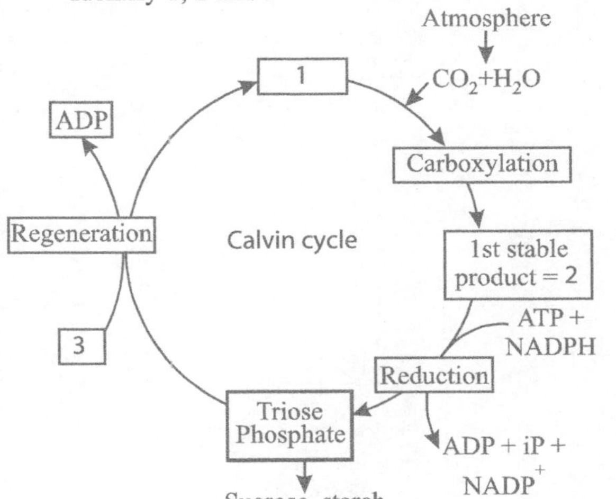

	1	2	3
(a)	ATP	3PGA	RuBP
(b)	3PGA	ATP	Sugar
(c)	RuBP	3-PGA	ATP
(d)	Sugar	RuBP	NADPH

68. Oxygenic photosynthesis occurs in: Critical Thinking
(a) *Oscillatoria*
(b) *Rhodospirillum*
(c) *Chlorobium*
(d) *Chromatium*

69. PGA as the first CO_2 fixation product was discovered in photosynthesis of:
(a) Bryophyte (b) Gymnosperm
(c) Angiosperm (d) Alga

70. 'Z-scheme' show the transfer of electrons involving:
(a) Bundle sheath cell
(b) PS I
(c) PS II
(d) Both (b) and (c)

71. In blue-green algae photosynthesis takes place in: Critical Thinking
(a) Chloroplasts (b) Chromatophore
(c) Heterocysts (d) Carotene

72. Which of the following element is a component of ferredoxin?
(a) Copper (b) Manganese
(c) Zinc (d) Iron

73. The family in which many plants are C_4 type:
(a) Malvaceae (b) Solanaceae
(c) Cruciferae (d) Graminae

74. As compared to a C_3 plant, how many additional molecules of ATP are needed for net production of one molecule of hexose sugar by C_4 plants ?

(a) Two (b) Six
(c) Twelve (d) Zero

75. Study the pathway given below. Which of the following option is correct?

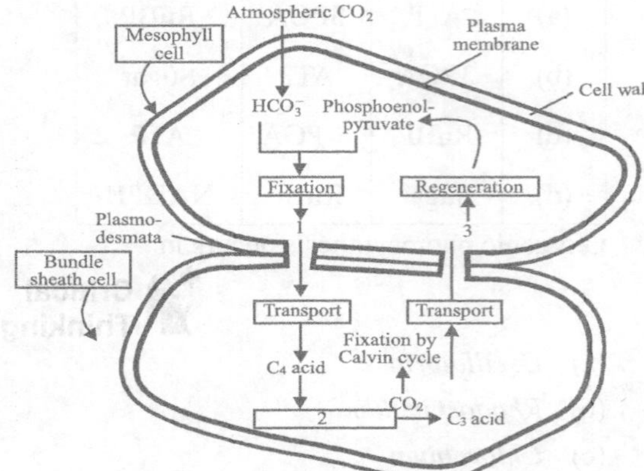

	1	2	3
(a)	C_3 acid	Reduction	C_4 acid
(b)	Fixation	C_4 acid	Regeneration
(c)	C_4 acid	Decarboxylation	C_3 acid
(d)	Carboxylation	C_3 acid	Reduction

76. Maximum CO_2 fixation is done by:

(a) Green plants
(b) Phytoplanktons
(c) Zooplanktons
(d) Bacteria

77. Which pigment system is inactivated in red drop?

(a) PS -I and PS -II (b) PS-I
(c) PS -II (d) None of these

78. Which of the following elements are essential for the photolysis of water?

(a) Ca and Cl (b) Mn and Cl
(c) Zn and I (d) Cu and Fe

79. In the leaves of C_4 plants, malic acid formation during CO_2 fixation occurs in the cells of:

(a) Epidermis (b) Mesophyll
(c) Bundle Sheath (d) Phloem

80. Algae and other submerged plants bounce in water during day time and sink at night, because:

Critical Thinking

(a) they come upto enjoy some time.
(b) they lose weight at night.
(c) they become buoyant due to accumulation of O_2 as a result of photosynthesis.
(d) they become light due to food material accumulation.

81. Energy transfer in photosystem occurs as:

Critical Thinking

(a) Phycoerythrin → Phycocyanin → Carotenoid → Chlorophyll *a*
(b) Chlorophyll *b* → Carotenoid → Phycoerythrin → Chlorophyll *a*
(c) Phycocyanin → Phycoerythrin → Carotenoid → Chlorophyll *a*
(d) Chlorophyll *b* → Carotenoid → Phycocyanin → Chlorophyll *a*

Answer KEYs

1	(b)	10	(b)	19	(c)	28	(a)	37	(a)	46	(d)	55	(a)	64	(c)	73	(d)
2	(d)	11	(a)	20	(c)	29	(b)	38	(a)	47	(a)	56	(b)	65	(c)	74	(c)
3	(a)	12	(b)	21	(b)	30	(a)	39	(d)	48	(d)	57	(a)	66	(c)	75	(c)
4	(c)	13	(b)	22	(d)	31	(b)	40	(b)	49	(d)	58	(c)	67	(c)	76	(b)
5	(c)	14	(d)	23	(d)	32	(a)	41	(d)	50	(b)	59	(a)	68	(a)	77	(c)
6	(b)	15	(b)	24	(a)	33	(b)	42	(b)	51	(b)	60	(b)	69	(d)	78	(b)
7	(c)	16	(d)	25	(b)	34	(c)	43	(a)	52	(c)	61	(a)	70	(d)	79	(b)
8	(b)	17	(d)	26	(a)	35	(b)	44	(c)	53	(d)	62	(a)	71	(b)	80	(c)
9	(d)	18	(b)	27	(a)	36	(d)	45	(b)	54	(a)	63	(a)	72	(d)	81	(a)

Hints & Solutions

1. **(b)**

2. **(d)** Xanthophylls are yellow coloured carotenoids also called xanthols or carotenols.

3. **(a)** **4.** **(c)** **5.** **(c)** **6.** **(b)**

7. **(c)** Water is split at photosystem–II to donate electrons to the reaction centre. The resulting protons are moved across the membrane to establish the proton motive force and oxygen is given off as a by-product.

8. **(b)** **9.** **(d)** **10.** **(b)** **11.** **(a)** **12.** **(b)**

13. **(b)**

14. **(d)** ATP is required at higher levels in the Calvin-Benson cycle than NADPH+H^+. Therefore, there must be a mechanism for producing additional ATP. Cyclic electron flow provides that mechanism. If non cyclic electron flow were to be speed up to meet ATP needs, an excess of NADPH+H^+ would result. Shifting between cyclic and non cyclic flow balances ATP/NADPH + H^+ ratios.

15. **(b)**

16. **(d)** Chloroplasts are the site of the photosynthetic reactions; chlorophyll is excited by photons of light and serve as reaction centers for the photosystems; visible light is necessary to excite chlorophyll and accessory pigments; water is the initial electron donor for the pathway and carbon dioxide is necessary to make energy storage precursor molecules.

17. **(d)** Carotenoids are accessory pigments and protect the chlorophyll from a lethal condition called photo-oxidation by a light of high intensity.

18. **(b)** **19.** **(c)** **20.** **(c)** **21.** **(b)** **22.** **(d)**

23. **(d)** **24.** **(a)** **25.** **(b)**

26. **(a)** Jan Ingenhousz showed that sunlight is essential to the plant process that somehow purifies the air fouled by burning candles or breathing animals. Ingenhousz in an experiment with an aquatic plant showed that in bright sunlight, small bubbles were formed around the green parts while in the dark they did not form any bubbles. Later he identified the bubbles as oxygen. Hence, he showed that it is only the green parts of the plants that could release oxygen.

27. **(a)**

28. (a) In 1930 C.B. Van Niel proved that, sulphur bacteria use H_2S (in place of water) and CO_2 to synthesise carbohydrates as follows :

$6CO_2 + 12H_2S \rightarrow C_6H_{12}O_6 + 6H_2O_2 + 12S$

This led Van Niel to the pastulation that in green plants, water (H_2O) is utilised in place of H_2S and O_2 is evolved in place of sulphur (S). He indicated that water is the electron donor in photosynthesis.

29. (b) 30. (a) 31. (b) 32. (a)

33. (b) Non cyclic photophosphorylation is involved both PS-I and PS-II. Flow of electrons is unidirectional. Here electrons are not cycled back and are used in the reduction of NADP to $NADPH_2$.

34. (c) By studying purple sulphur bacteria and green sulphur bacteria, Cornelius Van Niel was the first scientist to demonstrate that photosynthesis is a light-dependent redox reaction in 1931, in which hydrogen from an oxidisable compound reduces carbon dioxide to cellular materials. This can be expressed as: $2\ H_2A + CO_2 \rightarrow 2A + CH_2O + H_2O$, where A is the electron acceptor. His discovery predicted that H_2O is the hydrogen donor in green plant photosynthesis and is oxidised to O_2.

35. (b) The oxygen produced in photosynthesis is produced from the splitting of a water molecule by light energy, and not directly from the oxygen taken in during respiration.

36. (d) 37. (a) 38. (a)

39. (d) Damage to thylakoids have direct effect on the synthesis of ATP.

40. (b) 41. (d)

42. (b) Half leaf experiment proves that CO_2 is essential for photosynthesis.

43. (a) 44. (c)

45. (b) Photosynthesis occurs most efficiently with red or blue light because the chlorophyll pigments can absorb light maximally at these wavelengths and the least at green light.

46. (d) 47. (a) 48. (d) 49. (d) 50. (b)

51. (b) 52. (c) 53. (d) 54. (a)

55. (a) 56. (b)

57. (a) Light energy is required for the Calvin-Benson cycle to proceed. ATP synthesis is dependent on light energy and the Calvin-Benson cycle is dependent on ATP.

58. (c) 59. (a) 60. (b) 61. (a)

62. (a) Arnon (1965) used the term 'Assimilatory powers' for ATP and $NADPH_2$. ($NADPH_2$ alone is called reducing power).

63. (a) 64. (c) 65. (c) 66. (c) 67. (c)

68. (a) 69. (d) 70. (d)

71. (b) In blue-green algae photosynthesis takes place in chromatophores.

72. (d) 73. (d) 74. (c) 75. (c)

76. (b) 90% photosynthesis in world is carried out by algae (phytoplanktons).

77. (c) 78. (b)

79. (b) In the leaves of C_4 plants, malic acid formation during CO_2 fixation occurs in the cells of mesophyll.

80. (c) 81. (a)

14 Respiration in Plants

1. In plants energy is produced during the process of:
(a) Photosynthesis (b) Transpiration
(c) Respiration (d) Water absorption

2. How many ATP molecules are gained, when one molecule of glucose is degraded to pyruvic acid under anaerobic conditions? **Critical Thinking**
(a) 1 ATP is gain (b) 3 ATP is gain
(c) 2 ATP is gain (d) None of these

3. A eukaryotic cell can synthesise only 38 ATP instead of 40 ATP if the protons of cytosolic NADH+ H^+ are transported into mitochondria by: **Toughnut**
(a) Malate aspartate shuttle
(b) Glycerol phosphate shuttle
(c) ATPase
(d) Permease

4. Pyruvate dehydrogenase complex is used in converting:
(a) Pyruvate to glucose
(b) Glucose to pyruvate
(c) Pyruvic acid to lactic acid
(d) Pyruvate to acetyl Co-A

5. What is the final pathway followed by all carbon atoms derived from carbohydrates, lipids and proteins when they are oxidised during respiration?
(a) The Calvin cycle
(b) The Electron transport system
(c) The Kreb's cycle
(d) The Ornithine cycle

6. In which of the following reactions of glycolysis, a molecule of water is removed from the substrate?
(a) Fructose – 6 – phosphate → Fructose 1, 6 – bisphosphate
(b) 3 – phosphate glyceraldehyde → 1, 3 – bisphosphglyceric acid
(c) PEP → Pyruvic acid
(d) 2 – Phosphoglycerate → PEP

7. Hexose monophosphate shunt does not take place when:
(a) Oxygen is not enough
(b) Glucose is the substrate
(c) NAD is available
(d) NAD is not available due to shortage of oxygen

8. Which of the following molecules may be used as respiratory substrates?
(i) Fats (ii) Carbohydrates
(iii) Proteins (iv) Nucleic acid
(a) (i), (ii) and (iii) (b) (i) and (ii)
(c) (ii) and (iv) (d) (i) and (iii)

9. The graph shows the relation between light intensity and the giving off and taking up of carbon dioxide by the leaves of a plant. Why is most carbon dioxide given off when the light intensity is zero unit? **Critical Thinking**

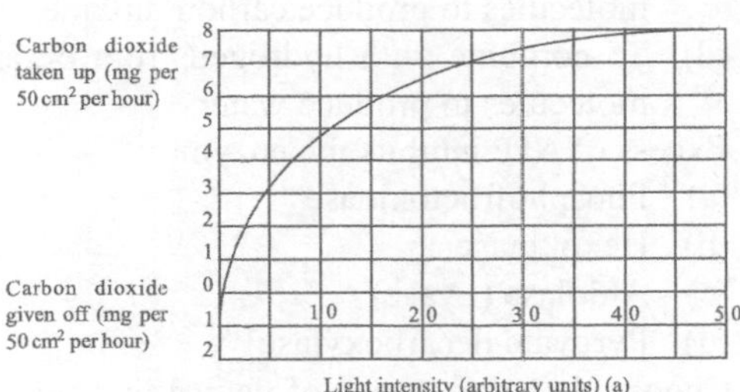

(a) Because it is just the start of the experiment.
(b) Only respiration is taking place at this intensity of light.
(c) Only photosynthesis is taking place at this intensity of light.
(d) The rate of photosynthesis is equivalent to the rate of respiration.

10. When proteins are the respiratory substrate, R.Q. will be:
(a) 0.7 (b) 1.0
(c) 0.9 (d) More than one

11. Which statement concerning ATP synthesis is **correct**?
(a) ATP can be synthesised through substrate level phosphorylation, photophosphorylation and oxidative phosphorylation.
(b) The proton-motive force is the establishment of proton gradients and electrochemical potentials across the inner membrane.
(c) Proton-motive force is essential for back flow of H^+ from outer chamber of matrix of mitochondria through proton channel (F_0) of $F_0 - F_1$ particle to produce ATP.
(d) All of the above

12. The first dicarboxylic acid in Krebs' cycle is:
(a) Isocitric acid (b) Pyruvic acid
(c) Oxalo acetic acid (d) α-ketoglutaric acid

13. Maximum usable energy per mole of glucose metabolised will be generated during:

(a) aerobic respiration by germinating seeds.
(b) production of methanol by enteric bacteria.
(c) fermentation into ethanol by yeast.
(d) glycolysis in the skeletal muscle of a sprinter performing a hundred metre dash

14. What is the function of molecular oxygen in cellular respiration? Tricky
(a) It causes the breakdown of citric acid.
(b) To combine with glucose to produce carbon dioxide.
(c) To combine with carbon from organic molecules to produce carbon dioxide.
(d) To combine with hydrogen from organic molecules to produce water.

15. Excess of ATP inhibits the enzyme:
(a) Phosphofructokinase
(b) Hexokinase
(c) Aldolase (Lyase)
(d) Pyruvate decarboxylase

16. Choose the correct pair of names that refer to one and the same process.
(a) Citric acid cycle and Calvin cycle
(b) Tricarboxylic acid cycle and urea cycle
(c) Kreb's cycle and Calvin cycle
(d) Tricarboxylic acid cycle and citric acid cycle

17. F_1 particle of oxysome:
(a) releases proton energy
(b) utilises proton energy
(c) has no role in energy consumption
(d) lies in outer chamber

18. Which of the following types of mammalian cell does not carry out oxidative phosphorylation? Critical Thinking
(a) Cardiac muscle cells
(b) Erythrocytes
(c) Neurons
(d) Oxyntic cells

19. During which stage in the complete oxidation of glucose are the highest number of energy deriving molecules (ATP) formed from ADP:
(a) Conversion of pyruvic acid to acetyl Co A
(b) Electron transport chain
(c) Glycolysis
(d) Kreb's cycle

20. In glycolysis, enyzme playing key role in splitting 6C compound into 3-C compound is:
(a) Hexokinase (b) Aldolase
(c) Isomerase (d) None of these

21. In some succulent plants like *Opuntia* the R.Q. is zero because:
(a) CO_2 is released without any absorption of O_2.
(b) O_2 is absorbed but CO_2 is not released.
(c) There is often compensation point in these plants due to thick phylloclades.
(d) None of the above

22. In Electron transport chain, $FADH_2$, is oxidised to FAD, the electrons are passed to cytochrome proteins whereas hydrogen ions are released into:
(a) Matrix
(b) Cytosol
(c) Outer membrane
(d) Perimitochondrial space

23. Electron transport chain is inhibited by:
(a) Rotenone and amytal
(b) Antimycin-A
(c) Cyanide (CN^-), azide (N_3^-) and carbon monoxide (CO)
(d) All of the above

24. Refer the figure showing major pathway of anaerobic respiration. Identify 1, 2, 3 and 4.

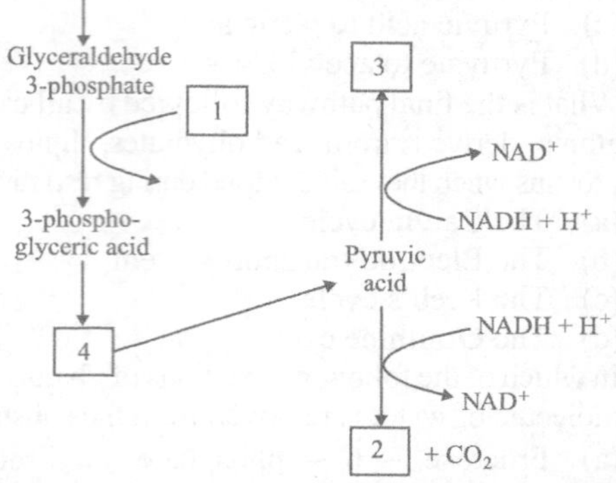

	1	2	3	4
(a)	NAD^+	Ethanol	Lactic acid	PEP
(b)	Ethanol	NAD^+	Lactic acid	ATP
(c)	Lactic acid	Ethanol	Glucose	ADP
(d)	NAD	Lactic acid	Ethanol	DHAP

25. Which of the following conversions represents a reduction reaction? **Tricky**
(a) Pyruvate → Acetyl CoA
(b) $C_6H_{12}O_6 \rightarrow 6CO_2$
(c) $NADH + H^+ \rightarrow NAD^+ + 2H$
(d) Acetaldehyde (C_2H_4O) → Ethanol (C_2H_6O)

26. Glycolysis consists of:
(a) ten enzymatic reactions that convert a six-carbon molecule to a four carbon pyruvate and result in a net gain of 4 ATP molecules.
(b) eight enzymatic reactions that convert a six-carbon molecule to a four carbon pyruvate and result in a net gain of 4 ATP molecules.
(c) ten enzymatic reactions that convert a six-carbon molecule to a three carbon pyruvate and result in a net gain of 2 ATP molecules.
(d) eight enzymatic reactions that convert a six-carbon molecule to a three carbon pyruvate and result in net gain of 2 ATP molecules.

27. During the early stages of alcoholic fermentation there is a high rate of growth of yeast. After some time the rate decreases. Which of the following conditions in the culture medium is least likely to have caused this?
(a) Depletion of glucose
(b) Depletion of oxygen
(c) Depletion of mineral salts
(d) Accumulation of waste products

28. The tissue of highest respiratory activity is:
(a) Meristems
(b) Ground tissue
(c) Phloem
(d) Mechanical tissue

29. Oxidative phosphorylation involves simultaneous oxidation and phosphorylation to finally form:

(a) Pyruvate (b) NADP
(c) DPN (d) ATP

30. In an electron transport chain in terminal oxidation the cytochrome which donates electrons to O_2 is:
(a) Cytochrome *b* (b) Cytochrome *c*
(c) Cytochrome a_3 (d) Cytochrome *a*

31. R.Q. in anaerobic respiration is:
(a) 0 (b) ∞
(c) 1 (d) 1

32. Two enzymes common to EMP pathway and C_3 cycle are: **Critical Thinking**
(a) Aldolase and triose phosphate isomerase
(b) Aldolase and enolase
(c) Cytochrome oxidase and enolase
(d) Phosphoglycerate mutase and triose phosphate isomerase

33. Protein is used as respiratory substrate only when:
(a) Carbohydrates are absent
(b) Fats are absent
(c) Both carbohydrates and fats are exhausted
(d) Fats and carbohydrates are abundant

34. Which of the following is coenzyme-II?
(a) NAD (b) NADP
(c) FAD (d) None of these

35. The Electron transport system in cells is coupled directly with the: **Tricky**
(a) Hydrolysis of hexose phosphates
(b) Polymerisation of hexoses
(c) Production of pyruvate
(d) Synthesis of ATP

36. Which of the following is link between carbohydrate and fat metabolism?
(a) CO_2 (b) Acetyl Co-A
(c) Pyruvic acid (d) Citric acid

37. Which is formed in oxidative phosphorylation?
(a) ATP in respiration
(b) ATP in photosynthesis
(c) NADH in respiration
(d) NADH in photosynthesis

38. Choose the correct combination of labelling of the molecules involved in the pathway of anaerobic respiration, in yeast:

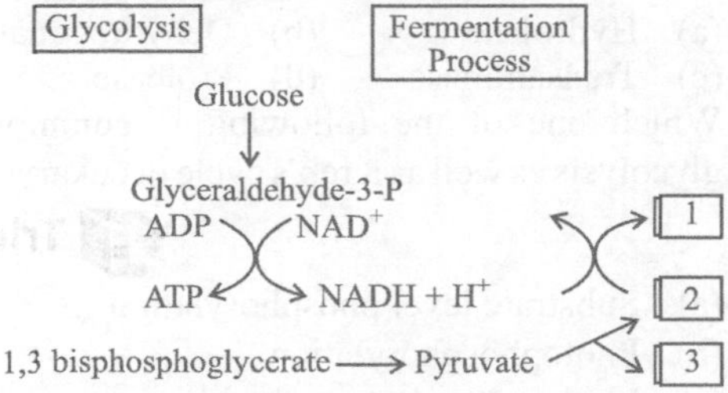

(a) 1-Ethanol, 2-CO_2, 3-Acetaldehyde
(b) 1-Ethanol, 2- Acetaldehyde, 3-CO_2
(c) 1-CO_2, 2-Ethanol, 3- Acetaldehyde
(d) 1-CO_2, 2-Acetaldehyde, 3- Ethanol

39. R.Q. of fatty substances is generally:
(a) Unity (b) Less than one
(c) Greater than one (d) Zero

40. Which of the following is used as a cellular respiration indicator? *Critical Thinking*
(a) Tetrazolium chloride
(b) Ethanol
(c) Schiff's reagent
(d) Lactic acid

41. The activity of succinate dehydrogenase is inhibited by:
(a) Pyruvate (b) Glycolate
(c) Malonate (d) Phosphoglycerate

42. How many moles of carbon dioxide are produced by the complete oxidation of 1 mole of pyruvate?

(a) 1 (b) 2 (c) 3 (d) 4

43. Why is glycolysis considered one of the first metabolic pathways to have evolved?
(a) It relies on fermentation, which is characteristic of the archaebacteria and eubacteria.
(b) It produces much less ATP than does the electron transport chain and chemiosmosis.
(c) It relies totally on enzymes that are produced by free ribosomes, and bacteria have only free ribosomes and no bound ribosomes.
(d) It is nearly universal, is located in the cytosol, and does not involve O_2.

44. During the formation of bread, it becomes porous due to release of CO_2 by the action of:
(a) Yeast (b) Bacterial
(c) Virus (d) Protozoans

45. Respiration is an:
(a) Endothermic process
(b) Exothermic process
(c) Anabolic process
(d) Endergonic process

46. Enzyme concerned with transfer of electrons are:
(a) Hydrolase (b) Dehydrogenase
(c) Transaminase (d) Protease

47. Which one of the following is common to glycolysis as well as Kreb's cycle in eukaryotes?

(a) Substrate level phosphorylation
(b) Photophosphorylation
(c) Localisation in mitochondria
(d) Production of $FADH_2$

48. Enzyme cytochrome oxidase can be inhibited by: *Tricky*
(a) Iodoacetate
(b) Azides & Cyanides
(c) Oligomycins
(d) Dinitrophenol

49. If R. Q. is less than 1.0 in a respiratory metabolism, it would mean that: *Critical Thinking*
(a) Carbohydrates are used as respiratory substrate.
(b) Organic acids are used as respiratory substrate.
(c) The oxidation of the respiratory substrate consumed more oxygen than the amount of CO_2 released.
(d) The oxidation of the respiratory substrate consumed less oxygen than the amount of CO_2 released.

50. Which group of the following scientists discovered the EMP pathway of glycolysis?
(a) Embden, Meyerhof and Parnas
(b) Emerson, Hoffman and Peterson
(c) Embden, Morrison and Pitcher
(d) Avery, MacLeod and McCarthy

51. The diagram below refers to mitochondrial electron transport chain. Identify P, Q, R, S, T. *Tricky*

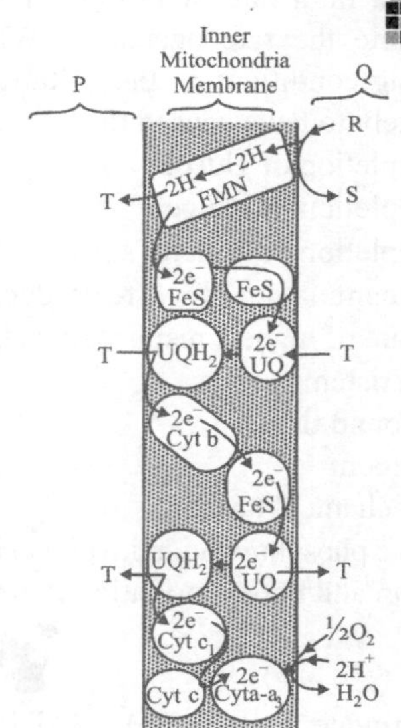

(a) P-Matrix; Q-Outer membrane R-$FMNH_2$; S-NAD_2; T-2H
(b) P-Outerchamber; Q-Matrix; R-$NADH+H^+$; S-NAD^+; T-$2H^+$
(c) P-Outer membrane; Q-Cristae; R-NAD^+; S-NAD^+; T-$2H^+$
(d) P-Cristae; Q-Outerchamber; R-$NADH+H^+$; S-NAD^+; T-$2H^+$

52. ATP:
(i) has high energy containing first phosphate bond.
(ii) is produced by oxidative phosphorylation and photophosphorylation.
(iii) acts as energy mediator in the cell.
(iv) has low energy containing second and third phosphate bond.
(a) (i), (ii) and (iii) (b) (i) and (ii)
(c) (ii) and (iv) (d) (i) and (iii)

53. Match Column-I with Column-II and select the correct answer using the codes given below.

Column-I	Column-II
A. 4C Compound	I. Acetyl CoA
B. 2C Compound	II. Pyruvate
C. 5C Compound	III. Citric acid
D. 3C Compound	IV. α-ketoglutaric acid
	V. Malic acid

(a) A–II; B–V; C–III; D–I
(b) A–V; B–I; C–IV; D–II
(c) A–III; B–I; C–IV; D–II
(d) A–V; B–III; C–I; D–II

54. In the process of respiration in plants 180 gms of sugar +192 gms of oxygen produce:

(a) large amount of CO_2, no water and no energy.
(b) 132 gms of CO_2, 54 gms of water and 337 kcals of energy.
(c) 264 gms of CO_2, 108 gms of water and 674 kcals of energy.
(d) 528 gms of CO_2, 216 gms of water and 1348 kcals of energy.

55. In mitochondrial electron transport system, for every two pairs of electrons that pass from NADH molecules through a sequential series of cytochrome enzymes to molecular oxygen generate: Tricky

(a) 3 ATP (b) 4 ATP (c) 6 ATP (d) 2 ATP

56. Oxidative phosphorylation refers to:
(a) Anaerobic production of ATP
(b) The citric acid cycle production of ATP
(c) Production of ATP by chemiosmosis
(d) Alcoholic fermentation

57. Connecting link between glycolysis and Krebs cycle is:
(a) Oxaloacetate (b) PEP
(c) Pyruvate (d) Acetyl CoA

58. Fermentation is essentially glycolysis plus an extra step in which pyruvic acid is reduced to form lactic acid or alcohol and CO_2. This last step: Toughnut

(a) removes poisonous oxygen from the environment.
(b) extracts a bit more energy from glucose.
(c) enables the cell to recycle NAD^+.
(d) inactivates toxic pyruvic acid.

59. Enzymes related with cristae are related with:
(a) Anaerobic respiration
(b) Aerobic respiration
(c) CO_2 formation
(d) Reduction of pyruvic acid

60. Sugars are not as good as fats as a source of energy for cellular respiration, because sugars:

(a) produce toxic amino groups when broken down.
(b) contain more hydrogen.
(c) usually bypass glycolysis and the Krebs cycle.
(d) contain fewer hydrogen atoms and electrons.

61. Aerobic respiratory pathway is appropriately termed:
(a) Parabolic (b) Amphibolic
(c) Anabolic (d) Catabolic

62. Number of oxygen atoms required for aerobic oxidation of pyruvate: Tricky

(a) 6 (b) 8
(c) 10 (d) 12

63. The energy-releasing process in which the substrate is oxidised without an external electron acceptor is called: Toughnut

(a) Fermentation (b) Photorespiration
(c) Aerobic respiration (d) Glycolysis

64. In glycolysis, during oxidation electrons are removed by:
(a) ATP
(b) Glyceraldehyde-3-phosphate
(c) NAD^+
(d) Molecular oxygen

65. Heme protein complexes that act as oxidising agents during plant respiration are known as:

(a) Chlorophyll (b) Haemoglobin
(c) Myoglobin (d) Cytochrome

66. If a starved plant is provided with glucose, the rate of respiration would: **Critical Thinking**
(a) Decrease
(b) Increase
(c) Become constant
(d) First rise and then fall

67. The common respiratory substrate is:
(a) Glucose (b) Fructose
(c) Sucrose (d) Fat

68. How many ATP molecules produced by aerobic oxidation of one molecule of glucose?

(a) 32 (b) 30 (c) 38 (d) 34

69. Kreb's cycle was discovered by Kreb in pigeon muscles in 1940. Which step is called gateway step/link reaction/transition reaction in respiration? **Critical Thinking**
(a) Glycolysis
(b) Formation of acetyl-coA
(c) Citric acid formation
(d) ETS terminal oxidation

70. $2(C_{51}H_{98}O_6) + 145O_2 \rightarrow 100CO_2 + 98H_2O$ + Energy
The R.Q. of above reaction is:
(a) 1 (b) 0.7 (c) 1.45 (d) 1.62

71. R.Q. value is less than 1 in:
(a) Lipids (b) Organic acids
(c) Carbohydrates (d) None of these

72. The diagram shows some of the reactions following glycolysis during aerobic respiration.

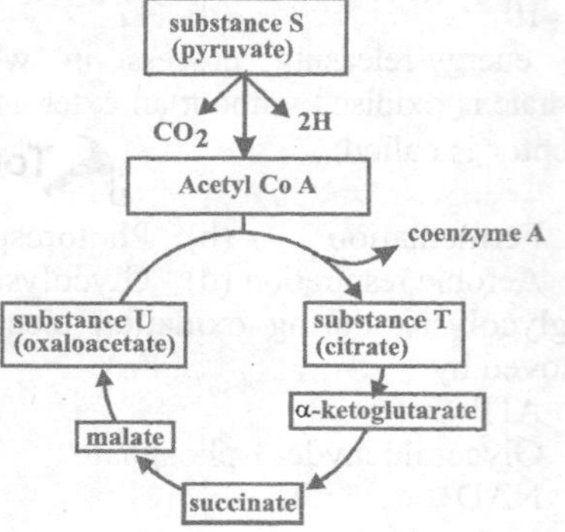

How many carbon atoms are in each of the substances S, T and U?

	S	T	U
(a)	2	5	3
(b)	2	6	4
(c)	3	4	6
(d)	3	6	4

73. Which graph represents the relationship between the relative energy intake of a person and temperature over a range of environmental temperatures from –40°C to +38°C?

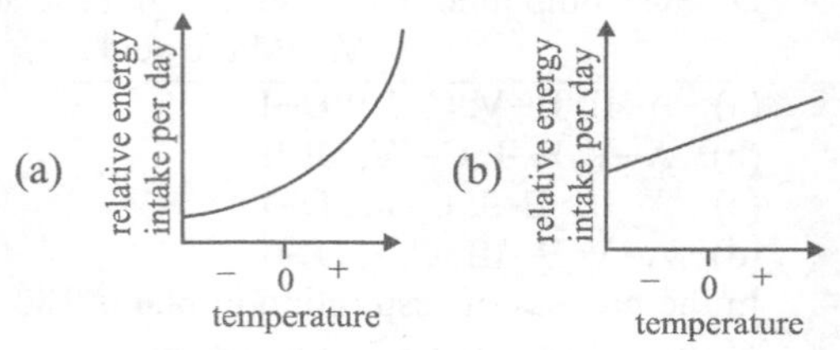

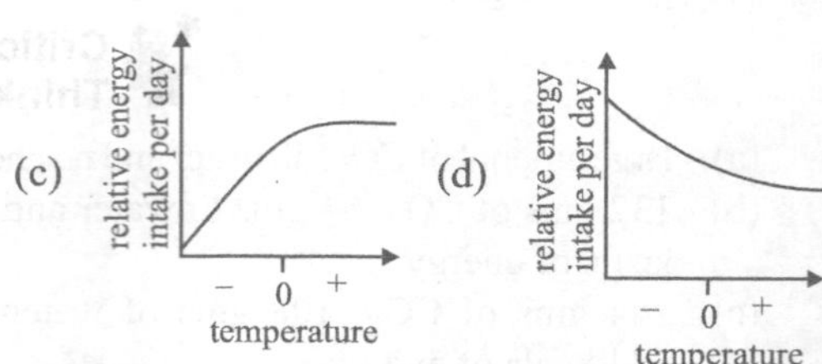

74. Conversion of fructose-6-phosphate to fructose-1, 6-diphosphate is catalysed by:
(a) Phosphofructokinase
(b) Aldolase
(c) Hexokinase
(d) None of the above

75. The principle of limiting factor was given by:
(a) Hill (b) Blackman
(c) Arnon (d) Calvin

Answer KEYs

1	(c)	9	(b)	17	(b)	25	(d)	33	(c)	41	(c)	49	(c)	57	(d)	65	(d)	73	(d)
2	(c)	10	(c)	18	(b)	26	(c)	34	(b)	42	(c)	50	(a)	58	(c)	66	(b)	74	(a)
3	(b)	11	(d)	19	(b)	27	(b)	35	(d)	43	(d)	51	(b)	59	(b)	67	(a)	75	(b)
4	(d)	12	(d)	20	(b)	28	(a)	36	(b)	44	(a)	52	(b)	60	(d)	68	(c)		
5	(c)	13	(a)	21	(b)	29	(d)	37	(a)	45	(b)	53	(b)	61	(b)	69	(b)		
6	(d)	14	(d)	22	(d)	30	(c)	38	(b)	46	(b)	54	(c)	62	(c)	70	(b)		
7	(d)	15	(a)	23	(d)	31	(b)	39	(b)	47	(a)	55	(c)	63	(d)	71	(a)		
8	(a)	16	(d)	24	(a)	32	(a)	40	(a)	48	(b)	56	(c)	64	(c)	72	(d)		

Hints & Solutions

1. (c) 2. (c) 3. (b) 4. (d)

5. (c) The Kreb's cycle is a complex of nine reactions where two molecules of acetyl CoA are oxidised to CO_2

6. (d)

7. (d) When oxygen is available $NADH_2$ produced in EMP-Krebs cycle pathway is released in the form of NAD. When there is shortage of oxygen NAD is not available and NADP takes up its place.

8. (a) 9. (b) 10. (c) 11. (d)

12. (d) 13. (a)

14. (d) The oxygen obtained from cellular respiration combines with the hydrogen obtained from the oxidation of organic molecules to form water.

15. (a)

16. (d) Tricaboxylic acid cycle is also known as citric acid cycle or Kreb's cycle. This is an aerobic process which takes place in the matrix of mitochondria. Kreb's discovered this cycle in 1937. So, this is also known as Hans Kreb's cycle.

17. (b)

18. (b) Oxidative phosphorylation is the addition of a phosphate group to ADP using the energy gained from electron transfer along a chain of electron carriers. This process generates energy in the form of ATP, and is absent in erythrocytes.

19. (b) 20. (b)

21. (b) In succulent plants like *Opuntia* there is an incomplete oxidation of the carbohydrates resulting in the formation of organic acids. CO_2 is not released.

$$2C_6H_{12}O_6 + 3O_2 \rightarrow 3C_4H_6O_5 + 3H_2O + 386 \text{ kcal.}$$

$$\text{R.Q.} = \frac{CO_2}{O_2} = \frac{0}{3} = 0$$

22. (d) 23. (d) 24. (a) 25. (d)

26. (c)

27. (b) Yeast causes alcoholic fermentation without using oxygen and hence depletion of oxygen has no effect on the metabolism or growth of yeast cells.

28. (a) Metabolic activity rate of meristematic tissue is very high.

29. (d)

30. (c) Cytochrome a_3 is terminal cytochrome of ETC. It has Fe^{+++} and Cu^{++} both. With Fe^{+++}, it picks up electrons and through Cu^{+} it passes electrons to oxygen.

31. (b) During anaerobic respiration, due to the absence of O_2 the value of R.Q. is infinite (or more than one).

32. (a)

33. (c) Primary respiratory substrate is carbohydrate and secondary respiratory substrate is fat.

34. (b)

35. (d) The electron transport system is coupled to proton transport across the inner membrane from the matrix, generating the proton-motive force, which powers the synthesis of ATP from ADP and iP.

36. (b) 37. (a) 38. (b) 39. (b)

40. (a) 41. (c)

42. (c) Pyruvate is a three-carbon compound, which on complete oxidation releases three moles of CO_2.

43. (d) 44. (a)

45. (b) Because it is an energy liberating process.

46. (b)

47. (a) Substrate-level-phosphorylation means synthesis of ATP by utilising energy released directly by the substrates. In glycolysis there are two steps in which ATP is synthesised. In Kreb's cycle, there is only one step where GTP is synthesised when succinyl CoA is converted to succinic acid.

48. (b) 49. (c) 50. (a) 51. (b)

52. (b) 53. (b)

54. (c) Reaction of respiration is

$$C_6H_{12}O_6 + 6O_2 \rightarrow 6CO_2 + 6H_2O + 674 \text{ kcal}$$

In this reaction considering molecular weight of each element, we get

$C_6H_{12}O_6 = 6 \times 12 + 12 \times 1 + 6 \times 16 = 180$ gm

$O_2 = 6 \times 32 = 192$ gm

Now we consider products of this reaction i.e. $6CO_2$ and $6H_2O$

$6CO_2 = 6\ (12 + 32) = 264$ gm

$6H_2O = 6\ (2 + 16) = 108$ gm

55. (c)

56. (c) Oxidative phosphorylation refers to the synthesis of ATP from ADP and inorganic phosphate by chemiosmosis. It occurs with the help of energy obtained from oxidation of reduced enzymes formed in cellular respiration.

57. (d)

58. (c) The last step regenerates NAD^+ from NADH and H^+.

59. (b)

60. (d) Fats make more electrons available to the electron transport system and more hydrogen ions available for chemiosmosis.

61. (b) 62. (c) 63. (d) 64. (c)

65. (d) Cytochromes are small proteins (intrinsic membrane proteins) that contain a cofactor, hence, which holds an iron atom. The iron carries electrons and cycles between + 2 and + 3 oxidation states. These form a part of electron transport chain in mitochondria and chloroplast and act as an electron transporter or electron acceptor in respiration and photosynthesis.

66. (b) 67. (a) 68. (c) 69. (b)

70. (b) 71. (a)

72. (d) Pyruvate is a three-carbon compound. Oxaloacetate, a four-carbon compound, combines with acetyl coenzyme A (two-carbon) to form citrate, a six-carbon compound.

73. (d) At lower temperatures, there is a need for a higher relative energy intake in order to carry on his metabolic processes and to generate heat in his body for enzyme activities. This need decreases as the temperature increases.

74. (a) 75. (b)

15 Plant Growth and Development

1. In plants, growth is:
   (a) Restricted to certain regions or structure
   (b) Irreversible
   (c) Change in size
   (d) All of the above
2. Growth is primarily affected by two climatic factors which are: Critical Thinking
   (a) Light and temperature
   (b) Temperature and relative humidity
   (c) Light and wind
   (d) Rainfall and temperature
3. Which of the following instrument can be used to record plant growth by seconds?
   (a) Arc auxanometer (b) Arc indicator
   (c) Space marker disc (d) Crescograph
4. Plant growth is ultimately regulated by:
   (a) Environmental cues
   (b) Hormones
   (c) Signal transduction pathways
   (d) The plant's genome
5. The process in which mature differentiated cells reverse to meristematic activity to form callus is called:
   (a) Cytodifferentiation (b) redifferentiation
   (c) dedifferentiation (d) differentiation
6. Which one of the following is primarily concerned with cell division?
   (a) GA_3 (b) IAA
   (c) Cytokinin (d) IBA
7. The discovery of gibberellins is related with which one of the following?
   (a) Blast disease of rice
   (b) Bakanae disease of rice
   (c) Early blight disease of potato
   (d) Rust disease of wheat
8. Which of the following light receptors is responsible for absorbing blue and ultraviolet light?

   (a) Phytochrome P_r (b) Photochrome P_{fr}
   (c) Cryptochrome (d) Phototropin
9. Match the column-I with column-II and choose the correct option.

	Column-I		Column-II
A.	Formative	I.	Plastic extension through the enzymatic loosening of microfibrils
B.	Phase of enlargement	II.	Cell develops into special or particular type of cells.
C.	Phase of differentiation	III.	New cells are produced by mitosis

   (a) A – I; B – II; C – III
   (b) A – I; B – III; C – II
   (c) A – III; B – II; C – I
   (d) A – III; B – I; C – II
10. In short day plants (SDP) flowering is induced by:
   (a) Photoperiod less than 12 hours.
   (b) Photoperiod shorter than initial value and uninterrupted long night.
   (c) Short photoperiod and interrupted long night.
   (d) Long night
11. Most appropriate temperature for vernalization is:
   (a) 1 – 6°C (b) 7°C
   (c) – 6°C (d) 0°C
12. Cell elongation in internode regions of the green plants takes place due to:
   (a) Ethylene (b) Indole acetic acid
   (c) Cytokinin (d) Gibberellins
13. An enzyme that can stimulate germination of barley seeds is:
   (a) Protease (b) Invertase
   (c) α-amylase (d) Lipase
14. The natural plant hormones were first isolated from:
   (a) Corn germ oil and human urine.
   (b) Cotton fruits, spinach leaves and rice plants.
   (c) Human urine and rice seedlings.
   (d) Spinach leaves and fungus *Gibberella*.

15. Cytokinins are known to:
(a) Inhibit cytoplasmic movement
(b) Help in retention of chlorophyll
(c) Influence water movement
(d) Promote abscission layer formation

16. Apical dominance means:
(a) Suppression of growth of apical bud by axillary buds.
(b) Suppression of growth of axillary buds by the presence of apical bud.
(c) Stimulation of growth of axillary buds by removal of apical bud.
(d) Inhibition of growth of axillary buds by removal of apical bud.

17. Maximum amount of growth in root occurs:

(a) in the presence of light.
(b) at its apex.
(c) behind the apex.
(d) in the presence of soil.

18. Auxin inhibits the growth of:
(a) Apical bud
(b) Lateral axillary buds
(c) Roots on stem cutting
(d) Parthenocarpic development of fruits

19. The growth rate is highest in:
(a) Red light (b) White light
(c) Green light (d) Blue light

20. Leaf movement in *Mimosa pudica* (Sensitive plant) due to:
(a) Nerve impulsion
(b) Delicate leaves
(c) Loss of tissue
(d) Turgor change in leaf base

21. Ageing is retarded by:
(a) ABA (b) CKN (c) GA (d) C_2H_4

22. Auxin causes cells to absorb water from hypertonic media because:
(a) it lowers transpiration rate.
(b) it increases rate of cellular respiration causing more production of ATP molecules thus facilitating active water absorption.
(c) it increases solute potential of the cells.
(d) it increase rate of transpiration.

23. The hormone reducing transpiration rate by inducing stomatal closure is:
(a) ABA (b) Ethylene
(c) Cytokinin (d) Auxin

24. Richmond-Lang effect is concerned with: Critical Thinking
(a) delay in senescence
(b) breaking dormancy
(c) suppression of apical dominance
(d) cell elongation

25. Which is employed for artificial ripening of banana fruits?
(a) Auxin (b) Coumarin
(c) Ethylene (d) Cytokinin

26. The movement of auxin is largely:
(a) Both acropetal and basipetal
(b) Basipetal
(c) Centripetal
(d) Acropetal

27. I.A.A. mainly inhibits growth of:
(a) Root
(b) Leaf
(c) Shoot
(d) Generally whole plant

28. Match column-I with column-II and select the correct answer using the codes given below.

	Column-I		Column-II
A.	*Xanthium*	I.	Its flower opens at higher temperature.
B.	*Crocus*	II.	Earlier and higher yield of crop can be obtained by keeping its seed at low temperature between 1 °C and 10 °C.
C.	Millet	III.	Growth regulator can be obtained from one of the parts of it.
D.	Coconut	IV.	It produces flower, when period of light available is less than critical period.

(a) A – IV; B – I; C – II; D – III
(b) A – IV; B – II; C – III; D – I
(c) A – IV; B – III; C – II; D – I
(d) A – I; B – II; C – III; D – IV

29. The names of many organisms are associated with classical studies in some areas of biology. Which of the following does not show such matching?
(a) Pea–Laws of heredity
(b) *Acetabularia*–Relationship between nucleus and cytoplasm
(c) *E. coli* – Genetic engineering
(d) Barley seeds– Role of phytochrome

30. Which of the following statements are **false**?
(i) C_2H_4 breaks seed and bud dormancy.
(ii) ABA stimulates the opening of stomata.
(iii) Cytokinin is primarily concerned with cell division.
(iv) ABA is synergistic to GA.
(a) (i), (ii), and (iii)
(b) (i) and (ii)
(c) (ii) and (iv)
(d) (i) and (iii)

31. The plants bend towards light because:
(a) the stem likes sunlight.
(b) the roots send the message to the stem to do so.
(c) due to secretion of growth promoting substances at the tip.
(d) the food material is translocated from leaves to stem causing bending.

32. Phytotron is device for: Tricky
(a) induction of mutations
(b) measuring wind velocity
(c) measuring electron flow
(d) growing plants under controlled conditions.

33. Which of the following hormones can replace vernalisation?
(a) Ethylene
(b) Gibberellins
(c) Cytokinins
(d) Auxins

34. When a dark period in SDP is interrupted by a brief exposure of light, the plant:
(a) Will not flower at all
(b) Flowers immediatly
(c) Gives more flowers
(d) Turns into LDP

35. Stem elongation is affected by:
(a) Gibberellin and florigen
(b) Auxin and gibberellin
(c) Florigen and kinin
(d) Kinin and auxin

36. Parthenocarpic tomato fruits can be produced by: Critical Thinking
(a) Raising the plants from vernalised seeds.
(b) Treating the plants with phenylmercuric acetate.
(c) Removing androecium of flowers before pollen grains are released.
(d) Treating the plants with low concentrations of gibberellic acid and auxins.

37. Induction of rooting in stem cutting by auxin treatment would be beneficial in which of the following? Critical Thinking
(a) *Marchantia* (b) Wheat
(c) *Cuscuta* (d) *Bougainvillea*

38. The growth of the plant body takes place by:
(a) Lateral meristems
(b) Apical meristems
(c) Both apical and lateral meristems
(d) None of the above

39. Match Column-I with Column-II and select the correct answer using the codes given below:

Column-I (Scientists)	Column-II (Discovery)
A. C. Darwin and F. Darwin	I. Cytokinin
B. Miller and Skoog	II. ABA
C. Addicott *et. al.*	III. C_2H_4.
D. Kurosawa	IV. Auxin
	V. GA

(a) A – IV; B – I; C – II; D – V
(b) A – IV; B – I; C – V; D – II
(c) A – III; B – I; C – II; D – IV
(d) A – V; B – IV; C – I; D – III

40. What will be the effect on phytochrome in a plant subjected to continuous red light?
(a) Level of phytochrome decreases
(b) Phytochrome is destroyed
(c) Phytochrome synthesis increases
(d) Destruction and synthesis of phytochrome remain in equilibrium.

41. Which one of the following is a long day plant?
(a) Bajra (b) Soyabean
(c) Tobacco (d) Wheat

42. Seedless fruits can be obtained by treating the unpollinated ovaries with:
(a) Colchicine (b) Sucrose solution
(c) Hormones (d) Pure lanolin

43. Which of the following test will you perform to demonstrate the presence of gibberellin?
(a) Bolting of cabbage
(b) Differentiation of shoots in tobacco callus culture
(c) Rapid division in carrot cells
(d) Elongation of *Avena* coleoptile

44. Which of the following controls flowering in long day plants and sex expression in certain species? *Critical Thinking*
(a) Auxin (b) Cytokinin
(c) Gibberellin (d) Ethylene

45. Which combination of gases is suitable for fruit ripening?
(a) 80% CO_2 and 20% CH_2
(b) 80% CH_4 and 20% CO_2
(c) 80% CO_2 and 20% O_2
(d) 80% C_2H_4 and 20% CO_2

46. To avoid excessive water loss during severe drought stress, the closure of stomata is signalled by the production of: *Critical Thinking*
(a) IAA (b) NAA (c) ABA (d) IBA

47. In tissue culture, roots can be induced by:
(a) no cytokinin and only auxins.
(b) higher concentration of cytokinin and lower concentration of auxins.
(c) lower concentration of cytokinin and higher concentration of auxins.
(d) only cytokinin and no auxins.

48. Which one of the following is a naturally occurring growth inhibitor?
(a) IAA (b) ABA (c) NAA (d) GA

49. "Foolish Seedling or Bakane" disease of rice led to the discovery of:
(a) IAA (b) GA
(c) ABA (d) 2, 4-D

50. Senescence as an active developmental cellular process in the growth and functioning of a flowering plant, is indicated in:
(a) Vessels and tracheid differentiation
(b) Leaf abscission
(c) Annual plants
(d) Floral parts

51. Which of the following triggers seed germination?
(a) Seeds imbibe in water.
(b) Seeds released from fruit.
(c) Seeds undergo chemical changes.
(d) All of the above

52. The stimulus for geotropism in roots is perceived by: *Critical Thinking*
(a) Root cap
(b) Quiescent centre
(c) Root tip
(d) None of the above

53. In callus culture, roots can be induced by the supply of:
(a) Auxin (b) Cytokinin
(c) Gibberellin (d) Ethylene

54. Which of the following shows the correct graph of arithmetic growth?

(a)
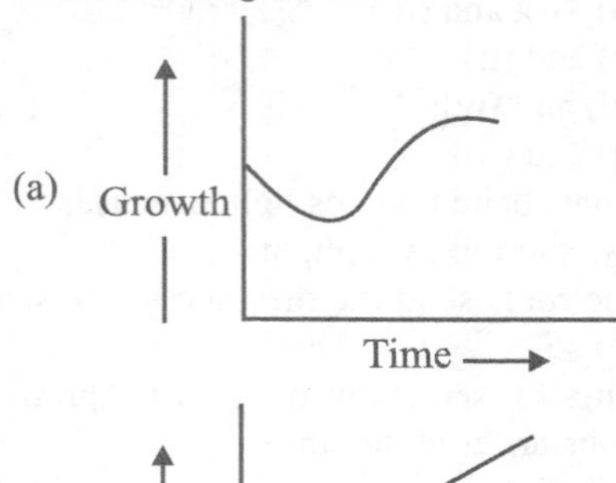

(b)
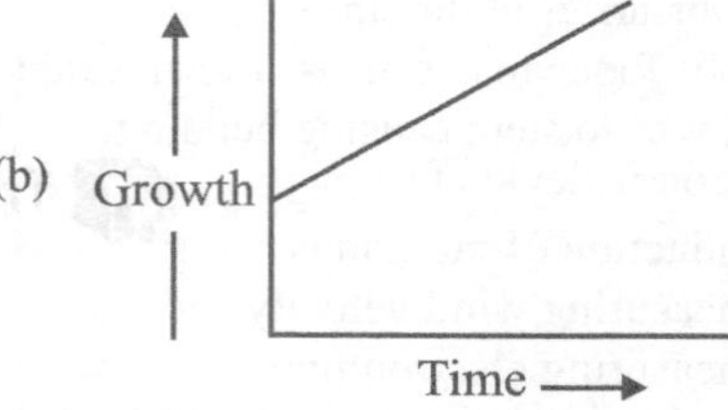

(c)
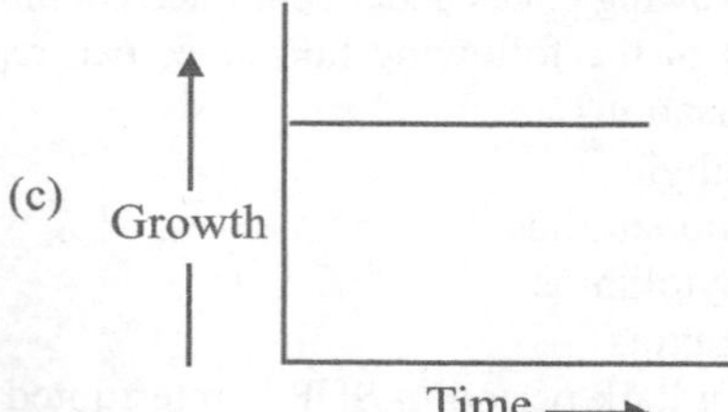

(d)
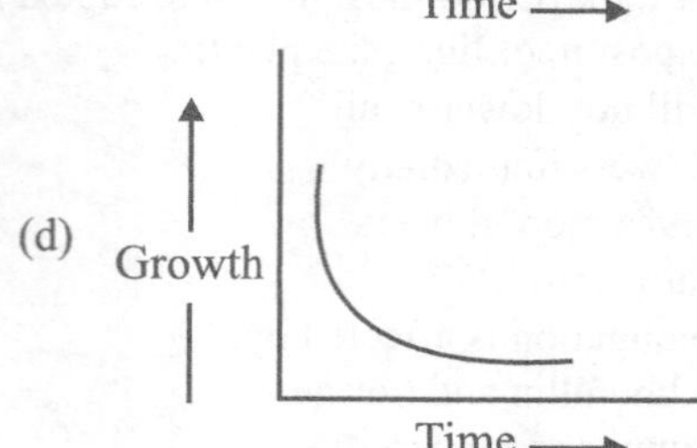

55. Exponential growth cannot be sustained for much time due to :
(i) limited space and nutrient.
(ii) accumulation of toxic agent.
(iii) unlimited space and nutrient.
(iv) accumulation of nutrient agent.
(a) (i) and (iii) (b) (iii) and (iv)
(c) (i) and (ii) (d) (iv) and (ii)

56. Which one is a short day plant?
(a) *Brassica campestris*
(b) *Raphanus sativus*
(c) *Glycine max*
(d) *Papaver somniferum*

57. One set of a plant was grown at 12 hours day and 12 hours night period cycles and it flowered while in the other set night phase was interrupted by flash of light and it did not produce flower. Under which one of the following categories will you place this plant? Critical Thinking

(a) Long day (b) Darkness neutral
(c) Day neutral (d) Short day

58. What is the site of vernalisation?
(a) Stem (b) Leaves
(c) Roots (d) Apical meristem

59. Maryland Mammoth Tobacco is a short day plant. Its critical duration of darkness is 10 hours. Under which of the following conditions will Maryland mammoth tobacco not flower?

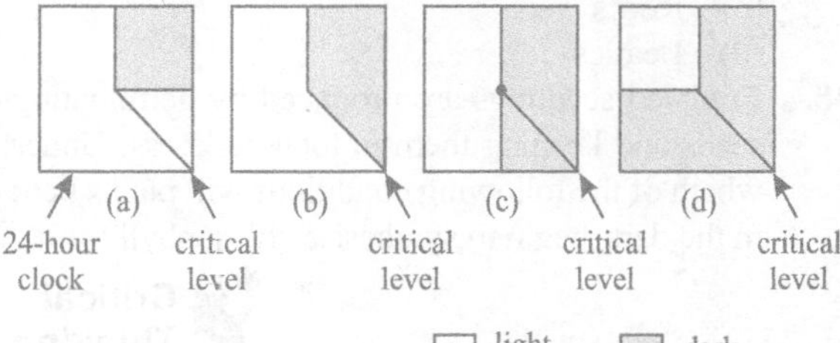

60. What will be the effect on phytochrome in a plant subjected to continuous red light?

(a) Level of phytochrome decreases
(b) Phytochrome is destroyed
(c) Phytochrome synthesis increases
(d) Destruction and synthesis of phytochrome remain in equilibrium.

61. If a flash of light is provided in the middle of dark period to a SDP then the plant is likely

(a) To flower
(b) Not to flower
(c) Produce more flowers
(d) None of the above

62. Which of the following statement regarding Gibberellin are **correct**?
(i) GA_3 is used to speed up the malting process in brewing industry.
(ii) GA_3 is a commercially available gibberellin.
(iii) Spraying juvenile conifers with GAs hastens maturity, thus leading to early seed production.
(iv) GA_3 cannot increase the length of internode of sugar cane.
(a) (i), (ii), and (iii) (b) (i) and (ii)
(c) (ii) and (iv) (d) (i) and (iii)

63. Which of the following statements are **incorrect** options.
(i) Primary growth involves increase in girth of plants.
(ii) Increase in girth of plants occurs due to apical meristem.
(iii) Secondary growth of plants occurs due to lateral meristem.
(iv) Vascular cambium and cork cambium are the lateral meristems of plants.
(v) Elongation of a plant along the axis is called primary growth.
(a) (i) and (ii) (b) (iii) and (iv)
(c) (iv) and (v) (d) (i) and (v)

64. Which of the following does not normally occur when an etiolated seedling is exposed to light?

(a) Chlorophyll synthesis begins.
(b) The rate of shoot elongation increases.
(c) The hypocotyl hook straightens (in species with a hypocotyl hook).
(d) Phytochrome P_r is converted to P_{fr}.

65. Which plant hormone promotes seed, bud dormancy and causes stomatal closure?
(a) IAA (b) Abscisic acid
(c) GA (d) Cytokinin

66. Which of the following hormone breaks bud dormancy of potato tuber? Critical Thinking
(a) Gibberellin (b) IAA
(c) ABA (d) Zeatin

67. Dwarfness can be controlled by treating the plant with:
(a) Cytokinin (b) Gibberellic acid
(c) Auxin (d) Antigibberellin

68. Florigen is produced in the region of:
(a) Leaves (b) Fruit
(c) Root (d) Trunk

69. Cell elongation in internodal regions of the green plants takes place due to:
(a) Indole acetic acid
(b) Cytokinins
(c) Gibberellins
(d) Ethylene

70. Which one increases in the absence of light? Critical Thinking

(a) Uptake of minerals
(b) Uptake of water
(c) Elongation of internodes
(d) Ascent of sap

71. The potent weed killer is:
(a) Gibberellin (b) 2,4-D
(c) Abscisic acid (d) Ethylene

72. Which one of the following acids is a derivative of carotenoids?
(a) Indole-3 -acetic acid
(b) Gibberellic acid
(c) Abscisic acid
(d) Indole butyric acid

73. Which of the following is not characteristic of changes associated with winter dormancy in many species of temperate plants?

(a) Development of wax-covered bud scales
(b) High concentration of abscisic acid in buds
(c) Decrease in solute concentration of sap
(d) Leaf abscission

74. Which of the following growth substance act as a stimulant during nodule formation in leguminous plant?
(a) Ethylene (b) ABA
(c) IAA (d) Morphactin

75. Primary precursor of I.A.A is:

(a) Phenylalanine
(b) Tyrosine
(c) Tryptophan
(d) Valine

76. Pineapple can be made to flower in off season by:
(a) Ethylene
(b) Zeatin
(c) Naphthalene Acetic Acid (NAA)
(d) Temperature

77. When a plant is not reproducing, most of the cytokinins are produced in its:
(a) Lateral buds
(b) Shoot apex
(c) Roots
(d) Leaves

78. Etiolated seedlings are produced by germinating seeds and keeping them in total darkness. Under which of the following conditions will plants kept in the dark-begin to synthesise chlorophyll?

(a) After being given a pulse of blue light.
(b) After being given a pulse of red light.
(c) After being given a pulse of red light followed by a pulse of far-red light.
(d) After being given a pulse of far-red light followed by a pulse of red light.

Answer KEYs

1	(d)	9	(d)	17	(c)	25	(c)	33	(b)	41	(d)	49	(b)	57	(d)	65	(b)	73	(c)
2	(a)	10	(b)	18	(b)	26	(a)	34	(a)	42	(c)	50	(b)	58	(d)	66	(a)	74	(c)
3	(d)	11	(b)	19	(a)	27	(a)	35	(b)	43	(a)	51	(a)	59	(a)	67	(b)	75	(c)
4	(d)	12	(d)	20	(d)	28	(a)	36	(d)	44	(c)	52	(a)	60	(a)	68	(a)	76	(c)
5	(c)	13	(c)	21	(b)	29	(d)	37	(d)	45	(d)	53	(a)	61	(b)	69	(c)	77	(b)
6	(c)	14	(a)	22	(b)	30	(c)	38	(c)	46	(c)	54	(b)	62	(a)	70	(c)	78	(c)
7	(b)	15	(b)	23	(a)	31	(c)	39	(a)	47	(c)	55	(c)	63	(a)	71	(b)		
8	(c)	16	(b)	24	(a)	32	(d)	40	(a)	48	(b)	56	(c)	64	(b)	72	(c)		

Hints & Solutions

1. (d) 2. (a) 3. (d)
4. (d) Though the environment influences development, final regulation is at the genomic level.
5. (c)
6. (c) Cytokinins induce cell division in plants and thus influence the plant growth by controlling cell division.
7. (b) 8. (c)
9. (d) Plant growth takes place in three phases;
   (i) Formative phase: It is also called the phase of cell formation or cell division. New cells are produced by mitotic division of the pre-existing cells. Mitosis adds new cells to the body.
   (ii) Phase of enlargement: The newly formed cells in formative phase undergoes enlargement. Cell wall of enlarging cell shows plastic extension through enzymatic loosing or microfilament and deposition of new material called intussusception.
   (iii) Phase of differentiation: The enlarged cells develop into special or particular type of cells by undergoing structural and physiological differentiation.
10. (b) Short-day plants flower, only if light periods are shorter than some critical length. In these plants the dark period is critical and must be continuous. If this dark period is interrupted even with a brief exposure of light, the short day plant will not flower. Strawberries, *Dahlias*, soyabeans and *Chrysanthemums* are short day plants.
11. (b) 12. (d) 13. (c)
14. (a) The first naturally occurring auxin was isolated by Kogl and Haagen-Smit (1931) from human urine. It was identified as auxin-a (auxentriolic acid). Later, in 1934 Kogl, Haagen-Smit and Erxleben obtained another auxin called auxin-b (auxenolonic acid) from corn germ oil (extracted from germinating corn seeds).
15. (b) Less chlorophyll and degeneration of proteins are the two important symptoms of senescence. Cytokinins delay these processes and, thus senescence is also delayed. This effect of cytokinin is known as Richmond-Lang effect.
16. (b)
17. (c) Root meristem is sub-apical in position due to presence of root cap.
18. (b) Auxin inhibits the growth of lateral axillary buds.
19. (a) 20. (d) 21. (b) 22. (b) 23. (a)
24. (a) 25. (c)
26. (a) The movement of auxin is polar and in stem it takes place in a basipetal direction, whereas in the roots it is in an acropetal direction.
27. (a) I.A.A. mainly inhibits growth of root.
28. (a)
29. (d) Barley seeds are associated with α-amylase bioassay of gibberellins.
30. (c) 31. (c) 32. (d)
33. (b) Gibberellin overcomes vernalisation requirement (low temperature) for flowering.
34. (a) 35. (b)
36. (d) Parthenocarpic tomato fruits can be produced by treating the plants with low concentrations of gibberellic acid and auxins.
37. (d) 38. (c) 39. (a) 40. (a)
41. (d) Long-day plants usually flower in the spring or early summer; they flower only if the light periods are longer than a critical length, which is usually 9-16 hours. For example, wheat plants flower only when light periods exceed fourteen hours.
42. (c)
43. (a) Bolting of cabbage can be induced artificially by the application of gibberellin under conditions that would normally maintain the rossette form.

44. (c) 45. (d) 46. (c) 47. (c) 48. (b)
49. (b) 50. (b)
51. (a) The uptake of water by the seed begins the processes that leads to seed germination.
52. (a) In root, geotropic response is perceived by root cap.
53. (a) 54. (b)
55. (c) Exponential or log phase cannot sustain for longer period because the nutrients and space are limited and there is competition as well. Microorganisms, when nutrients get exhausted, secrete toxic chemicals which inhibit the growth of other organisms.
56. (c) 57. (d) 58. (d)
59. (a) Short day plant generally require light period of less than 12 hours and continuous dark period of about 14-16 hours for subsequent flowering.
60. (a) 61. (b) 62. (a)
63. (a)
(i) Elongation of plant along the axis is called the primary growth.
(ii) Primary growth happens due to the presence of root apical meristem and shoot apical meristem. Root apical meristem (RAM), shoot apical meristem (SAM) and Intercalary meristem are responsible for the primary growth of the plants and they principally contribute to the elongation of the plants along their axis. In the dicotyledons and gymnosperms, the lateral meristems, vascular cambium and cork cambium appear later in life. These are the meristems that cause increase in the girth of the organ in which they are active. This is known as the secondary growth of the plant.
64. (b) The rate of shoot elongation is greatest when the seedling is in the dark, and decreases upon exposure to light.
65. (b) 66. (a) 67. (b) 68. (a) 69. (c)
70. (c) 71. (b) 72. (c)
73. (c) Solute concentration of the sap is increased to prevent freezing during winters.
74. (c) 75. (c) 76. (c)
77. (b) The ratio of cytokinins to auxins, control cell differentiation in plant tissue culture. If there is more of cytokinin than auxin, shoot buds develop from a callus. If there is relatively more auxins than cytokinins, roots develop.
78. (c) The pulse of red light converts P_r to P_{fr}. Subsequent pulses of far-red light stimulate changes that lead to chlorophyll synthesis.

16 Digestion and Absorption

1. The lamina propria and muscularis mucosa belong to:
(a) Serosa (b) Sub mucosa
(c) Mucosa (d) None of these

2. The centre of hunger or centre which regulates the amount of food we eat or our appetite is located in:
(a) Medulla (b) Cerebrum
(c) Hypothalamus (d) Alimentary canal

3. Which of the following is a freely movable muscular organ that is attached to the floor of the oral cavity by the frenulum?
(a) Jaws (b) Teeth
(c) Tongue (d) Salivary glands

4. The lymph vessels called lacteals, are found in:
(a) Mammary glands
(b) Pancreas
(c) Mucosa of intestine
(d) Villi of ileum

5. Pylorus occurs between:
(a) Ileum and rectum
(b) Duodenum and ileum
(c) Stomach and duodenum
(d) Oesophagus and stomach

6. The innermost layer of the digestive tract is the:
(a) Serosa membrane
(b) Mucosa membrane
(c) Submucosa membrane
(d) Lumen

7. The sphincter of Oddi is present between:
(a) Oesophagus and cardiac stomach.
(b) Pyloric stomach and duodenum.
(c) Hepatic duct and cystic duct.
(d) Hepatopancreatic duct and duodenum.

8. Parotid salivary glands are present:
(a) Below the tongue
(b) Below the ears
(c) In the angle between two jaws
(d) Below the eye orbits

9. The common bile duct in human is formed by the joining of: **Tricky**
(a) Pancreatic duct and bile duct
(b) Cystic duct and hepatic duct
(c) Cystic duct and pancreatic duct
(d) Hepatic duct and pancreatic duct

10. Which of the following is the correct chronological order for flow of food from mouth to anus?
(a) Oesophagus $\rightarrow$ Stomach $\rightarrow$ Small intestine $\rightarrow$ Large Intestine
(b) Large intestine $\rightarrow$ Oesophagus $\rightarrow$ Stomach $\rightarrow$ Small Intestine
(c) Small intestine $\rightarrow$ Large intestine $\rightarrow$ Oesophagus $\rightarrow$ Stomach
(d) Stomach $\rightarrow$ Small intestine $\rightarrow$ Large intestine $\rightarrow$ Oesophagus

11. If the epiglottis does not function correctly, then what will happen to the human being?

(a) One might congest.
(b) Peristalsis will discontinue.
(c) Acid reflux disease will damage the oesophagus.
(d) Swallowing will be difficult or impossible.

12. The enormously long tusks in elephant are:
(a) Upper incisors (b) Upper canines
(c) Lower canines (d) Lower incisors

13. Which of the following type of enzyme is not matched correctly with the molecule that it breaks down?
(a) Amylase–Starch
(b) Lipase–Starch
(c) Protease–Proteins
(d) Disaccharidase–Sugars

14. Vitamin needed for blood coagulation is:
(a) E (b) D (c) K (d) C

15. Cod liver oil is a good source of :
(a) Vitamin B (b) Vitamin C
(c) Vitamin A (d) Iodine

16. How does a gastrovascular cavity differ from an alimentary canal? The gastrovascular cavity

Critical Thinking

(a) stores food but does not digest it.
(b) is usually much larger.
(c) has only one opening.
(d) functions in digestion but not absorption.

17. True stomach in cud-chewing animals is:
(a) Abomasum (b) Reticulum
(c) Omasum (d) Rumen

18. Which of the following vitamin is essential for normal vision?
(a) Folic acid (b) Biotin
(c) Riboflavin (d) Niacin

19. Which one is essential for DNA replication or cell division?
(a) Vit. E (b) Folic acid
(c) Vit. K (d) Vit. D

20. Intestinal villi are more numerous and larger in posterior part of small intestine than in anterior part because:

Toughnut

(a) digestion is faster in posterior part.
(b) blood supply is poorer in anterior part.
(c) blood supply is poorer in posterior part.
(d) there is more of digested food in posterior part.

21. How would you expect the digestive system of a hawk, a carnivore, to be compared with that of a sparrow, a seed-eater?

Critical Thinking

(a) The hawk would have a larger gastrovascular cavity.
(b) The sparrow's digestive system would be longer.
(c) The hawk would have a gizzard, but the sparrow would not.
(d) The hawk's digestive system would be longer.

22. Liver cells secrete:
(a) Amylopsin (b) Trypsin
(c) Bile and no enzymes (d) Lipase

23. Physiologically, vitamin E is known to be:
(a) Anti-inflammatory
(b) A powerful antioxidant
(c) An important co-enzyme
(d) Anti-hypertensive factor

24. Vitamin B_2 is related with:
(a) FAD (b) NAD (c) NADP (d) CoA

25. Which is the best source for vitamin B_1?
(a) Cod liver oil (b) Egg
(c) Whole wheat bread (d) Curd

26. Ptyalin cannot work in stomach, because it is:
(a) inactivated by pepsin.
(b) inactivated by HCl.
(c) carbohydrate - digesting enzyme.
(d) inactivated by rennin.

27. The structure which prevents entry of food into windpipe during swallowing in mammals is:
(a) Larynx (b) Pharynx
(c) Glottis (d) Epiglottis

28. The deficiency of which vitamin is known to cause abortion in early pregnancy of rat?
(a) Retinol (b) Calciferol
(c) Tocopherol (d) Napthoquinone

29. The swallowing of food from buccal cavity into oesophagus is a reflex action controlled by:
(a) Brain
(b) Medulla oblongata
(c) Hormones
(d) Peripheral nerves

30. Digestion of proteins, fats and carbohydrate is completed in:
(a) Stomach (b) Duodenum
(c) Ileum (d) None of these

31. Glottis is a structure which:

Critical Thinking

(a) prevents entry of food into windpipe during swallowing food.
(b) allows food to pass into oesophagus.
(c) produces sound as air is forced through it.
(d) allows air to enter in trachea.

32. Succus entericus is secreted by:

Critical Thinking

(a) Villi of intestine
(b) Brunner's glands
(c) Crypts of Lieberkuhn
(d) Brunner's glands and Crypts of Lieberkuhn

33. Hydrolysis of phospholipids yield:

(a) Glycerol, phosphoric acid and fatty acids.
(b) Glycerol, phosphoric acid and nitrogen base.
(c) Glycerol & fatty acids.
(d) Acetyl coA.

34. Which of the following food reserves would be first used in a starving person?
(a) Muscle proteins
(b) Skin adipose tissue
(c) Liver glycogen
(d) Liver proteins

35. Secretion of hormone cholecystokinin is for:
(a) Controlling blood pressure
(b) Absorption of iron
(c) Bile functions
(d) Release of insulin

36. Trypsinogen is produced by:
(a) Liver (b) Duodenum
(c) Stomach (d) Pancreas

37. The below diagram repesents the T.S. of gut. Identify 1, 2, 3 and 4

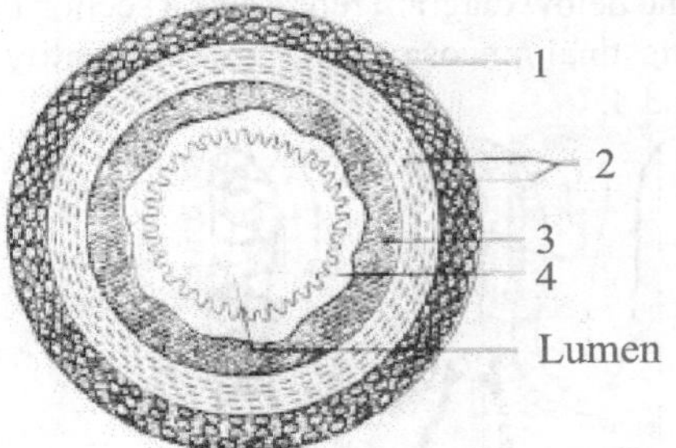

(a) 1 - Serosa; 2 - Muscularis; 3 - Submucosa; 4 - Mucosa
(b) 1 - Muscularis; 2 - Serosa; 3 - Submucosa; 4 - Mucosa
(c) 1 - Serosa; 2 - Muscularis; 3 - Mucosa; 4 - Submucosa
(d) 1 - Serosa; 2 - Submucosa; 3 - Muscularis; 4 - Mucosa

38. Which of the following statement is incorrect about chylomicrons? **Critical Thinking**
(i) Chylomicrons are produced in the epithelial cells of small intestinal.
(ii) It contains triglycerides, cholesterol and phospholipids.
(iii) It is protein coated small vesicles.
(iv) Chylomicrons are released from the epithelial cells into lacteals.
(a) (i) and (iv)
(b) (ii) and (iii)
(c) (i), (ii), (iii) and (iv)
(d) None of the above

39. When gall bladder of a man is removed:
(a) fat digestion is not possible. **Tricky**
(b) acidity continues in duodenum.
(c) effect of pancreatic juice upon food is impaired.
(d) All of the above

40. Contraction of gall bladder is induced by:
(a) Gastrin (b) Cholecystokinin
(c) Secretin (d) Enterogastrone

41. Meals which are rich in fat are not digested in the intestine in absence of:
(a) Pepsin (b) Enterokinase
(c) Insulin (d) Steapsin

42. When a piece of bread is chewed, it tastes sweet because: **Toughnut**
(a) the sugar contents are drawn out.
(b) saliva converts starch into maltose.
(c) it does not taste sweet.
(d) the taste buds are stimulated by chewing.

43. Functional units of absorption of digested food are:
(a) Peyer's patches
(b) Villi
(c) Crypts of Lieberkuhn
(d) Brunner's gland

44. Herbivorous animals can digest cellulose because: **Tricky**
(a) their molar and premolar teeth can crush and grind the food.
(b) bacteria present in their caecum help in digestion of cellulose.
(c) gastric juice has a digestive enzyme for cellulose digestion.
(d) alimentary canal is very long.

45. In which part of gut, proteins ultimately degraded to amino acids? **Critical Thinking**
(a) Small intestine (b) Colon
(c) Stomach (d) Caecum

46. Vermiform appendix is made up of: **Tricky**
(a) Digestive tissue (b) Respiratory tissue
(c) Excretory tissue (d) Lymphatic tissue

47. Pancreatic juice takes part in digestion of:
(a) Proteins and carbohydrates
(b) Proteins, fats and carbohydrates
(c) Proteins and fats
(d) Proteins only

48. A carbohydrate splitting enzyme is secreted by:
(a) Liver
(b) Zymogen cells of gastric glands
(c) Salivary glands
(d) Crypts of Lieberkuhn

49. Which group contains biocatalysts?
(a) Peptidase, amylase, rennin
(b) Myosin, oxytocin, adrenaline
(c) Rhodopsin, pepsin, steapsin
(d) Glucose, amino acids, fatty acids

50. Duct leading from parotid gland and opening into vestibule is: Tricky
(a) Wharton's duct (b) Stensen's duct
(c) Wolffian duct (d) Infra-orbital

51. Pancreatic lipase acts upon:
(a) Glycogen (b) Triglycerides
(c) Disaccharides (d) Polypeptides

52. Small intestine is distinguishable into 3 parts, a 'U' shaped ________, a long coiled middle portion ________ and a highly coiled ______.
(a) Ileum, jejunum,duodenum
(b) Jejunum, duodenum, ileum
(c) Duodenum, jejunum, ileum
(d) Caecum, duodenum, ileum

53. Man cannot digest cellulose, whereas cow and other herbivorous animals can because: Critical Thinking
(a) they have enzyme cellulase in their stomach.
(b) they masticate it well by chewing teeth.
(c) they have bacteria in their alimentary canal which digest cellulose.
(d) none of the above

54. DNase and RNase are enzymes secreted by:
(a) Pancreas (b) Stomach
(c) Gall bladder (d) Intestine

55. Teeth of adult man, not present in milk dentition are:
(a) Molars (b) Premolars
(c) Canines (d) Incisors

56. Diastema is a toothless gap in the jaws of rabbit between: Critical Thinking
(a) Premolars and molars
(b) Incisors and molars
(c) Canines and molars
(d) Incisors and premolars

57. Which of the following are reabsorbed in the alimentary canal as such? Critical Thinking
(a) Albumen of egg
(b) Polysaccharides
(c) Proteins
(d) Fat soluble vitamins

58. The pH of succus entericus is:

(a) 7.6 (b) 6.6 (c) 5.6 (d) 2.0

59. Mammalian teeth are:
(a) Acrodont (b) Homodont
(c) Thecodont (d) Polyphyodont

60. Which of the following control the amount of enzymes in pancreatic juice?
(a) Cholecystokinin (b) Secretin
(c) Pancreozymin (d) Gastrin

61. The below diagram represents a section of small intestinal mucosa showing villi. Identify 1, 2, 3 and 4.

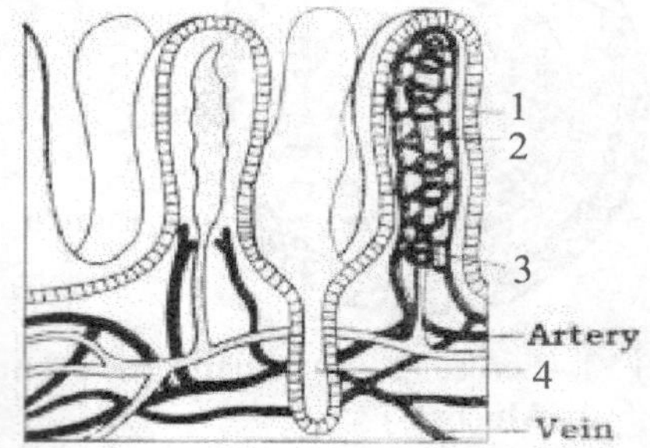

(a) 1-Villi, 2-Lacteal, 3-Capillaries, 4-Crypts
(b) 1-Lacteal, 2-Villi, 3-Capillaries, 4-Crypts
(c) 1-Villi, 2-Lacteal, 3-Crypts, 4-Capillaries
(d) 1-Crypts, 2-Lacteal, 3-Capillaries, 4-Villi

62. Which of the following is/are **correct** statements regarding beri-beri?
(i) A crippling disease prevalant among the native population of sub-Saharan Africa.
(ii) A deficiency disease caused by lack of thiamine (vitamin B_1).
(iii) A nutritional disorder in infants and young children when the diet is persistently deficient in essential protein.
(iv) Occurs in those countries where the staple diet is polished rice.
(v) The symptoms are pain, paralysis, muscle, cramps, progressive oedema, mental deterioration and finally heart failure.
(a) (i), (ii) and (v) (b) (i), (ii) and (iv)
(c) (i), (iii) and (v) (d) (ii), (iii) and (v)

63. A "1" flap called "2" prevents the entry of food into the glottis – opening of the "3" – during "4". Identify 1, 2, 3, and 4.

	1	2	3	4
(a)	Cartilaginous	Epiglottis	Wind pipe	Swallowing
(b)	Bony	Glissons's capsule	Stomach	Digestion
(c)	Bony	Pharynx	Oral cavity	Ingestion
(d)	Cartilaginous	Oesophagus	Trachea	Mastication

64. Rennin acts on milk protein and changes:
(a) Caseinogen into casein
(b) Casein into paracasein
(c) Caseinogen into paracasein
(d) Paracasein into caseinogen

65. Gall bladder is attached to liver in the region of:
(a) Quadrate lobe (b) Caudate lobe
(c) Right lobe (d) Left lobe

66. If pH of stomach is made 7, which component of food would be affected? **Critical Thinking**
(a) Starch (b) Protein
(c) Fat (d) Sucrose

67. If the chyme of a person who had orally consumed only starch as food is analysed before it enters the duodenum, it will show the presence of:
(a) Maltose and glucose
(b) Dextrin and maltose
(c) Starch, dextrin and maltose
(d) Starch, dextrin and glucose

68. Sodium taurocholate aids in:
(a) Absorption (b) Defecation
(c) Assimilation (d) Emulsification

69. How many teeth in human, erupt only once in the life time? **Tricky**
(a) 4 (b) 12 (c) 20 (d) 16

70. Under which of the following circumstances will insulin be secreted?
(a) The blood sugar level in the liver is low.
(b) The blood sugar level in the hepatic portal vein is low.
(c) The blood sugar level in the islets of Langerhans is high.
(d) The glycogen level in the skeletal muscle is high.

71. What will happen if the secretion of parietal cells of gastric glands is blocked with an inhibitor? **Toughnut**
(a) Enterokinase will not be released from the duodenal mucosa and so trypsinogen is not converted into trypsin.
(b) Gastric juice will be deficient in chymosin.
(c) Gastric juice will be deficient in pepsinogen.
(d) In the absence of HCl secretion, inactive pepsinogen is not converted into the active enzyme pepsin.

72. Which one of the given option represents lophodont?
(a) Incisors and canines
(b) Canines and premolars
(c) Premolars and molars
(d) Incisors and premolars

73. How do nutrients, absorbed by the small intestine, travel to the individual cells of the human body? **Critical Thinking**
(a) The nutrients are absorbed from the small intestine into the blood and move through the circulatory system to the body cells.
(b) The nutrients move from the small intestine directly to the liver and then move through the lymphatic system to the body cells.
(c) The small intestine forces the nutrients into the kidneys, where the nutrients are then dissolved in fluids used by the body cells.
(d) The body cells send nerve impulses indicating a lack of nutrients to the small intestine, and the small intestine sends the nutrients back to the cells.

74. Match column-I with column-II and select the correct answer using the codes given below.

Column-I	Column-II
A. Hepatic lobule	I. Base of Villi
B. Crypts of Lieberkuhn	II. Glisson's capsule
C. Sphincter of Oddi	III. Gall bladder
D. Cystic duct	IV. Hepato-pancreatic duct

(a) A – II; B – I; C – IV; D – III
(b) A – I; B – II; C – IV ; D – III
(c) A – I; B – II; C – III; D – IV
(d) A – III; B – IV; C – II; D – I

75. Match column-I with column-II and select the correct answer using the codes given below.

Column-I	Column-II
A. Neck cells	I. HCl, Intrinsic factor
B. Peptic / Chief cells	II. Mucus
C. Parietal / Oxyntic	III. Pepsinogen

(a) A – II; B – III; C – I
(b) A – III; B – II; C – I
(c) A – I; B – II; C – III
(d) A – II; B – I; C – III

76. What is common among amylase, renin and trypsin?
(a) All are proteins.
(b) All are proteolytic enzymes.
(c) Produced in stomach.
(d) Act at pH lower than 7.

77. Valves of Kerckring (also called plicae circulares or valvulae conniventes) occur in: Tricky
(a) between right auricle and right ventricle.
(b) left auricle and left ventricle.
(c) stomach and duodenum.
(d) intestine.

78. In pancreatic juice, which of the followings are secreted in inactive form as proenzymes?
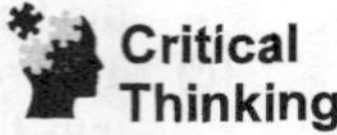

(a) Trypsin, chymotrypsin, & carboxypeptidases
(b) Pepsin, trypsin and chymotrypsin
(c) Trypsin only
(d) Trypsin and chymotrypsin only

79. If the bile-pancreatic duct is blocked then which of the following will not be affected?
(a) Digestion of proteins
(b) Emulsification of fats
(c) Level of blood glucose
(d) Digestion of starch

80. If for some reason our goblet cells becomes non-functional then this will adversely affect: Critical Thinking
(a) production of somatostatin.
(b) secretion of sebum from the sebaceous glands.
(c) maturation of sperms.
(d) smooth movement of food, down the intestine.

81. Match column-I (organs) with column-II (functions) and choose the correct option.

Column-I (Organs)	Column-II (Functions)
A. Mouth	I. Reclaims water and salts.
B. Stomach	II. Carries out most of the digestion and absorption of nutrients.
C. Small intestine	III. Releases amylase enzyme that break down carbohydrates.
D. Large intestine	IV. An acidic compartment that begins to break proteins into larger polypeptides.

(a) A – I; B – II; C – III; D – IV
(b) A – III; B – IV; C – II; D – I
(c) A – II; B – III; C – I; D – IV
(d) A – IV; B – I; C – II; D – III

82. The given pair represent substrate–enzyme pair. Which of the following pair is (are) correct?
(i) Maltose – Lactase
(ii) Protein – Pepsin
(iii) Carbohydrate – Lipase
(iv) Casein – Rennin
(a) (i) and (ii) (b) (i) and (iii)
(c) (i) and (iv) (d) (ii) and (iv)

83. During prolonged fasting,

(a) first fats are used up, followed by carbohydrates from liver and muscles and proteins in the end.
(b) first carbohydrates are used up, followed by fat and proteins towards end.
(c) first lipids, followed by proteins and carbohydrates towards end.
(d) None of the above

84. Dental formula shows type and number of teeth in:
(a) Upper jaw
(b) Lower jaw
(c) Both jaws
(d) One half of both jaws

85. It is important to get some vitamin B_1 every day, but it is all right if intake of vitamin A varies a bit. Why?

Critical Thinking

(a) Vitamin B_1 is an essential nutrient, and vitamin A is not.
(b) Vitamin A can be stored by the body, but vitamin B_1 cannot.
(c) The body needs much larger amounts of vitamin B_1 than vitamin (A).
(d) The body requires vitamin B_1, but vitamin A is just an extra vitamin.

86. Bow legged condition in children is due to deficiency of vitamin:
(a) D (b) A (c) K (d) C

87. In ruminants, bacterial action occurs in:
(a) Reticulum (b) Omasum
(c) Abomasum (d) Rumen

88. Osteomalacia in adults is caused due to the deficiency of:
(a) Vitamin D (b) Vitamin C
(c) Vitamin A (d) Vitamin K

89. Riboflavin deficiency causes which disease?
(a) Scurvy (b) Pellagra
(c) Beriberi (d) Cheilosis

90. Ampulla of Vater is guarded by:
(a) Sphincter of Oddi
(b) Pyloric sphincter
(c) Cardiac sphincter
(d) Anal sphincter

91. Alimentary canal of herbivores differ from carnivores having:
(a) Short alimentary canal
(b) Long alimentary canal
(c) Wide alimentary canal
(d) Thin alimentary canal

92. Certain amino acids are essential to the diet of animals because:
(a) they prevent over-nourishment.
(b) they are cofactors and coenzymes that are required for normal physiological function.
(c) an animal cannot directly synthesise them through the transfer of an amino group to an appropriate carbon skeleton.
(d) animals need these substances in order to make the stored fats that are used during hibernation and migration.

93. Mental retardation in children suffering from galactosemia can be avoided by:

Toughnut

(a) giving them more milk.
(b) giving them milk free diet.
(c) giving them milk fortified with vitamin.
(d) giving them more proteinaceous diet.

94. Excessive simulation of vagus nerve in humans may lead to:
(a) Hoarse voice
(b) Peptic ulcers
(c) Efficient digestion of proteins
(d) Irregular contractions of diaphragm

95. A dental disease characterised by mottling of teeth due to ingredient in drinking water, namely:
(a) Fluorine (b) Chlorine
(c) Boron (d) Mercury

96. Gaucher's disease is related to:
(a) Abnormal fat metabolism
(b) Vitamin deficiency
(c) Disturbed carbohydrate metabolism
(d) Abnormal protein metabolism

97 Lathyrism due to consumption of khesri dal is characterised by:

Critical Thinking

(a) skeletal deformation and thinning of collagen fibres.
(b) skeletal abnormalities, diabetes mellitus and reproductive failure.
(c) retarded growth, precocious puberty and renal dysfunction.
(d) cardiovascular abnormalities mental retardation and delayed puberty.

98. When breast feeding is replaced by less nutritive food, low in proteins and calories; the infants below the age of one year are likely to suffer from:

Critical Thinking

(a) Pellagra (b) Marasmus
(c) Rickets (d) Kwashiorkor

99. Pancreatic cancer is an especially dangerous disease in people because the pancreas is

(a) the organ that produces and stores bile.
(b) the site of synthesis for all of the essential amino acids.
(c) one of the organs through which food must pass on its way to the colon.
(d) an organ in which many different kinds of digestive enzymes are manufacture.

100. Colostomy is the removal of large parts of the colon. The principal effect of a colostomy is:

(a) Diarrhoea
(b) Constipation
(c) Increased loss of bile pigments
(d) Kwashiorkor and marasmus

101. Jaundice is yellowing of the skin due to accumulation of bilirubin and/or other products of haemoglobin breakdown. Bilirubin is formed in the blood but is further transformed by the liver and excreted in the bile. Which of the following is a cause of jaundice? Critical Thinking

(a) Excessive fragility of RBC
(b) Hepatitis and gall stones
(c) Cancer of pancreas
(d) All of the above

102. Which of the following statement is true?

(a) Pepsin cannot digest casein.
(b) Trypsin can digest collagen.
(c) Pepsin cannot digest collagen.
(d) Chymotrypsin can digest casein.

Directions (Qs. 103 and 104): The given diagram shows the duct systems of liver, gall bladder and pancreas in which few structures are marked as 1, 2, 3 and 4. On the basis of this figure, answer the questions.

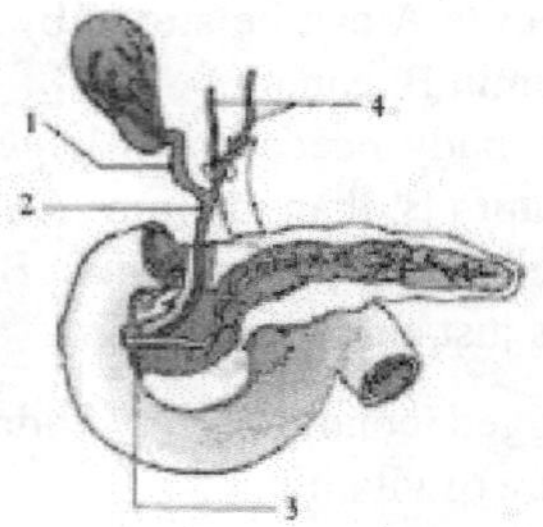

103. Sphincter of Oddi controls the flow of digestive juice by guarding which duct?

(a) 1 (b) 2 (c) 3 (d) 4

104. Which two ducts are responsible for the formation of a duct that carry bile from the gall bladder and conduct it into the first section of the small intestine?

(a) 1 and 2 (b) 2 and 3
(c) 3 and 4 (d) 4 and 1

Answer KEYs

1	(c)	12	(a)	23	(b)	34	(c)	45	(a)	56	(d)	67	(c)	78	(a)	89	(d)	100	(a)
2	(c)	13	(b)	24	(a)	35	(c)	46	(d)	57	(d)	68	(d)	79	(c)	90	(a)	101	(d)
3	(c)	14	(c)	25	(c)	36	(d)	47	(b)	58	(a)	69	(b)	80	(d)	91	(b)	102	(d)
4	(d)	15	(c)	26	(b)	37	(a)	48	(c)	59	(c)	70	(c)	81	(b)	92	(c)	103	(c)
5	(c)	16	(c)	27	(d)	38	(d)	49	(a)	60	(c)	71	(d)	82	(d)	93	(b)	104	(d)
6	(b)	17	(a)	28	(c)	39	(d)	50	(b)	61	(a)	72	(c)	83	(b)	94	(b)		
7	(d)	18	(c)	29	(b)	40	(b)	51	(b)	62	(a)	73	(a)	84	(d)	95	(a)		
8	(b)	19	(b)	30	(c)	41	(d)	52	(c)	63	(a)	74	(a)	85	(b)	96	(a)		
9	(b)	20	(d)	31	(d)	42	(b)	53	(c)	64	(b)	75	(a)	86	(a)	97	(a)		
10	(a)	21	(b)	32	(c)	43	(b)	54	(a)	65	(c)	76	(a)	87	(d)	98	(b)		
11	(a)	22	(d)	33	(a)	44	(b)	55	(b)	66	(b)	77	(d)	88	(a)	99	(d)		

Hints & Solutions

1. (c) 2. (c)
3. (c) Tongue is a freely movable muscular organ that is attached to the floor of the oral cavity by the frenulum.
4. (d) 5. (c)
6. (b) The membranes of the digestive tract are from the inside to the outside: mucosa, submucosa, circular longitudinal muscles and serosa.
7. (d) The opening of hepato-pancreatic duct in the duodenum is guarded by sphincter of Oddi.
8. (b)
9. (b) Bile is drained from the liver by a bile duct which is formed by the joining of a cystic duct from the gall bladder and a common hepatic duct from different liver lobes.
10. (a) Alimentary canal is a long continuous tube having muscular wall and glandular epithelium extending from mouth to anus. Arranged sequentially, it includes the following parts: Buccopharyngeal cavity $\rightarrow$ Oesophagus $\rightarrow$ Stomach $\rightarrow$ Small intestine $\rightarrow$ Large intestine.
11. (a) During swallowing, the epiglottis covers the opening to the airways to prevent the entry of food or liquid, therefore, if the epiglottis does not function properly then one might get congestion.
12. (a) 13. (b) 14. (c) 15. (c)
16. (c) A gastrovascular cavity has a single opening.
17. (a) 18. (c) 19. (b) 20. (d)
21. (b) Herbivores require longer digestive tracts than do carnivores.
22. (d) 23. (b) 24. (a) 25. (c) 26. (b)
27. (d) 28. (c) 29. (b) 30. (c) 31. (d)
32. (c) 33. (a) 34. (c) 35. (c) 36. (d)
37. (a) 38. (d) 39. (d)
40. (b) Contraction of gall bladder is induced by cholecystokinin.
41. (d) Pancreatic lipase (formely called steapsin) hydrolyses fats into glycerol and fatty acids.
42. (b) Because saliva converts starch into maltose.
43. (b) 44. (b) 45. (a)
46. (d) The appendix is the blind ended–tube located at the junction of the small and large intestines. The tissues that make up the appendix and the surrounding gut region are immune tissues and is lymphatic in origin.
47. (b)
48. (c) 49. (a) 50. (b) 51. (b) 52. (c)
53. (c) 54. (a) 55. (b) 56. (d) 57. (d)
58. (a) 59. (c) 60. (c) 61. (a)
62. (a) Deficiency of vitamin B_1 causes beri-beri disease. Beri beri disease was discovered by Eijkman in 1897. Eijkman observed that the disease beri-beri occured mainly in people feeding on a polished rice diet. This disease is rare in the western world because of food habits and food enrichment. Beri-beri disease is characterised by loss of appetite and weight and vigour, retarted growth, degeneration of nerves, muscle atrophy, even paralysis, weakened heart beat and failure.
63. (a) A cartilaginous flap called epiglottis prevents the entry of food into the glottis - opening of the wind pipe - during swallowing.
64. (b) 65. (c)
66. (b) 67. (c) 68. (d) 69. (b)
70. (c) Insulin is secreted by the Islets of Langerhans in the pancreas when the blood sugar level is high. This increases the rate of glucose uptake from the blood into the muscle cells.
71. (d) Parietal cells secrete hydrochloric acid (HCl). HCl makes the medium acidic, which is required for activation of protein digesting enzyme called pepsin.

72. (c)

73. (a) The nutrients are absorbed from the small intestine into the blood and move through the circulatory system to the body cells.

74. (a) 75. (a)

76. (a) Amylase, renin and trypsin are all proteins.

77. (d) Valves of kerckring are large valvular flaps projecting into the lumen of small intestine.

78. (a) Pancreatic juice contains variety of inactive enzymes like trypsinogen, chymotrypsinogen, and carboxypeptidases.

79. (c) The level of glucose will not be affected because glucose level can be maintained by the insulin hormone secreted from pancreas and hormones do not pass through duct.

80. (d) Goblet cells, found in the intestinal mucosal epithelium, secretes mucus. The mucus lubricates the food for its easy passage. So, if for some reason, goblet cells become non-functional, it will adversely affect the smooth movement of food, down the intestine.

81. (b) Mouth releases saliva (secreted from salivary glands) containing salivary amylase enzyme that breaks down carbohydrates. Stomach is the most distensible and widest organ of the alimentary canal. It is an acidic environment where digestion of protein takes place (by breaking proteins into larger polypeptides). Small intestine carries out most of the digestion and absorption of nutrients. Large intestine reclaims water and salts.

82. (d) The correct substrate for lactase is lactose and for lipase is fats.

83. (b) During prolonged fasting, first carbohydrate are used up, followed by fat and proteins towards the end.

84. (d)

85. (b) Excess fat-soluble vitamins can be stored; excess water-soluble vitamins are secreted.

86. (a) 87. (d) 88. (a) 89. (d) 90. (a)

91. (b)

92. (c) Essential amino acids must be acquired through diet because an animal cannot directly synthesise all of the amino acids needed for protein.

93. (b) 94. (b) 95. (a) 96. (a) 97. (a)

98. (b)

99. (d) The pancreas produces a number of important digestive enzymes, without which digestion and nutrient absorption are greatly hampered.

100. (a) One of the major functions of the large intestine is the reabsorption of large quantities of water used in the digestive process. Diarrhoea or excessive watery stool is the principal effect of colostomy.

101. (d) Both gall stones and enlargement of pancreas can produce jaundice by blocking the entry of bile into the duodenum. Hepatitis decreases the liver's ability to excrete bilirubin. Excessive fragile RBC may release so much haemoglobin in the blood that bilirubin is formed in very high concentration, producing jaundice.

102. (d) Milk protein can be digested by pepsin and chymotrypsin.

103. (c) Hepato – pancreatic duct (marked as 3) is guarded by sphincter of Oddi. The sphincter of Oddi (or Hepatopancreatic sphincter) is a muscular valve that controls the flow of digestive juices (bile and pancreatic juice) through the ampulla of vater into the second part of the duodenum.

104. (d) Ducts [marked as 1(duct from gall bladder) and 4 (duct from liver)] are responsible for the formation of common bile duct (marked as 2)].The common bile duct is a small, tube-like structure formed where the common hepatic duct (duct from liver) and the cystic duct (duct from gall bladder) join. Its function is to carry bile from the gallbladder and conduct it into the upper part of the small intestine (the duodenum). The common bile duct is part of the biliary system. Bile, a yellow-brown fluid which helps break down fats in food, is produced by the liver. It is stored and concentrated in the gallbladder until it is needed to help digest food.

17 Breathing and Exchange of Gases

1. Much developed larynx of human male is called:
   (a) Aristole's lantern
   (b) Syrinx
   (c) Adam's apple
   (d) Muller's organ
2. Which of the following feature is common in the alveoli of lungs and villi of intestine in mammals?
   Critical Thinking
   (a) Provide a large surface area.
   (b) Have ciliated epithelium.
   (c) Are suited for diffusion of gases.
   (d) Have rich supply of blood vessels and lymph ducts.
3. Left lung of human has:
   (a) 2 lobes (b) 4 lobes
   (c) 3 lobes (d) 5 lobes
4. The alveoli of the lungs do not contain "air" because:
   (a) we normally do not ventilate our lungs at a high enough rate.
   (b) the lungs have too many alveoli to ventilate.
   (c) there is "dead space" in the trachea and bronchi.
   (d) the trachea and bronchi are too small in volume.
5. The process of migration of chloride ions from plasma to RBC and carbonate ions from RBC to plasma is:
   (a) Chloride shift (b) Ionic shift
   (c) Atomic shift (d) Na^+ pump
6. At rest, most people require an alveolar ventilation for about 4 L/minute. Suppose an emphysema victim with a tidal volume of 0.5 L has suffered alveolar damage that has produced a respiratory dead space of 0.3 L, what will be that person's frequency of breathing at rest?

   (a) 5 breaths/min (b) 8 breaths/min
   (c) 20 breaths/min (d) 50 breaths/min
7. A zoologist compared the respiratory efficiency and swimming speed of different fish. He found that less efficient fish tended to have:
   Tricky
   (a) greater ventilation
   (b) a thicker respiratory surface
   (c) more haemoglobin
   (d) a faster heart rate
8. Ciliated epithelium in trachea of mammals helps in:
   (a) sucking inspired air in.
   (b) perceiving sense of smell.
   (c) pushing expired air out.
   (d) pushing mucus out.
9. When, under certain conditions, the p50 value of haemoglobin rises, the affinity of the pigment of combining with O_2 will:
   (a) Remain same
   (b) Rise
   (c) Fall
   (d) First rise and then fall
10. Partial pressure of oxygen in inspired and expired air is ……. and ……. mm of Hg.
    (a) 100, 46 (b) 158, 40
    (c) 158, 90 (d) 100, 95
11. The given diagram represents the human respiratory system with few structures labelled as 1, 2, 3 and 4.

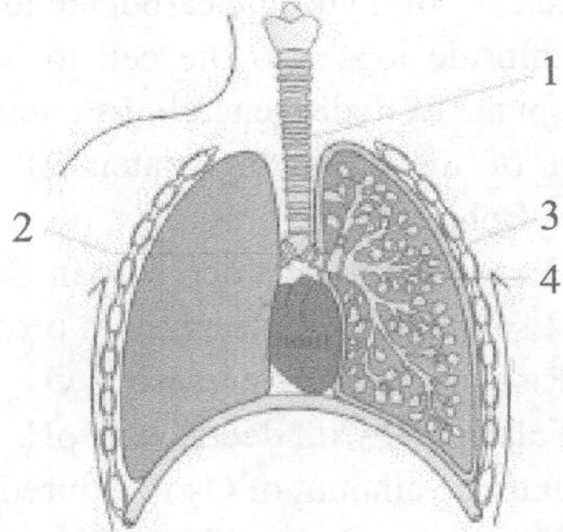

The exchange of gases takes place in which labelled structure?
    (a) 1 : Trachea (b) 2 : Bronchi
    (c) 3 : Bronchioles (d) 4 : Alveoli

12. Which of the following statement sequences is correct to initiate expiration?
    (i) Relaxation of external intercostal muscles and return of diaphragm and sternum to their normal position.
    (ii) Air expelled from lungs.
    (iii) Volume of thorax decreases.
    (iv) Intrapulmonary pressure increases.
    (a) (i), (iii), (iv), (ii) (b) (ii), (iv), (iii), (i)
    (c) (iv), (iii), (ii), (i) (d) (i), (ii), (iii), (iv)
13. The graph shows four dissociation curves.

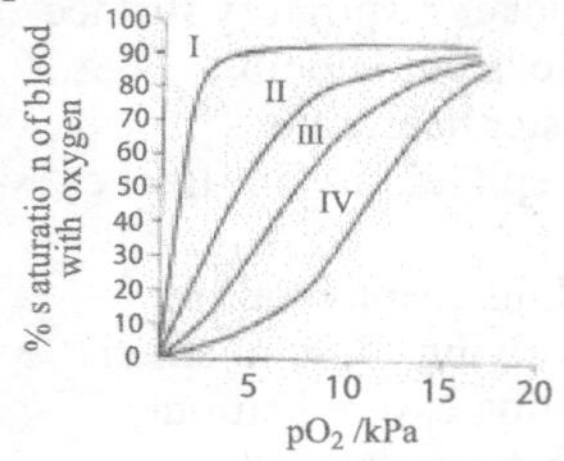

Which curve represents the oxygen dissociation curve for myoglobin?
    (a) I (b) II
    (c) III (d) IV
14. The chloride shift in red blood cells is the movement of: **Critical Thinking**
    (a) cations into the cell to balance its uptake of chloride ions.
    (b) chloride ions from the blood into the glomerular filtrate for excretion.
    (c) chloride ions out of the cell to balance the release of hydrogencarbonate ions.
    (d) chloride ions into the cell to balance its uptake of hydrogencarbonate ions.
15. Which of the following statement **correctly** defines Bohr effect?
    (a) Rise in p50 with a decrease in CO_2 conc.
    (b) Rise in p50 with decrease in pH.
    (c) Rise in p50 with increase in O_2.
    (d) Fall in p50 with decrease in pH.
16. Approximate amount of O_2 in expired air is:
    (a) 16% (b) 4.6%
    (c) 19% (d) < 1%
17. R.Q. is less than 1 for:
    (a) Starch (b) Organic acid
    (c) Protein (d) Carbohydrate
18. The alveoli contain specialised cells that secrete a phospholipoprotein complex known as pulmonary surfactant. The function of pulmonary surfactant is: **Tricky**
    (a) to provide some rigidity to the alveoli to prevent lung collapse.
    (b) to decrease the surface tension of the alveoli, making it easier to expand them.
    (c) to facilitate the diffusion of O_2 and CO_2 by providing protein channels through which these gases flow.
    (d) to provide a sticky surface upon which dust and microbes are trapped and disposed of.
19. Because of the relatively high altitude of Antonito, Colorado, the town has a normal barometric pressure of about 600 mm Hg rather than 760 mm Hg as at sea level. The partial pressure of oxygen in Antonito's air is approximately: **Critical Thinking**
    (a) 75 mm Hg (b) 126 mm Hg
    (c) 160 mm Hg (d) 760 mm Hg
20. Smoking destroys the cilia in the respiratory passage ways. This:
    (a) makes it harder to move air in and out of the lungs.
    (b) decreases the surface area for respiration.
    (c) slows blood flow through lung blood vessels.
    (d) makes it harder to keep the lungs clean.
21. Breathing with the help of ribs is more pronounced in:
    (a) Males (b) Females
    (c) Pregnant females (d) None of these
22. The function of nasal cavity and nasopharynx is to:
    (a) warm the inspired air.
    (b) moisten the inspired air.
    (c) filter out the dust particles from the inspired air.
    (d) All of the above
23. Switch off centre for breathing lies in:
    (a) Medulla oblongata (b) Hypothalamus
    (c) Pons (d) Carotid bodies
24. The tissue respiration refers to:
    (a) Inspiration
    (b) Expiration
    (c) External respiration
    (d) Internal respiration
25. Total lungs capacity is approximately:
    (a) 1200 ml (b) 2400 ml
    (c) 4000 ml (d) 5800 ml

26. Voice in mammal is produced:
(a) by syrinx (b) during inhalation
(c) by bronchus (d) during exhalation

27. The vocal cords are attached to which of the following paired cartilage of larynx?
(a) Cricoid (b) Thyroid
(c) Arytenoid (d) All of these

28. Which of the following statements sequences is correct to initiate inspiration?
(i) The contraction of external intercoastal muscles raises the ribs and sternum.
(ii) Volume of thorax increases in the dorso-ventral axis.
(iii) Increase in pulmonary volume.
(iv) Intrapulmonary pressure decreases.
(v) Volume of thorax increases in the anterior-posterior axis.
(a) (i), (ii), (iv), (v), (iii)
(b) (i), (ii), (iii), (iv), (v)
(c) (i), (ii), (iv), (iii), (v)
(d) (v), (i), (ii), (iii), (iv)

29. The 'blue baby' syndrome results from:
(a) Methaemoglobin
(b) Excess of dissolved oxygen
(c) Excess of TDS (total dissolved solids)
(d) Excess of chloride

30. A person can hold his breath longer if he gradually exhales than if he keeps his lungs fully expanded. This phenomenon is due to the fact that:

Critical Thinking

(a) concentration of CO_2 in the blood decreases, lessening the excitatory input to the carotid bodies.
(b) he is then using less O_2 as his thoracic muscles are not working as hard.
(c) stretch receptors in his lungs are then not firing as much thus lessening the excitatory input to the expiratory centres.
(d) he is preventing the pressure in his lungs from increasing too much.

31. Hamburger phenomenon is also known as:
(a) Calcium shift (b) Bohr effect
(c) Chloride shift (d) Na^+-K^+ pump

32. What would happen if human blood becomes acidic (low pH)?
(a) Oxygen-carying capacity of haemoglobin decreases.
(b) Oxygen-carrying capacity of haemoglobin increases.
(c) RBC count increases.
(d) RBC count decreases.

33. Amount of O_2 normally carried by 100 mL. of pure blood is:
(a) 40 mL (b) 20 mL
(c) 10 mL (d) 30 mL

34. Although much CO_2 is carried in blood, yet blood does not become acidic, because:

Critical Thinking

(a) it is absorbed by the leucocytes.
(b) blood buffers play an important role in CO_2 transport.
(c) it combines with water to form H_2CO_3which is neutralised by $NaCO_3$.
(d) it is continuously diffused through tissues and is not allowed to accumulate.

35. If the thoracic wall of the lungs is punctured then:
(a) the lungs get inflated.
(b) the man dies as the lungs get collapsed.
(c) the breathing rate decreases.
(d) the breathing rate increases.

36. The covering of lungs is called:
(a) Pleura (b) Pericardia
(c) Peritoneum (d) Mediastinum

37. Alveoli increase the surface area by:

(a) 10 m^2 (b) 100 m^2
(c) 150 m^2 (d) 200 m^2

38. The percentage of carbon dioxide, oxygen and nitrogen in exhaled air of man are about:

Tricky

(a) 0.04: 16: 79 (b) 0.4: 20: 70
(c) 4.0 : 20: 79 (d) 4.0 : 16: 79

39. A biochemist mixed 10 drops of acid with 100 mL of water, and the pH dropped from 7.4 to 5.0. She, then mixed 10 drops of acid with 100 mL of blood. The pH dropped from 7.4 to 7.2. What is the reason for this difference?

Critical Thinking

(a) Blood is thicker than water.
(b) Blood is already very acidic, so the acid has less effect.
(c) Blood is saturated with oxygen; there is little room for acid.
(d) Blood contains buffers that reduce pH change.

40. Forced deep breathing for a few minutes by a person at rest may be followed by a temporary cessation of breathing, because of:

Critical Thinking

(a) too much O_2 in blood.
(b) too much CO_2 in blood.
(c) little CO_2 in blood.
(d) both too much O_2 and very little CO_2 in blood.

41. In the given figure, label A represents __________ while label B represents __________.

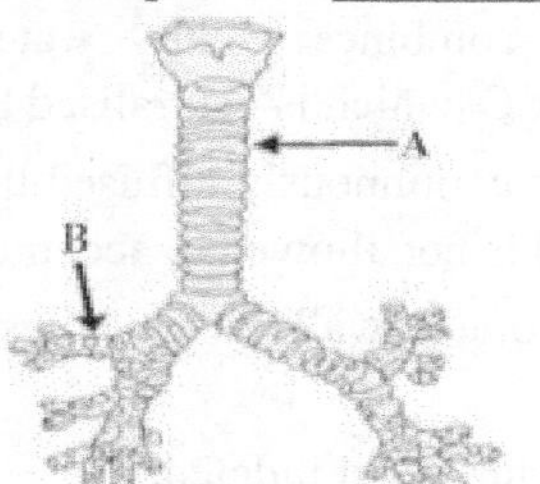

(a) A- Trachea, B- Bronchus
(b) A- Alveolus, B- Bronchiole
(c) A-Bronchiole, B- Trachea
(d) A- Trachea, B- Bronchiole

42. Patients with chronic lung disease and difficulty breathing often adapt to the high concentration of CO_2 in their blood. The breathing centres stop responding to CO_2 level. If such a patient has difficulty in breathing, medical personnel are reluctant to give the patient pure oxygen. Based on what you know about control of breathing, why do you think this is the case?

Toughnut

(a) The patient's body would use the oxygen to make even more CO_2.
(b) The oxygen would increase concentration of bicarbonate, altering pH.
(c) Increased oxygen in the blood might slow or stop breathing.
(d) The body is not used to the oxygen, and the patient would overdose.

43. The graph shows dissociation curves for four respiratory pigments.

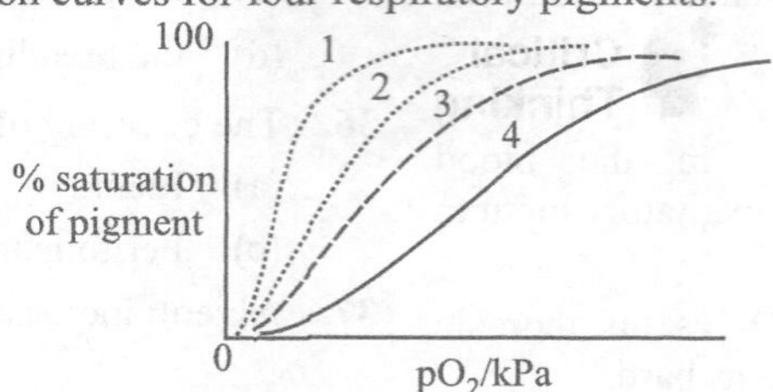

Which combination identifies the four curve?

	Adult Haemoglobin	Haemoglobin with high CO_2	Foetal Haemoglobin	Adult myoglobin
(a)	1	2	3	4
(b)	2	1	4	3
(c)	3	2	1	4
(d)	3	4	2	1

44. Match the disorders given in column-I with symptoms under column-II. Choose the answer which gives the correct combination of alphabets with numbers.

	Column-I		Column-II
A.	Asthma	I.	Inflammation of nasal tract
B.	Bronchitis	II.	Spasm of bronchial muscles
C.	Rhinitis	III.	Fully blown out alveoli
D.	Emphysema	IV.	Inflammation of bronchi
		V.	Cough with blood strained sputum

(a) A – IV, B – II, C – V, D – I
(b) A – V, B – III, C – II, D – I
(c) A – III, B – I, C – V, D – IV
(d) A – II, B – IV, C – I, D – III

45. After taking a long deep breath, we do not respire for some seconds due to: Critical Thinking

(a) more CO_2 in blood.
(b) more O_2 in blood.
(c) less CO_2 in blood.
(d) less O_2 in blood.

46. Bicarbonate ions can be generated in:

(a) Lymphocytes (b) Neutrophil
(c) Basophil (d) RBCs

47. A person receives a serious cut on the back of the neck that is deep enough to severely damage the brain stem in the lower medulla. The respiratory effect of this type of wound would likely be:

(a) catastrophic and it might result in a complete cessation of breathing.
(b) to reduce respiration to an irregular rhythm.
(c) to shift the oxygen-binding curve of haemoglobin to the left.
(d) to stimulate the production of additional diphosphoglyceric acid.

48. Which fact suggests that most O_2 is transported from lungs to the tissues combined with haemoglobin rather than dissolved in blood plasma? Toughnut

(a) O_2-carrying capacity of whole blood is much higher than that of plasma and O_2 content of blood leaving the lungs is greater than that of blood entering the lungs.
(b) Haemoglobin can combine with O_2.
(c) Oxyhaemoglobin can dissociate into haemoglobin and O_2.
(d) Increase in CO_2-concentration decreases the O_2-affinity of haemoglobin.

49. Presence of large number of alveoli around alveolar ducts opening into bronchioles in mammalian lungs is: Critical Thinking

(a) inefficient system of ventilation with little of residual air.
(b) inefficient system of ventilation with high percentage of residual air.
(c) an efficient system of ventilation with no residual air.
(d) an efficient system of ventilation with little residual air.

50. In expiration, diaphragm becomes:

(a) Flattened (b) Contract
(c) Straightened (d) Arched

51. Common feature of human and insect trachea is:

(a) Non-collapsible wall
(b) Supporting rings
(c) Ectodermal origin
(d) Endodermal origin

52. The impulse for voluntary (forced) breathing starts in:

(a) Medulla
(b) Vagus
(c) Cerebral hemisphere
(d) Spinal cord

53. Membrane separating air in pulmonary alveoli from blood capillaries is:

(a) Alveolar epithelium
(b) Cardiac epithelium
(c) Endothelium of blood capillaries
(d) Both (a) and (c)

54. During CO_2 transport, diffuses from erythrocytes to plasma and in turn upsets the ionic equilibrium momentary, in order to keep the ionic balance, and equal number of Cl^- ions pass into the erythrocytes from plasma. This process is known as:

(a) Bicarbonate shift
(b) Carbonation
(c) Hamburger phenomenon
(d) Carbo-chlorosis

55. Arrange the given steps of respiration in the sequence of event they occur.

Choose the correct options given below.

(i) Diffusion of gases, O_2 and CO_2 across the alveolar membrane.
(ii) Transport of gases by the blood.
(iii) Utilisation of O_2 by the cells for catabolic reactions and the resultant releases of CO_2.
(iv) Pulmonary ventilation by which atmospheric air is drawn in and CO_2 rich alveolar air is released out.
(v) Diffusion of O_2 and CO_2 between the blood and tissue.

(a) (iii), (v), (ii), (i), (iv)
(b) (iii), (ii), (v), (i), (iv)
(c) (v), (iv), (iii), (ii), (i)
(d) (i), (ii), (iii), (iv), (v)

56. During oxygen transport, the oxyhaemoglobin at the tissue level liberates oxygen to the cells because in tissue:

Critical Thinking

(a) O_2 concentration is high and CO_2 is low.
(b) O_2 concentration is low and CO_2 is high.
(c) O_2 tension is high and CO_2 tension is low.
(d) O_2 tension is low and CO_2 tension is high.

57. Asthma is caused due to:

(a) Infection of lungs
(b) Spasm in bronchial muscles
(c) Bleeding into pleural cavity
(d) Infection of trachea

58. The Bohr shift describes:

(a) the outward movement of Cl^- from the blood cell in exchange for HCO_3^- moving into the cell.
(b) the leftward shift of the entire oxygen equilibrium curve when temperature rises.
(c) the rightward shift of the entire oxygen equilibrium curve when pH rises.
(d) the rightward shift of the entire oxygen equilibrium curve when pH falls.

59. Which of the following gases makes the most stable combination with the haemoglobin of RBCs?

(a) CO_2 (b) CO
(c) O_2 (d) N

60. When the carbon dioxide concentration in blood increases, the rate of breathing will:

(a) Remain unaffected (b) Decrease
(c) Stop (d) Increase

61. Which of the following statements are true/false?

(i) The blood transports CO_2 comparatively easily because of its higher solubility.
(ii) Approximately 8.9% of CO_2 is transported being dissolved in the plasma of blood.
(iii) The carbon dioxide produced by the tissues, diffuses passively into the blood stream and passes into red blood corpuscles and react with water to form H_2CO_3.
(iv) The oxyhaemoglobin (HbO_2) of the erythrocytes is basic.
(v) The chloride ions diffuse from plasma into the erythrocytes to maintain ionic balance.

(a) (i), (iii) and (v) are true, (ii) and (iv) are false.
(b) (i), (iii) and (v) are false, (ii) and (iv) are true.
(c) (i), (ii) and (iv) are true, (iii) and (v) are false.
(d) (i), (ii) and (iv) are false, (iii) and (v) are true.

62. Which of the given statements are **incorrect**?

(i) A healthy human on an average breathes 12-16 times/minute.
(ii) The volume of air involved in the breathing movements can be estimated by spirometer.
(iii) Diaphragm is very useful in both inspiration and expiration.

(a) (i) and (ii)
(b) (ii) and (iii)
(c) (i) and (iii)
(d) None of the above

63. Gas exchange between maternal and foetal blood takes place in the placenta. Foetal and adult haemoglobin differ slightly in composition, and their dissociation curves are different so that oxygen can be transferred from maternal blood to foetal blood. If curve III represents the dissociation curve of adult haemoglobin, which curve would represent foetal haemoglobin ? (Hint : Must foetal haemoglobin have a higher or lower affinity for oxygen than maternal haemoglobin has ?)

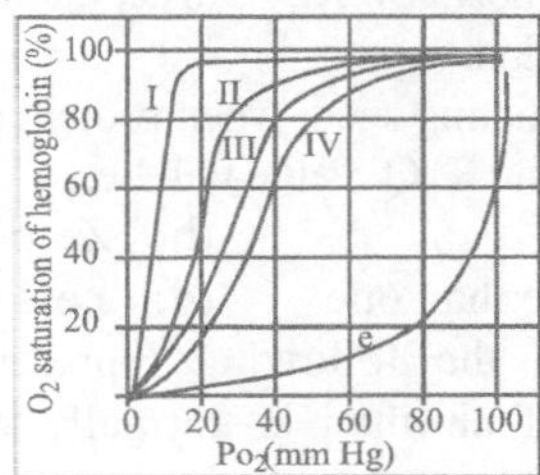

(a) I (b) II
(c) III (d) IV

64. The breathing rate gets increased by the increase in the content of:
(a) Nitrogen (b) Oxygen
(c) Carbon dioxide (d) Carbon monoxide

65. In an old science fiction movie, the hero tried to drown a giant ant by holding its head under water. Would this work? Why? Toughnut

(a) Yes, Ants use lungs to breath much as we do.
(b) Yes, The skin surface, covered with water, could not get O_2 from the air.
(c) No, Ants use gills for respiration, like crabs do.
(d) No, Ants breathe through holes in the sides of their bodies.

66. Respiratory system is derived from:
(a) Ectoderm (b) Mesoderm
(c) Endoderm (d) None of these

67. When you exhale, the diaphragm:
(a) Relaxes and arches
(b) Relaxes and flattens
(c) Contracts and arches
(d) Contracts and flattens

68. About how many times per minute a man normally respires? Tricky
(a) 12 to 15 (b) 10
(c) 72 (d) 100

69. Which of the following is not involved in the neural control of ventilation?

(a) Neurons in the medulla
(b) The vagus nerve
(c) The contraction state of the diaphragm
(d) Chemosensors on the surface of the medulla

70. In which of the following animals respiration occurs without a respiratory organ?
(a) Fish (b) Cockroach
(c) Tadpole (d) Earthworm

71. Book lungs are respiratory organs of:
(a) Molluscs (b) Mammals
(c) Arachnids (d) Earthworm

72. Which one of the following is called inspiratory muscle in mammals?
(a) Radial muscle of diaphragm
(b) External intercostal muscle
(c) Internal intercostal muscle
(d) Pleural muscle

73. The respiratory centre, which regulates respiration, is located in: Critical Thinking
(a) Cerebral peduncle
(b) Vagus nerve
(c) Pons
(d) Medulla oblongata

74. Which one of the following can respire in the absence of oxygen? Critical Thinking
(a) *Amoeba*
(b) Tapeworm
(c) House fly
(d) *Hydra*

75. The disease that occurs when the haemoglobin content of the blood goes down is:
(a) Pleurisy
(b) Emphysema
(c) Anaemia
(d) Pneumonia

76. Refer the given figure and answer the question.

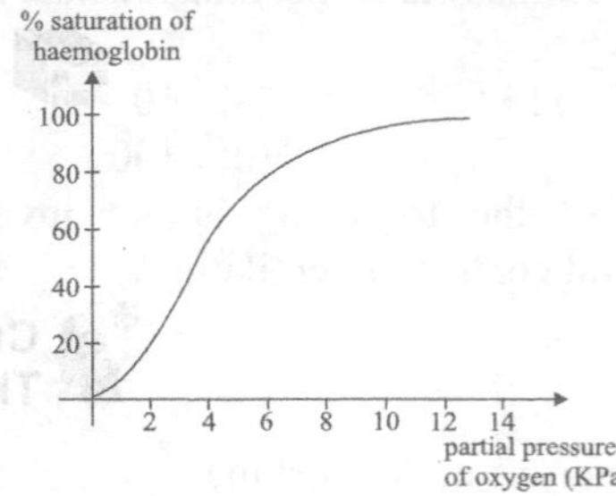

Which of the following statement is **correct** regarding the above figure?
(a) When percentage saturation of haemoglobin is plotted against the partial pressure of oxygen, a sigmoid curve is obtained.
(b) Binding of oxygen with haemoglobin is primarily related to partial pressure of carbon monoxide.
(c) The given graph illustrates the amount of HbO_2 as similar to Hb at different pO_2.
(d) None of the above

77. When CO_2 concentration in blood increases, breathing becomes:
(a) Slow and deep
(b) Faster and deeper
(c) Shallower and slow
(d) There is no effect on breathing

78. Exposure to carbon monoxide (from coal gas) is extremely dangerous and can kill a patient because:

Critical Thinking

(a) the compound carboxy-haemoglobin is formed with haemoglobin which can gradually clot the blood resulting in circulatory failure.
(b) carboxy-haemoglobin reduces the ability of blood to transport oxygen by rupturing a vast majority of erythrocytes.
(c) carboxy-haemoglobin greatly modifies the structure of haemoglobin, thus making it lose its affinity for oxygen.
(d) the compound formed, carboxy-haemoglobin does not allow RBCs to act for their respiratory function.

79. In a germinating seed, when protein is aerobically oxidised, the R. Q. value will be:
(a) One (b) Zero
(c) More than one (d) Less than one

80. Which of the following represents a larger volume of air than is normally found in the resting tidal volume of a human lung?

(a) Residual volume
(b) Inspiratory reserve volume
(c) Expiratory reserve volume
(d) All of the above

81. The total lung capacity is represented by:

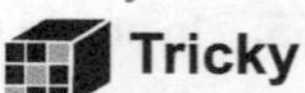

(a) Tidal volume + vital capacity
(b) Tidal volume + residual volume
(c) Vital capacity + residual volume
(d) Inspiratory and expiratory reserve volumes

Answer KEYs

1	(c)	10	(b)	19	(b)	28	(c)	37	(b)	46	(d)	55	(a)	64	(c)	73	(d)
2	(a)	11	(d)	20	(d)	29	(a)	38	(d)	47	(a)	56	(d)	65	(d)	74	(b)
3	(a)	12	(a)	21	(c)	30	(c)	39	(d)	48	(a)	57	(b)	66	(c)	75	(c)
4	(c)	13	(a)	22	(d)	31	(c)	40	(d)	49	(d)	58	(d)	67	(a)	76	(a)
5	(a)	14	(c)	23	(c)	32	(a)	41	(d)	50	(d)	59	(b)	68	(a)	77	(b)
6	(c)	15	(b)	24	(d)	33	(b)	42	(c)	51	(a)	60	(d)	69	(c)	78	(d)
7	(b)	16	(a)	25	(d)	34	(b)	43	(d)	52	(c)	61	(a)	70	(d)	79	(d)
8	(d)	17	(c)	26	(d)	35	(b)	44	(d)	53	(d)	62	(d)	71	(c)	80	(d)
9	(c)	18	(b)	27	(c)	36	(a)	45	(c)	54	(c)	63	(b)	72	(b)	81	(c)

Hints & Solutions

1. (c) Much developed larynx of human male is called Adam's apple.
2. (a) 3. (a)
4. (c) The alveoli do not contain "air" with 20.9% O_2 because incoming air is mixed with air left in the "dead space" of the trachea and bronchi that has had some of the O_2 removed by the lungs.
5. (a) During the transport of CO_2 through the blood, bicarbonate ions diffuse out of RBCs while chloride ions from plasma, enter the RBCs to maintain ionic equilibrium. This is called chloride shift.
6. (c) $f = \frac{A}{T-D} = \frac{4000\,ml/min}{500\,ml - 300\,ml}$

$\frac{4000}{200} = 20$ breaths/min
7. (b) 8. (d) 9. (c) 10. (b)
11. (d) In the diagram, label 1, 2, 3 and 4 represents trachea, bronchi, bronchioles and alveoli respectively. Alveoli are sac like structure present in lungs. They are the site of exchange of gases.
12. (a)
13. (a) Myoglobin has only one haem group and there can be non co-operative binding, hence the shape of the dissociation curve is hyperbolic rather than sigmoidal. Also, myoglobin has a very high affinity for oxygen and does not release oxygen until a very low oxygen tension.
14. (c) Most of the hydrogencarbonate ions formed in the red blood cells diffuse out of the plasma along a concentration gradient and combine with sodium in the plasma to form sodium hydrogencarbonate. The loss of negatively-charged hydrogencarbonate ions leaves them with a more positive charge that is balanced by chloride ions diffusing into the red blood cells from the plasma.
15. (b) 16. (a) 17. (c)
18. (b) Thin layer of water on the inner surface of alveoli serves to contract the alveoli and makes them behave like unfilled balloons that resist inflation. This would mean tremendous effort to inhale and expand the alveoli and therefore to overcome this, cells of the alveoli secrete pulmonary surfactant to decrease the surface tension of water lining the alveoli.
19. (b) Air at sea level with an atmospheric pressure of 700 mm Hg has a partial pressure of oxygen of 160 mm Hg (700 x 21). Therefore, the partial pressure of O_2 at Antonito, Colorado, is 600 x 21 = 126 mm Hg.
20. (d) Debris normally gets trapped in mucus lining of the respiratory passage ways. This mucus is then swept up and away from the lungs by the action of cilia.
21. (c) 22. (d) 23. (c)
24. (d) 25. (d) 26. (d) 27. (c)
28. (c) 29. (a)
30. (c) When the chest is expanded, stretch receptors in the lungs relax and excite the expiratory centre in the mdeulla which tends to produce expiration. Partial exhalation lessens the intensity of this reflex, allowing breath to be held for a longer time. Exhalation does not change the blood CO_2 concentration unless fresh air (low in CO_2) is brought into the lungs, allowing more CO_2 to diffuse out of the blood.
31. (c) 32. (a) 33. (b)
34. (b) CO_2 enters RBC and reacts with water to form carbonic acid. Carbonic acid dissociates to form bicarbonate and hydrogen ions. Some bicarbonate ions are transported in erythrocytes while some diffuse into the blood plasma. Exit of bicarbonate ions change the ionic balance between the plasma and erythrocytes. To restore this balance chloride ions diffuse from plasma into erythrocytes. Due to this the pH of blood is maintained.
35. (b) 36. (a)
37. (b) Alveoli are tiny air sacs within the lungs where exchange of oxygen and carbon dioxide takes place. These tiny alveoli, structures taken all together form a very large surface area to do the work of breathing. The alveoli cover a surface that measures more than 1,076.4 square feet (100 square metre).

38. (d)
39. (d) The blood contains several buffering systems.
40. (d)
41. (d) The label 1 represents trachea while label 2 represents bronchiole. Trachea (wind pipe) carries air between larynx and bronchi and bronchioles are the terminal part of bronchi.
42. (c) Since carbon dioxide levels are not being responded to, the signal to inhale and exhale is dependent upon oxygen levels.
43. (d) Myoglobin has the greatest affinity for oxygen and does not unload the oxygen until very low partial pressures of oxygen (below 20 mmHg). Foetal haemoglobin has a higher affinity for oxygen than adult haemoglobin, thus facilitating the transfer of oxygen from maternal haemoglobin to foetal, haemoglobin at the placenta. When respiration is occurring at fast rate, more oxygen is needed. High levels of respiration are associated with high levels of carbon dioxide, hence the affinity of haemoglobin for oxygen decreases at higher carbon dioxide partial pressures so that oxygen can be released to respiring tissues.
44. (d) 45. (c) 46. (d)
47. (a) Cutting the brain stem below the medulla eliminates neural contact with the respiratory mechanism and could easily result in a complete stoppage of breathing.
48. (a) 49. (d) 50. (d) 51. (a)
52. (c) 53. (d) 54. (c)
55. (a) Major steps involving respiration are

Step 1: Utilisation of O_2 by cell of catabolic reactions.

Step 2: Diffusion of O_2 and CO_2 between blood and tissues.

Step 3: Transportation by blood.

Step 4: Diffusion of gases (O_2 and CO_2) through alveolar membrane.

Step 5: CO_2 goes out and atmospheric air is drawn in.

56. (d) 57. (b)
58. (d) The Bohr effect describes the action of pH on the oxygen-binding curve. Decreases in pH result in a net right shift of the entire oxygen equilibrium curve.
59. (b) 60. (d) 61. (a)
62. (d) Diaphragm is very important in both expiration and inspiration. On an average, a healthy human breathes 12-16 times/minute. The volume of the air involved in breathing movements can be calculated by using a spirometer which helps in the clinical assessment of pulmonary functions.
63. (b) 64. (c)
65. (d) Openings to the tracheal system are found along the sides of the body.
66. (c)
67. (a) When you exhale, the diaphragm relaxes and arches. This decreases the size of the thoracic cavity, which increases the pressure within the thoracic cavity, forcing air out of the lungs.
68. (a)
69. (c) The contraction state of the diaphragm does not influence the neural control of ventilation.
70. (d) 71. (c) 72. (b) 73. (d)
74. (b)
75. (c) The disease that occurs when the haemoglobin content of the blood goes down is anaemia.
76. (a) The given graph shows the oxygen – haemoglobin dissociation curve. The relationship between the pO_2 and the percent saturation of haemoglobin when represented on a graph is termed as oxygen – haemoglobin dissociation curve. It is S-shaped because of the behaviour of the Hb in different pO_2.
77. (b) 78. (d) 79. (d)
80. (d) All of the listed choices have a larger volume than the resting tidal volume.
81. (c)

18 Body Fluids and Circulation

1. Which of the following is the largest lymphatic vessels of the human body?
(a) Lacteal duct
(b) Thoracic duct
(c) Cisterna chyli
(d) Right lymphatic duct

2. Which of the following cations is required for the conversion of prothrombin into active thrombin by thromboplastin? **Critical Thinking**
(a) Cu^{2+} (b) Fe^{3+}
(c) Fe^{2+} (d) Ca^{2+}

3. The normal pH of the arterial blood is:
(a) 6.8 (b) 7.8
(c) 7.4 (d) 8.8

4. If cardiac supply of vagus nerve is cut then: **Tricky**
(a) Rate of heart beat increases.
(b) Rate of heart beat decreases.
(c) There is no change in heart beat.
(d) Heart stops beating.

5. Blood pressure increases and heart rate decreases in response to:
(a) Exercise
(b) Haemorrage
(c) Exposure to high altitude
(d) Increased intracranial pressure

6. Systolic pressure of heart is higher than diastolic pressure, because **Critical Thinking**
(a) blood is forcefully pumped into arteries by the heart during systole and not during diastole.
(b) arteries offer resistance to the flowing of blood during systole only.
(c) arteries contract during systole only.
(d) volume of blood in heart is greater during systole than during diastole.

7. Match column-I with column-II and choose the correct option.

	Column-I		Column-II
A.	Basophils	I.	Phagocytes
B.	Neutrophils	II.	Secrete histamine, serotonin, heparin and involved in inflammatory response
C.	Monocytes	III.	Resist infections and are also involved in allergic reaction
D.	Eosinophils	IV.	Immunity
E.	Lymphocytes		

(a) A – II; B and C – I; D – III; E – IV
(b) A – II; B and C – III; D – I; E – IV
(c) A – III; B and C – I; D – II; E – IV
(d) A – IV; B and C – III; D – I; E – II

8. The given diagram represents human heart with four chambers labelled as 1, 2, 3 & 4?

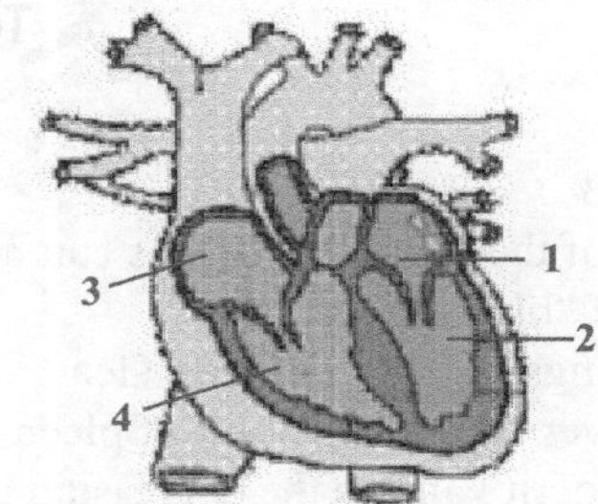

Which labelled structure receives carbon dioxide rich blood from the body?
(a) 1- Left atrium (b) 2- Left ventricle
(c) 3- Right atrium (d) 4- Right ventricle

9. Anaemia due to failure of maturation of RBC is:
(a) Von Kluger's disease
(b) Microcytic anaemia
(c) Normocytic anaemia
(d) Pernicious anaemia

10. Which of the following accelerates extrinsic or tissue thromboplastin formation? Tricky
(a) Factor VII (proconvertin)
(b) Factor XI (plasma thromboplastin antecedent)
(c) Heparin
(d) Calcium

11. Which of the following structures of the lymphatic system acts primarily as a filter for detecting and destroying microorganisms in lymph traveling through major lymph vessels? Critical Thinking
(a) Lymph nodes
(b) Thymus
(c) Lymph capillaries
(d) Tonsil, but not the appendix

12. pH of blood in arteries and veins is:
(a) Same
(b) More in arteries and less in veins
(c) More in veins and less in arteries
(d) None of the above

13. The blood protein involved in blood coagulation is:
(a) Heparin (b) Prothrombin
(c) Thrombin (d) All of these

14. An open circulatory system occurs in:
(a) Reptiles (b) Man
(c) Animals (d) Insects

15. In the ABO system of blood groups, if both antigens are present but no antibody, the blood group of the individual would be: Toughnut
(a) B (b) O
(c) AB (d) A

16. Which of the following organs can be called as a sort of "blood bank"?
(a) Lungs (b) Heart
(c) Liver (d) Spleen

17. Haematocrit value is the expression of:
(a) ratio of oxyhaemoglobin and reduced haemoglobin in blood.
(b) ratio of normal and abnormal haemoglobin in blood.
(c) ratio of red blood corpuscles to plasma in blood.
(d) None of the above.

18. Which one of the following blood vessel is without valves?
(a) Artery (b) Pulmonary aorta
(c) Vein (d) Systemic aorta

19. A vein possesses a large lumen because:
(a) tunica media and tunica externa form a single coat.
(b) tunica media and tunica interna form a single coat.
(c) tunica interna, tunica media and tunica externa are thin.
(d) tunica media is a thin coat.

20. The purpose of the AV node is to ____ and the purpose of the Purkinje fibres is to___. Toughnut
(a) create simultaneous atrial and ventricular depolarisation; speed up transmission of the cardiac impulse into the ventricle.
(b) delay ventricular depolarisation relative to atrial depolarisation; insulate the cardiac impulse from the general ventricular fibres.
(c) delay ventricular depolarisation relative to atrial depolarisation; transmit the cardiac impulse to very small localised groups of ventricular fibres.
(d) delay atrial depolarisation relative to ventricular depolarisation; transmit the cardiac impulse to very small localised groups of ventricular fibres.

21. If cardiac output is 6.0 L/min and 1.44 L of urine is formed daily, what percentage of flow in the nephrons and collecting ducts is reabsorbed (kidneys receive 20% of cardiac output)? Tricky
(a) 0.1% (b) 0.02%
(c) 99.9% (d) Data insufficient

22. The pacemaker of the human heart is:
(a) SA node (b) Tricuspid valve
(c) AV node (d) All of these

23. Papillary muscles are associated with:
(a) Ventricles (b) Auricles
(c) Dorsal aorta (d) Sinus venosus

24. If nerves of heart are cut, then heart will:
(a) Beat rhythmically
(b) Stop working
(c) Beat arhythmically
(d) Shrink

25. Cardiac output is:
(a) the product of heart rate and stroke volume.
(b) the product of auricular and ventricular volume.
(c) the blood pumped in one minute.
(d) Both (a) and (c)

26. In the human body, the highest blood pressure can be recorded in:
(a) arteries near the heart.
(b) veins returning to the heart.
(c) capillaries of the body tissue.
(d) the lungs.

27. The following statement contains some of the events that lead to a 'heart attack'.
(i) Blood clot that breaks, loose and passes along blood vessel.
(ii) Region of heart muscle suffers a myocardial infarction.
(iii) Narrow branch of coronary artery becomes blocked by thrombus.
(iv) Thrombus forms on inner surface of coronary artery.
The correct sequence of these events is:
(a) (i), (iv), (ii), (iii)
(b) (i), (iv), (iii), (ii)
(c) (iv), (i), (ii), (iii)
(d) (iv), (i), (iii), (ii)

28. Which of the following term is correctly matched with its description?
(a) Plasma – Straw coloured fluid that consists of suspended blood cells
(b) Heart – Site where exchange of nutrients and gases occur
(c) Capillary – A red colour fluid that provides food to the cells
(d) Blood – Pumping station of the body

29. Carotid artery carries:
(a) impure blood from the brain.
(b) impure blood to kidneys.
(c) oxygenated blood to the brain.
(d) oxygenated blood to the heart.

30. Pulmonary aorta in mammalian embryo communicates with carotico-systemic aorta by a thin vessel called ductus arteriosus which later closes and remains in adult as: Critical Thinking
(a) Fossa ovalis
(b) Ligamentum arteriosus
(c) Carotico-pulmonary aperture
(d) None of the above

31. The heart sound 'murmur' is heard during:
(a) closing of bicuspid and tricuspid valves.
(b) closing of semiluner valves.
(c) leaking of blood through valves.
(d) All of the above

32. Myocardial Infarction is caused by:

(a) hardening of arteries.
(b) lumpy thickness develop in the inner walls of arteries.
(c) clot which may occur in the lumen of a coronary artery.
(d) sudden interruption in blood flow towards a portion of heart.

33. In arteriosclerosis:
(a) arterial walls become very thin and weak so that the blood oozes out of the walls.
(b) sex linked heredity is involved.
(c) blood coagulates even in the arteries.
(d) arterial walls become inelastic and thickened.

34. The accompanying diagram shows a small part of a normal electrocardiogram. Which region represents a wave of excitation passing through the ventricles?

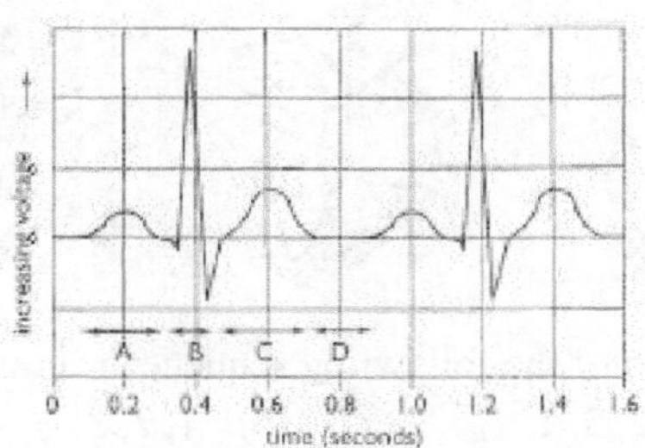

(a) A (b) B
(c) C (d) D

35. Which layer of the wall of blood vessels is made up of circular smooth muscles?
(a) Outer (b) Middle
(c) Inner (d) Both (a) and (b)

36. First heart sound (Lubb) coincides with the ________ wave of ECG:
(a) R (b) T
(c) P (d) Q

37. QRST is related with:
(a) Ventricular contraction or depolarisation
(b) Auricular contraction
(c) Auricular relaxation
(d) Cardiac cycle

38. Child death may occur in the marriage between:
(a) Rh^+ man and Rh^+ woman.
(b) Rh^+ man and Rh^- woman.
(c) Rh^- man and Rh^- woman.
(d) Rh^- man and Rh^+ woman.

39. Artery differs from vein in having:
(a) Narrow wall
(b) Thick walls
(c) Valves
(d) None of the above

40. 'P' wave of ECG occurs before the:
(a) onset of ventricular contraction.
(b) end of arterial contraction.
(c) beginning of atrial contraction.
(d) None of the above.

41. In human, the opening of post caval in the right auricle is guarded by:
(a) Bicuspid valve
(b) Tricuspid valve
(c) Eustachian valve
(d) Sino-auricular valve

42. The accompanying diagram shows three stages in the cardiac cycle.

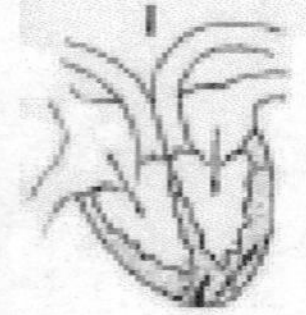

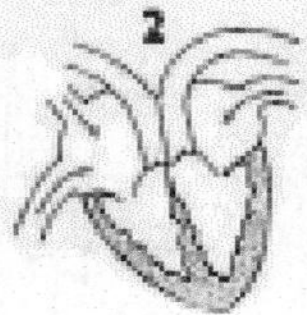

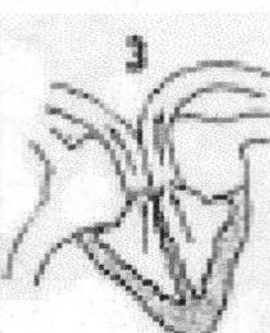

Which of the followiing sequence is correct?
(a) 2, 3, 1 (b) 1, 2, 3
(c) 2, 1, 3 (d) 3, 1, 2

43. Different factors play important roles in coagulation of blood, some of the factors are listed in column-I and their nomenclatures are given in column-II. Find out the accurate matching.

	Column-I		Column-II
A.	Factor II	I.	Thromboplastin
B.	Factor III	II.	Prothrombin
C.	Factor VIII	III.	Hageman factor
D.	Factor XII	IV.	Antihaemophilic globulin

(a) A – II; B – I; C – IV; D – III
(b) A – I; B – II; C – III; D – IV
(c) A – III; B – IV; C – II; D – I
(d) A – IV; B – III; C – II; D – I

44. Wall of blood capillary is formed of:
(a) Haemocytes
(b) Parietal cells
(c) Endothelial cells
(d) Oxyntic cells

45. Heart beats are accelerated by:
(a) Cranial nerves and acetylcholine
(b) Sympathetic nerves and acetylcholine
(c) Cranial nerves and adrenaline
(d) Sympathetic nerves and epinephrine

46. In mammals, veins differs from arteries in having:
(a) Thicker walls
(b) Deeply present
(c) Carry blood away from heart
(d) Internal valves

47. Which of the following best describes an artery?
(a) Carries blood away from the heart.
(b) Carries oxygenated blood.
(c) Contains valves.
(d) Has thin walls.

48. Blood clotting can be prevented by the use of potassium oxalate. It is due to: Tricky
(a) Checking of conversion of fibrinogen into fibrin.
(b) Inactivation of all factors for blood coagulation.
(c) Neutralisation of acidic medium.
(d) Removal of free Ca^{2+} ions.

49. You are required to draw blood from a patient and to keep it in a test tube for analysis of blood corpuscles and plasma. You are also provided with the following four types of test tubes. Which of them will you not use for the purpose?

(a) Test tube containing calcium bicarbonate
(b) Chilled test tube
(c) Test tube containing heparin
(d) Test tube containing sodium oxalate

50. A drop of each of the following, is placed separately on four slides. Which of them will not coagulate?
(a) Whole blood from pulmonary vein
(b) Blood plasma
(c) Blood serum
(d) Sample from the thoracic duct of lymphatic system.

51. Blood circulation that starts in capillaries and ends in capillaries is called:

(a) Portal circulation
(b) Hepatic circulation
(c) Lymphatic circulation
(d) Renal circulation

52. Which of the following substances, if introduced into the blood stream, would cause coagulation of blood at the site of its introduction?

(a) Prothrombin (b) Fibrinogen
(c) Thromboplastin (d) Heparin

53. Match the blood vessels of human heart listed under column-I with the functions given under column-II; Choose the answer which gives the correct combination of both columns.

	Column-I (Blood vessel)		Column-II (Function)
A.	Superior vena cava	I.	Carries deoxygenated blood to lungs
B.	Inferior vena cava	II.	Carries oxygenated blood to lungs
C.	Pulmonary artery	III.	Brings deoxygenated blood from lower parts of the body to the right atrium
D.	Pulmonary vein	IV.	Brings oxygenated blood to the left atrium
		V.	Brings deoxygenated blood from upper parts of the body into the right atrium

(a) A – V; B – I; C – III; D – II
(b) A – V; B – III; C – I; D – IV
(c) A – IV; B – V; C – III; D – I
(d) A – V; B – I; C – II; D – III

54. Which of the following statements (s) is/are **incorrect**?

(i) The AV node and the bundle of His constitute the electrical link between the atria and the ventricles.
(ii) The bundle of His is a bundle of electrical nodes which allows the ventricles to contract.
(iii) The bundle of His is a group of fibres that carry the electrical impulses through the centre of the heart.
(iv) The bundle of His is located in the arterial region.

(a) (ii), (iii) and (iv) (b) (i), (iii) and (iv)
(c) (i), (ii) and (iv) (d) (i), (ii) and (iii)

55. A heart attack occurs when: **Tricky**

(a) a heart valve malfunctions.
(b) a coronary artery is blocked.
(c) the heart is weakened by overwork.
(d) the aorta is blocked.

56. Emphysema damages the tissues of the lungs and slows pulmonary blood flow. This causes blood to back up, stretching and weakening the walls of the heart and blood vessels. Which of the following do you think would be most affected by this backup of blood from the lungs?

(a) Aorta (b) Right atrium
(c) Left atrium (d) Right ventricle

57. In the human heart, blood from the lungs enters the heart through the left atrium, pumps into the left ventricle, out the aorta and through the body, and then returns into the right atrium, pumps into the right ventricle and exits to the lungs.

Using the diagram, which set of letters (1, 2, 3, 4, 5) correctly represents the process describe above?

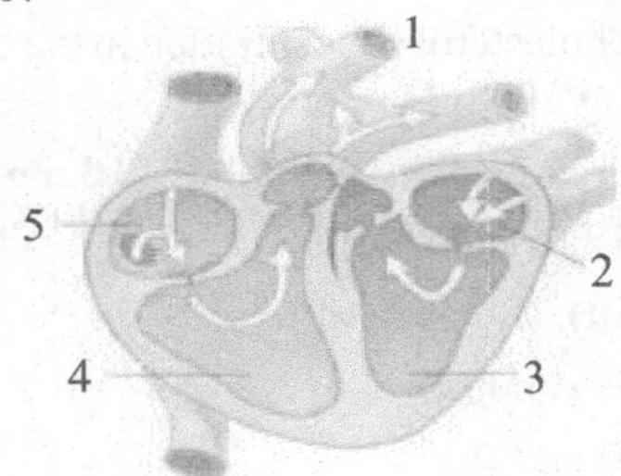

(a) 5, 4, 1, 2, 3 (b) 2, 3, 1, 5, 4
(c) 3, 4, 1, 2, 5 (d) 4, 3, 1, 5, 2

58. Which of the following vein has least amount of urea?

(a) Pulmonary vein
(b) Hepatic portal vein
(c) Hepatic vein
(d) Renal vein

59. We feel sleepy just after taking meals because:

(a) Blood pressure increases
(b) Blood pressure decreases
(c) Body weight increases
(d) We feel lethargic

60. Which of the following statements about the control of circulation in humans is **incorrect**?

(a) Sympathetic nerve input to skeletal muscle causes the blood vessels in the muscle to dilate.
(b) Blood flow can be regulated by autonomic nerve signals emanating from cardiovascular control centres in the medulla of the brain.
(c) Carotid artery chemosensors detect low O_2 levels in the blood and promote, increased blood pressure.
(d) Hormones such as angiotensin and vasopressin cause venules to constrict.

61. Identify the **correct** statements regarding the human blood.

(i) It constitutes 30-35% of the total extracellular fluid.
(ii) The volume of the blood in an adult is 5L.
(iii) Glucose concentration in the blood is 50 mg/100 mL.
(iv) Cholesterol concentration in the blood is 30 mg/100 mL.
(v) Urea level in the blood is 10 mg/ 100 mL.

(a) (i), (ii) and (iii)
(b) (iii), (iv) and (v)
(c) (iv) and (v)
(d) (i) and (ii)

62. Persons suffering from high blood pressure should take the following precaution to avoid excessive rise in their blood pressure:

Critical Thinking

(a) Sleep as much as possible
(b) Avoid standing
(c) Increase their weight
(d) Avoid emotional disturbances and excitement.

63. If heart of a mammal is injected with 2% $CaCl_2$ solutions, then:

(a) Heart beat will increase.
(b) Heart beat will first increase then decrease.
(c) Heart beat will stop.
(d) No effect

64. Tunica adventitia is the:

(a) outermost fibrous layer of the artery.
(b) middle vascular layer of the artery.
(c) outermost covering of the stomach.
(d) innermost layer of fine blood capillaries.

65. Which one of the following will be the cardiac output (in litres per minute) of a heart that has a stroke volume of 0.07 litres and is beating at a rate of 90 per minute?

Tricky

(a) 63.30 (b) 63.00
(c) 00. 63 (d) 06.30

66. Chordae tendinae are found in:

(a) Joints of legs
(b) Atria of heart
(c) Ventricles of brain
(d) Ventricles of heart

67. The heartbeat of a person increases at the time of an interview due to secretion of:

(a) Renin (b) Adrenaline
(c) ADH (d) ACTH

68. Mitral valve in mammals guards the opening between:

(a) Left auricle and left-ventricle
(b) Pulmonary vein and left auricle
(c) Stomach and intestine
(d) Liver and spleen

69. PR interval in a normal ECG, at heart beat of 70/min is approximately:

(a) 180 ms (b) 80 ms
(c) 8 ms (d) 320 ms

70. Which of the following correctly traces the electrical impulses that trigger each heartbeat?

(a) Ventricles–Pacemaker–AV node–Atria
(b) Pacemaker–AV node–Atria–Ventricles
(c) Atria–Pacemaker–AV node–Ventricles
(d) Pacemaker–Atria–AV node–Ventricles

71. Which of the following is not one of the reasons that closed circulatory systems are more efficient than open circulatory systems? Critical Thinking

(a) Closed systems rely exclusively on simple diffusion for transport, whereas open systems rely on pumping mechanisms.
(b) Transport within closed systems is more rapid than in open systems.
(c) Blood can easily be directed to specific areas in closed systems, but not in open systems.
(d) Closed systems operate better under higher pressure than do open systems.

72. Consider the statements and choose the correct option.

(i) Proteins contributes 6-8% of the blood plasma.
(ii) Plasma contain very high amount of minerals.
(iii) Plasma without the clotting factors is called serum.
(iv) Glucose, amino acids, lipids, etc., are also present in the plasma as they are always in transit in the body. Critical Thinking

(a) All are correct
(b) Only (ii) is false
(c) Only (i) is correct
(d) All are false

73. Cardiac cycle in man takes about: 

(a) 0.5 seconds (b) 1.0 second
(c) 1.2 seconds (d) 0.8 seconds

74. The left atrio-ventricular aperture in man is guarded by:

(a) Tricuspid valve (b) Semilunar valve
(c) Bicuspid valve (d) Sphinctor valve

75. What happens when the pacemaker is non functional? Critical Thinking

(a) Only the auricles will contract rhythmically.
(b) The cardiac muslces do not contract in a coordinated manner rhythmically.
(c) Only ventricles will contract rhythmically.
(d) Auricles and ventricles contract simultaneously.

76. A recording of the electrical activity of a patient's heart, shows that the atria are contracting regularly and normally, but every few beats the ventricles fail to contract. Which of the following is probably not functioning properly? Critical Thinking

(a) AV node (b) Semilunar valve
(c) Coronary artery (d) Pacemaker

77. Which of the following statements is **incorrect** about lymph.

(i) Lymph is colourful as it has haemoglobin but no RBC.
(ii) The fluid present in lymphatic system is called lymph.
(iii) It contains specilised lymphocytes which are responsible for immunity of the body.
(iv) Lymph is an important carrier for nutrients and hormones.
(v) Fats are absorbed through lymph in the lacteals present in the intestinal villi.

(a) Only (i) (b) (iii) and (iv)
(c) (ii) and (iii) (d) Only (iv)

78. Which one of the statement is **correct** with reference to the circulation of blood in a mammal?

(a) Left auricle receives oxygenated blood.
(b) Pulmonary artery carries oxygenated blood to the lungs.
(c) Blood from the lungs is returned to heart through 2-veins, one from each lung.
(d) None of the above

Answer KEYs

1	(b)	**9**	(d)	**17**	(c)	**25**	(d)	**33**	(d)	**41**	(c)	**49**	(a)	**57**	(b)	**65**	(d)	**73**	(d)
2	(d)	**10**	(a)	**18**	(a)	**26**	(a)	**34**	(b)	**42**	(c)	**50**	(c)	**58**	(d)	**66**	(d)	**74**	(c)
3	(c)	**11**	(a)	**19**	(d)	**27**	(d)	**35**	(b)	**43**	(a)	**51**	(a)	**59**	(b)	**67**	(b)	**75**	(b)
4	(a)	**12**	(b)	**20**	(c)	**28**	(a)	**36**	(a)	**44**	(c)	**52**	(c)	**60**	(d)	**68**	(a)	**76**	(a)
5	(d)	**13**	(b)	**21**	(d)	**29**	(c)	**37**	(a)	**45**	(d)	**53**	(b)	**61**	(d)	**69**	(a)	**77**	(a)
6	(a)	**14**	(d)	**22**	(a)	**30**	(b)	**38**	(b)	**46**	(d)	**54**	(d)	**62**	(d)	**70**	(d)	**78**	(a)
7	(a)	**15**	(c)	**23**	(a)	**31**	(c)	**39**	(b)	**47**	(a)	**55**	(b)	**63**	(c)	**71**	(a)		
8	(c)	**16**	(d)	**24**	(a)	**32**	(d)	**40**	(c)	**48**	(d)	**56**	(d)	**64**	(a)	**72**	(b)		

Hints & Solutions

1. (b) 2. (d)

3. (c) The pH of venous blood during rest is lower by 0.02 than that of arterial blood. The average pH of blood is 7.4 but it varies between 7.36 to 7.45.

4. (a)

5. (d) With the increase in intra-cranial pressure, the cerebral blood flow is reduced. This increased pressure also stimulates the vasomotor centre and increases systemic blood pressure.

6. (a) 7. (a)

8. (c) The label 3 in the given figure represents right atrium. The right atrium receives carbon dioxide rich blood from the body. The label 1 represents left atrium. It receives oxygenated blood from the lungs. The label 2 represents left ventricle. It pumps oxygenated blood to rest of the body. The label 4 represents right ventricle. It pumps deoxygenated blood to the lungs.

9. (d) Deficiency of RBCs in the blood causes anaemia. It may be due to excessive blood loss or increased destruction of RBCs (haemolysis). In the absence of vitamin B_{12} and folic acid (extrinsic factor), pernicious (Addison's) anaemia results. It is also called megaloblastic anaemia.

10. (a) Factor VII or proconvertin or plasma is not used up during clotting. During clotting, it is changed into convertin. It accelerates tissue or extrinsic thromboplastin formation. Factor IX is necessary for intrinsic thromboplastin formation. Calcium is for both intrinsic and extrinsic thromboplastin formation and also in the conversion of prothrombin into thrombin. Heparin is the anticoagulant in the normal blood circulation.

11. (a) The lymph nodes filter and destroy microorganisms that are traveling through the lymph system.

12. (b) 13. (b) 14. (d)

15. (c) Blood group AB is also known as the universal recipient.

16. (d)

17. (c) The ratio of RBCs to plasma is known as haematocrit value. Hematocrit is the instrument used for finding it.

18. (a) 19. (d)

20. (c) The AV node delays the ventricular depolarisation relative to atrial depolarisation, so atrial contraction occurs before ventricular contraction. The Purkinje fibres transmit the cardiac impulse to very small localised groups of ventricular fibres.

21. (d) Data is insufficient because it is not given what percentage of the flow through the kidneys is filtered through the glomerulus and enters the rest of the nephron. Glomerular filtration rate and urine flow rate must have been given to calculate the reabsorption rate.

22. (a) 23. (a) 24. (a) 25. (d)

26. (a) Contraction of the ventricle produces pressure in the human circulatory system. Blood leaves the ventricles and then immediately enters the arteries leading away from the heart. Blood pressure will have its highest recording in these arteries.

27. (d)

28. (a) Plasma is a straw coloured fluid that consists of suspended blood cells. Heart is the pumping station of the body. Capillary is the site where exchange of nutrients and gases takes place while blood is a red colour fluid that provides food to the cells.

29. (c) 30. (b) 31. (c) 32. (d)

33. (d) 34. (b) 35. (b) 36. (a)

37. (a) It is also called ventricular complex.

38. (b) Rh factor was discovered by Karl Landsteiner. A child of Rh^+ man will be Rh^+ whether the mother is Rh^+ or Rh^-. If the mother is Rh^+ then there will be no problem but if mother is Rh^- so when the blood of Rh^+ child (in womb) mixes with the blood of Rh^- mother then some antibodies in mother's blood are formed against Rh+ factor which coagulate the womb blood causing death. If birth takes place then there is a possibility of child death in early years. This in known as erythroblastosis foetalis. In most cases, the Ist pregnancy may succeed but after that it fails.

39. (b) 40. (c) 41. (c) 42. (c)

43. (a)

44. (c) The walls of blood capillaries do not possess tunica externa (connective tissue layer containing lymph vessels), tunica media (a thick network of smooth muscle fibres) but only tunica interna (a single layer endothelium of squamous cells).

45. (d) 46. (d)

47. (a) Arteries are defined as vessels that carry blood away from the heart.

48. (d)

49. (a) Presence of calcium will remove heparin, a blood anti-coagulant and will promote blood clotting. Sodium oxalate and heparin containing test tubes will not allow the blood to clot.

50. (c) 51. (a) 52. (c) 53. (b)

54. (d) Bundle of His is present in the intraventricular septum connected to AV bundle and its branches reach the Purkinje fibres in the ventricles. AV bundles provide the only route for the transmission of wave of excitation from the atria to ventricles.

55. (b) A heart attack occurs when coronary artery is blocked. In it, the heart muscles is suddenly damaged by inadequate blood supply.

56. (d) The right ventricle pumps blood to the lungs. Such backup of blood flow would force the right ventricle to work harder.

57. (b) In the given diagram of human heart, the parts marked as 1, 2, 3, 4 and 5 are respectively aorta, left atrium, left ventricle, right ventricle and right atrium. The correct sequence of the letters which represent the process of blood flow described in the question is 2, 3, 1, 5, 4.

58. (d) Least concentration of urea is found in renal vein. This is because urea is excreted through urine formed in kidney.

59. (b) Blood is diverted towards alimentary canal, therefore the pressure of blood falls in brain.

60. (d) Angiotensin and vasopressin act to control constriction of the arterioles and not the venules.

61. (d) Blood measures about 5-5.5 L in an adult man, constituting 30-35% of the total extracellular fluid.

Glucose value is 80-100 mg/100 mL of blood.

Cholesterol is 50-180 mg/100 mL of blood.

Urea Normal level is 17-30 mg/100 mL.

62. (d)

63. (c) Because excess of Ca^{++} will increase the force of contraction, thereby causing spasm of cardiac muscles.

64. (a) 65. (d) 66. (d) 67. (b)

68. (a) 69. (a)

70. (d) The pacemaker initiates an impulse through the atria, which is then conducted to the ventricles by the AV node.

71. (a) Both closed systems and open systems rely on pumping mechanisms to distribute fluid throughout the body.

72. (b) 73. (d) 74 (c) 75. (b)

76. (a) The AV node conducts the impulse from the atria to the ventricles.

77. (a) 78. (a)

19 Excretory Products and their Elimination

1. Waste products of adenine and guanine metabolism are excreted by man as:
(a) Ammonia (b) Urea
(c) Uric acid (d) Allantois

2. A man takes large amount of protein. He is likely to excrete more amount of: Tricky
(a) Water (b) Glucose
(c) Urea and uric acid (d) Salts

3. Aquatic reptiles are:
(a) Ammonotelic
(b) Ureotelic over land
(c) Ureotelic
(d) Ureotelic in water

4. In mammalian kidneys, Bowman's capsules or Malpighian corpuscles occur in:
(a) Cortex (b) Medulla
(c) Pelvis (d) All of these

5. Reabsorption of chloride ions from glomerular filtrate in kidney tubule occurs by:
(a) Active transport
(b) Diffusion
(c) Osmosis
(d) Brownian movement

6. Synthesis of urea in liver takes place by:
(a) Nitrogen cycle (b) Kreb's cycle
(c) Ornithine cycle (d) Glycolysis

7. Uric acid gets deposited in small joints to produce:
(a) Rheumatoid arthritis
(b) Gout
(c) Osteoarthritis
(d) All of the above

8. Diabetes insipidus is a condition in which a person is unable to produce sufficient levels of the hormone ADH. The hormone increases the permeability of water of the second (distal) convoluted tubule and collecting duct in the kidney nephrons. What is produced as a result?

(a) Large volumes of concentrated urine
(b) Large volumes of dilute urine
(c) Small volumes of concentrated urine
(d) Small volumes of dilute urine

9. The loops of Henle in the kidneys of a desert kangaroo rat are much longer than those in a white laboratory rat because the:

(a) kangaroo rat lives in an environment where water is scarce.
(b) white rat's diet is much less varied than the kangaroo rat's.
(c) kangaroo rat cannot always find food.
(d) kangaroo rat produces more body wastes.

10. Blood plasma contains proteins but glomerular filtrate does not. Why does this difference in composition occur?
(a) Blood osmotic pressure is maintained by the presence of plasma proteins.
(b) Proteins are actively transported from the kidney tubule back into the blood capillaries.
(c) Proteins cannot pass through the membranes of the glomerular capillaries.
(d) There is a high hydrostatic pressure in the blood within the glomerular capillaries.

11. In human beings the capsular urine entering the Proximal Convoluted Tubule (PCT) is:
(a) Isotonic to blood
(b) Hypotonic to blood
(c) Hypertonic to blood
(d) Isotonic to sea water

12. If the diameter of afferent renal arteriole is decreased and that of efferent renal arteriole increased, ultra filtration will:
(a) Be faster
(b) Be slower
(c) Not take place
(d) Take place in the same speed

13. The voluntary response to the distension of urinary bladder is:
(a) Polyurea (b) Micturition
(c) Mellitus (d) Menstruation

14. Refer the following diagram and identify the parts of a kidney indicated as 1, 2, 3, 4 and 5.

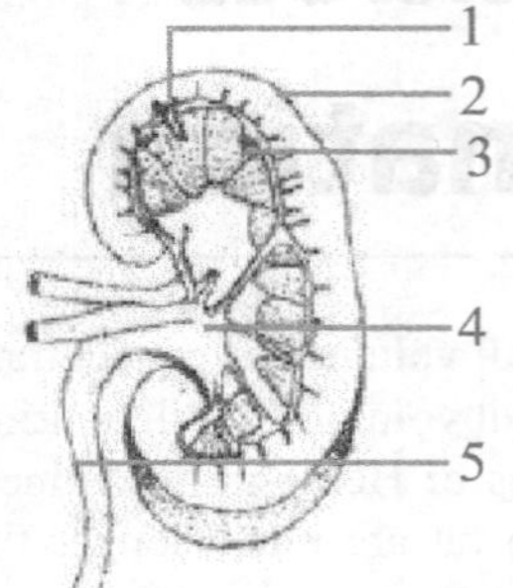

(a) 1 – cortex, 2 – nephron, 3 – pelvis, 4 – medulla, 5 – ureter
(b) 1 – cortex, 2 – medulla, 3 – nephron, 4 – pelvis, 5 – ureter
(c) 1 – nephron, 2 – cortex, 3 – medulla, 4 – ureter, 5 – pelvis
(d) 1 – nephron, 2 – cortex, 3 – medulla, 4 – pelvis, 5 – ureter

15. Match the column-I with column-II and choose the correct option.

Column-I (Excretory organs)	Column-II (Animals)
A. Nephridia	I. *Hydra*
B. Malpighian tubules	II. Leech
C. Protonephridia	III. Shark
D. Kidneys	IV. Round worms
	V. Cockroach

(a) A – II; B – V; C – IV; D – III
(b) A – IV; B – II; C – I; D – V
(c) A – V; B – II; C – IV; D – III
(d) A – II; B – IV; C – V; D – I

16. If the afferent arteriole that supplies blood to the glomerulus becomes dilated: Critical Thinking

(a) the protein concentration of the filtrate decreases.
(b) hydrostatic pressure in the glomerulus decreases.
(c) the glomerular filtration rate increases.
(d) All of the above

17. In which one of the following ways does the blood leaving the glomerulus of mammalian kidney tubule differ from the blood entering the glomerulus? Tricky

(a) It has a lower concentration of plasma proteins.
(b) It contains fewer corpuscles per unit volume.
(c) It has a lower concentration of crystalloids.
(d) It has a higher concentration of crystalloids.

18. The appearance of albumin in the urine is most likely due to: Critical Thinking

(a) increase in the blood pressure.
(b) decrease in the blood osmotic corpuscles.
(c) damage to the Malpighian corpuscles.
(d) damage to the proximal convoluted tubules.

19. Mesonephric kidney is found in:

(a) Aves (b) Reptilia
(c) Fishes (d) All of these

20. Which of the following is totally reabsorbed in renal tubules?

(a) Na (b) K
(c) H_2O (d) $C_6H_{12}O_6$

21. In man, the two kidneys lie:

(a) at the level of ovaries.
(b) at the same level.
(c) left kidney at a higher level than the right one.
(d) right kidney at a higher level than the left one.

22. Vasa rectae are peritubular capillaries around:

(a) Posterior part of alimentary canal
(b) PCT
(c) Loop of Henle
(d) DCT

23. Which vitamin is excreted out in high quantity through urine in man?

(a) Vit. C (b) Vit. D
(c) Vit. E (d) Vit. K

24. Urine contains how much of water?

(a) 96% (b) 90%
(c) 70% (d) None of these

25. The glomerular filtrate consists of:

(a) Urea, sodium chloride, fibrinogen and water
(b) Glucose, amino acids, urea, oxytocin and calcitonin
(c) Both (a) and (b)
(d) Urea, glucose, salts and water

26. "Columns of Bertini" in the kidney of mammals are formed as the extension of:

(a) Medulla into cortex
(b) Cortex into medulla
(c) Medulla into pelvis
(d) Pelvis into ureter

27. Filtration pressure in human kidneys is about:

Tricky

(a) + 15 mm Hg (b) + 70 mm Hg
(c) + 45 mm Hg (d) +55 mm Hg

28. Bowman's capsule is lined by:
(a) Ciliated cuboidal epithelium
(b) Squamous epithelium
(c) Non-ciliated cuboidal epithelium
(d) Non-ciliated columnar epithelium

29. Excretory products of mammalian embryo are eliminated by:
(a) Placenta (b) Amniotic fluid
(c) Allantois (d) Ureters

30. Accessory excretory organs of man are:
(a) Only skin
(b) Only skin and liver
(c) Only skin and lungs
(d) Skin, lungs, liver and intestine

31. Brush border epithelium is characteristic of:
(a) Neck of nephron
(b) Collecting tube
(c) Proximal convoluted tubule
(d) All of the above

32. Healthy human does not excrete out in his urine:
(a) Uric acid
(b) Alanine
(c) Creatinine
(d) B-complex vitamins

33. Consider the following statements and choose the correct option.
(i) Flame cells are excretory structures in flatworms.
(ii) Green glands are excretory organs in annelids.
(iii) Columns of Bertini are the conical projections of renal pelvis into renal medulla between the renal pyramids.
(a) (i) and (ii) are correct
(b) (ii) and (iii) are incorrect
(c) (i) and (iii) are correct
(d) (i), (ii) and (iii) are correct

34. Which of the following statements about the excretory system of insects is **not true**?
(a) Active transport moves materials from the coelomic fluid into the Malpighian tubules.
(b) The Malpighian tubules can produce a highly concentrated urea solution, allowing insects to inhabit some of Earth's driest habitats.
(c) Reabsorption of salts takes place mostly in the gut.
(d) Water reabsorption is by osmotic movement only.

35. A single load of insulin is injected into the bloodstream. If the urine flow is constant, which graph best expresses the relationship between urine-insulin concentration and time?

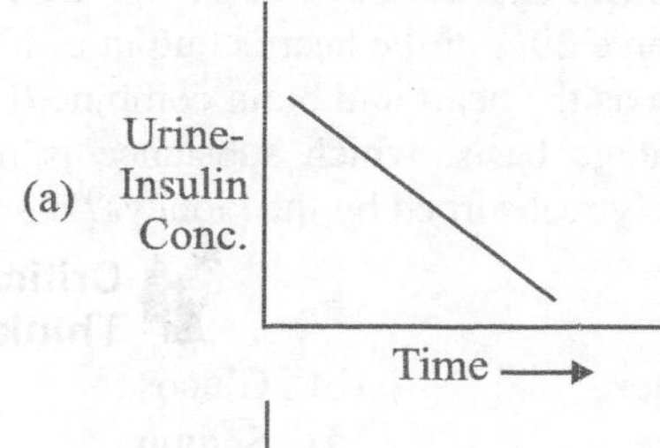

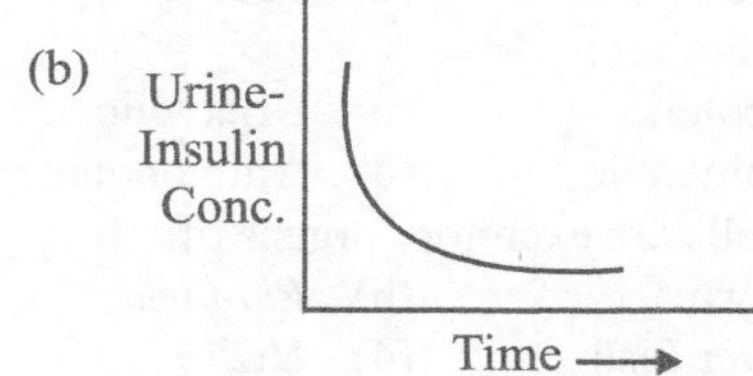

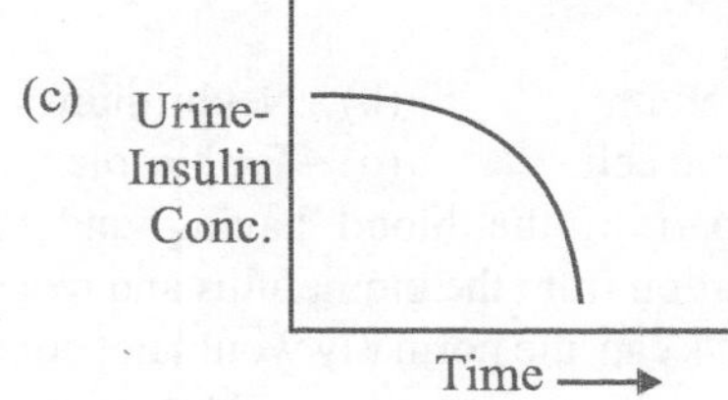

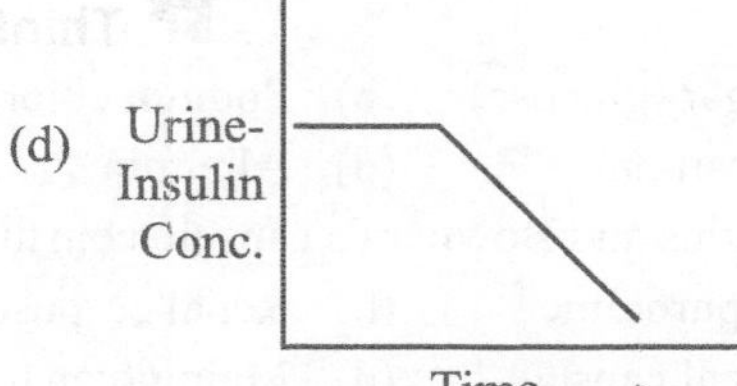

36. On a cold day, blood vessels in the skin:

(a) dilate, allowing blood to keep the skin warm.
(b) constrict, forcing blood to flow through vessels in the skin.
(c) constrict, reducing heat loss from blood at the surface.
(d) dilate, causing blood to pass through the cold skin more quickly.

37. Osmoreceptors in the brain detect changes in blood ion concentration which can reflexively result in:
(a) water retention by the kidneys.
(b) water loss by the kidneys.
(c) ion retention by the kidneys.
(d) All of the above

38. Kidneys help in the conservation of useful materials and excretion of wastes and therefore they receive 20% of the heart's output of blood (as much as the heart and brain combined). On a percentage basis, which substance is most completely reabsorbed by the kidneys?

(a) Water (b) Glucose
(c) Urea (d) Sodium

39. Bony fishes are:
(a) Ureotelic (b) Uricotelic
(c) Aminotelic (d) Ammonotelic

40. Flame cells are excretory organs of:
(a) Prawn (b) *Planaria*
(c) Silver Fish (d) *Hydra*

41. Which organ of earthworm is analogous to our kidney?
(a) Clitellum (b) Nephridium
(c) Flame cell (d) Typhlosole

42. Liquid part of the blood having undergone ultrafiltration from the glomerulus and reaching Bowman's capsule normally would not contain:

(a) Sugar (glucose) (b) Sodium chloride
(c) Creatinine (d) Albumin

43. Glomerulus and Bowman's capsule constitute:
(a) Nephrotome (b) Renal corpuscle
(c) Renal capsule (d) Malpighian tubule

44. Human sweat does not contain:
(a) Lactic acid (b) Glucose
(c) Amino acids (d) Uric acid

45. Workers in deep mines usually suffer from dehydration because:
(a) water is lost due to evaporation.
(b) water is lost due to defecation.
(c) water is lost in the form of urine.
(d) water is lost along with salts in the form of sweat.

46. Glomerular filtration rate would be decreased by: **Tricky**
(a) constriction of the efferent arteriole.
(b) an increase in afferent arteriolar pressure.
(c) compression of the renal capsule.
(d) an increase in renal blood flow.

47. Kidney crystals are solid clusters of:
(a) Calcium nitrate and uric acid
(b) Phosphate and uric acid
(c) Calcium carbonate and uric acid
(d) Calcium metabisulphite and uric acid

48. The hormone that promotes reabsorption of water from glomerular filtrate is:
(a) Oxytocin (b) Vasopressin
(c) Calcitonin (d) Relaxin

49. Select the correct match of the types of organs given in column-I with their role in excretion given in column-II.

	Column-I (Types of organ)		Column-II (Role in excretion)
A.	Lungs	I.	Secretes bile containing substances like bilirubin, biliverdin, cholesterol, degraded steroid hormones, vitamins and drugs.
B.	Liver	II.	Eliminates water and salts in sweat and substances like sterols, hydrocarbons and waxes through sebum.
C.	Skin	III.	Remove large amounts of CO_2 (18 litres/day) and also significant quantities of water every day.
D.	Kidney	IV.	Remove wastes (metabolic by products) and regulate pH, ion concentration, volume and osmolarity of blood.

(a) A – I; B – II; C – III; D – IV
(b) A – III; B – I; C – II; D – IV
(c) A – III; B – I; C – IV; D – II
(d) A – IV; B – II; C – III; D – I

50. Which of the following hormone/enzyme is/are correctly paired with its function?

(i) Renin – Enzyme that catalyses the formation of angiotensin I.
(ii) Aldosterone – Regulates water reabsorption at the distal convoluted tubule.
(iii) Anti-diuretic hormone (ADH) – It is a powerful vasoconstrictor that stimulates the secretion of aldosterone.
(iv) Angiotensin II – Promotes reabsorption of sodium at distal convoluted tubule.

(a) Only (i) (b) Only (iii)
(c) (i), (ii) and (iii) (d) (ii), (iii) and (iv)

51. The diagram shows a simplified nephron.

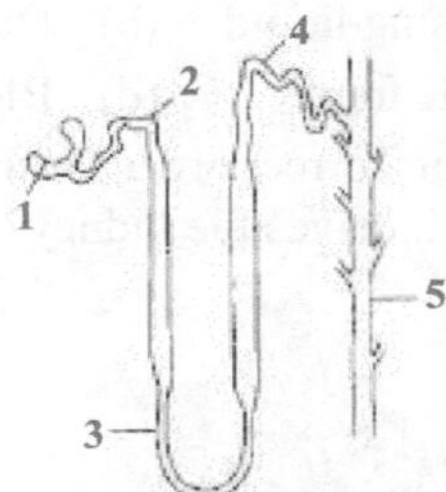

In which regions (marked as 1, 2, 3, 4 and 5) would cells with many microvilli and mitochondria be found?

(a) 1 and 5 (b) 2 and 4
(c) 2 and 3 (d) 3 and 4

52. If the human kidneys filter 150 litres of plasma in a 24 hour period, what is the typical amount of urine produced and eliminated in that time period?

Tricky

(a) 0.15 litres (b) 1.5 litres
(c) 15 litres (d) 30 litres

53. Out of the four parts given below, which parts plays significant role in forming concentrated urine in human?

Tricky

(i) Loop of Henle
(ii) Glomerulus
(iii) Bowman's capsule
(iv) Vasa recta

(a) (i) and (ii) (b) (iii) and (iv)
(c) (ii) and (iii) (d) (i) and (iv)

54. Colloidal osmotic pressure in blood plasma is mainly due to:

(a) Albumin (b) Globulin
(c) Fibrinogen (d) Sodium chloride

55. The number of nephrons in a kidney is equal to:

(a) the number of Bowman's capsules.
(b) sum of Bowman's capsules and glomeruli.
(c) double the number of Bowman's capsules.
(d) sum of Bowman's capsules and Malpighian corpuscles.

56. Expulsion of urine become voluntary when it is present in:

(a) Urethra (b) Ureters
(c) Urinary bladder (d) Pelvis of kidneys

57. Renin is secreted from:

(a) Stomach
(c) Juxta cortical cell
(b) Juxta glomerular body
(d) Juxta medullary cell

58. ADH influences water permeability in the:

(a) Proximal tubule
(b) Distal tubule
(c) Collecting tubule
(d) Both (a) and (b)

59. A terrestrial animal must be able to:

(a) conserve water.
(b) excrete large amounts of salts in urine.
(c) excrete large amounts of water in urine.
(d) actively pump salts out through the skin.

60. The diagram below shows a nephron of the mammalian kidney.

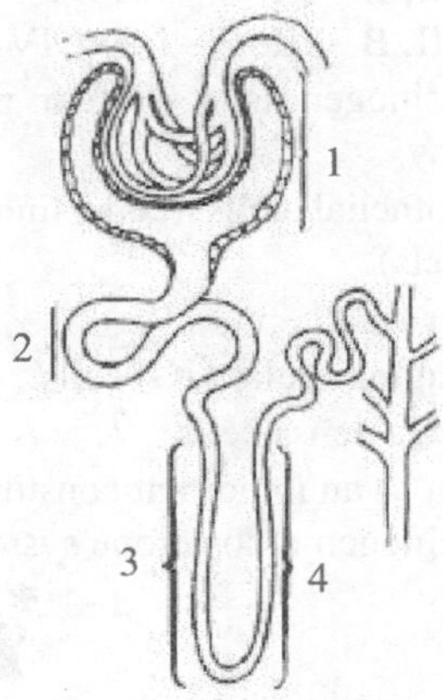

Which one of the regions (marked as 1, 2, 3 and 4) is the main region responsible for restoring the metabolic level of the blood plasma to its pre-renal level?

(a) Region 1 (b) Region 2
(c) Region 3 (d) Region 4

61. Which of the following statement(s) is/are **correct**?

(i) Urine is hypertonic in distal convoluted tubule.
(ii) When the urine passes into collecting tubule, it becomes hypotonic.
(iii) Urine is isotonic in proximal convoluted tubule.
(iv) Urine becomes more and more hypotonic as it passes through Henle's loop.

(a) (i) and (iv) (b) (i), (ii) and (iii)
(c) (ii) and (iii) (d) (iii) only

62. Match the column-I with column-II and choose the correct option.

Column-I	Column-II
A. Delivers blood to glomerulus	I. Proximal convoluted tubules
B. Carries urine to pelvis	II. Renal artery
C. Collects filtrate from Bowman's capsule	III. Collecting duct
D. Loop of Henle	IV. Ascending and descending limb

(a) A – IV; B – III; C – II; D – I
(b) A – I; B – II; C – III; D – IV
(c) A – II; B – III; C – IV; D – I
(d) A – II; B – III; C – I; D – IV

63. Angiotensinogen is a protein produced and secreted by:

(a) Endothelial cells (cells lining the blood vessels)
(b) Liver cells
(c) Juxtaglomerular (JG) cells
(d) Macula densa cells

64. Which one is an important constituent of renin-angiotensinogen-aldosterone system?

(a) JGA cell (b) Macular cell
(c) Erythropoetin (d) Plasma cell

65. Which is finally reabsorbed in distal convoluted tubule?

(a) Calcium (b) Potassium
(c) Bicarbonate (d) Water

66. Na^+ and Cl^- are actively transported out of the tubules to help set up the countercurrent multiplier. Which of the following are sites of active Na^+ and Cl^- transport in the nephron?

(a) Proximal tubule, ascending limb of the loop of Henle
(b) Descending limb of the loop of Henle, ascending limb of the loop of Henle
(c) Ascending limb of the loop of Henle, proximal tubule
(d) Collecting duct, descending limb of the loop of Henle

67. Which of the following is acting as an artificial kidney in haemodialysis?

(a) Dialysing liquid (b) Dialyser
(c) Bubble trap (d) Blood pump

68. Which is the correct order for the path taken by urine after it leaves the kidney?

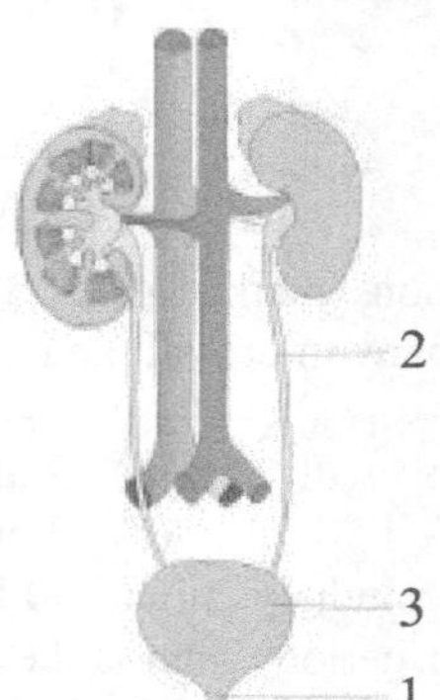

(a) 1, 2, 3 (b) 3, 1, 2
(c) 2, 3, 1 (d) 2, 1, 3

69. Which of the following statements is/are **correct**?

(i) Urine is hypertonic in distal convoluted tubule.
(ii) When the urine passes into the collecting tubule, it becomes hypotonic.
(iii) Urine is isotonic in proximal convoluted tubule.
(iv) Urine becomes more and more hypotonic as it passes through the Henle's loop.

(a) (i) and (iv) only
(b) (i), (ii), and (iii) only
(c) (ii) and (iii) only
(d) (iii) only

70. Identify the **correct** statements.
(i) The outer layer of the kidney is called capsule.
(ii) Cortex is divided into outer cortex and inner medulla.
(iii) Medulla is divided into medullary pyramids.
(iv) The cortex extends in between the medullary pyramids which is called as columns of Bertini.
(a) (i), (iii) and (iv) (b) (i) and (iv)
(c) (i), (ii) and (iii) (d) (i), (ii), (iii) and (iv)

71. The blood constituents that remain unchanged in quantity after circulating through the kidney are:
(a) Urea and glucose
(b) Glucose and proteins
(c) Urea and proteins
(d) Urea and uric acid

72. Coxal glands are excretory organs in:
(a) Spiders and scorpions
(b) Insects
(c) Annelids
(d) Molluscs

73. A fall in glomerular filtration rate (GFR) activates:
(a) juxtaglomerular cells to release renin.
(b) adrenal cortex to release aldosterone.
(c) adrenal medulla to release adrenaline.
(d) posterior pituitary to release vasopressin.

74. Antennal glands are excretory organs of:
(a) Moths (b) Scorpions
(c) Spiders (d) Cray fishes

75. A marine fish lives in a hypertonic environment. What mechanisms does a marine fish use to maintain homeostasis? **Toughnut**
(a) It drinks copious amounts of water, excretes only small amounts of water, and pumps sodium out of the body.
(b) It drinks little water, excretes large amounts of water, and pumps sodium into the body.
(c) It drinks little water, excretes only small amounts of water, and pumps sodium out of the body.
(d) None of the above

76. A person who is on a long hunger strike and is surviving only on water, will have: **Tricky**
(a) less amino acids in his urine
(b) more glucose in his blood
(c) less urea in his urine
(d) more sodium in his urine

77. In micturition:
(a) Urethra relaxes (b) Ureters contract
(c) Ureters relax (d) Urethra contracts

78. Match column-I with column-II regarding human excretory system and choose the correct option given below.

	Column-I		Column-II
A.	Epithelial cells of Bowman's capsule	I.	Juxtamedullary nephron
B.	Extension of cortex between the medullary pyramids as renal columns	II.	Vasa recta
C.	Nephrons with long loop of Henle running deep into the medulla	III.	Juxtaglomerular Apparatus
D.	A fine vessel of the peritubular capillaries running parallel to Henle's loop	IV.	Podocytes
E.	A special sensitive region in the DCT and afferent arteriole at the location of their contact	V.	Columns of Bertini
		VI.	Cortical nephron

(a) A – III; B – II; C – I; D – IV; E – V
(b) A – V; B – I; C – II; D – III; E – IV
(c) A – IV; B – III; C – I; D – II; E – V
(d) A – II; B – IV; C – VI; D – I ; E – III

79. The part of the nephron impermeable to water is: **Tricky**
(a) Proximal tubule
(b) Distal tubule
(c) Ascending limb of Henle's loop
(d) Collecting duct

80. In the renal tubules, the permeability of the distal convoluted tubule and collecting duct to water is controlled by: **Tricky**
(a) Aldosterone (b) Vasopressin
(c) Growth hormone (d) Renin

81. In which part of excretory system of mammals can you first use the term 'urine' for contained fluid?

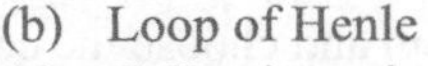

(a) Bowman's capsule
(b) Loop of Henle
(c) Collecting tubule
(d) Urinary bladder

82. In distal convoluted tubule of the nephrons:

(a) Na^+ reabsorption requires energy.
(b) secretion of K^+ ions does not require energy.
(c) water reabsorption requires energy.
(d) ammonia is secreted.

83. Most aquatic animals excrete ammonia, while land animals excrete urea or uric acid. What is the most likely explanation for this difference?

(a) They have different diets.
(b) Land animals can get the energy needed to make urea or uric acid.
(c) Ammonia is very toxic, and it takes lots of water to dilute it.
(d) Land animals cannot afford the energy needed to make ammonia.

84. Osmoregulation in *Paramoecium* is a function of:
(a) Contractile vacuole (b) Trichocysts
(c) Cytopyge (d) Cyostome

85. Common excretory product of insects is:
(a) NH_3 (b) Urea
(c) Uric acid (d) Hippuric acid

86. In prawn, excretion is carried out by:
(a) Nephrons (b) Malpighian tubules
(c) Flame cells (d) Green glands

87. Hippuric acid, creatinines and ketones are added to urine through:

Critical Thinking

(a) Reabsorption
(b) Glomerular filtration
(c) Tubular secretion
(d) Both (b) and (c)

88. If Henle's loop were absent from mammalian nephron then which of the following is to be expected?
(a) There will be no urine formation.
(b) There will be hardly any change in the quality and quantity of urine formed.
(c) The urine will be more concentrated.
(d) The urine will be more diluted.

89. The given diagram represents a kidney nephron.

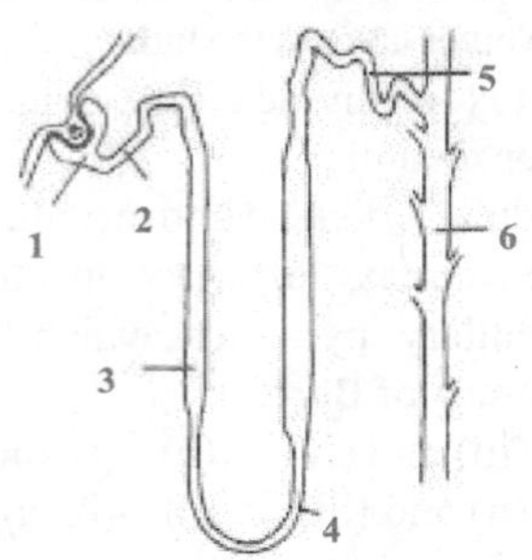

At which position (marked as 1, 2, 3, 4, 5 and 6) does antidiuretic hormone (ADH) affect the permeability of membranes to water?
(a) 1 and 2
(b) 3 and 4
(c) 2 and 3
(d) 5 and 6

90. Following are four statements. Find out the correct combination.

Critical Thinking

(i) Glucose has high threshold value.
(ii) Urine is concentrated in Henle's loop.
(iii) Haemodialyser removes urea, uric acid, glucose and proteins.
(iv) In glomerulus, urea, uric acid, water, glucose and plasma proteins are filtered out.

(a) (i), (iii), (iv)
(b) (ii), (iii), (iv)
(c) (i), (ii)
(d) (i), (iii)

91. The Urine under normal conditions does not contain glucose because:

Critical Thinking

(a) the normal blood sugar is fructose.
(b) glucose of blood is not filtered in the glomerulus.
(c) glucose in glomerular filtrate is reabsorbed in the uriniferous tubules.
(d) glucose in glomerular filtrate is converted into glycogen.

92. Na^+ and Cl^- both are removed by:
(a) Ascending limb of Henle's loop
(b) Proximal convoluted tubule
(c) Both (a) and (b)
(d) Descending limb of Henle's loop

93. Due to insufficient filtration in the Bowman's capsule, all are likely to happen except:

Toughnut

(a) Accumulation of fluid in the body
(b) Increase in blood pressure
(c) Increase in blood urea level
(d) Loss of glucose through urine

94. The kidneys resemble contractile vacuoles of protozoans in expelling:

(a) Excess water (b) Excess glucose
(c) Urea & uric acid (d) All salts

95. Dialysis (used for the treatment of chronic kidney diseases) differs from the process of osmosis in the respect that :

Critical Thinking

(a) both solvent and solute pass through the membrane.
(b) only solute selectively passes through the membrane.
(c) only solvent selectively passes through the membrane.
(d) gases are the only substances that pass through the membrane and the blood is cleaned.

Directions for (Qs. 96 to 99): Refer the given figure of nephron showing blood vessels and duct and answer the questions.

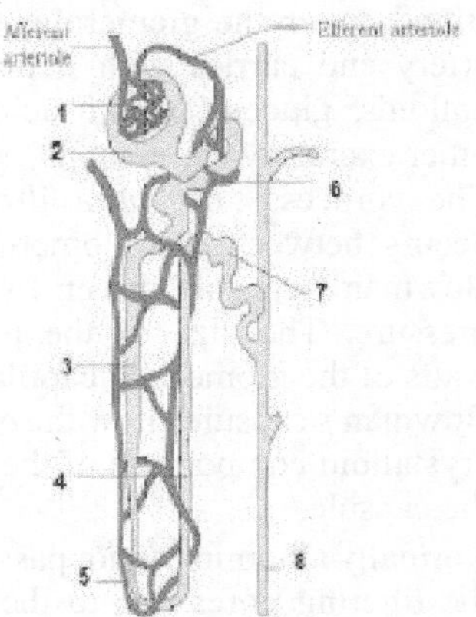

96. Which parts in the given figure have minimum reabsorption and play a significant role in the maintenance of high osmolarity of medullary interstitial fluid?

(a) 1 and 2
(b) 3 and 4
(c) 5 and 6
(d) 7 and 8

97. Cells of which part is lined by simple cuboidal brush border epithelium that increases the surface area for reabsorption? Identify the name also.

(a) 3–Descending limb of loop of Henle
(b) 5–Vasa recta
(c) 6–Proximal convoluted tubule
(d) 7–Collecting duct

98. Which part is capable of reabsorption of HCO_3^- and selective secretion of hydrogen and potassium ions and NH_3 to maintain the pH and sodium-potassium balance in blood?

(a) 1 (b) 3
(c) 5 (d) 7

99. Which structures have an ability to produce the concentrated urine?

(a) 1, 2, 3
(b) 3, 4, 5
(c) 6, 7, 8
(d) 2, 4, 7

100. In hot weather the urine becomes more concentrated and darker in colour, due to:

(a) Dehydration
(b) Rehydration
(c) Absorption of aldosterone
(d) Absence of renin

Answer KEYs

1	(c)	11	(a)	21	(c)	31	(c)	41	(b)	51	(c)	61	(d)	71	(b)	81	(c)	91	(c)
2	(c)	12	(c)	22	(c)	32	(b)	42	(d)	52	(b)	62	(d)	72	(a)	82	(a)	92	(c)
3	(c)	13	(b)	23	(a)	33	(b)	43	(b)	53	(d)	63	(b)	73	(a)	83	(c)	93	(c)
4	(a)	14	(d)	24	(a)	34	(b)	44	(d)	54	(a)	64	(a)	74	(d)	84	(a)	94	(a)
5	(b)	15	(a)	25	(d)	35	(c)	45	(d)	55	(a)	65	(c)	75	(a)	85	(c)	95	(a)
6	(c)	16	(c)	26	(b)	36	(c)	46	(c)	56	(c)	66	(a)	76	(c)	86	(d)	96	(b)
7	(b)	17	(c)	27	(a)	37	(d)	47	(b)	57	(d)	67	(c)	77	(a)	87	(d)	97	(c)
8	(b)	18	(c)	28	(b)	38	(b)	48	(b)	58	(b)	68	(c)	78	(d)	88	(d)	98	(d)
9	(a)	19	(c)	29	(a)	39	(d)	49	(b)	59	(a)	69	(d)	79	(c)	89	(d)	99	(b)
10	(c)	20	(d)	30	(d)	40	(b)	50	(a)	60	(b)	70	(d)	80	(b)	90	(c)	100	(a)

Hints & Solutions

1. **(c)** Main sources of uric acid are purines. Purines like adenine and guanine change into xanthine which changes into uric acid

   Guanine
   ↓
   Adenine → xanthine → uric acid

2. **(c)** Because they are the ultimate products of protein catabolism.
3. **(c)** Ureotelic animals excrete urea as nitrogenous waste products, and includes mainly amphibians and mammals besides annelids (earthworm), elasmobranch fishes (shark) and aquatic animals.
4. **(a)** 5. **(b)** 6. **(c)** 7. **(b)**
8. **(b)** If the walls of the collecting duct are water-permeable, water leaves the ducts to pass into the hyperosmotic surrounding and concentrated urine is produced. Thus, when there is insufficient ADH, less water is reabsorbed and more dilute urine is produced in copious amounts.
9. **(a)** A longer loop of Henle promotes water conservation.
10. **(c)** The filtration slits in the epithelium of the renal corpuscle are about 25nm wide, too small for most of the proteins to pass through. Only small solute molecules like glucose may be filtered out.
11. **(a)** 12. **(c)** 13. **(b)** 14. **(d)**
15. **(a)**
16. **(c)** Changes in the afferent arteriole pressure affects glomerular filtration rate. Increases in pressure will increase filtration rate and decreases in pressure will decrease filtration rate.
17. **(c)** Blood enters the glomerulus by the renal artery and carries with it plasma, blood colloids, glucose, amino acids, urea and other excretory compounds, and minerals. The process of ultra-filtration, then occurs between the glomerulus and the Bowman's capsule driven by hydrostatic pressure. The size of the pores on the walls of the glomerular capillaries and the Bowman's capsule allow the passage of all crystalloid components of the plasma into the capsule.
18. **(c)** Normally albumin cannot pass out through the filtering pores due to their large size but in some pathological conditions *viz.* anoxia or heart failure, the filtration membrane becomes damaged and filtering pores enlarge in size, so that serum albumin passes out in the largest amount and appears in the urine.
19. **(c)** 20. **(d)** 21. **(c)** 22. **(c)**
23. **(a)** 24. **(a)** 25. **(d)**
26. **(b)** The renal columns of Bertini is the part of cortex continued inside medulla between

pyramids.

27. **(a)** Effective filtration pressure in man is + 15 mm Hg. In case of horse and some other mammals, it is + 25 mm Hg.

28. **(b)** **29.** **(a)** **30.** **(d)** **31.** **(c)**

32. **(b)** **33.** **(b)**

34. **(b)** Insects except that they excrete uric acid rather than urea.

35. **(c)** At any given moment, the concentration of insulin in the blood will be approximately equal in all parts of the bloodstream. It follows then, that insulin will disappear from the blood at a constant rate, meaning that a plot of blood concentration versus time will be exponential in nature. The plot of the urine insulin concentration versus time will be the reverse plot.

36. **(c)** Reducing blood flow to the body surface will reduce heat loss.

37. **(d)** Osmoreceptors in the brain detect changes in blood ion concentration, decreases in both ion and water retention, depending on the status of blood osmolarity.

38. **(b)** Glucose is 100% reabsorbed and thus, a healthy person will excrete no glucose in the urine. Sodium and water are usually over 99% reabsorbed. Urea is the main excretory product formed as a result of protein breakdown.

39. **(d)** **40.** **(b)** **41.** **(b)** **42.** **(d)**

43. **(b)** **44.** **(d)** **45.** **(d)** **46.** **(c)**

47. **(b)** **48.** **(b)** **49.** **(b)**

50. **(a)** Renin is an enzyme that catalyses the formation of angiotensin I.

Aldosterone promotes reabsorption of sodium at distal convoluted tubule. Anti-diuretic hormone (ADH) regulates water reabsorption at the distal convoluted tubule. Angiotensin II is a powerful vasoconstrictor that stimulates the secretion of aldosterone.

51. **(c)**

1 – Bowman's capsule : Mainly involved in ultrafiltration. This is a passive process that does not require an energy source like the mitochondria.

2 – Proximal convoluted tubule : The large surface area (microvilli) and mitochondria are adaptations for selective reabsorption of substances from the glomerular filtrate.

3 – Descending limb of the loop of Henle : Creates and maintains an increasing osmotic gradient for sodium, chloride and urea.

4 – Distal convoluted tubule : Brush borders and mitochondria in this site for the fine control of salt, water and pH balance of blood.

5 – Collecting duct.

52. **(b)** About 1 percent of the filtrate is excreted as urine, so about 1.5 litres of the original 150 litres would be urinated.

53. **(d)** The flow of filtrate in the two limbs of Henle's loop is in opposite directions and thus, forms a counter current mechanism. (it is a process due to which the urine is made hypertonic). Vasa-recta also plays a significant role in counter-current mechanism.

54. **(a)** **55.** **(a)** **56.** **(c)** **57.** **(d)**

58. **(b)** Antidiuretic hormone (ADH) produced by neurosecretory cells in the paraventricular and supraoptic nuclei of the hypothalamus, stimulates water reabsorption from (in DCT part of nephron) kidney tubules into the blood and vasocontraction of arterioles.

59. **(a)**

60. **(b)** Region 2 represents the proximal convoluted tubule of the kidney which is responsible for selective reabsorption, restoring the level of metabolites.

61. **(d)** The filtrate is isotonic to blood plasma (in proximal convoluted tubule) and the filtrate becomes hypertonic to blood plasma (in descending limb of loop of Henle). The filtrate is hypotonic to blood plasma (in ascending limb of loop of Henle). In distal convoluted tubule ADH make the filtrate isotonic to blood plasma.

62. **(d)** **63.** **(b)** **64.** **(a)** **65.** **(c)**

66. **(a)** The proximal tubule is the site of active transport of Na^+ out of the tubule. Na^+ also moves out of the tubule at the ascending limb of the loop of Henle, but this is a passive transport with Cl^- being actively transported out.

67. **(c)**

68. **(c)** The nephron carries the urine into the collecting tubule of the kidney from where it is carried to ureter. From ureter, the urine passes into the urinary bladder. The urinary bladder stores the urine until it is forced out of the body, through an opening known as the urethra. The label 2 represents ureter, label 3 represents urinary bladder and label 1 represents urethra.

69. (d)

70. (d) The outer layer of the kidney is a tough capsule. Inside the kidney, there are two zones-cortex and medulla. Medulla is divided into few conical masses called medullary pyramids. The cortex extends in between the medullary pyramids as renal columns called columns of Bertini.

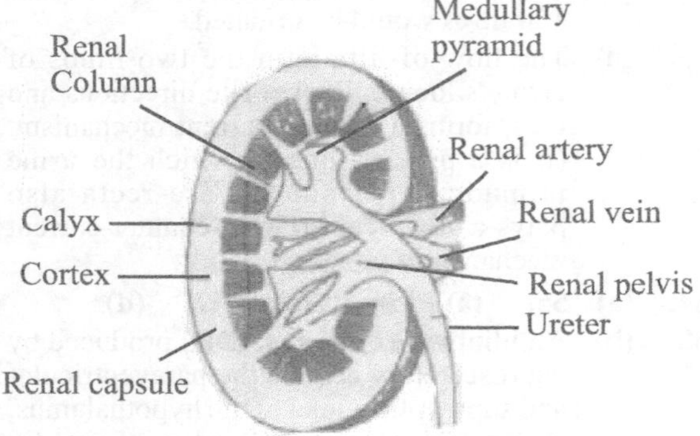

71. (b) Proteins are not filtered through glomerulus and glucose is totally reabsorbed in renal tubule.

72. (a) 73. (a) 74. (d)

75. (a) Marine fishes live in an environment in which salts try to enter the body and water leave the body. To counter these effects marine fishes drink lots of water, excrete small amounts of waters and actively pump sodium out of the body across the gills.

76. (c)

77. (a) Process of passing out of urine is called micturition. During discharge of urine, urethral sphincters relaxes and smooth muscles of bladder wall contract gradually.

78. (d) 79. (c) 80. (b) 81. (c)

82. (a) Reabsorption of Na^+ in the tubules is an active process.

83. (c) The animal does not expend energy forming ammonia, but it is highly toxic and must be diluted in large amounts of water.

84. (a) 85. (c) 86. (d) 87. (d)

88. (d)

89. (d) ADH increases the permeability of the distal convoluted tubules and the collecting duct to water by increasing the number of water channels in membranes lining tubules. It also increases the permeability of the collecting duct to urea, which increases the concentration of medullary tissue fluid, resulting in more water being, removed in the descending limb.

90. (c) 91. (c) 92. (c)

93. (c) Insufficient filtration will increase the blood urea level because all the quantity of urea produced by liver would not be filtered through glomerular capsule. Loss of glucose through urine always takes place due to insufficient reabsorption.

94. (a)

95. (a) Osmosis is a process in which solvent passes from an area of lower solute concentration to an area of higher solute concentration. In dialysis, solvent and solute both pass through the membrane.

96. (b) 3 and 4 are respectively descending limb and ascending limb of loop of Henle. They have minimum reabsorption and play a significant role in the maintenance of high osmolarity of medullary interstitial fluid.

97. (c) Part marked as 6 is proximal convoluted tubule (PCT). PCT is the convoluted portion of the vertebrate nephron that lies between Bowman's capsule and the loop of Henle. Cells of PCT are lined by simple cuboidal brush border epithelium that increases the surface area for reabsorption. The cells lining the PCT, absorb organic nutrients, plasma proteins & ions from the tubular fluid & release them into the interstitial fluid surrounding the renal tubule where this reabsorbed fluid enters the peritubular capillaries.

98. (d) Distal convoluted tubule (DCT) is marked as 7. The DCT is an important site for the active secretion of ions, acids, drugs, & toxins and the selective reabsorption of sodium ions from the tubular fluid. In the final portions of the DCT, an osmotic flow of water assists in concentrating the tubular fluid.

99. (b) 3, 4 and 5 are respectively descending limb, ascending limb of loop of Henle and vasa recta. These three structures have an ability to produce the concentrated urine through the process of countercurrent exchange system. Counter current mechanism (the process due to which urine is made hypertonic) is regular exchange of Na^+ ions between the ascending and descending limbs of kidney.

100. (a) In hot weather the urine becomes more concentrated and darker in colour due to dehydration (water loss). In hot weather, an individual need more water to cool themselve by evaporation of perspiration, so body compensates by cutting down the water in urine.

20 Locomotion and Movement

1. What is sarcomere?
(a) Part between two H-lines
(b) Part between two A-band
(c) Part between two I-bands
(d) Part between two Z-lines

2. Which one of the following is a skull bone?
(a) Atlas (b) Coracoid
(c) Arytenoid (d) Pterygoid

3. During muscle contraction:
(a) ATP is broken down
(b) ATP is formed
(c) GTP is broken down
(d) None of the above

4. Which is a part of pectoral girdle?
(a) Glenoid cavity (b) Sternum
(c) Ilium (d) Acetabulum

5. Which of the following is striped muscle but not voluntary in action ? **Critical Thinking**
(a) Cardiac muscle (b) Abdominal
(c) Shank (d) Wrist muscles

6. Which of the following are involved in muscle contraction? **Critical Thinking**
(a) Ca^{2+} and Mg^{2+}
(b) Ca^{2+} and Na^{+}
(c) Na^{+} and K^{+}
(d) Mg^{2+} and K^{+}

7. Part of the body having a single pair of bones is:
(a) Pelvic girdle (b) External ear
(c) Wrist (d) Lower jaw

8. Cardiac muscles are different from that of skeletal muscles as the former are:
(a) Striated but involuntary.
(b) Non-striated and involuntary.
(c) Smooth or unstriated.
(d) Voluntary in action.

9. Which of the following joint is found between sternum and the ribs in humans?
(a) Angular joint
(b) Fibrous joint
(c) Cartilaginous joint
(d) Gliding joint

10. Which of the following sets of antagonistic muscles is incorrect? **Critical Thinking**
(a) Flexor - Relaxor
(b) Abductor - Adductor
(c) Pronator - Supinator
(d) Levator - Depressor

11. Which one yields ATP, required for muscle contraction? **Tricky**
(a) Myoglobin
(b) Creatine phosphate
(c) Creatinine phosphate
(d) Myosin

12. Ankle joint is:
(a) Pivot joint (b) Ball and socket joint
(c) Hinge joint (d) Gliding joint

13. Which of the following is not exclusively supplied with involuntary muscles?
(a) Muscles of iris
(b) Muscles of the ducts of gland
(c) Muscles of urethra
(d) Muscular coats of blood vessel

14. ATPase activity is found in:
(a) Myosin filament (b) Actin filament
(c) Both (a) and (b) (d) None of these

15. Biceps are attached with:
(a) Radius (b) Ulna
(c) Femur (d) Humerus

16. Example of an imperfect joint is:
(a) Suture (b) Pubic symphysis
(c) Elbow joint (d) Ball and socket joint

17. Elbow joint is:
(a) Ball and socket joint
(b) Hinge joint
(c) Suture joint
(d) Gliding joint

18. Which set clearly identify striated muscles?
(a) Cylindrical, syncytial and unbranched
(b) Spindle, unbranched and uninucleated
(c) Cylindrical, striped and nucleated
(d) Cylindrical, striped and branched

19. The joint between incus and stapes is:
(a) Hinge joint (b) Gliding joint
(c) Pivot joint (d) Ball and socket joint

20. Which is a high energy compound in muscles?
(a) Glucose (b) Starch
(c) Glycogen (d) Creatine phosphate

21. Muscles must have something rigid to pull against if they are to cause movement. In case of humans, the rigidity is supplied by ________; in the case of earthworms, the rigidity comes from ________ . **Critical Thinking**
(a) a hydrostatic skeleton; an endoskeleton.
(b) opposing muscles; an exoskeleton.
(c) an exoskeleton; an endoskeleton.
(d) an endoskeleton; a hydrostatic skeleton.

22. Sutures present between various bones of skull are:
(a) Cartilaginous joints
(b) Synovial joints
(c) Hinge joints
(d) Fibrous joints

23. Astragalus and calcaneum are part of:
(a) Forelimb (b) Hindlimb
(c) Scapula (d) Clavicle

24. The lactic acid generated during muscle contraction is converted to glycogen in:
(a) Muscle (b) Kidney
(c) Pancreas (d) Liver

25. Hinge joints:
(a) are synovial joints
(b) permit movement in one direction
(c) are found in knee
(d) All of the above

26. In mammals, the number of cervical vertebrae is:
(a) 7 (b) 9
(c) 1 (d) 12

27. Joint between atlas and axis is:
(a) Pivot joint (b) Saddle joint
(c) Angular joint (d) Hinge joint

28. Muscles are arranged in pairs, ________. **Tricky**
(a) enabling them to perform opposing movements.
(b) doubling their strength.
(c) so they can take turns contracting and resting.
(d) so if one is injured, the other can take over.

29. The strength of contraction of an entire skeletal muscle is dependent on the: **Critical Thinking**
(a) number of muscle fibres that contract simultaneously.
(b) frequency of contraction of each muscle fibre.
(c) number of active cross bridges in each muscle fibre.
(d) All of the above

30. Part of the body having a single pair of bones is:
(a) Pelvic girdle (b) External ear
(c) Wrist (d) Lower jaw

31. The functional unit of contractile system of a striated muscle is:
(a) Sarcomere (b) Z-band
(c) Sarcosome (d) Myofibril

32. During muscle contraction: **Tricky**
(a) Size of A-bands remains the same.
(b) Size of H-zone becomes smaller.
(c) Size of I-bands decreases.
(d) All of the above

33. Quadriceps and gastrocnemius muscles lie in:
(a) Legs (b) Hands
(c) Wrist (d) Shoulder

34. Pneumatic bones are found in:
(a) House lizard (b) Pigeon
(c) Flying fish (d) Frog's tadpole

35. Intercostal muscles occur in:
(a) Abdomen (b) Thigh
(c) Ribs (d) Diaphragm

36. Which of the following would you not expect to find in both a skeletal muscle cell and a cardiac muscle cell?
(a) Actin filaments
(b) Myosin binding sites
(c) Potassium channels that close slowly after an action potential.
(d) Large numbers of mitochondria

37. Striated muscle fibres are found in:
(a) Urinary bladder (b) Lungs
(c) Gall bladder (d) Leg muscles

38. Which of the following animal has no skeleton?
(a) Sponge (b) Butterfly
(c) Jelly fish (d) Star fish

39. Which of the following muscle bends a part over another?
(a) Flexor (b) Extensor
(c) Abductor (d) Adductor

40. Sliding filament theory of muscle contraction was given by:
(a) Arnon and Hill
(b) Huxley and Pullman
(c) Huxley and Huxley
(d) None of these

41. A deltoid ridge occurs in:
(a) Radius (b) Ulna
(c) Femur (d) Humerus

42. Multi-unit smooth muscles are found in the wall of:
(a) Large blood vessels
(b) Intestine
(c) Stomach
(d) Urinary bladder

43. Phalangeal formula of hand of man is:
(a) 1, 2, 2, 2, 2 (b) 2, 1, 1, 1, 1
(c) 2, 3, 3, 3, 3 (d) 2, 3, 3, 2, 2

44. Sprain is pulling of:
(a) Bone (b) Ligament of muscles
(c) Cartilage (d) All of these

45. Foramen magnum occurs on:
(a) Lateral side of skull
(b) Posterior side of skull
(c) Anterior side of skull
(d) Neural canal

46. Creatine phosphate is present in:
(a) Nervous tissue
(b) Connective tissue
(c) Muscles of vertebrates
(d) Muscles of invertebrates

47. Joint between bones of human skull is:
(a) Hinge joint
(b) Synovial joint
(c) Cartilaginous joint
(d) Fibrous joint

48. In the given figure of pelvic girdle of man, structure marked as 1, 2, 3, 4 and 5 represents:

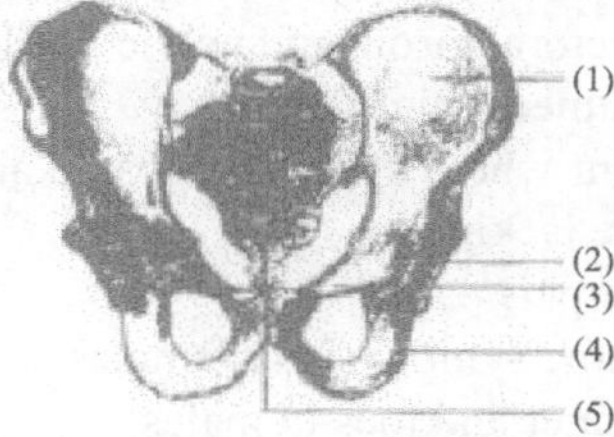

(a) 1-Pubis, 2-Acetabulum, 3-Ilium, 4-Ischium, 5-Pubic symphysis
(b) 1-Ilium, 2-Acetabulum, 3-Pubis, 4-Ischium, 5-Pubic symphysis
(c) 1-Ischium, 2-Acetabulum, 3-Pubis, 4-Ilium, 5-Pubic symphysis
(d) 1-Ilium, 2-Pubis, 3-Acetabulum, 4-Pubic symphysis, 5-Ischium

49. Red muscle fibres are rich in:
(a) Golgi bodies (b) Mitochondria
(c) Lysosomes (d) Ribosomes

50. A person is suffering from an age related disorder "X". X is characterised by decreased bone mass and increased chances of fractures. Identify X and its common cause.

(a) Tetany, Increased levels of estrogen
(b) Osteoporosis, Decreased levels of estrogen
(c) Myasthenia gravis, Decreased levels of estrogen
(d) Muscular dystrophy, Increased levels of estrogen

51. In the given diagram of skull, what does "1" represent?

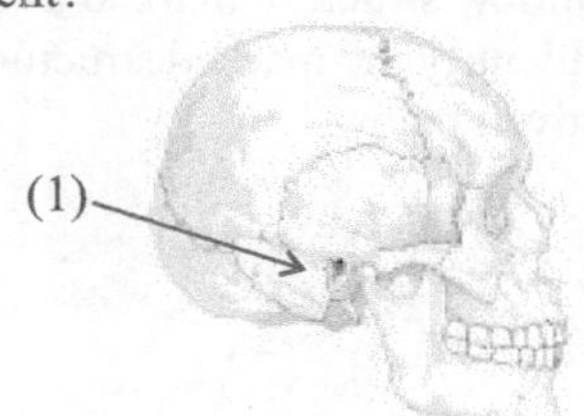

(a) Frontal bone (b) Temporal bone
(c) Occipital bone (d) Parietal bone

52. Long uninucleate muscles are found in:
(a) Diaphragm (b) Alimentary canal
(c) Tongue (d) Eye

53. Match the name of bone given in column-I with their numbers given in column-II.

	Column-I (Bone)		Column-II (Number)
A.	Thoracic	I.	8
B.	Wrist bones	II.	2 pairs
C.	False ribs	III.	12
D.	Metatarsal	IV.	3 pairs
E.	Skull (cranial & facial)	V.	26
		VI.	5
		VII.	22

(a) A – I; B – VI; C – II; D – V; E – III
(b) A – III; B – I; C – IV; D – VI; E – VII
(c) A – I; B – VII; C – III; D – V; E – II
(d) A – II; B – V; C – VII; D – I; E – IV

54. Obturator foramen occurs in:
(a) Pelvic girdle (b) Pectoral girdle
(c) Cranium (d) Vertebrae

55. Which of the following match is **incorrect**?
(a) 8th, 9th and 10th pairs of ribs – Do not articulate directly with the sternum but join the sixth rib with the help of hyaline cartilage.
(b) Glenoid cavity – Articulates with the head of the humerus to form the shoulder joint.
(c) Fibrous joint – Flat skull bones which fuse end-to-end with the help of dense fibrous connective tissues in the form of sutures to form the cranium.
(d) Increase in Ca^{++} level – Leads to the binding of calcium with a subunit of troponin on actin filaments and thereby remove the masking of active sites for myosin.

56. In the given diagram of right pectoral girdle and upper arm few structure marked as 1, 2, 3, 4, 5, 6 and 7. Identify the marked structures from the options given below.

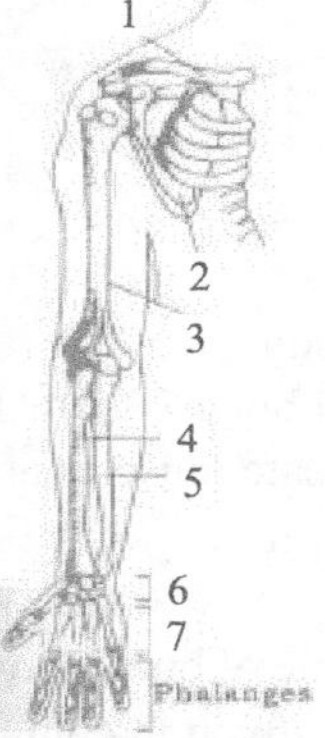

(a) 1-1st Vertebra, 2-Scapula, 3-Humerus, 4-Radius, 5-Ulna, 6-Carpals, 7-Metacarpals
(b) 1-Scapula, 2-Clavicle, 3-Humerus, 4-Radius, 5-Ulna, 6-Carpals, 7-Metacarpal
(c) 1-Ilium, 2-Scapula, 3-Humerus, 4-Radius, 5-Ulna, 6-Carpals, 7-Metacarpals
(d) 1-Clavicle, 2-Scapula, 3-Humerus, 4-Radius, 5-Ulna, 6-Carpals, 7-Metacarpals

57. Upon preventing acetylcholine from diffusing across a neuromuscular junction, which of the following will not occur? **Tricky**
(a) No action potential will be produced in the affected muscle fibres.
(b) The endoplasmic reticulum releases calcium ions (Ca^{2+}) into the cytoplasm.
(c) Myosin will not bind to actin in the affected muscle fibre.
(d) The affected muscle fibre will fail to contract.

58. In human beings, the second cervical vertebra helps in rotatory movements of head through knob-like process called: **Critical Thinking**
(a) Prezygapophysis
(b) Olecranon process
(c) Odontoid process
(d) Metaphysis

59. Latissimus dorsi muscles are:
(a) Muscles of fore arm
(b) Muscles of lower jaw
(c) Muscles of the chest
(d) Muscles of the shoulder

60. The condition in which a number of nuclei are present in a muscle fibre is called:
(a) Coenocytic (b) Syncytial
(c) Polykaryon (d) Endoduplication

61. A stronger muscle contraction occurs when the brain ____________.
(a) send signals to larger number of motor units to contract.
(b) activates muscle cells more quickly.
(c) activates the motor units of the muscle one at a time.
(d) sends nerve impulses to the muscle one at a time instead of in bursts.

62. The vertebral column is connected to the pelvic girdle in the:
(a) Coccygeal region
(b) Sacral region
(c) Lumbar region
(d) Cervical region

63. Which pair does not possess corresponding bones?
(a) Carpals and tarsals
(b) Atlas and coccyx
(c) Pectoral girdle and pelvic girdle
(d) Humerus and femur

64. The joint where synovial capsule and synovial fluid are lacking is:
(a) Carpals
(b) Pubic symphysis in females
(c) Finger and toes in males
(d) Femur and pelvis in females

65. Which one of the following shortens when a muscle fibre contracts?
(a) Thick filament (b) Sarcomere
(c) Myosin molecule (d) Thin filament

66. The number of 'true ribs' in a man is:
(a) 12 pair (b) 7 pair
(c) 3 pair (d) 2 pair

67. Muscle contraction is triggered:

(a) when high levels of oxygen and sugar are released by the sarcolemma.
(b) when a surplus of ATP is released by a nerve motor unit.
(c) by release of a neurotransmitter at a synapse that directly causes actin and myosin to slide.
(d) by the nerve releasing a neurotransmitter, which triggers a flow of calcium that attaches to actin filaments and exposes the myosin binding sites.

68. A cricket player is chasing a fast ball in the field. Which one of the following groups of bones are directly contributing in this movement?

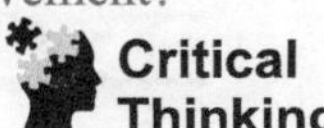

(a) Femur, malleus, tibia, metatarsals
(b) Pelvis, ulna, patella, tarsals
(c) Sternum, femur, tibia, fibula
(d) Tarsals, femur, metatarsals, tibia

69. Which of the following muscles draw the lower jaw, tongue and the head backward?
(a) Proctor (b) Retractor
(c) Abductor (d) All of these

70. An example of gliding joint is:
(a) Humerus and glenoid cavity
(b) Femur and tibio-fibula
(c) Occipital condyle and odontoid process
(d) Zygapophyses of adjacent vertebrae

71. Which of the following human skeletal parts are correctly matched with their respective category?

Pairs of skeletal parts		Category
A. Humerus and ulna	–	Appendicular skeleton
B. Malleus and stapes	–	Ear ossicles
C. Sternum and ribs	–	Axial skeleton
D. Clavicle and glenoid cavity	–	Pelvic girdle

(a) A and B only
(b) A, B and C only
(c) A, B, and D only
(d) All of these

72. A motor unit is best described as:
(a) all the nerve fibres and muscle fibres in a single muscle bundle.
(b) one muscle fibre and its single nerve fibre.
(c) a single motor neuron and all the muscle fibres that it innervates.
(d) the neuron that provides the CNS with information on the state of contraction of the muscle.

73. Children sing a song that goes 'head, shoulders, knees, and toes...' If they sang the names of the bones, it would be (in order) **Critical Thinking**
(a) Hyoid, sternum, patella, and tarsals.
(b) Cranial, clavicle, tibia, and carpals.
(c) Cranial, scapula, patella, and phalanges.
(d) Clavicle, carpals, phalanges, and tarsals.

74. Synovial fluid is found in:
(a) Cranial cavity
(b) Spinal cavity
(c) Immovable joints
(d) Freely movable joints

75. The given diagram shows a human skeleton. Which of the following correctly identifies the bones marked as 1, 2, 3 & 4?

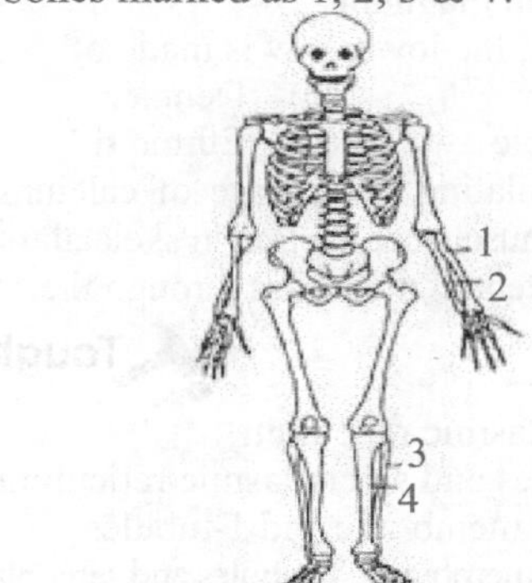

	1	2	3	4
(a)	Tibia	Fibula	Radius	Ulna
(b)	Radius	Ulna	Tibia	Fibula
(c)	Ulna	Radius	Tibia	Fibula
(d)	Fibula	Tibia	Radius	Ulna

76. Which of the following is **incorrect**?

(a) Cardiac muscle is striated.
(b) Smooth muscle does not contain actin.
(c) Skeletal muscle is considered voluntary.
(d) Smooth muscle is found in the iris of the eye and the walls of the bladder.

77. Smallest bone in the human body is:
(a) Stapes (b) Malleus
(c) Patella (d) Navicular bone

78. Lower jaw of man is made up of:
(a) 1 bone
(b) 2 bones
(c) No bones, only muscles
(d) 3 bones

79. The action potential that triggers a muscle contraction, travels deep within the muscle cell by means of: **Tricky**
(a) Sarcoplasmic reticulum
(b) Transverse tubules
(c) Synapses
(d) Motor end plates

80. Which of the following does not reveal in the detailed analysis of a cilium from the human respiratory tract?
(a) Tubulin (b) Plasmagel
(c) Microtubules (d) Axoneme

81. Spinal cord passes through:
(a) Foramen of Monro
(b) Foramen ovale
(c) Obturator foramen
(d) Foramen magnum

82. In mammals, the lower jaw is made of:
(a) Maxilla (b) Dentary
(c) Mandible (d) Ethmoid

83. Before stimulating the release of calcium ions that causes muscle contraction in skeletal muscle, an action potential must pass through the:

(a) Sarcoplasmic reticulum
(b) T-tubules and sarcoplasmic reticulum
(c) Plasma membrane and T-tubules
(d) Plasma membrane, T-tubules and sarcoplasmic reticulum

84. Which of the following statements about the molecular arrangement of actin and myosin in myofibrils is **incorrect**? **Critical Thinking**
(i) Each actin (thin filament) is made of 2F (filamentous) actins.
(ii) F-actin is the polymer of G (globular) actin.
(iii) 2F-actins are twisted into α-helix.
(iv) Two strands of tropomyosin (protein) lie in the grooves of F-actin.
(v) Troponin molecules (complex proteins) are distributed at regular intervals on the tropomysin.
(vi) Troponin forms the head of the myosin molecule.
(vii) The myosin is a polymerised protein.
(a) (i), (iii), (vii) (b) (ii), (iv) and (v)
(c) Only (vi) (d) Only (iii)

85. Patella bone is derived from:
(a) Cartilage (b) Areolar tissue
(c) Tendon (d) Ligament

86. In humans, coccyx is formed by the fusion of:
(a) 3 vertebra (b) 4 vertebra
(c) 5 vertebra (d) 6 vertebra

87. The intercalated discs of____muscle____.
(a) smooth; provide strong mechanical adhesion and rapid electrical communication
(b) skeletal; are the basis for all voluntary muscle action
(c) skeletal; make possible both fast twitches and slow twitches.
(d) cardiac; provide strong mechanical adhesion and rapid electrical communication.

88. Choose the letter from the figure that most appropriately corresponds to the structure.

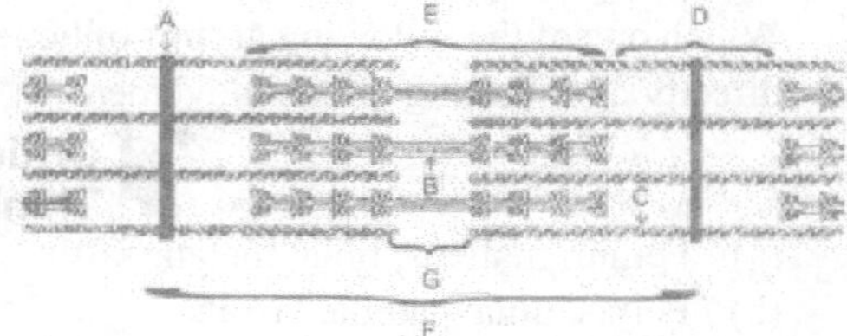

(i) A-band (ii) I-band
(iii) Sarcomere (iv) H-zone
(v) Myosin
(vi) Actin, Troponin, Tropomyosin
(vii) Z-line
(a) (i) – E, (ii) – D, (iii) – F, (iv) – G, (v) – B, (vi) – C, (vii) – A
(b) (i) – E, (ii) – D, (iii) – C, (iv) – G, (v) – B, (vi) – A, (vii) – F
(c) (i) – E, (ii) – D, (iii) – F, (iv) – G, (v) – C, (vi) – A, (vii) – B
(d) (i) – E, (ii) – D, (iii) – F, (iv) – A, (v) – B, (vi) – C, (vii) – G

89. Sigmoid notch is found in:
(a) Tibio-fibula (b) Femur
(c) Radio-ulna (d) Humerus

90. Which of the following statements about the striated muscles is **incorrect**?
(i) In the centre of each I-band, is an elastic fibre (Z-line) which bisects it.
(ii) Thin filaments are firmly attached to the Z-line.
(iii) M-line is a fibrous membrane in the middle of A-bands.
(iv) A sarcomere comprises one full A-bands and 2 half I-bands.
(a) All of these (b) Only (iv)
(c) (i) and (ii) (d) None of these

91. In the given diagramatic representation of muscle contraction, identify the process marked as 1, 2, 3, 4 and 5.

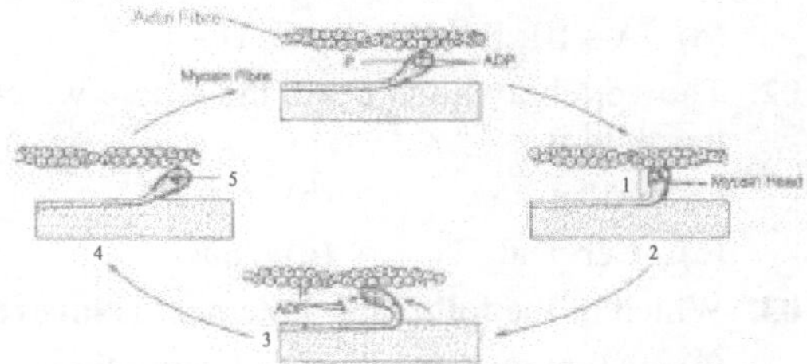

(a) 1-Cross bridge, 2-Cross bridge formation, 3-Breaking of cross bridge, 4-Sliding (rotation), 5-ATP
(b) 1-Cross bridge, 2-Cross bridge formation, 3-Sliding / rotation, 4-Breaking of cross bridge, 5-ATP
(c) 1-Cross bridge, 2-Breaking of cross bridge, 3-Sliding/rotation, 4-Cross bridge formation, 5-AMP
(d) 1-Cross bridge, 2-Cross bridge formation, 3-Sliding/rotation, 4-ADP, 5-Breaking of cross bridge.

92. Zygomatic arch in Rabbit is formed by:
(a) Premaxilla, Maxilla and Palatine
(b) Maxilla, Jugal and Squamosal
(c) Periotic, Maxilla & Premaxilla
(d) All of the above

93. Lower jaw in mammals articulates with a bone called:
(a) Jugal
(b) Alisphenoid
(c) Sqamosal
(d) Parietal

94. The given figure shows the diagrammatic cross sectional view of a muscle with their parts marked as 1, 2, 3 and 4. Which part is held together by a common collagenous connective tissue layer?

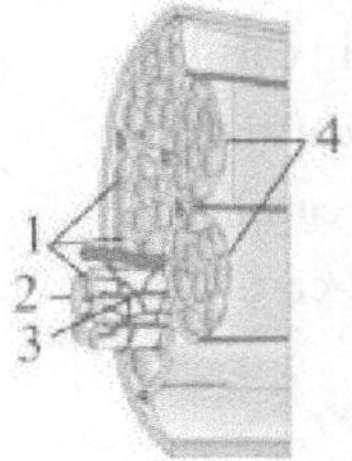

(a) 1 (b) 2
(c) 3 (d) 4

95. Put the following phrases in proper order to describe what occurs at the neuromuscular junction to trigger muscle contraction.
(i) Receptor sites on sarcolemma.
(ii) Nerve impulse
(iii) Release of Ca^{+2} from sarcoplasmic reticulum.
(iv) The neurotransmitter acetylcholine is released.
(v) Sarcomere shortern.
(vi) Synaptic cleft.
(vii) Spread of impulses over sarcolemma on T-tubules.
(a) (ii), (iv), (i), (vi), (vii), (iii), (v)
(b) (ii), (iv), (vi), (i), (vii), (iii), (v)
(c) (i), (ii), (iii), (iv), (v), (vi), (vii)
(d) (vii), (vi), (v), (iv), (iii), (ii), (i)

96. Just as, there are pseudopodia *in Amoeba*, there are:
(a) Teeth in rabbit
(b) Suckers upon scolex of tapeworm
(c) Legs in cockroach
(d) Spicules *in Leucosolenia*

97. Which of the following statement is not the feature of red muscle fibres?
(a) They have plenty of mitochondria.
(b) They have high content of myoglobin.
(c) They have high amount of sarcoplasmic reticulum.
(d) They are called aerobic muscles.

98. Consider the diagram given below and identify the parts labelled as 1, 2, 3, 4, and 5 respectively.

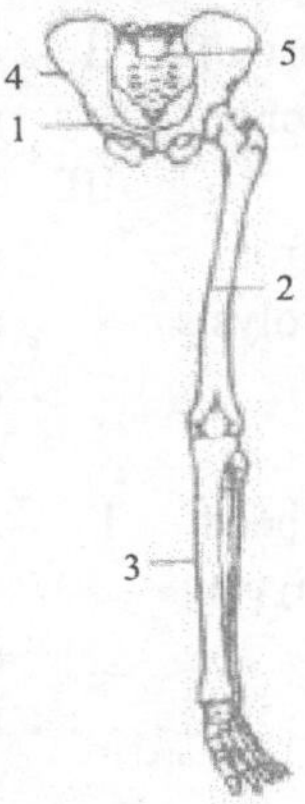

(a) Femur, Ilium, Tibia, Pubis and Sacrum
(b) Pubis, Tibia, Femur, Ilium and Sacrum
(c) Ilium, Femur, Tibia, Pubis and Sacrum
(d) Pubis, Femur, Tibia, Ilium and Sacrum

99. A student was given sample of two muscles marked as 1 and 2. When he compared the muscles he found that muscle 1 contains large amount of myoglobin and utilise large amount of stored oxygen for ATP production whereas muscle 2 contains few myoglobin, mitochondria and high sarcoplasmic reticulum.
Identify the correct statement regarding the muscle 1 and 2 from the option given below.

(a) Both the muscles are called aerobic muscles.
(b) Both the muscles are called red fibres and depend on aerobic process for energy.
(c) Muscle 1 is called white fibres and whereas muscle 2 depends on aerobic process for energy.
(d) Muscle 1 is called red fibres and aerobic muscles whereas muscle 2 is called white fibres and depend on anaerobic process for energy.

100. Locomotory organs in starfishes are:
(a) Parapodia (b) Pseudopodia
(c) Suckers (d) Tubefeet

101. Match the column-I with column-II and select the correct option.

	Column-I		Column-II
A.	Striated appearance of myofibril	I.	Distribution pattern of actin and myosin
B.	Store house of calcium ions	II.	Sarcoplasmic reticulum
C.	Energy released from ATP hydrolysis	III.	Helps myosin head to bind to exposed active sites on actin to form a cross bridge
D.	Globular head of meromyosin	IV.	Active AT Pase enzyme has binding sites for ATP and active sites for actin.

(a) A – I; B – II; C – III; D – IV
(b) A – III; B – I; C – IV; D – II
(c) A – II; B – I; C – III; D – IV
(d) A – III; B – IV; C – II; D – I

102. The vertebra which bears the whole weight of the skull is:
(a) Atlas (b) Axis
(c) Cervical (d) Sacral

103. Which of the following statement is **incorrect**?
(a) All movements lead to locomotion.
(b) Ciliary movement help in passage of ova through female reproductive tract.
(c) Microfilaments are involved in amoeboid movement.
(d) In *Paramoecium*, the cilia help in movement of food through cytopharynx and in locomotion as well.

104. Match column-I with column-II and select the correct answer using the codes given below.

	Column-I (Bone)		Column-II (Number of bones)
A.	Cranium / Brainbox	I.	29
B.	Skull (cranial and facial bones)	II.	8
C.	Face	III.	14
D.	Hind limb	IV.	12 pairs
E.	Ribs	V.	30

(a) A – I; B –II; C – III; D – V; E – IV
(b) A – II; B – I; C – III; D – V; E – IV
(c) A – I; B – II; C – III; D – IV; E – V
(d) A – V; B – IV; C – III; D – II; A – I

105. In the larynx of rabbit, which one of the following is the biggest cartilage?

(a) Arytenoid cartilage
(b) Cricoid cartilage
(c) Thyroid cartilage
(d) Tracheal cartilage

106. Which of the following bones are not found in the human appendicular skeleton?
(a) Finger bones
(b) Upper arm bone
(c) Jaw bone
(d) Ankle bones

107. Match column-I with column-II and select the correct answer using the codes given below.

	Column-I		Column-II
A.	Atlas and Axis	I.	Cartilaginous joint
B.	Two parietal bones	II.	Fibrous joint
C.	Two pubic bones	III.	Saddle joint
D.	First carpal and first metacarpal	IV.	Pivot joint

(a) A – II; B – I
(b) B – II; C – III
(c) C – II; D – III
(d) D – III; A – IV

108. Match column-I with column-II and select the correct answer using the codes given below.

	Column-I		Column-II
A.	Ribs are attached to the sternum ventrally and to the vertebrae dorsally.	I.	True ribs
B.	Ribs are attached to sternum through coastal cartilage (hyaline) of 7th rib	II.	False ribs
C.	Ribs are not attached to sternum	III.	Floating ribs

(a) A – I; B – II; C – III
(b) A – I; B – III; C – II
(c) A – II; B – I; C – III
(d) A – III; B – II; C – I

Answer KEYs

1	(d)	**12**	(c)	**23**	(b)	**34**	(b)	**45**	(b)	**56**	(d)	**67**	(d)	**78**	(a)	**89**	(c)	**100**	(d)
2	(d)	**13**	(c)	**24**	(d)	**35**	(c)	**46**	(c)	**57**	(a)	**68**	(d)	**79**	(b)	**90**	(d)	**101**	(a)
3	(a)	**14**	(a)	**25**	(d)	**36**	(c)	**47**	(d)	**58**	(c)	**69**	(b)	**80**	(b)	**91**	(b)	**102**	(a)
4	(a)	**15**	(d)	**26**	(a)	**37**	(d)	**48**	(b)	**59**	(d)	**70**	(d)	**81**	(d)	**92**	(b)	**103**	(a)
5	(a)	**16**	(b)	**27**	(a)	**38**	(c)	**49**	(b)	**60**	(b)	**71**	(b)	**82**	(c)	**93**	(c)	**104**	(b)
6	(a)	**17**	(b)	**28**	(a)	**39**	(a)	**50**	(b)	**61**	(a)	**72**	(c)	**83**	(c)	**94**	(d)	**105**	(c)
7	(a)	**18**	(a)	**29**	(d)	**40**	(c)	**51**	(b)	**62**	(b)	**73**	(c)	**84**	(c)	**95**	(b)	**106**	(c)
8	(a)	**19**	(a)	**30**	(a)	**41**	(d)	**52**	(b)	**63**	(b)	**74**	(d)	**85**	(c)	**96**	(c)	**107**	(d)
9	(c)	**20**	(d)	**31**	(a)	**42**	(a)	**53**	(b)	**64**	(b)	**75**	(b)	**86**	(b)	**97**	(c)	**108**	(a)
10	(a)	**21**	(d)	**32**	(d)	**43**	(c)	**54**	(a)	**65**	(b)	**76**	(b)	**87**	(d)	**98**	(d)		
11	(b)	**22**	(d)	**33**	(a)	**44**	(b)	**55**	(a)	**66**	(b)	**77**	(a)	**88**	(a)	**99**	(d)		

Hints & Solutions

1. (d) Dark bands on the myofibril form A-band. Middle part of A-band contains light zone-Hensen line (H-lines). Light band on myofibril is called I-band. Middle part of I-band contains Z-line. Sarcomere is part between two Z-lines.
2. (d) Atlas is the first cervical vertebra. Coracoid is part of the pectoral girdle. Pterygoid is a small bone articulated with palatine.
3. (a)
4. (a) Glenoid cavity is a shallow concavity on the lateral side of pectoral girdle in which the head of humerus fits making the shoulder joint.
5. (a)
6. (a) Ca^{2+} and Mg^{2+} are involved in muscle contraction.
7. (a) 8. (a)
9. (c) In angular joint, articulating surface of one bone is oval and convex whereas that of the other is elliptical and concave, it allows movement only in two directions, e.g., wrist. Fibrous joints are inmovable joints, e.g., sutures of skull. Cartilaginous joints have little motility due to disc of fibrocartilage between the articular ends. In gliding joints articulate end of both the bones are either flat or curved to allow gliding movements, e.g., between carpals and wrist.
10. (a) 11. (b) 12. (c) 13. (c)
14. (a) 15. (d) 16. (b)
17. (b) Elbow joint is hinge joint.
18. (a) 19. (a) 20. (d)
21. (d) An exoskeleton is a hard encasement deposited on the surface of the animal; an endoskeleton consists of hard supporting elements, such as bones, buried within the soft tissues of an animal.
22. (d) 23. (b) 24. (d) 25. (d)
26. (a)
27. (a) Joint between atlas and axis is pivot joint.
28. (a) Since muscles can only shorten, an opposing muscle pair allows a joint to move in different directions.
29. (d) 30. (a) 31. (a) 32. (d)
33. (a) 34. (b) 35. (c)
36. (c) Only cardiac muscle cells have the unique potassium channels that give this type of muscle its myogenic properties.
37. (d) 38. (c) 39. (a) 40. (c)
41. (d) 42. (a)
43. (c) 1 humerus, 1 radius, 1 ulna, 8 carpal bones 5 metacarpal bones, 5 digits (14 phalanges), therefore phalangeal formula : 2, 3, 3, 3, 3
44. (b) 45. (b) 46. (c) 47. (d)
48. (b) 49. (b)
50. (b) Decreased levels of estrogen is a common cause of osteoporosis. Osteoporosis is a condition in which the bones become brittle and fragile from loss of tissue, typically as a result of hormonal changes or deficiency of calcium or vitamin D. Osteoporosis is characterised by low bone mass and deterioration of bone tissue. This leads to increased bone fragility and risk of fracture (broken bones), particularly of the hip, spine, wrist and shoulder. Osteoporosis is often known as "the silent thief" because bone loss occurs without symptoms.
51. (b) 52. (b) 53. (b) 54. (a)
55. (a) 8th, 9th and 10th pairs of ribs (called false ribs) do not articulate directly with the sternum but join the seventh rib with the help of hyaline cartilage.
56. (d) 57. (a) 58. (c)
59. (d) Latissimus dorsi is one of the pair of large triangular muscles on the thoracic and lumbar areas of the back. The latissimus dorsi extends, adducts and rotates the arm medially; draws the shoulder back and down.
60. (b)

61. **(a)** The strength of a contraction for a given muscle is proportional to the number of motor units activated.

62. **(b)** **63.** **(b)** **64.** **(b)**

65. **(b)** The sarcomere is the functional unit of muscle. As actin myofilaments slide over myosin, the sarcomere shortens, but not the individual components that make up the sarcomere.

66. **(b)**

67. **(d)** Muscle contraction is triggered by a nerve releasing a neurotransmitter, which in turn triggers the sarcoplasmic reticulum to release calcium ions into the muscle interior where they bind to troponin, thus causing tropomyosin to shift from the face of the actin filament to which myosin heads need to bind to produce contraction.

68. **(d)** **69.** **(b)** **70.** **(d)**

71. **(b)** Clavicle and glenoid cavity belongs to pectoral girdle rather than pelvic girdle. Hence, it is incorrectly paired.

72. **(c)** A motor unit is a single motor neuron and all the muscle fibres that it innervates.

73. **(c)** The scapula is the shoulder blade, patella is the kneecap, and phalanges are the fingers and toes and the head is the cranial bone.

74. **(d)**

75. **(b)** In the given figure of human skeletal system, the marked bones (1, 2, 3 and 4) are identified as radius, ulna, tibia and fibula respectively.

Radius and ulna are two large bones of the forearm. Radius is a long bone, prism-shaped and slightly curved longitudinally. It articulates with the capitulum of the humerus, and with the ulna at two points: the radial notch (lower) and the head (upper part) of the ulna. Ulna runs parallel to the radius, which is shorter and smaller. The tibia and fibula are two bones of hind limbs. Tibia, also known as the shin bone, is the larger and stronger of the two lower leg bones. It forms the knee joint with the femur and the ankle joint with the fibula and tarsus. The support and movement of the tibia is essential to many activities performed by the legs, including standing, walking, running, jumping and supporting the body's weight. Fibula or calf bone is located on the lateral side of the tibia, with which it is connected above and below. It is shorter, thinner and slender.

76. **(b)** Although contraction of smooth muscle is controlled differently from that of skeletal muscle, smooth muscle does not contain actin and myosin.

77. **(a)** **78.** **(a)**

79. **(b)** The action potential arriving to the muscle travels into the muscle through the transverse tubules.

80. **(b)** Plasmagel is a special category of cytoplasm that influences the amoeboid movement of some cells. It is not a structural component of a cilium.

81. **(d)** **82.** **(c)**

83. **(c)** The action potential is initiated on the plasma membrane at the neuromuscular junction. It then spreads through the cell via the T-tubule system before reaching the triads that initiate the opening of the Ca^{2+} channels of the sarcoplasmic reticulum.

84. **(c)** **85.** **(c)** **86.** **(b)**

87. **(d)** Both of these characteristics are important for the function of cardiac muscle.

88. **(a)** **89.** **(c)** **90.** **(d)** **91.** **(b)**

92. **(b)** **93.** **(c)**

94. **(d)** 1, 2, 3 and 4 are respectively muscle fibre, sarcolemma, blood capillary and fascicle. Each organised skeletal muscle in human body is made of a number of muscle bundles or fascicles held together by a collagenous connective tissue layer called fascia.

95. **(b)** **96.** **(c)**

97. **(c)** Red muscle fibres are one of the two main types of skeletal muscle which contains abundant mitochondria and myoglobin. Red muscle fibres contract and fatigue more slowly than white fibres and it generate ATP by aerobic catabolism of glucose and fats utilising myoglobin-bound O_2.

Sarcoplasmic reticulum is a system of membrane-bound tubules that surrounds muscle fibrils, releasing calcium ions during contraction and absorbing them during relaxation.

98. **(d)**

99. (d) Muscle 1 and 2 are respectively red muscle fibre and white muscle fibre. Red muscle fibre (or Type I, slow twitch muscle fibre) is dense with capillaries and is rich in mitochondria and myoglobin, giving the muscle tissue its characteristic red colour. It can carry more oxygen and sustain aerobic activity using fats or carbohydrates as fuel. Slow twitch fibres contract for long periods of time but with little force.

White muscle fibre (or Type II, fast twitch muscle fibre) contract quickly and powerfully but fatigue very rapidly, sustaining only short, anaerobic bursts of activity before muscle contraction becomes painful. They contribute most to muscle strength and have greater potential for increase in mass. Type II is anaerobic, glycolytic, "white" muscle that is least dense in mitochondria and myoglobin.

100. (d) Tubefeet are the characteristic organs of echinoderms, serving for locomotion, food capture, respiration, etc.

101. (a)

102. (a) According to Greek mythology, atlas is the Titan who born the heavens on his shoulders. If the skull can be considered as the heavens, the first vertebra which bears the whole weight of the skull can be regarded as atlas.

103. (a) All movements do not lead to locomotion. Locomotion is the voluntary movements resulting in a change in location.

104. (b)

105. (c) The larynx of rabbit is made up of thyroid cartilage, arytenoid cartilage and cricoid cartilage. Cricoid cartilage is C-shaped while the biggest cartilage of larynx is thyroid.

106. (c) The appendicular skeleton includes all bones associated with the appendages. The jaw bone is part of the skull, which belongs to the axial skeleton.

107. (d) **108. (a)**

21 Neural Control and Co-ordination

1. Which of the following controls reflex action?
(a) Central nervous system
(b) Sympathetic nervous system
(c) Parasympathetic nervous system
(d) Sensory nerves

2. Perception of colour is possible among:

(a) Birds (b) Reptiles
(c) Men (d) Both (a) and (c)

3. Match the column-I (various phase of an action potential) with column-II (ionic activity associated) and choose the correct option.

	Column-I		Column-II
A.	Resting stage of a neuron	I.	Opening and then closing of the sodium channels
B.	Depolarisation phase in the generation of an action potential.	II.	All voltage gated sodium and potassium channels are closed.
C.	Repolarisation phase in the generation of action potential	III.	The sodium channels remain opened.
D.	Absolute refractory phase.	IV.	Opening of potassium gates and the rushing of potassium

(a) A – II; B – I; C – IV; D – III
(b) A – I; B – II; C – III; D – IV
(c) A – III; B – IV; C – I; D – II
(d) A – IV; B – II; C – III; D – I

4. In the given diagram of brain what does "1" represent?

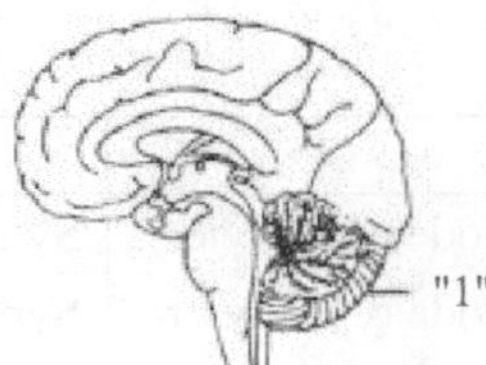

(a) Pons (b) Midbrain
(c) Cerebellum (d) Medulla oblongata

5. Nerve fibres are surrounded by an insulating fatty layer called:
(a) Adipose sheath (b) Myelin sheath
(c) Hyaline sheath (d) Peritoneum

6. Which of the following disorder of eyes caused by UV radiations?
(a) Cataract
(b) Glaucoma
(c) Dilation pupil
(d) Some defect of retina

7. The human eye is sensitive only to light having wavelength ranging from: Tricky
(a) 80 to 280 nanometre
(b) 380 to 760 nanometre
(c) 780 to 870 nanometre
(d) 880 to 980 nanometre

8. "X" is spiral shaped structure (given below) consisting of hair cells that serve as receptors for auditory stimuli.
Identify "X" and its label & location (marked as 1, 2, 3 and 4) from the given diagrammatic representation of the sectional view of cochlea.

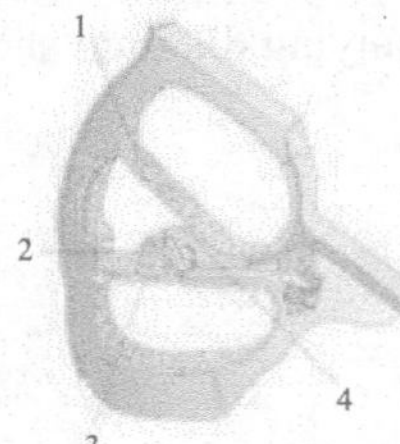

(a) X- Organ of corti, 2, 3.
(b) X- Eustachian tube, 1, 2.
(c) X- Semicircular canal, 3, 4.
(d) X- Crista ampullaris, 1, 4.

9. Three layers in eye ball from inside out are:
(a) Retina, choroid, sclerotic
(b) Choroid, retina, sclerotic
(c) Sclerotic, choroid, retina
(d) Sclerotic, retina, choroid

10. Eye muscles are innervated by:

(a) Oculomotor, abducens and vagus
(b) Oculomotor, trochlear and abducens
(c) Oculomotor, abducens and facial
(d) Oculomotor, facial and vagus

11. The neural phenomenon depicted by the given diagram is best described as: **Critical Thinking**

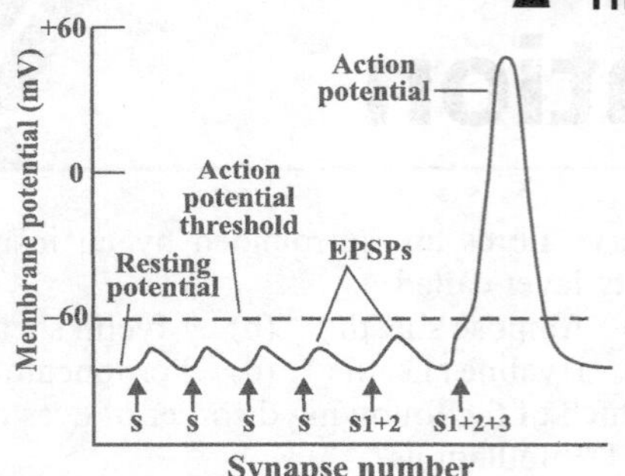

(a) Spatial summation
(b) Temporal summation
(c) Presynaptic excitation
(d) Presynaptic inhibition

12. Which of the following is devoid of blood supply?
(a) Retina (b) Choroid
(c) Cornea (d) Scleroid

13. Clusters of neuron cell bodies embedded in the white matter of the brain are referred to as:
(a) Nuclei (b) Gyri
(c) Sulci (d) Ganglia

14. Which of the following statements is **correct** regarding a myelinated nerve fibre?
(i) It is always associated with an axon.
(ii) It allows rapid conduction of nerve impulses.
(iii) It allows slow conduction of nerve impulses.
(iv) It has nodes of Ranvier.
(a) Only (i) and (ii)
(b) Only (i), (ii), and (iii)
(c) Only (i), (ii), and (iv)
(d) Only (i), (iii), and (iv)

15. The resting membrane potential for neuron A is –70mV, while the resting potential for neuron B is –50mV. The threshold voltage for the production of an action potential is –35mV for both neurons. Which of the following statements is **incorrect**? **Critical Thinking**
(a) Neuron A must depolarise by 35mV to reach the threshold voltage.
(b) Neuron B must hyperpolarise by 15mV to reach the threshold voltage.
(c) The inside of both neurons is negatively charged with respect to the outside.
(d) A single EPSP received by neuron A would cause it to depolarise slightly.

16. Which of the following substances can cure Parkinson's disease?
(a) GABA (b) Acetylcholine
(c) Dopamine (d) Glutamic acid

17. Eustachian tube connects:
(a) External ear with middle ear
(b) External ear with internal ear
(c) Middle ear with pharynx
(d) Internal pharyax

18. The accompanying diagram shows the structure of neuron. Identify 1 to 5.

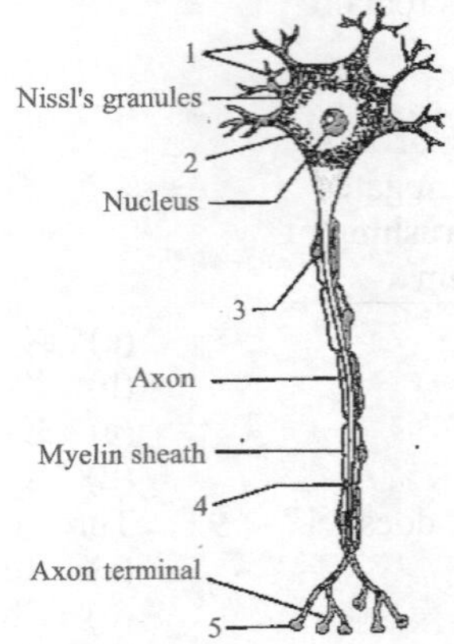

	1	2	3	4	5
(a)	Nerve fibre	Cyton or cell body	Schwann cell	Node of Ranvier	Synaptic knob
(b)	Dendrites	Cyton or cell body	Schwann cell	Node of Ranvier	Synaptic knob
(c)	Dendrites	Nerve cell	Schwann cell	Node of Ranvier	Synaptic knob
(d)	Dendrites	Cyton or cell body	Nerve cell	Node of Ranvier	Synaptic knob

19. In the given figure of synapse few structure are marked as 1, 2, 3 and 4. Study the figure carefully and identify the structure which transmits signal across a chemical synapse.

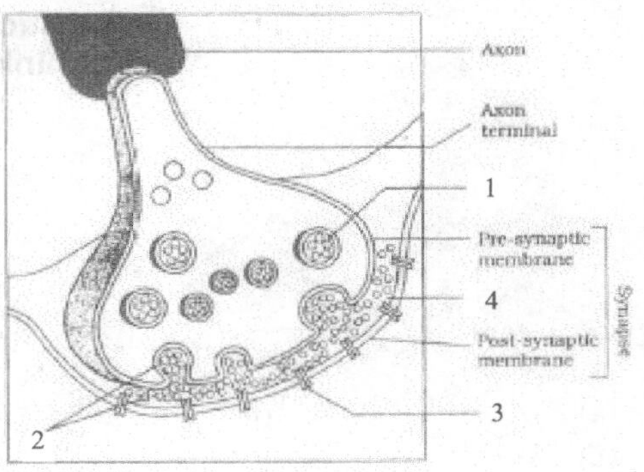

(a) 1 (b) 2
(c) 3 (d) 4

20. The primary cause for the existence of resting membrane potential can be attributed to:

(a) Potassium diffusion potential
(b) Presence of electrogenic pump
(c) Diffusion potentials of Na^+, K^+ and Cl^-
(d) Anions like proteins, phosphates and sulphates.

21. The given diagram shows the axon terminal and synapse with few part labelled as 1, 2, 3 & 4. Choose the correct combination of labelling from the given options.

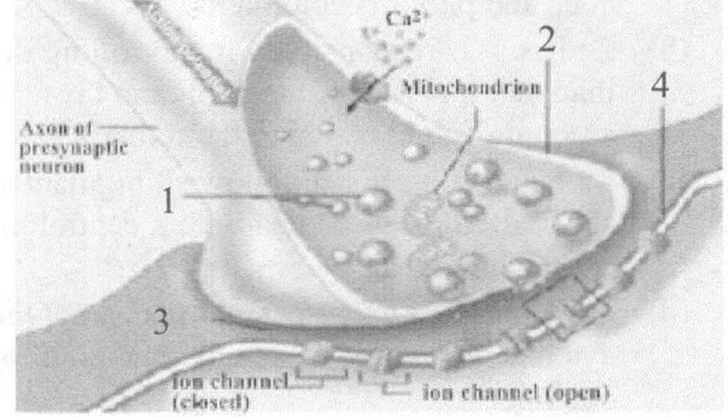

(a) 1-Synaptic vesicle, 2-Axon terminal, 3-Synaptic cleft, 4-Postsynaptic membrane
(b) 1-Axon terminal, 2-Synaptic vesicle, 3-Postsynaptic membrane, 4-Synaptic cleft.
(c) 1-Synaptic vesicle, 2-Synaptic cleft, 3-Axon terminal, 4-Post synaptic membrane
(d) 1-Post synaptic membrane, 2-Axon terminal, 3-Synaptic vesicle, 4-Synaptic cleft

22. Match the column-I with column-II and select the correct option.

	Column-I		Column-II
A.	Cervical nerves	I.	5 pairs
B.	Thoracic nerves	II.	1 pair
C.	Lumbar nerves	III.	12 pairs
D.	Coccygeal nerves	IV.	8 pairs

(a) A – II; B – IV; C – I; D – III
(b) A – IV; B – III; C – I; D – II
(c) A – III; B – I; C – II; D – IV
(d) A – IV; B – I; C – II; D – III

23. The mandibular nerve is the branch of which cranial nerve?

(a) II (b) III
(c) V (d) VI

24. Select the correct match of the types of neuron given in column-I with its location given in column-II.

	Column-I		Column-II
A.	Sclera	I.	The visible coloured portion of the eye.
B.	Choroid	II.	Its diameter is regulated by the muscle fibres of iris.
C.	Pupil	III.	Composed of a dense connective tissue
D.	Fovea	IV.	Portion of the retina where only the cones are densely packed
E.	Iris	V.	Contains many blood vessels and looks bluish in colour

(a) A – I; B – II; C – III; D – IV; E – V
(b) A – III; B – V; C – II; D – IV; E – I
(c) A – IV; B – I; C – V; D – II; E – III
(d) A – V; B – IV; C – III; D – I; E – II

25. The given diagram shows different parts of a human eye with one part labeled as X.

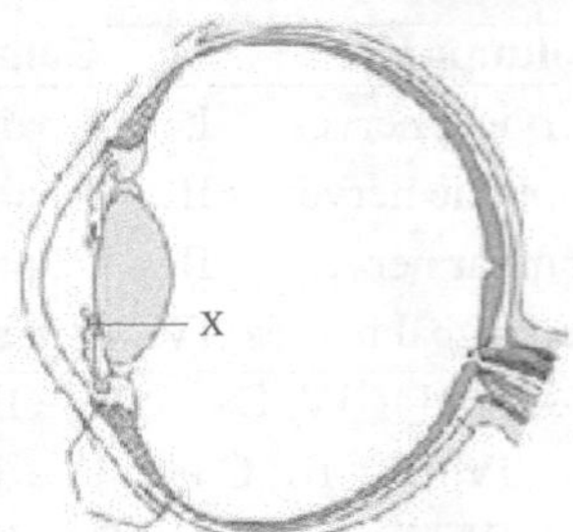

Which of the following statement is **correct** regarding label X?
(a) It is the opening in lens that permits light into the inner chambers of the eye.
(b) It is the coloured portion of vascular tunic.
(c) It is a biconcave structure that changes shape to bring objects into focus.
(d) It is thick, jelly-like substance in the posterior compartment of the eye.

26. The size of aperture of the pupil of one eye is controlled by:
(a) Iris (b) Retina
(c) Cornea (d) Conjunctiva

27. Refer the given figure of ear with few structure marked as 1, 2, 3 & 4. Which labelled structure converts sound waves into mechanical vibrations?

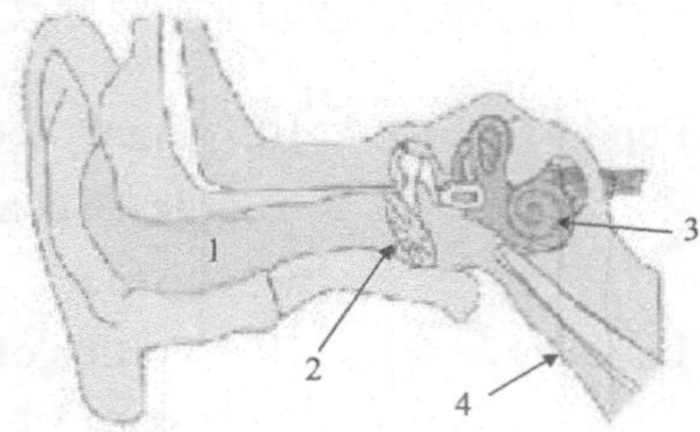

(a) 1 (b) 2
(c) 3 (d) 4

28. Part of mammalian brain controlling muscular coordination is:
(a) Cerebrum
(b) Corpus callosum
(c) Medulla oblongata
(d) Cerebellum

29. Cranial nerve showing maximum branching is:
(a) Trigeminal (b) Vagus
(c) Optic (d) Facial

30. The given diagram represents the sectional view of cochlea with few part labelled as 1, 2, 3 & 4. The movement of which marked label causes hair cell microvilli to bend back and forth.

Critical Thinking

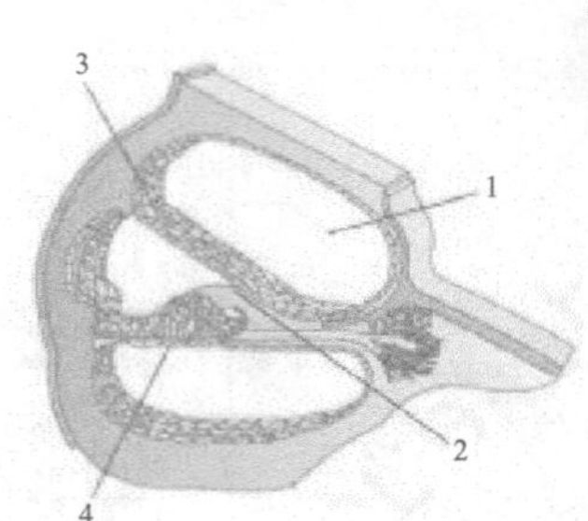

(a) 1 (b) 2
(c) 3 (d) 4

31. In the given figure of eye, few parts are labelled as 1, 2, 3 and 4. Select the option which shows the correct identification of the part with its characteristics.

Tricky

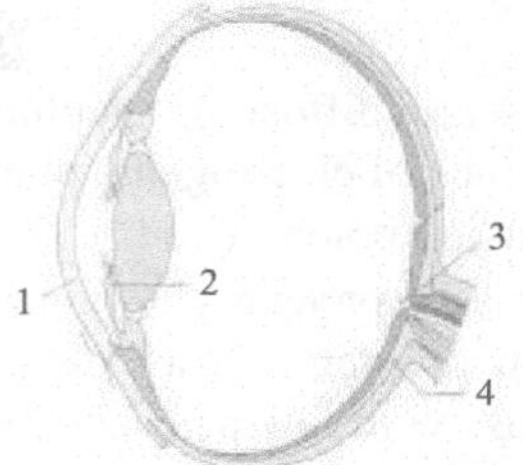

(a) 1 : Choroid, it contains ganglion cells, bipolar cells and photoreceptor cells.
(b) 2 : Iris, it is responsible for controlling the diameter and size of the pupil and thus the amount of light reaching the retina.
(c) 3 : Blind spot, it is a yellowish pigmented spot called macula lutea with a central pit called the fovea.
(d) 4 : Cornea, it is a transparent front part of the eye that covers the iris, pupil, and anterior chamber.

32. Glands of Swammerdam are found associated with:
(a) Cranial nerves (b) Pancreas
(c) Spinal nerves (d) Liver

33. Which part of the brain is more develop in human?
(a) Medulla (b) Cerebellum
(c) Cerebrum (d) Optic lobes

34. When a membrane is at the resting potential, the concentration of: Tricky

(a) Sodium and potassium ions is higher on the inside of its membrane.
(b) Sodium and potassium ions is higher on the outside of its membrane.
(c) Sodium ions is higher on the inside of its membrane and of potassium ions is higher on the outside.
(d) Sodium ions is higher on the outside of its membrane and of potassium ions is higher on the inside.

35. How are neurons structurally adapted to chemically transmit impulses to neighboring neurons? Tricky

(a) The trigger zone is near the cell body.
(b) Axon terminals contain neurotransmitter substances within synaptic vesicles.
(c) Schwann cells surround axons.
(d) They have numerous dendrites.

36. Dopamine acetylcholine and GABA are:
(a) Hormone (b) Enzyme
(c) Synthetic vitamin (d) Neurotransmitters

37. Four events in the transmission of nerve impulses across synapses are given below.

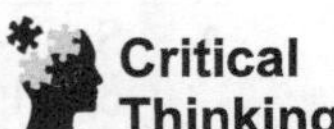

(i) Depolarisation of the presynaptic membrane.
(ii) Propagation of postsynaptic action potential.
(iii) Hydrolysis of transmitter substance.
(iv) Rupturing of synaptic vesicles.

In which sequence do these events occur?
(a) (i) → (iii) → (ii) → (iv)
(b) (i) → (iv) → (ii) → (iii)
(c) (iv) → (i) → (iii) → (ii)
(d) (iv) → (iii) → (i) → (ii)

38. Area of cerebral cortex controlling vision is:
(a) Frontal lobe (b) Parietal lobe
(c) Temporal lobe (d) Occipital lobe

39. Absolute refractory period during nerve impulse conduction is the period of: Tricky
(a) repolarisation
(b) depolarisation
(c) both repolarisation and depolarisation
(d) neither repolarisation nor depolarisation

40. The anterior choroid plexus forms the roof of:
(a) Optocoel
(b) 4th ventricle
(c) Diocoel
(d) Olfactory ventricle

41. A thin vascular membrane of brain is:
(a) Arachnoid (b) Piameter
(c) Durameter (d) Epineurium

42. Seventh cranial nerve enters:
(a) Jugular ganglion
(b) Geniculate ganglion
(c) Gasserian ganglion
(d) Femoral ganglion

43. Several nervous system diseases result from a loss of support cells or of substances produced by support cells (such as myelin). The primary function of these support cells is to:
(a) act as supporting structures within nervous tissue.
(b) produce insulating material around axons.
(c) assist in the conduction of impulses along neurons.
(d) All of the above

44. The set of mixed nerve is: Critical Thinking
(a) Optic, auditory, olfactory
(b) Auditory, trigeminal, hypoglossal
(c) Occulomotor, pathetic, abducens
(d) Vagus, facial, trigeminal

45. In static condition, the body balance is sensed by:
(a) Crista (b) Macula
(c) Both (a) and (b) (d) Cochlear canal

46. Given below is a table comparing the effects of sympathetic and parasympathetic nervous system. Which of the following feature is correctly described?

	Feature	Sympathetic nervous system	Parasympathetic nervous system
(a)	Salivary glands	Stimulates secretion	Inhibits secretion
(b)	Pupil of eye	Dilates	Constricts
(c)	Heart rate	Decreases	Increases
(d)	Intestinal peristalsis	Stimulates secretion	Inhibits secretion

47. Eye lens focus light on the retina containing rod cells, which are receptors. Number of rod cells in eyes are: Tricky

(a) 120 million (b) 105 million
(c) 150 million (d) 118 million

48. A particular disease of the nervous system specifically involves the Ca^{2+} ion channels at the chemical synapses of motor neurons where neurotransmitter is stored and released. In other words, this disease affects the

Critical Thinking

(a) axon terminals of the presynaptic cell and the release of acetylcholine.
(b) axon terminals of the postsynaptic cell and the release of K^+ ions.
(c) electrical synapses.
(d) axon terminals of the presynaptic cell and the release of K^+ ions.

49. Organs of Ruffini are receptors of:
(a) Cold (b) Pressure
(c) Heat (d) Touch

50. Ora serrata is:
(a) oral cavity of protochordates.
(b) anterior edge of sensory portion of the retina.
(c) taste receptor.
(d) a part of utriculus of ear.

51. Which of the following statements is **false** about the electrical synapse?

Tricky

(i) At electrical synapses, the membranes of pre and post synaptic neurons are in very close proximity.
(ii) Electrical current can flow directly from one neuron into the other across the synapses.
(iii) Transmission of an impulse across electrical synapses is very similar to impulse conduction along single axon.
(iv) Electrical synapses pass electrical signal between cells with the use of acetylcholine (Ach).
(v) Electrical synapses are fast rare in our system.
(a) (i) and (ii) (b) Only (ii)
(c) Only (iv) (d) Only (v)

52. Which of the following statement is **correct** regarding the organ of eye-sight?

(a) The space between cornea and lens is filled with transparent gel.
(b) When all cones are stimulated equally, a sensation of no light (dark) is produced.
(c) Rhodopsin is purplish red protein, hence called visual purple.
(d) The anterior transparent portion of choroid is called cornea.

53. Which of the following is the largest nerve?
(a) Vagus (b) Olfactory
(c) Trigeminal (d) Glossopharyngeal

54. Which one is the spinal nerve?
(a) Trigeminal (b) Hypoglossal
(c) Olfactory (d) None of these

55. The following statements about nerve metabolism, excitation and conduction are true except:
(a) Nerve fibre can be excited under anaerobic conditions.
(b) Nerve conduction proceeds in the absence of oxygen and development of action potential does not depend on metabolism immediately.
(c) In the absence of oxygen, action potential can be generated and nerve conduction proceeds, but later, it fails as the Na^+ pump cannot operate.
(d) At rest, the oxygen consumption in the brain is lesser than what is seen in a nerve.

56. Match the given terms in column-I with its definition given in column-II and choose the correct options.

	Column-I (Terms)		Column-II (Definition)
A.	Semi-circular canal	I.	Spiral organ of Corti
B.	Vestibule	II.	Fluid found in the scala vestibule and scala tympani
C.	Cochlea	III.	Evaluates rotational motion
D.	Perilymph	IV.	Fluid found within the Organ of Corti
E.	Endolymph	V.	Responds to gravity and movements of the head

(a) A – III; B – V; C – I; D – II; E – IV
(b) A – I; B – II; C – III; D – IV; E – V
(c) A – II; B – III; C – IV; D – V; E – I
(d) A – IV; B – I; C – V; D – II; E – III

57. A fall in ECF, K^+ concentration would cause:

(a) hyperpolarisation of cell membrane.
(b) decrease in membrane potential.
(c) increases Na^+ current into the cell.
(d) stimulation of electrogenic pump.

58. 'Ishihara charts' are used by ophthalmologists to detect: **Tricky**
(a) Visual acuity
(b) Night blindness
(c) Colourblindness
(d) Microbial infection

59. One group of neurotransmitter receptors is the metabotropic receptors. Which one of the following statements about metabotropic receptors is **false**? **Critical Thinking**
(a) Both muscanaric and β-adrenergic receptors are metabotropic receptors.
(b) Metabotropic receptors are coupled with A proteins.
(c) Metabotropic receptors can stimulate gene expression.
(d) Muscanaric receptors bind with acetylcholine.

60. Which of the following statements is **correct**?

(a) The internal ear receives sound waves and directs them to the ear drum.
(b) The membranous canals are suspended in the endolymph of the bony canals.
(c) The ear ossicles increase the efficiency of transmission of sound waves to the inner ear.
(d) The malleus is attached to the oval window and the stapes is attached to the temporal membrane of the cochlea.

61. Which of the following terms is not correctly matched with its feature? **Critical Thinking**
(a) Efferent neurons - Carries signals from CNS to the effector.
(b) Axon terminal - Possess neurotransmitter containing vesicle.
(c) Limbic system - Along with the hypothalamus, it is involved in the regulation of sexual behaviour, expression of emotional reactions and motivation.
(d) Association areas - Present in cerebellum and responsible for functions like intersensory associations, memory and communication.

62. Which of the following statement is **incorrect**?
(a) The ear ossicle attached to tympanic membrane is malleus.
(b) Opsin (of rhodopsin) develops from vitamin A.
(c) The pressure on ear drum is equalised by Eustachian tube.
(d) Otolith organ consists of saccule and utricle.

63. Identify the correct match of types of neurons with its location.
(i) Unipolar neuron - Embryonic stage
(ii) Bipolar neuron - Cornea of eyes
(iii) Multipolar neuron - Cerebral cortex
(a) (i) and (ii) only
(b) (ii) and (iii) only
(c) (i) and (iii) only
(d) All of the above

64. Refer the given figure which shows the axon terminal and synapse with their parts marked as 1 to 7. Identify the correct parts whose constitution forms the structure of synapse. **Critical Thinking**

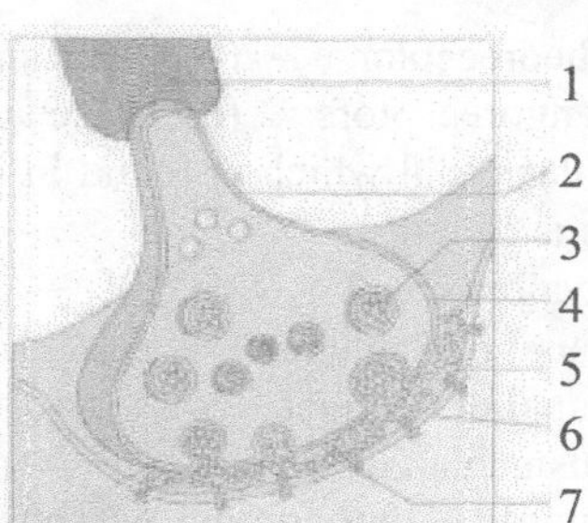

(a) 1, 2, 3 (b) 3, 4, 5
(c) 4, 5, 6 (d) 5, 6, 7

65. Trochlear nerve supplies:
(a) Nasal epithelium
(b) Superior oblique eye muscles
(c) Inferior oblique eye muscle
(d) Superior rectus eye muscle.

66. An axon has four terminal ends connected with dendrites of four different neurons. Its nerve impulse will:
(a) Become weak due to distribution into four.
(b) Travel in all the four neurons with equal strength.
(c) Pass on to one neuron only.
(d) Travel to none because the movement of impulse is from dendrite to axon.

67. The diagram shows the sequence of events occurring as an action potential arrives at a synapse.
The numbered arrows represent the movement of substances across the membranes.

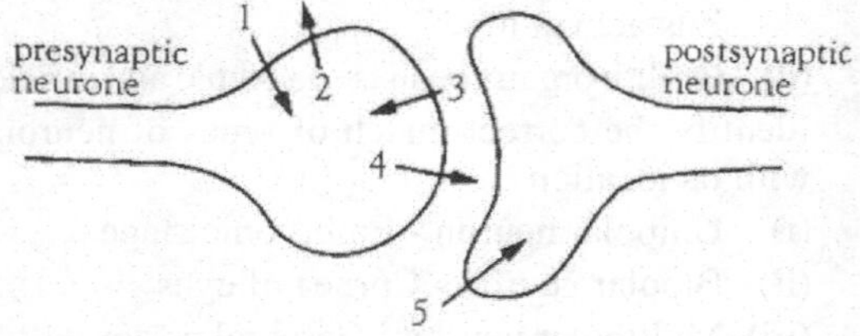

What are the substances moving across the membranes?

	1	2	3	4	5
(a)	K^+	Na^+	Ach.	Ca^{2+}	K^+
(b)	K^+	Na^+	K^+	Ca^{2+}	Ach.
(c)	Na^+	K^+	Ca^{2+}	Ach.	Na^+
(d)	Na^+	K^+	Na^+	Ach.	Ca^{2+}

68. Receptors which are sensitive to pain are known as:
(a) Rheoreceptor (b) Algesireceptors
(c) Tangoreceptors (d) Mechanoreceptors

69. A circular canal which is found in limbus part of eyes is called:
(a) Hyaloid canal
(b) Canal of Schlemm
(c) Canal of Cloquet
(d) Eustachian tube

70. An example of autonomous nervous system is:
(a) Swallowing food
(b) Pupillary reflex
(c) Peristalsis of intestine
(d) Knee-jerk response

71. The eyes of tigers and wolves shine (reflect back the light focused on them) because:
Critical Thinking
(a) cornea is opaque
(b) retina does not have enough rods
(c) retina has only rods and no cones
(d) None of the above

72. Our ear can hear the frequency of sound waves
Tricky
(a) 20 to 20,000 cycles/sec
(b) 1000 to 2000 cycles/sec
(c) 5000 to 7000 cycles/sec
(d) 5000 to 10,000 cycles/sec

73. In the blind spot, where the optic nerves leave the eye:
(a) Rods and cones are absent.
(b) Only cones are present.
(c) Only rods are present.
(d) Special neurons are present.

74. Part of brain responsible for hearing is:
(a) Cerebellum (b) Cerebrum
(c) Medulla (d) Hypothalamus

75. Fenestra ovalis is the opening of:
(a) Cranium (b) Tympanum
(c) Tympanic cavity (d) Brain

76. Which is not a reflex action?
(a) Swallowing of food
(b) Shivering in cold
(c) Salivation at choicest food
(d) Closure of eyelid by flashing light

77. No image formation occurs on blind-spot of retina because:
Critical Thinking
(a) It is not present on the optical axis of the eye.
(b) Here cones and rods are absent.
(c) On this part only cones are present.
(d) The nerve fibres of this region do not contribute in the formation of optic chiasma.

78. Center for sense of smell is:
(a) Cerebellum (b) Olfactory lobes
(c) Cerebrum (d) Midbrain

79. Route of reflex arc is:
Tricky
(a) Receptors, effectors, grey mater and motor fibres
(b) Receptors, sensory fibres, grey mater, motor fibres and effectors
(c) Sensory fibres, grey mater, motor-fibres, receptors and effectors
(d) Effectors, grey mater, motor, sensory fibres and receptors

80. The small canal leading from pinna into the head in mammals is:
(a) Haversian canal
(b) Ext. auditory meatus
(c) Inguinal canal
(d) Eustachian tube

81. Ear drum is known as:
(a) Tympanic membrane
(b) Tensor tympani
(c) Scala tympani
(d) Scala vestibuli

82. The nodes of Ranvier are:
(a) the points in which the axons is exposed.
(b) the contact points found over the non-myelinated nerve fibre.
(c) areas in which the axons swell up.
(d) areas where the neurilemma touches the axon.

83. Which of the following statement is **correct**?
(a) Electrical synapses are more common in our neural system than chemical synapses.
(b) The new potential in post-synaptic neuron may be either excitatory or inhibitory.
(c) Hypothalamus is the major coordination centre for sensory and motor signaling.
(d) The tracts of nerve fibres that connect two cerebral hemispheres are called corpora bigemina.

84. "Jumping of the action potential" at the nodes of Ranvier is known as:
(a) Saltatory conduction
(b) Neurotransmission
(c) Recovery phase
(d) Active phase

85. The rods and cones of the eye retinal layer are modified:
(a) Hair
(b) Unipolar neurons
(c) Bipolar neurons
(d) Multipolar neurons

86. Which of the following statement is **correct**?

(a) The space within cochlea called scala media is filled with endolymph.
(b) The vestibular apparatus is composed of two semi-circular canals and the otolith organ consisting of the saccule and utricle.
(c) The external auditory meatus helps in equalising the pressures on either sides of the ear drum.
(d) The membranes constituting cochlea, the Reissner's and basilar, divide the surrounding perilymph filled bony labyrinth into an upper scala tympani and a lower scala vestibuli.

87. Neurons receive signals through their ________ and send signals to other neurons through their ________.
(a) dendrites ... receptors
(b) end feet ... cell bodies and dendrites
(c) cell bodies and dendrites ... axons
(d) transmitter vesicles ... axons

88. Which of the following statements is **incorrect** about cones?
(i) The daylight vision and colour vision are function of cones.
(ii) In human eye, there are three types of cones having characteristic photopigments that respond to red, green and blue light.
(iii) The sensations of different colours are produced by various combination of these 3 types of cones.
(iv) When these 3 types of cones are stimulated equally, a sensation of white light is produced.
(a) Only (iv) (b) Only (i) and (iv)
(c) Only (iii) (d) None of these

89. When a physician elicits the knee-jerk reflex by tapping deep tendons in the knee, the normal response is for the lower leg to swing forward. When this happens:

Critical Thinking

(a) muscles in the back of the lower leg are contracting but that in the front are relaxing.
(b) muscles in the front of the lower leg are contracting but that in the back are relaxing.
(c) muscles in the back of the thigh are contracting but that in the front are relaxing.
(d) muscles in the front of the thigh are contracting but that in the back are relaxing.

90. Which of these is an example of conditioned reflex?
(a) Watering of mouth at the taste of food.
(b) Withdrawal of hand on touching a hot plate.
(c) Cycling.
(d) Flowing of tears while peeling and cutting onions.

91. In snakes, eye lids are:
(a) Immovable (b) Movable
(c) Absent (d) All of these

92. In mammalian eye, the power of accommodation is due to:
(a) Regulating light through iris
(b) Focussing by cornea
(c) Changing the focal length of eye lens by ciliary body
(d) Changing the diameter of pupil

93. Depolarisation of nerve impulse involves:

Tricky

(a) Entry of Na^+ only
(b) Entry of K^+ only
(c) Entry of Na^+ and exit of K^+
(d) Entry of K^+ and exit of Na^+

94. The autonomic nervous system has control over:
(a) Reflex action (b) Sense organs
(c) Internal organs (d) Skeletal muscles

95. Nerve impulses are initiated by nerve fibres only when the membrane shall become more permeable to:
(a) Adrenaline (b) Phosphorus
(c) Sodium ions (d) Potassium ions

96. The parts of the neurons that perform basic cellular functions such as protein synthesis are the:
(a) Axons (b) Dendrites
(c) Synaptic knobs (d) Soma

97. Which of the following correctly traces the energy of sound waves into the ear?

Critical Thinking

(a) Auditory canal-eardrum-ear bones-cochlea
(b) Eardrum-auditory canal-cochlea-ear bones
(c) Auditory canal-ear bones-eardrum-cochlea
(d) Eardrum-auditory canal-ear bones-cochlea

98. The sense of balance is maintained with the help of:
(a) Hammer
(b) Stirrup
(c) Semicircular canals
(d) Cochlea

99. The enzyme required for the conduction of nerve impulse across synapse is:
(a) Peroxidase
(b) Choline acetylase
(c) Ascorbic acid oxidase
(d) Succinic dehydrogenase

100. Which is activated in stress condition?
(a) Somatic (b) Sympathetic
(c) Parasympathetic (d) Whole ANS

101. Ear is most sensitive to sound waves of:

Tricky

(a) 1000 cycles/sec (b) 10000 cycles/sec
(c) 6000 cycles/sec (d) 20 cycles/sec.

102. Find out the right matching from the following pairs.
(a) Accommodation .... Pupil
(b) Colour perception .... Cones
(c) Night blindness .... Rods
(d) Binocular vision .... Iris

103. The pinna of the ear is composed of:
(a) Elastic cartilage
(b) Hyaline cartilage
(c) White fibrocartilage
(d) Bone

104. Reflex arcs, which involve no processing in the brain, are beneficial because they permit:

(a) instantaneous signaling to motor neurons.
(b) avoidance of dangerous stimuli, even without a previously "learned" response.
(c) immediate response to dangerous stimuli.
(d) All of the above

105. A bipolar neuron has:

Tricky

(a) 2 dendrites and 1 axon
(b) 2 axons and 1 dendrite
(c) 1 dendrite and 1 axon
(d) 2 axons and 2 dendrites

106. Vermis is a part of:
(a) Optic lobe (b) Midbrain
(c) Cerebellum (d) Medulla oblongata

107. An abnormal production of antidiuretic hormone (ADH) could result from a dysfunction of:

(a) Hypothalamus
(b) Choroid plexus
(c) Medulla oblongata
(d) Reticular activation system

108. In a nerve, if sodium pump is blocked, which of the following is most likely to happen?

(a) Na^+ and K^+ will increase outside the cell.
(b) Na^+ outside the nerve will increase.
(c) Na+ inside the nerve will increase.
(d) K^+ inside the nerve will increase.

109. Brain waves common to a healthy sleeping person and a brain-damaged awake person are called:

(a) Alpha waves (b) Beta waves
(c) Theta waves (d) Delta waves

110. Which of the following options correctly represents the name of 1, 2, 3 and 4 in the given diagram of neuron?

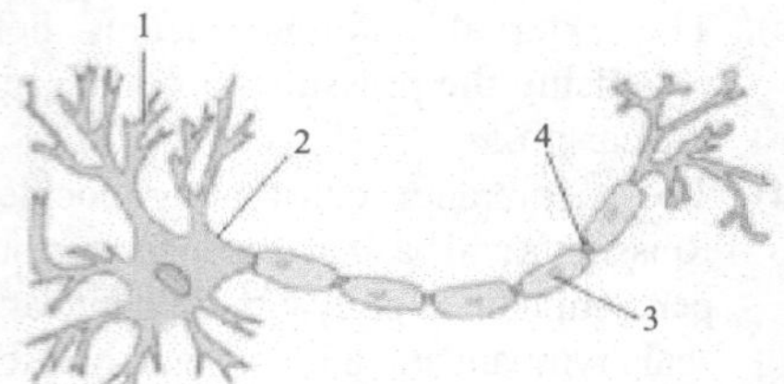

(a) 1- Axon, 2- Dendrites, 3- Node of Ranvier, 4- Myelin sheath
(b) 1- Dendrites, 2- Axon, 3- Node of Ranvier, 4- Myelin sheath
(c) 1- Dendrites, 2- Cell body, 3- Myelin sheath, 4- Node of Ranvier
(d) 1- Axon, 2- Cell body, 3- Dendrites, 4- Node of Ranvier

111. Match the column-I with column-II and select the correct option.

Column-I		Column-II
A. Unipolar neuron	I.	In cerebral cortex
B. Bipolar neuron	II.	In embryonic stage
C. Multipolar neuron	III.	In retina of eye

(a) A – I; B – II; C – III
(b) A – I; B – III; C – II
(c) A – II; B – I; C – III
(d) A – II; B – III; C – I

112. Which of the following statements is **correct** about rods compared to cones?
(i) Rods are most numerous in the fovea.
(ii) Rods contain rhodopsin.
(iii) Rods produce general outlines of objects rather than sharp images.
(iv) Rods produce black and white but not colour images.
(a) (i), and (ii)
(b) (ii), and (iii)
(c) (i), (ii) and (iii)
(d) (ii), (iii) and (iv)

113. Which of the following processes occurs during repolarisation of nerve fibre?
(i) Open Na^+ channel
(ii) Closed Na^+ channel
(iii) Closed K^+ channel
(iv) Open K^+ channel
(a) (ii) and (iv) (b) (i) and (iii)
(c) (ii) and (iii) (d) (i) and (ii)

114. The amount of neurotransmitter released from a presynaptic cell is a function of:
(a) The voltage of the individual action potential in the presynaptic neuron.
(b) The frequency of arriving action potentials is the presynaptic neuron.
(c) Whether the action potential is excitatory or inhibitory.
(d) The receptors in the postsynaptic cell.

115. Which is absent in aqueous humor?
(a) Carbon dioxide (b) Oxygen
(c) Hyaluronic acid (d) Glucose

116. Which of the following statement is **correct** regarding cerebellum of brain?
(a) It is concerned with the maintenance of posture/equilibrium.
(b) It is responsible for olfactory functions.
(c) It controls optic functions.
(d) Both (a) and (c)

117. Which is not connected with autonomic system?
(a) Memory & learning
(b) Peristalsis
(c) Digestion
(d) Excretion

118. In mammalian cochlea, the thin-walled sloping roof of the scala media is referred to as:
(a) Organ of Corti
(b) Scala tympani
(c) Basilar membrane
(d) Reissner's membrane

119. Which has H - shaped grey matter?
(a) Cerebrum
(b) Medulla oblongata
(c) Cerebellum
(d) Spinal cord

120. The function of cerebrospinal fluid does not include:
(a) protection of delicate brain and spinal cord from shock.
(b) protection of brain and spinal cord by containing antibody.
(c) buoyancy to brain.
(d) as a medium for excretion of waste product.

121. The one-way flow of information between presynaptic and postsynaptic cells is a characteristic of most kinds of neuron interactions. A notable exception to this rule is the: **Critical Thinking**
(a) Neuromuscular junction
(b) Nicotinic receptor system of vertebrate neurons
(c) Motor end plate
(d) Gap junction

122. Scala vestibuli is connected with:
(a) Fenestra rotundus (b) Fenestra ovalis
(c) Scala media (d) Eustachian tube

123. Which of the following functions as a signal filter and relays signals to the appropriate portion of the brain?
(a) Thalamus (b) Hypothalamus
(c) Cerebrum (d) Cerebellum

124. Once the threshold level is reached, _____. **Critical Thinking**
(a) the Na^+ gates open.
(b) an action potential will be generated.
(c) the membrane is depolarised.
(d) the refractory period is over.

125. Sequence of meninges from inner to outside is:
(a) Duramater – Arachnoid – Piamater
(b) Duramater – Piamater – Arachnoid
(c) Arachnoid – Duramater – Piamater
(d) Piamater – Arachnoid – Duramater

Answer KEYs

1	(a)	14	(c)	27	(b)	40	(c)	53	(c)	66	(b)	79	(b)	92	(c)	105	(c)	118	(d)
2	(d)	15	(b)	28	(d)	41	(b)	54	(d)	67	(c)	80	(b)	93	(c)	106	(c)	119	(d)
3	(a)	16	(c)	29	(a)	42	(b)	55	(d)	68	(b)	81	(a)	94	(c)	107	(a)	120	(b)
4	(c)	17	(c)	30	(d)	43	(d)	56	(a)	69	(b)	82	(a)	95	(c)	108	(c)	121	(d)
5	(b)	18	(b)	31	(b)	44	(d)	57	(a)	70	(c)	83	(b)	96	(d)	109	(d)	122	(b)
6	(a)	19	(b)	32	(c)	45	(b)	58	(c)	71	(d)	84	(a)	97	(a)	110	(c)	123	(a)
7	(b)	20	(a)	33	(c)	46	(b)	59	(b)	72	(a)	85	(c)	98	(c)	111	(d)	124	(b)
8	(a)	21	(a)	34	(d)	47	(a)	60	(c)	73	(a)	86	(a)	99	(b)	112	(d)	125	(d)
9	(a)	22	(b)	35	(b)	48	(a)	61	(d)	74	(b)	87	(c)	100	(b)	113	(a)		
10	(b)	23	(c)	36	(d)	49	(c)	62	(b)	75	(c)	88	(d)	101	(a)	114	(b)		
11	(a)	24	(b)	37	(b)	50	(b)	63	(c)	76	(b)	89	(d)	102	(b)	115	(c)		
12	(c)	25	(b)	38	(d)	51	(c)	64	(c)	77	(b)	90	(c)	103	(a)	116	(d)		
13	(a)	26	(a)	39	(a)	52	(c)	65	(b)	78	(c)	91	(c)	104	(d)	117	(a)		

Hints & Solutions

1. **(a)** **2.** **(d)** **3.** **(a)** **4.** **(c)**

5. **(b)** **6.** **(a)** **7.** **(b)**

8. **(a)** The marked label 1, 2, 3 and 4 are respectively Reissner's membrane, organ of Corti, basilar membrane and tectorial membrane.

Organ of Corti (X) is a sensitive element in the inner ear. It is a spiral shaped structure located on the basilar membrane (3) which contains hair cells that acts as auditory receptors. The hair cells are present in rows on the internal side of the organ of corti.

9. **(a)** **10.** **(b)**

11. **(a)** The diagram shows that three different stimuli arrive at the same time and evoke a large EPSP that leads to an action potential, while less than three do not. This is spatial summation. Temporal summation involves the same stimulus arriving in rapid succession.

12. **(c)**

13. **(a)** Nuclei are areas of grey matter within the white matter, where nerve impulses are processed.

14. **(c)** Conduction of impulses is very slow in unmyelinated fibres.

15. **(b)** Hyperpolarising neuron B by 15 mV would make its membrane potential –65mV, actually taking it further away from the threshold voltage needed to fire an action potential.

16. **(c)** **17.** **(c)** **18.** **(b)**

19. **(b)** When an impulse, or nerve signal reaches the end of axon, the vesicle release a neurotransmitter into the small space between the adjoining cells (synaptic gap). Neurotransmitters diffuse across the synapse and bend to receptors in the recieving cell that are specific for the neurotransmitter.

20. (a) The resting membrane potential is primarily due to diffusion potential of K^+. The E_k is closer to resting membrane potential, than K_{Na} or E_{Cl}. Presence of electrogenic pump will help to maintain the potential difference, but not a primary cause for resting membrane potential. The Na^+ pump contributes only 4 mv to the resting membrane potential.

21. (a) In the given figure of axon terminal and synapse, the correct labelling of 1, 2, 3 and 4 are respectively synaptic vesicles, axon terminal, synaptic cleft and postsynaptic membrane.

22. (b) 23. (c)

24. (b) Sclera is the external layer composed of dense connective tissue. It is the tough, opaque tissue that serves as protective outer coat for eye. Choroid contains many blood vessels and looks bluish in colour. Pupil is a hole located in the centre of the iris of the eye that allows light to strike the retina. It appears black because light rays entering the pupil are either absorbed by the tissues inside the eye directly, or absorbed after diffuse reflections within the eye that mostly miss exiting the narrow pupil. Fovea is a portion of retina where only the cones are densely packed. Iris is the visible coloured portion of the eye.

25. (b) In the given figure of human eye, the part marked as "X" is iris. The iris is a thin, circular structure in the eye which is responsible for controlling the diameter and size of the pupil and thus the amount of light reaching the retina. The colour of the iris gives the eye its colour.

26. (a)

27. (b) The label (2) represents tympanic membrane. Tympanic membrane converts sound waves into mechanical vibrations.
The structure (1), (2), (3) and (4) are external auditory canal, tympanic membrane, cochlea and eustachian tube respectively.

28. (d) 29. (a)

30. (d) The label (4) represents basilar membrane. The movement of the basilar membrane is most important to hearing. Once the hair cell microvilli bend, the hair cells depolarize, thus inducing action potentials in the cochlear nerve.

31. (b) The parts marked as 1, 2, 3 and 4 in the given figure of human eye are respectively cornea, iris, blind spot and choroid.
Cornea (1), the anterior portion of sclera, is the transparent, dome-shaped window covering the front of the eye. It is a powerful refracting surface. Blind spot (3) is the area on the retina without receptors that respond to light. Therefore, an image that falls on this region will not be seen. It is in this region that the optic nerve exits the eye on its way to the brain. Choroid (4) is the layer of blood vessels and connective tissue between the sclera (white of the eye) and retina.

32. (c) In frog, white soft chalky masses called calcareous bodies or glands of Swammerdam cover the dorsal root ganglia of spinal nerves.

33. (c)

34. (d) The resting potential of a neuron membrane occurs when the sodium ion concentration is higher on the outside and the potassium ion concentration is higher on the inside.

35. (b) Numerous sacs called synaptic vesicles located in the cytoplasm at the tip of the presynaptic axon release molecules of a neurotransmitter across the synaptic cleft, which bind to receptors imbedded in the membrane of the postsynaptic cell.

36. (d)

37. (b) The presynaptic membrane is first depolarised, stimulating the release of acetylcholine, the transmitter substance. This diffuses to the postsynaptic membrane, propagating the nervous stimulus. Acetylcholine is then hydrolysed.

38. (d) 39. (a) 40. (c) 41. (b)

42. (b)

43. (d) Support cells have a variety of roles in the nervous system, but they all tend to involve maintaining the electrical impulse.

44. (d)

45. (b) In static condition, the body balance is sensed by macula. Macula provide information on head position (static equilibrium), as well as linear acceleration and deceleration, a type of dynamic equilibrium. The macula consist of hair cells with hair bundles and supporting cells.

46. (b)

	Features	Sympathetic nervous system	Parasympathetic nervous system
(a)	Salivary glands	Inhibits secretions	Stimulates secretions
(b)	Pupil of eye	Dilates	Constricts
(c)	Heart rate	Increases strength and rate of heart beat	Decreases strength and rate of heart beat
(d)	Intestinal peristalsis	Inhibits secretions	Stimulates secretions

47. (a)

48. (a) If the disease acts on a chemical synapse where the neurotransmitter is stored and released, it is affecting the axon terminals of the presynaptic cell. Ca^{2+} channels are involved in regulating the release of acetylcholine by allowing Ca^{2+} to enter the presynaptic cell and promoting the fusion of acetylcholine containing vesicles to the membrane.

49. (c) 50. (b) 51. (c)

52. (c) The space between the cornea and the lens is called the aqueous chamber and contains a thin watery fluid called the aqueous humor. When all the cones are stimulated equally, a sensation of white light is produced. The anterior transparent portion of sclera is called cornea.

53. (c)

54. (d) Hypoglossal is spinal nerve in frog.

55. (d) Oxygen consumption is greater in brain than in a nerve fibre.

56. (a) Semi-circular canal is interconnected tubes located inside each ear. They are lined with cilia and filled with a fluid known as endolymph. With every movement of head, the endolymph moves the cilia. This works as a type of motion sensor, as the movements of the cilia are communicated to the brain. The vestibule is the central part of the osseous labyrinth and responds to gravity and movements of head. Cochlea is the auditory portion of the inner ear. It is a spiral-shaped cavity in the bony labyrinth. Perilymph is an extracellular fluid located within the cochlea (part of the inner ear) in two of its three compartments: the scala tympani and scala vestibuli. The major cation (positively charged ion) in perilymph is sodium. Endolymph is the fluid contained in the membranous labyrinth of the inner ear.

57. (a) There is more K^+ exit taking place and the cell interior becomes more negative. That is, the membrane is hyperpolarised.

58. (c)

59. (b) Metabotropic receptors are coupled with G proteins that bind GDP. When the receptor binds a neurotransmitter, the GDP is turned into GTP and a subunit of the G protein separates and activates effector proteins.

60. (c) The external ear (pinna) collects sound waves and directs them to the ear drum. The membranous canals are suspended in the perilymph of the bony canals. The malleus is attached to the tympanic membrane and the stapes is attached to the oval window of the cochlea.

61. (d) Association areas are the regions of the cerebral cortex of the brain connected by numerous nerve fibres to all parts of both cerebral hemispheres and coordinating such higher activities as learning and reasoning. Association areas are responsible for functions like intersensory associations, memory and communication.

62. (b) The photopigments of the human eye are composed of opsin (protein) and retinal. Upon light absorption, opsins can change their conformation from a resting state to a signalling state. Then it activates the G protein, thereby resulting in a signalling cascade that produces physiological responses. Retinal, also called retinaldehyde or vitamin A aldehyde, is a polyene chromophore which is bound to proteins called scotopsins and photopsins, and is the chemical basis of animal vision.

63. (c) Bipolar neurons is a type of neuron which has two extensions - one axon and one dendrite. Bipolar cells are specialised sensory neurons for the transmission of special senses. As such, they are part of the sensory pathways for smell, sight, taste, hearing and vestibular functions. They are found in the retina of eye.

64. (c) The parts labelled as 4, 5 and 6 are respectively pre-synaptic membrane, synaptic cleft and post-synaptic membrane. These parts constitute the structure of synapse. A synapse is a structure that permits a neuron (or nerve cell) to pass an electrical or chemical signal to another cell (neural or otherwise). Synapses are essential to neuronal function. At a synapse, the plasma membrane of the signal-passing neuron (the presynaptic neuron) comes into close apposition with the membrane of the target (postsynaptic) cell. Both the presynaptic and postsynaptic sites contain extensive arrays of molecular machinery that link the two membranes together and carry out the signaling process. In many synapses, the presynaptic part is located on an axon, but some postsynaptic sites are located on a dendrite or soma.

65. (b) 66. (b)

67. (c) As the action potential reaches the synapse, the pre-synaptic membrane is depolarised due to the entry of sodium ions (1). The membrane is subsequently repolarised by the exit of potassium ions (2). The arrival of the action potential causes calcium channels in the membrane to open, thus calcium ions rush into the synaptic knob (3). This causes the synaptic vesicles to fuse with the pre-synaptic membrane, releasing acetylcholine into the synaptic cleft (4). Acetylcholine binds to specific receptor sites on the post-synaptic membrane, opening ion channels and causing sodium ions to enter (5), depolarising the membrane.

68. (b) 69. (b) 70. (c) 71. (d)

72. (a) These are called the limits of audibility. Actual value is 20 to 20000 cycles/sec.

73. (a) Blind spot in the eye is the place where neither rods nor cones present.

74. (b) 75. (c) 76. (b) 77. (b)

78. (c) Olfactory lobes receive the sensation of smell but relay the same to temporal lobe of cerebrum.

79. (b)

80. (b) The small canal leading from pinna into the head in mammals is ext. auditory maetus

81. (a) Ear drum is also known as tympanic membrane which separates middle ear from external ear.

82. (a) The nodes of ranvier are the points in which the axons is exposed.

83. (b) An electrical synapse is a mechanical and electrically conductive link between two adjoining neurons that is formed at a narrow gap between the pre- and postsynaptic neurons known as a gap junction. Electrical synapses are rare in our neural system. Hypothalamus contains a number of centres which control body temperature, urge for eating and drinking. It also contains several groups of neurosecretory cells which secretes hormone called hypothalamic hormones. Thalamus is a major coordinating centre for sensory and motor signalling. The tracts of nerve fibres that connect two cerebral hemispheres are called corpus callosum.

84. (a) 85. (c)

86. (a) The vestibular apparatus is composed of three semi-circular canals and the otolith organ consisting of the saccule and utricle. The Eustachian tube helps in equalising the pressures on either sides of the ear drum. The membranes constituting cochlea, the Reissner's and basilar, divide the surrounding perilymph filled bony labyrinth into an upper scala vestibuli and a lower scala tympani.

87. (c) Dendrites generally receive inputs and conduct signals toward the cell body, whereas axons conduct signals away from the cell body.

88. (d)

89. (d) When any muscle contracts, its antagonist relaxes by rule. The leg swings forward because the muscles in the front of the thigh (quadriceps femoris) contract. At the same time the hamstrings in the back of the thigh, relax and are lengthened.

90. (c) 91. (c) 92. (c) 93. (c)

94. (c) 95. (c) 96. (d) 97. (a)

98. (c) 99. (b) 100. (b) 101. (a)

102. (b) Cone cells are responsible for colour vision. They function best in relatively bright light, as compared to rod cells that work better in dim light. Cone cells are densely packed in the fovea centralis, a rod-free area with very thin, densely packed cones which quickly reduce in number towards the periphery of the retina.

Accommodation of vision occurs by contraction and relaxation of ciliary muscles. Pupil determines the amount of light that enters the eye. Night blindness occurs due to vitamin A deficiency. It is due to a disorder of the cells in the retina that are responsible for vision in dim light. Binocular vision is a vision using two eyes with overlapping fields of view, allowing good perception of depth. Iris is responsible for controlling the diameter and size of the pupil and thus the amount of light reaching the retina.

103. (a)

104. (d) Reflex arcs allow for rapid processing and response.

105. (c) **106. (c)**

107. (a) The hypothalamus influences the production of ADH by the posterior pituitary.

108. (c)

109. (d) Delta waves have a low frequency of 1 to 5 Hz and are normal during sleep.

110. (c) In the given diagram of neuron, the part marked as 1, 2, 3 and 4 are respectively dendrites, cell body, myelin sheath and nodes of Ranvier. Neuron is a specialised, impulse-conducting cell that is the functional unit of the nervous system, consisting of the cell body and its processes, the axon and dendrites.

111. (d)

112. (d) Rods and cones are light receptors present in retina. Rods are absent in the fovea while cones are numerous in the fovea.

113. (a)

114. (b) Neurotransmitter is released from the presynaptic cell when an action potential reaches the synapse. Increase in the frequency of arriving action potentials in the presynaptic neuron will increase the release of neurotransmitter into the synapse.

115. (c) **116. (d)** **117. (a)** **118. (d)**

119. (d) **120. (b)**

121. (d) Gap junctions are electrical synapses that physically connect neurons via membrane proteins. Electrical signals can thus travel in either direction.

122. (b)

123. (a) The thalamus may be thought as the "junction box" of the brain.

124. (b) If a depolarisation reaches a threshold, a different type of response, an action potential, will be triggered.

125. (d)

22 Chemical Co-ordination & Integration

1. According to the accepted concept of hormone action, if receptor molecules are removed from target organs, then the target organ will:

(a) not respond to the hormone.
(b) continue to respond to hormone without any difference.
(c) continue to respond to the hormone but in the opposite way.
(d) continue to respond to the hormone but will require higher concentration.

2. Endocrine glands produce or action of endocrine glands is mediated through:
(a) Hormones (b) Enzymes
(c) Minerals (d) Vitamins

3. Ovulation in humans is controlled by:
(a) FSH and LTH
(b) FSH and GH
(c) LTH and LH
(d) FSH and LH

4. Deficiency of vasopressin primarily results in: Tricky
(a) Increased volume of urine
(b) Decreased volume of urine
(c) Excessive secretion of urochrome
(d) Change in pH from acidic to alkaline range.

5. The first hormone artificially produced by genetically engineered bacteria is:
(a) Insulin (b) Somatotropin
(c) Testosterone (d) Adrenaline

6. Which endocrine gland stores its secretion in extracellular spaces before discharging it into blood? Tricky
(a) Testis (b) Pancreas
(c) Thyroid (d) Adrenal

7. Pineal body develops from:
(a) Dorsal part of diencephalon
(b) Ventral part of diencephalon
(c) Ventral side of cerebellum
(d) Lateral side of cerebrum

8. If thyroxine is added in a beaker which has some small tadpoles then: Tricky
(a) all tadpoles die.
(b) they metamorphose very fast.
(c) they develop small body.
(d) they develop a giant body.

9. Which of the following hormones is a cholesterol-derivative?
(a) Estrogen (b) Prostaglandin
(c) Epinephrine (d) All of these

10. "Upregulation" of hormone receptors refers to:
(a) increase in hormone receptor numbers with low hormone levels.
(b) increase in hormone receptor numbers with high neurotransmitter levels.
(c) increase in hormone levels produced by increase in hormone receptor numbers.
(d) decrease in hormone levels produced by decrease in hormone receptor numbers.

11. The hormone released by "X" marked in the given figure helps to restore Y. Identify X and Y. Critical Thinking

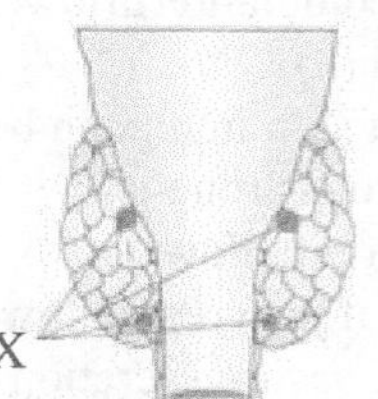

	X	Y
(a)	Thyroxine	Too much calcium in the blood.
(b)	PTH	Lowered levels of calcium in blood.
(c)	Thymosin	Decreased level of blood sugar.
(d)	Adrenaline	Excessive loss of sodium in extracellular fluid.

12. The effect of prolactin hormone is on:
(a) Liver (b) Pancreas
(c) Mammary glands (d) Bones

13. Which of the following is an example of negative feedback?
(a) Sucking reflex
(b) Uterine contractions during labour.
(c) Secretion of CRH in response to ACTH.
(d) Hair erection due to low body temperature.

14. What is the effect on metamorphosis of tadpole of frog when there is less amount of iodine in water ?
(a) Accelerated (b) Delayed
(c) Stopped (d) Not affected

15. Which one of the following four glands is correctly matched with the accompanying description?
(a) Thyroid – Hyperactivity in young children causes cretinism.
(b) Thymus – Starts undergoing atrophy after puberty.
(c) Parathyroid – Secretes parathormone which promotes movement of sodium ions from blood into bones during calcification.
(d) Pancreas – Delta cells of Islets of Langerhans secrete a hormone which stimulates glycolysis in liver.

16. MSH of pars intermedia of pituitary is responsible for: **Critical Thinking**
(a) darkening of skin in lower verterbrates.
(b) lighten skin colouration in lower vertebrates.
(c) Both (a) and (b)
(d) darkening of skin in human beings.

17. Function of enterogastrone is:
(a) regulation of bile flow.
(b) stimulation of pancreatic flow.
(c) inhibition of gastric secretion.
(d) stimulation of gastric secretion.

18. A hormone circulating in the blood with target tissue 1, 2, and 3 could act selectively on only target tissue 1 if: **Critical Thinking**
(a). target tissue 1 had fewer hormone receptors than 2 and 3.
(b) target tissues 2 and 3 had hormone receptors with a higher hormone affinity than 1.
(c) target tissue 1 had hormone receptors with a higher hormone-affinity than 2 and 3.
(d) the hormone was released only in the vicinity of target tissue 1.

19. Anterior lobe of pituitary secretes:
(a) ACTH, TSH and oxytocin
(b) STH, GH and ADH
(c) TSH, ADH and prolactin
(d) FSH,GH and LH

20. Thymosin stimulate secretion of:
(a) Milk
(b) Erythrocytes
(c) T-lymphocytes
(d) Melanocytes

21. Secretin and cholecystokinin are digestive hormones. They are secreted in:
(a) Pyloric stomach
(b) Duodenum
(c) Ileum
(d) Oesophagus

22. Hormone connected with increased rate of glycogenolysis, blood pressure and heart beat is:
(a) Insulin (b) Glucagon
(c) Adrenaline (d) FSH

23. Which of the following statements is **not correct** for the hormones?
(a) They act on target organs away from the source glands.
(b) They are secreted directly into the blood.
(c) They are used again and again like catalysts.
(d) They are produced in very minute quantities and are biologically very active.

24. Which of the following statements is **correct** in relation to the endocrine system?
(a) Organs in the body like gastrointestinal tract, heart, kidney and liver do not produce any hormones.
(b) Non-nutrient chemicals produced by the body in trace amounts that act as intercellular messenger are known as hormones.
(c) Releasing and inhibitory hormones are produced by the pituitary gland .
(d) Adenohypophysis is under direct neural regulation of the hypothalamus.

25. Which one of the following flows directly into the blood from of its production to act on an organ away from it? **Critical Thinking**
(a) Glucagon
(b) Renin
(c) Cholesterol
(d) Cholesterase

26. Which of the following is related to obesity, low plasma Na^+, high K^+ and increased blood pressure?

Critical Thinking

(a) Growth hormone
(b) Cortisol
(c) Thyroxine
(d) Adrenaline

27. Uterine contraction at the child birth is stimulated by:
(a) Prolactin
(b) Progesterone
(c) Adrenaline
(d) Oxytocin

28. Match column-I with column-II and select the correct answer using the codes given below.

Column-I	Column-II
A. Hypothalamus	I. Relaxin
B. Anterior pituitary	II. Estrogen
C. Testis	III. FSH and LH
D. Ovary	IV. Androgens
	V. Gonadotropin releasing hormone

(a) A – V; B – III; C – IV; D – II
(b) A – V; B – III; C – II; D – IV
(c) A – I; B – II; C – IV; D – III
(d) A – III; B – V; C – IV; D – II

29. The urine of a man is very dilute and the quantity of urine is too much and dehydration has started in his body and he is very thirsty by the cause of:

Tricky

(a) Hypersecretion of ADH
(b) Hyposecretion of ADH
(c) Both (a) and (b)
(d) None of the above

30. Addison's disease results from:
(a) Hyposecretion of adrenal cortex
(b) Hypersecretion of adrenal cortex
(c) Hypertrophy of gonads
(d) Hyperactivity of cells of Leydig

31. Injections of a hormone are sometimes given to strengthen contractions of the uterus during childbirth. What hormone might this be?

(a) Adrenocorticotropic hormone (ACTH)
(b) Thyroxine
(c) Oxytocin
(d) Insulin

32. Which of the following statement is **correct** for thyrocalcitonin?

Critical Thinking

(a) Produced by parathyroid and decreases Ca^{++} in ECF.
(b) Produced by thyroid and decreases Ca^{++} in ECF.
(c) Produced by parathyroid and increases Ca^{++} in ECF.
(d) Produced by thyroid and increases Ca^{++} in ECF.

33. A man suffering from diabetes mellitus drinks water more frequently as he has to eliminate from blood, the extra:
(a) Salts (b) Sugar
(c) Insulin (d) Protein

34. Which of the following statements about glucocorticoids is **correct**?
(a) The major glucocorticoid in humans is cortisol.
(b) They are secreted by the zona fasciculata of the adrenal cortex.
(c) Secretion of these hormones is decreased in Addison's disease.
(d) All of the above

35. Match column-I with column-II and select the correct answer using the codes given below.

Column-I	Column-II
A. Hypothalamus	I. Sperm lysins
B. Acrosome	II. Estrogen
C. Graafian follicle	III. Relaxin
D. Leydig cells	IV. GnRH
E. Parturition	V. Testosterone

(a) A – IV; B – I; C – II; D – III; E – V
(b) A – II; B – I; C – IV; D – III; E – V
(c) A – II; B – I; C – V; D – IV; E – III
(d) A – IV; B – I; C – II; D – V; E – III

36. Which of the following hormones have antagonistic (opposing) effects?
(a) Thyroxine and calcitonin
(b) Insulin and glucagon
(c) Growth hormone and epinephrine
(d) ACTH and glucocorticoids

37. Which of the following radioactive isotope is used in the detection of thyroid cancer?

(a) Iodine-131 (b) Carbon-14
(c) Uranium-238 (d) Phosphorus-32

Direction (For Qs. 38 and 39): The given figure shows two endocrine glands (marked as A and B) and the hormones (marked as 1 and 2) secreted from them.

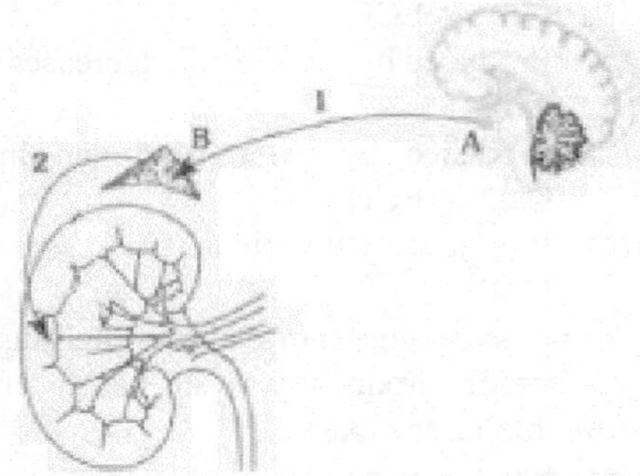

38. Identify (A) and (B).
(a) A- Pituitary gland , B- Adrenal gland
(b) A- Hypothalamus, B- Pituitary gland
(c) A- Hypothalamus, B- Adrenal gland
(d) A- Pituitary gland, B- Hypothalamus

39. Identify the hormone represented by lines 1 and 2. **Tricky**
(a) 1- ACTH, 2- Aldosterone
(b) 1- ACTH, 2- Adrenaline
(c) 1- TSH, 2- Thyroxine
(d) 1- TSH, 2- Aldosterone

40. Jet lag occurs when a person moves rapidly from one time zone to another, causing conflict between the body's biological rhythm and the new cycle of light and dark. Some scientists suspect that jet lag may result from disruption of a daily hormone cycle. Which of the following hormones do you think is the most likely suspect? **Toughnut**

(a) Epinephrine (b) Insulin
(c) Melatonin (d) Estrogen

41. During pregnancy corpus luteum:
(a) Degenerates
(b) Changes to corpus albicans
(c) Persists until parturition
(d) Persists upto the middle of pregnancy

42. Blood calcium levels are maintained because the bones act as a calcium sink in which calcium can either be stored or procured based on the body's needs. If a person secretes too much parathyroid hormone (one possible factor in the bone disease osteoporosis), What happens to calcium levels in the body? **Critical Thinking**
(a) Bone stores are depleted and the kidneys move calcium.
(b) Bone stores are depleted and the kidneys retain calcium.
(c) Blood levels of calcium may read high.
(d) Both (b) and (c)

43. The graph below illustrates the changes in blood sugar concentration after a healthy man has drunk a glucose solution.

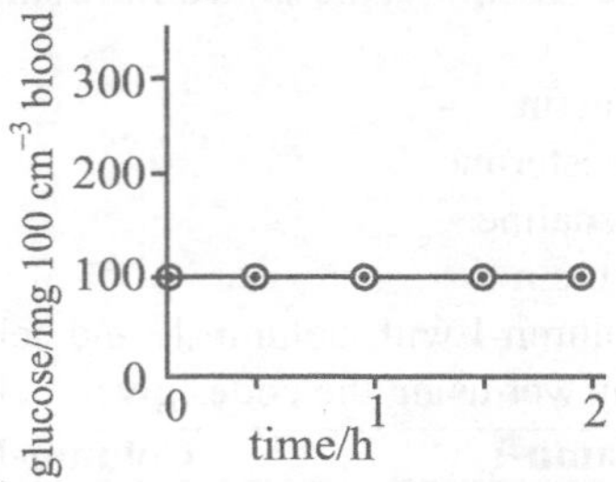

Which one of the following graphs would apply to a diabetic man in similar circumstances?

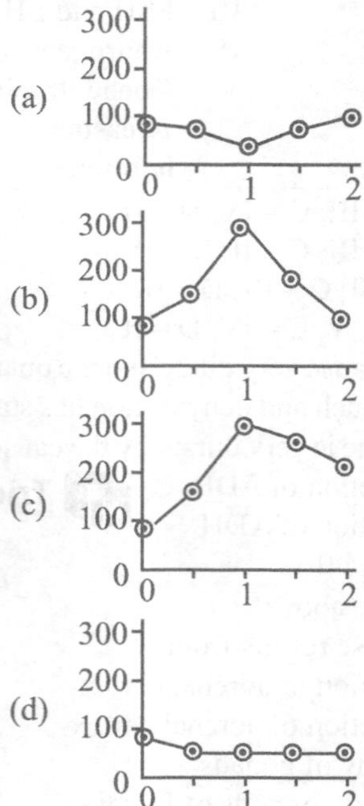

44. Which of the following does not secrete any hormone?
(a) Spleen (b) Ovary
(c) Testes (d) Pancreas

45. 'Brain sand' is associated to:
(a) Pineal body (b) Thymus gland
(c) Thyroid gland (d) Pituitary gland

46. The given figure is related with principal endocrine glands in human. Identify 1 to 7.

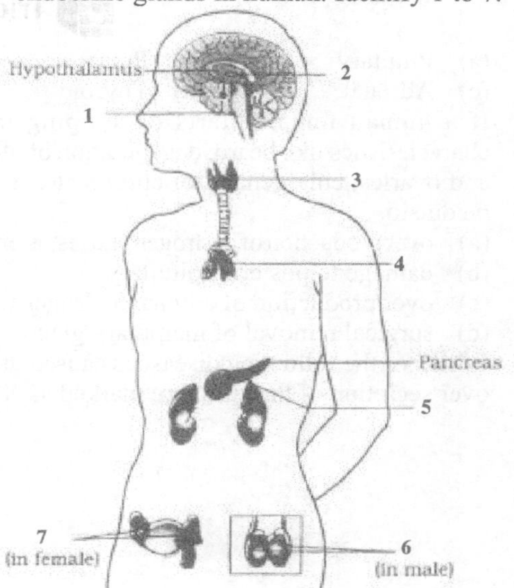

(a) 1- Pineal, 2- Pituitary, 3- Thyroid and parathyroid, 4- Thymus, 5- Adrenal, 6- Testis, 7- Ovary
(b) 1- Pituitary, 2- Pineal, 3- Thyroid and parathyroid, 4- Thymus, 5- Adrenal, 6- Ovary, 7- Testis
(c) 1- Pituitary, 2- Pineal, 3- Thyroid and parathyroid, 4- Thymus, 5- Kidney, 6- Testis, 7- Ovary
(d) 1- Pituitary, 2- Pineal, 3- Thyroid and parathyroid, 4- Thymus, 5- Adrenal, 6- Testis, 7- Ovary

47. Which of the following statement is **correct** regarding glucagon hormone?
(a) It has opposite effect to that of insulin.
(b) It converts glucose to glycogen.
(c) It is given to diabetic patients.
(d) It is formed by β-cells of pancreas.

48. If both the ovaries of rat are removed then which of the following hormones will be deficient?
(a) Prolactin
(b) Oestrogen
(c) Oxytocin
(d) Gonadotrophic hormone

49. Which one is different from the category of other three?
(a) Gastrin (b) Ptyalin
(c) Secretin (d) Glucagon

50. Match column-I with column-II and select the correct answer using the codes given below.

Column-I	Column-II
A. Adenohypophysis	I. Epinephrine
B. Adrenal medulla	II. Somatotropin
C. Parathyroid gland	III. Thymosin
D. Thymus gland	IV. Calcitonin

(a) A – III; B – I; C – IV; D – II
(b) A – I; B – II; C – III; D – IV
(c) A – II; B – I; C – IV; D – III
(d) A – IV; B – III; C – II; D – I

51. Antiageing hormone is:
(a) Thyroxine (b) Melatonin
(c) Estrogen (d) Testosterone

52. Which of the following statement(s) is/are **correct**? **Critical Thinking**
(i) Maximum iodine is stored in thyroid gland.
(ii) Calcitonin is non-iodinised hormone secreted by parafollicular cells of thyroid gland.
(iii) Calcitonin (TCT) regulates the blood Ca^{+2} level.
(iv) TCT is hypocalcemic factor.
(a) All of these
(b) None of these
(c) (i), (ii) and (iii)
(d) Only (iv)

53. A decrease in the level of oestrogen and progesterone causes:
(a) Growth and dilation of myometrium.
(b) Growth of endometrium.
(c) Constriction of uterine blood vessels leading to sloughing of endometrium or uterine epithelium.
(d) Release of ovum from the ovary.

54. Complete failure of the anterior lobe of pituitary causes:
(a) Conn's disease
(b) Acromegaly
(c) Cushing's disease
(d) Simmond's disease

55. If a person has protruding eyes, tachycardia and higher body temperature, he is suffering from:
(a) Cretinism
(b) Hyperthyroidism
(c) Myxoedema
(d) Acromegaly

56. When an egg is fertilised and pregnancy occurs, which of the following is **correct** regarding hormone concentrations?
(a) Progesterone levels drop, where as oestrogen levels remain high.
(b) Both progesterone and oestrogen levels drop.
(c) Both progesterone and oestrogen levels rise.
(d) Progesterone levels remain high, whereas oestrogen levels drop.

57. What is the significance of the discovery of different hormone receptor types for the hormone epinephrine?
(a) Two types of epinephrine hormone response elements were identified.
(b) Different tissues responded differently to the presence of epinephrine.
(c) It was discovered that slightly different forms of epinephrine (agonists) were secreted by the adrenal glands.
(d) All target tissues for epinephrine possess all four types of epinephrine receptors.

58. Match column-I with column-II and select the correct answer using the codes given below.

	Column-I		Column-II
A.	Adrenaline	I.	Myxoedema
B.	Hyperparathyroidism	II.	Accelerates heart beat
C.	Oxytocin	III.	Salt-water balance
D.	Hypothyroidism	IV.	Childbirth
E.	Aldosterone	V.	Demineralisation

(a) A – II; B – V; C – IV; D – I; E – III
(b) A – III; B – IV; C – V; D – I; E – II
(c) A – V; B – III; C – II; D – IV; E – I
(d) A – II; B – III; C – IV; D – V; E – I

59. Hormone relaxin is meant for: Tricky
(a) release of ova from ovary.
(b) relaxing uterus to prevent abortion.
(c) relaxing pubic symphysis.
(d) the contraction of uterus for child birth.

60. The primary difference between sex hormones in general and gonadotropins specifically is that gonadotropins: Critical Thinking
(a) are produced before birth only.
(b) are present only in one sex.
(c) cease functioning at puberty
(d) influence the production of gametes specifically, while sex hormones have a wide variety of primary and secondary roles in sex development and determination.

61. Sella tursica is a depression lodging:

(a) Pituitary (b) Pineal
(c) Adrenal (d) Thyroid

62. If a human female starts developing male characteristics like beard, degeneration of uterus and ovaries, enlargement of clitoris etc. It may be due to:
(a) over production of oestrogen and testosterone.
(b) damage to posterior pituitary.
(c) over production of adrenal androgens.
(d) surgical removal of mammary gland.

63. Which of the following disease is caused due to over secretion of the structure marked as X?

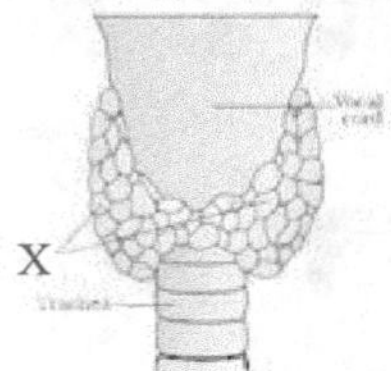

(a) Gigantism (b) Diabetes mellitus
(c) Diabetes insipidus (d) Grave's disease

64. Match column-I with column-II and select the correct answer using the codes given below.

	Column-I		Column-II
A.	Gastrin	I.	Acts on gastric gland and stimulates secretion of HCl and pepsinogen.
B.	Secretin	II.	Acts on the exocrine part of pancreas and stimulates secretion of water and HCO_3^- ions.
C.	CCK	III.	Acts on both pancreas and gall bladder and stimulates the secretion of both pancreatic enzymes and bile juice respectively.
D.	GIP	IV.	Inhibits gastric secretion and motility.

(a) A – IV; B – III; C – II; D – I
(b) A – II; B – IV; C – III; D – I
(c) A – I; B – II; C – III; D – IV
(d) A – I; B – III; C – II; D – IV

65. Which of the following hormone stimulates pancreas to secrete enzymes?
(a) Enterocrinin (b) CCK-PZ
(c) Gastrin (d) Secretin

66. Most of the contraceptive pills contain:
(a) Oestrogen + FSH
(b) Progesterone + LH
(c) FSH + LH
(d) Oestrogen + Progesterone

67. Heterocrine glands are the glands, which:
(a) work as exocrine glands.
(b) work as endocrine glands.
(c) have dual (exo- and endocrine) mode of function.
(d) None of the above

68. Mainly which type of hormones control the menstrual cycle in woman?
(a) FSH
(b) LH
(c) FSH, LH, oestrogen
(d) Progesterone

69. Which of the followings is the more scientific definition of hormone?
(a) They are extracellular messengers.
(b) They always act at distantly located target organs.
(c) They are the products of well organised endocrine glands.
(d) They are non-nutrient chemicals that act as intercellular messengers.

70. Growth hormone of pituitary is more effective in:
(a) Presence of thyroxine
(b) Absence of thyroxine
(c) Absence of insulin
(d) Presence of adrenaline

71. Restlessness, intolerance to temperature and increased metabolism is due to:
(a) Thyrotoxicosis
(b) ACTH deficiency
(c) Pituitary deficiency
(d) Calcitonin deficiency

72. Match the following columns. 

	Column-I		Column-II
A.	Protein hormones	I.	Epinephrine
B.	Steroid hormones	II.	Testosterone, progesterone
C.	Iodothyronines	III.	Thyroid hormones
D.	Amino acid	IV.	Insulin and derivatives glucagon hormones

(a) A – I; B – II; C – III; D – IV
(b) A – IV; B – III; C – II; D – I
(c) A – IV; B – II; C – III; D – I
(d) A – IV; B – II; C – I; D – III

73. The given figure shows feedback relationship between three glands and their hormones. Toughnut

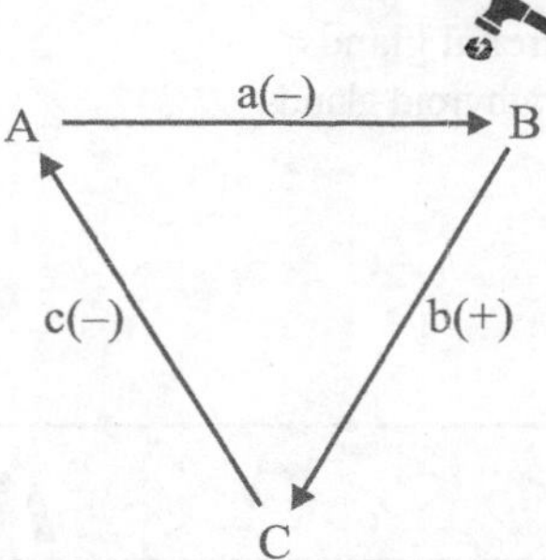

The secretion of '*b*' is decreased by '*a*', secretion of '*c*' is increased by '*b*' while the secretion of '*a*' is decreased by '*c*'. Increased or decreased amounts of a hormone cause an increase or decrease in the amount of the following hormone in the system. Now if '*a*' is added to the system, the secretion of '*c*' is
(a) decreased
(b) increased
(c) first decreased and then increased
(d) first increased and then decreased

74. A hormone that acts upon the renal tubule and blood capillaries is:
(a) Glucagon (b) Aldosterone
(c) Vasopressin (d) Glucocorticoids

75. Which one of the following hormone stimulates the "let-down" (release) of milk from the mother's breasts when the baby is sucking? Tricky
(a) Progesterone (b) Oxytocin
(c) Prolactin (d) Relaxin

76. Three hormones control the metamorphosis of a caterpillar into a butterfly. Which statement about these hormones is false?
(a) Molting hormone promotes the changes in the epidermis (external skeleton).
(b) Prothoracicotropic hormone stimulates the prothoracic gland to release juvenile hormone.
(c) Levels of juvenile hormone control the time at which an insect turns into an adult.
(d) Molting hormone promotes the development of imaginal disks.

77. A tumor in an endocrine gland caused Jennifer to have weakened bones and unusually high levels of blood calcium. Which of the following was affected? **Tricky**

(a) Anterior pituitary
(b) Pancreas
(c) Adrenal glands
(d) Parathyroid glands

78. In human, adult females oxytocin:
(a) is secreted by anterior pituitary.
(b) stimulates growth of mammary glands.
(c) stimulates pituitary to secrete vasopressin.
(d) causes strong uterine contractions during parturition.

79. Which of the following exerts control over all the others?
(a) Adrenal cortex (b) Hypothalamus
(c) Thyroid gland (d) Anterior pituitary

Answer KEYs

1	(a)	9	(a)	17	(c)	25	(b)	33	(b)	41	(d)	49	(b)	57	(b)	65	(b)	73	(a)
2	(b)	10	(a)	18	(c)	26	(b)	34	(d)	42	(d)	50	(c)	58	(a)	66	(d)	74	(c)
3	(d)	11	(b)	19	(d)	27	(d)	35	(d)	43	(c)	51	(b)	59	(c)	67	(c)	75	(b)
4	(a)	12	(c)	20	(c)	28	(a)	36	(b)	44	(a)	52	(a)	60	(d)	68	(c)	76	(b)
5	(a)	13	(c)	21	(b)	29	(b)	37	(a)	45	(a)	53	(c)	61	(a)	69	(d)	77	(d)
6	(c)	14	(c)	22	(c)	30	(a)	38	(a)	46	(d)	54	(d)	62	(c)	70	(a)	78	(d)
7	(a)	15	(b)	23	(c)	31	(c)	39	(a)	47	(a)	55	(b)	63	(d)	71	(a)	79	(b)
8	(b)	16	(a)	24	(b)	32	(b)	40	(c)	48	(b)	56	(d)	64	(c)	72	(c)		

Hints & Solutions

1. **(a)** The hormones in blood act on specific sites or receptors of target organs. So if we remove the receptor molecule from the target organs, there will be no response.
2. **(b)** The molecules of amino acid derivatives, peptides or polypeptide (protein) hormones bind to specific receptor molecules located on the plasma membrane of target cells.
3. **(d)** 4. **(a)** 5. **(a)** 6. **(c)**
7. **(a)** 8. **(b)** 9. **(a)**
10. **(a)** "Upregulation" of hormone receptors on a cell is the production of more receptors when the hormone is present in low concentration over time in the blood or other fluids surrounding the cell.
11. **(b)** The label X represents parathyroid gland. It helps to restore lowered levels of calcium in blood. Too much calcium in the blood is restored by calcitonin, secreted by thyroid gland. Decreased level of blood sugar is restored by insulin, secreted by pancreas. Excessive loss of sodium in extracellular fluid is restored by aldosterone, secreted by adrenal gland.
12. **(c)** 13. **(c)** 14. **(c)**
15. **(b)** Thymus starts functioning in the embryonic stage itself, becomes active during childhood and undergoes regression and gradually stops functioning in old individuals. So, production of thymosin decreases. As a result, immune responses of old persons become weak.
16. **(a)** 17. **(c)**
18. **(c)** Because hormones are typically distributed throughout the blood, all potential target tissues will experience the same hormone concentration. Only by having either higher numbers of receptors, or receptors with greater hormone affinity, can one tissue be stimulated preferentially over other tissues by the same hormone.
19. **(d)** 20. **(c)** 21. **(b)** 22. **(c)**
23. **(c)**
24. **(b)** Hormones are non-nutrient chemicals which act as intercellular messengers and are produced in trace amounts.

    Gastrointestinal tracts secrete four major peptide hormones – gastrin, secretin, cholecystokinin (CCK) and gastricinhibitory peptide (GIP) while juxtaglomerular cells of kidney secrete erythropoietin, a peptide hormone. Releasing and inhibitory hormones are produced by hypothalamus. Neurohypophysis or posterior pituitary is under the direct neural regulation of the hypothalamus.
25. **(b)** 26. **(b)** 27. **(d)** 28. **(a)**
29. **(b)** Due to deficiency of ADH, water absorption by kidney tubules is decreased leading to excessive urine formation *i.e.* diuresis, so the person feels thirsty due to dehydration.
30. **(a)**
31. **(c)** Oxytocin stimulates muscle contractions during childbirth and milk letdown.
32. **(b)** Thyrocalcitonin (TCT) is secreted by parafollicular cells of thyroid gland. It is a protein hormone that regulates (lowers) the blood calcium levels in blood plasma.
33. **(b)** 34. **(d)** 35. **(d)**
36. **(b)** Insulin lowers blood sugar levels. Glucagon raises blood sugar levels.
37. **(a)** Iodine is an element used by thyroid gland for the synthesis of thyroid hormones. Thus, radioactive isotope Iodine -131 is used as radioactive material for detection of thyroid cancer.
38. **(a)** (A) and (B) are respectively pituitary gland and adrenal gland.
39. **(a)** The hormone representing by lines 1 and 2 are respectively ACTH and aldosterone. ACTH (adrenocorticotrophic hormone) is secreted by anterior pituitary gland. It controls the structure and function of adrenal cortex especially secretion of glucocorticoids and sex corticoids. Aldosterone is a type of mineralocorticoid, secreted from adrenal cortex. It is a salt retaining hormone which is secreted in response to increased potassium levels or decreased blood flow and sodium to the kidney.
40. **(c)** Melatonin is involved in the regulation of circadian rhythms.

41. (d)
42. (d) In response to an excess of parathyroid hormone, the body works to conserve calcium. Liberation of calcium from the bones and retention by the kidneys may result in high blood levels.
43. (c) In diabetics, the rise in blood glucose above the normal level of 80–90 mg/100ml fails to stimulate the production of insulin from the pancreatic β-cells of the Islets of Langerhans. As a result, the glucose is not removed from the blood to be stored in muscle cells as glycogen, resulting in a high blood glucose level for a long period of time.
44. (a) 45. (a) 46. (d)
47. (a) Glucagon is secreted from the alpha cells of pancreas when the blood sugar level is low. It stimulates glycogen breakdown and glucose synthesis in the liver by increasing blood glucose concentration. It has opposite effect to that of insulin because insulin is released when the level of blood sugar is high.
48. (b) 49. (b) 50. (c) 51. (b)
52. (a) 53. (c) 54. (d) 55. (b)
56. (d) Progesterone is necessary for maintenance of the uterine lining during pregnancy.
57. (b) 58. (a) 59. (c)
60. (d) Gonadotropins are activated at puberty and necessary for the production and maturation of gametes.
61. (a) 62. (c)
63. (d) The label X represents thyroid gland. The over-secretion of hormone from thyroid gland may lead to Grave's disease. Grave's disease is categorised as an autoimmune disorder.
64. (c) 65. (b)
66. (d) Contraceptive pills for women contain female sex hormones oestrogen and progesterone. These prevent development of eggs and ovulation by inhibiting secretion of FSH. Some pills contain progesterone only in such cases, ovulation may occur but cervical mucus is thickened preventing the entry of sperm.
67. (c) Heterocrine glands are the glands which have dual (exo- and endocrine) mode of function.
68. (c) Oestrogens are steroid hormones secreted by growing ovarian follicles. During menstrual cycle a negative feedback prevents the over secretion of oestrogen. FSH stimulates maturation of Graafian follicles. LH stimulates ovulation and development of corpus luteum.
69. (d) Hormones are chemical messengers of the body that transfer information from one set of cells to another.
70. (a) 71. (a)
72. (c) On the basis of nature, hormones can be divided into four groups
(i) Peptide, polypeptide, protein hormone (e.g., Insulin, glucagon, pituitary hormones, hypothalamic hormones).
(ii) Sterols (e.g., Cortisol, testosterone, estradiol and progesterone).
(iii) Iodothyronine (e.g., Thyroid hormones).
(iv) Amino acid derivatives (e.g., Epinephrine).
73. (a) Increasing the amount of '*a*' decreases the secretion of '*b*'. Since increasing '*b*' increases the secretion of '*c*', decreased '*b*' leads to decreased '*c*' which in turn leads to the increased secretion of '*a*'.
74. (c)
75. (b) Prolactin is a proteinaceous hormone released by anterior lobe of pituitary. Prolactin supplements the action of gonadal hormones in stimulating the growth and activity of female mammary gland during pregnancy and lactation. Oxytocin released by the posterior lobe of pituitary gland, is important for the mechanism of ejection of milk from the mammary glands due to sucking of breasts by infant. Progesterone, released by corpus luteum brings about pregnancy changes such as uterine growth, implantation of embryo etc. Relaxin is a proteinaceous hormone secreted by corpus luteum towards the end of gestation period for loosening and softening of pelvic ligaments, relaxing of uterus for easing parturition.
76. (b) Prothoracicotropic hormone stimulates the prothoracic gland to release juvenile hormone.
77. (d) Parathyroid hormone stimulates the movement of calcium from bone to blood.
78. (d) Oxytocin is an important hormone which is released in neurohypophysis of pituitary. It stimulates the contraction of smooth muscles. It stimulates and increases labour pain. Hence, it is a hormone for quick birth, whose high secretion results in parturition.
79. (b) By controlling the pituitary, the hypothalamus exerts control over these glands.

23 Reproduction in Organisms

1. A potato tuber is sown in a pot and pot is placed in the freezer of refrigerator. New plants will:
   (a) Emerge normally
   (b) Not emerge
   (c) Be without leaves
   (d) Be without flowers
2. Soil around stem of potato plant is constantly removed from very young stage so that only roots remain in the soil. Potato tubers in plant will be:
   (a) Larger (b) Smaller
   (c) Formed normally (d) Not formed
3. Banana fruits are seedless, because:

   (a) auxins are sprayed for rapid development of fruit.
   (b) vegetative propagation of plant.
   (c) they are triploid plants.
   (d) fruits are artificially ripened.
4. Reproducing new plants by cells instead of seeds is known as:
   (a) Biofertiliser (b) Mutation
   (c) Tissue culture (d) Antibiotics
5. The formation of gametophyte from the sporophyte (without meiosis) is called:
   (a) Apospory (b) Apogamy
   (c) Agamospermy (d) None of these
6. A small portion of cane-sugar stem between the two successive nodes is cut off and sown in the soil normally. New plants will be:
   (a) formed normally (b) not formed
   (c) without juice (d) without nodes
7. Apomixis in plant means development of a plant:
   (a) from root cuttings.
   (b) without fusion of gametes.
   (c) from cuttings of stem.
   (d) from fusion of gametes.
8. After culturing the anther of a plant, a few diploid plants were found along with haploid plants. The diploid plants could have arisen from:
   (a) Generative cell of pollen
   (b) Cells of anther wall
   (c) Vegetative cell of pollen
   (d) Exine of pollen wall
9. Carrot is micropropagated through:
   (a) Embryo (b) Embryoids
   (c) Shoot culture (d) Callus
10. Modified stem present in *Gladiolus* is:
   (a) Bulb (b) Rhizome
   (c) Corm (d) Bulbil
11. Examine the figures given below and select the right options out of (a- d); in which all the 4 items A, B, C and D are identified correctly

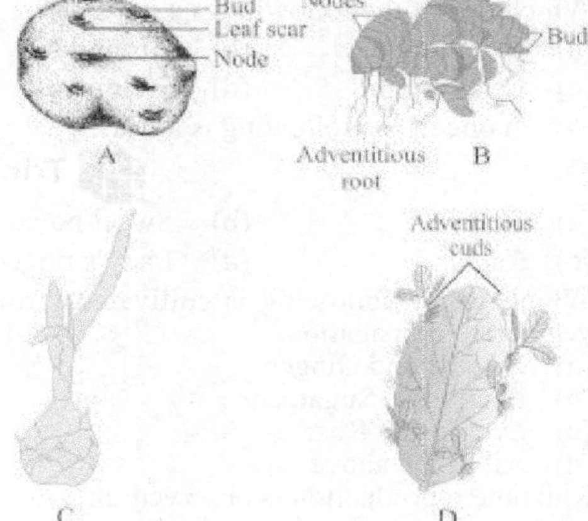

	A	B	C	D
(a)	Tuber	Rhizome	Bulb	Leaf buds
(b)	Offset	Sucker	Stolon	Leaf buds
(c)	Offset	Stolon	Sucker	Leaf buds
(d)	Tuber	Rhizome	Bulbil	Leaf buds

12. Match column-I with column- II and select the correct answer using the codes given below.

Column-I (Name of the organism)	Column-II (Haploid chromosome number in gamete)
A. *Ophioglossum (fern)*	I. 23
B. Rice	II. 24
C. Potato	III. 12
D. Man	IV. 630

   (a) A – I; B – II; C – III; D – IV
   (b) A – II; B – III; C – IV; D – I
   (c) A – III; B – IV; C – II; D – I
   (d) A – IV; B – III; C – II; D – I

13. Which one of the following pair is correctly matched?
(a) Onion – Bulb
(b) Ginger – Sucker
(c) *Chlamydomonas* – Conidia
(d) Yeast – Zoospores

14. Sweet potato is a modification of:
(a) Stem (b) Adventitious root
(c) Hypocotyl (d) Flowering axis

15. In Onion, the edible swollen portion is:
(a) Aerial stem
(b) Underground stem
(c) Internodes
(d) Leaf bases

16. Which one of the following is a nut fruit?
(a) Betel nut (b) Ground nut
(c) Cashew nut (d) Walnut

17. Which one of the following is a fruit?

(a) Ginger (b) Sweet potato
(c) Radish (d) Lady's finger

18. Which of the following is cultivated through vegetative propagation?
(a) Banana and Ginger
(b) Potato and Sugarcane
(c) *Dahlia* and Rose
(d) All of the above

19. Rhizome reproduction is observed in:
(a) *Solanum tuberosum*
(b) Ginger
(c) *Achyranthes*
(d) All of the above

20. Roses and banana are usually propagated through
(a) Parthenocarpy (b) Cutting
(c) Grafting (d) Layering

21. .... are thick walled spores produced by *Rhizopus* which are capable of withstanding unfavourable conditions.
(a) Chlamydospores (b) Zoospores
(c) Oidia (d) Sporangiospores

22. Vegetative reproduction in *Cycas* occurs by:
(a) Sporophyll (b) Bulbils
(c) Scaly leaves (d) Fragmentation

23. Artificial induction of roots on stems before it is separated from the parent plant for propagation is called:
(a) Root-stem joint
(b) Plant tissue culture
(c) Grafting
(d) Layering

24. Match the following columns and choose the correct option.

	Column-I		Column-II
A.	Clone	I.	Agamospermy
B.	Plant apomixis	II.	Scion
C.	Grafting	III.	Callus
D.	Tissue culture	IV.	Flower
E.	Inflorescence	V.	Ramet

(a) A – V; B – I; C – II; D – III; E – IV
(b) A – IV; B – I; C – III; D – II; E – IV
(c) A – I; B – II; C – III; D – IV; E – V
(d) A – V; B – II; C – III; D – IV; E – I

25. Which the following statements are **correct**?
(i) Stamens are male reproductive part whereas carpels are female reproductive parts.
(ii) In *Spirogyra*, asexual reproduction takes place by fragmentation.
(iii) Vegetative propagation by leaves occurs in sweet potato.
(a) (i) and (ii) (b) (ii) and (iii)
(c) (i) and (iii) (d) All of the above

26. Identify the correct option that describes all the three diagrams correctly.

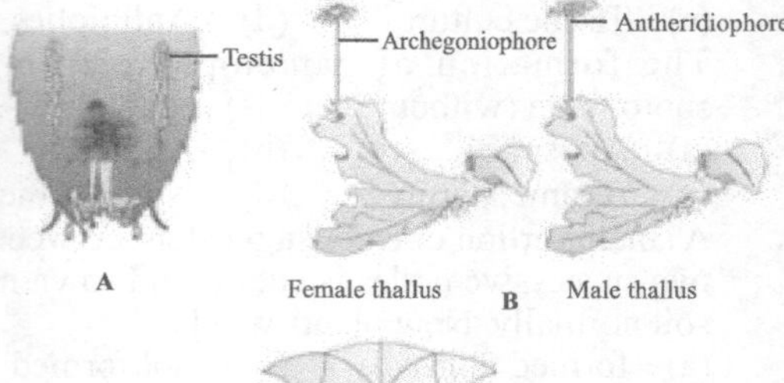

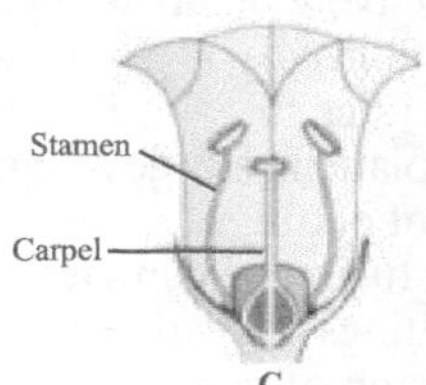

(a) A-Male cockroach, B-*Funaria*, C-Unisexual flower
(b) A-Male cockroach, B-*Marchantia*, C-Bisexual flower
(c) A-Female cockroach, B-*Cycas*, C-Unisexual flower
(d) A-Female cockroach, B-*Marchantia*, C-Bisexual flower

27. Common mode of reproduction in *Penicillium* is:
(a) Conidia
(b) Gemmule formation
(c) Buds
(d) Zoospore

28. For vegetative propagation, the vegetative part of the plant should have:
(a) A growing point (b) Reserved food
(c) Leaf (d) Both (a) and (b)

29. In a grafted plant, stock has 48 chromosomes while scion has 24 chromosomes. The chromosome number for root cells and eggs are: Toughnut
(a) 48 and 24 (b) 24 and 24
(c) 24 and 12 (d) 48 and 12

30. Which is the most common method of reproduction in majority of fungi and bacteria?
(a) Binary fission (b) Multiple fission
(c) Budding (d) Spore formation

31. Which of the following definitions about oviparity is **false**? Critical Thinking
(a) Only birds and reptiles are oviparous.
(b) The large amount of yolk provides the nutrients for the developing embryo.
(c) The shell protects the egg from dehydration.
(d) Both oxygen and carbon dioxide can diffuse through the shell.

32. During spermatogenesis, a single male germ cell produces _____ sperm cell(s). Tricky
(a) 1 (b) 2 (c) 3 (d) 4

33. Ploidy of ovary, anther, egg, pollen, male gamete and zygote are respectively: Tricky
(a) 2n, 2n, n, 2n, n, 2n
(b) 2n, 2n, n, n, n, 2n
(c) 2n, n, n, n, n, n
(d) 2n, 2n, n, 2n, 2n, 2n

34. Oestrous cycle occurs in females of:
(a) Monkeys (b) Cows
(c) Apes (d) Humans

35. Which of the following statements regarding the given terms are **correct**? Tricky

Runners, Tubers, Offsets

(a) These all are incapable of giving rise to new individuals and formation of these structures does not involve two parents.
(b) These all are capable of giving rise to new individuals and formation of these structures does not involve two parents.
(c) These all are capable of giving rise to new individuals but formation of these structures involves two parents for reproduction.
(d) These are all incapable of giving rise to new individuals but formation of these structures involves two parents for reproduction.

36. Match column-I with column- II and select the correct answer using the codes given below.

	Column-I (Name of the organism)		Column-II (Chromosome number in meiocyte) (2*n*)
A.	House fly	I.	20
B.	Fruit fly	II.	34
C.	Apple	III.	8
D.	Maize	IV.	12

(a) A – II; B – III; C – IV; D – I
(b) A – III; B – IV; C – II; D –I
(c) A – IV; B – III; C – I; D – II
(d) A – IV; B – III; C – II; D – I

37. The terms homothallic and monoecious are used to denote:
(a) Bisexual condition
(b) Unisexual condition
(c) Staminate flowers
(d) Pistillate flowers

38. If cotyledons are brought above the soil, the germination is:
(a) Hypogeal (b) Epigeal
(c) Vivipary (d) None of these

39. Some organisms are capable of asexual or sexual reproduction. Under favourable conditions, reproduction proceeds asexually. When conditions become more stressful, reproduction switches to a sexual mode. Why? Tricky
(a) Sexual reproduction is simple and more rapid allowing larger numbers of offsprings to be produced.
(b) Sexual reproduction requires two separate individuals, who can mutually provide nutrient support during stress.
(c) Sexual reproduction produces individuals with new combinations of recombined chromosomes increasing diversity.
(d) Asexual reproduction requires more energy.

40. Meiocytes is observed in all except:
(a) Human egg (b) Blue-green algae
(c) *Equisetum* (d) Dicot plant

41. In the given figure of water hyacinth, a structure is marked as "X". This structure is involved in vegetative propagation as a unit of vegetative propagules. Identify the type of unit.

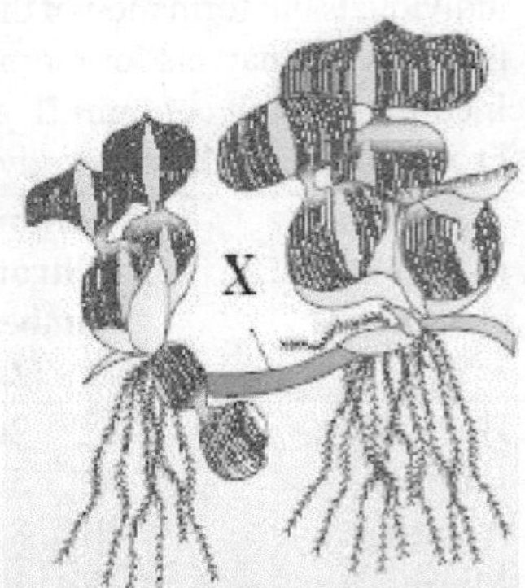

(a) Tuber (b) Offsets
(c) Sucker (d) Rhizome

42. Match column-I with column-II and select the correct answer using the codes given below.

	Column - I		Column - II
A.	Sponge	I.	Tuber
B.	Yeast	II.	Offset
C.	Potato	III.	Gemmules
D.	Water hyacinth	IV.	Budding

(a) A – IV; B – I; C – II; D – III
(b) A – III; B – I; C – IV; D – II
(c) A – III; B – IV; C – I; D – II
(d) A – III; B – IV; C – II; D – I

43. Which of the following statements are **correct**?
(i) *Euglena* divides by longitudinal binary fission.
(ii) Cytokinesis takes place along transverse axis in *Paramoecium*.
(iii) *Amoeba* divides by irregular binary fission.
(iv) In multiple fission, parental body divides into many daughter cells under favourable conditions.

(a) (i) and (iv) (b) Only (iv)
(c) (i), (ii) and (iii) (d) (ii) and (iv)

44. Which of the following statements regarding the asexual reproduction is **incorrect**?
(a) Both mitotic and meiotic division occurs.
(b) It does not contribute to evolution and speciation.
(c) It is uniparental and usually occurs in unicellular organisms.
(d) There is no variation and the offsprings have the same phenotype and genotype.

45. Offsprings of oviparous animals are at greater risk as compared to offsprings of viviparous animals because: **Tricky**
(a) proper embryonic care and protection is lesser.
(b) embryo is not developed.
(c) progenies are with more variation.
(d) progenies are larger.

46. Agamospermous embryo sac develops from:
(a) Pollen
(b) Egg
(c) Ovum
(d) Megaspore mother cell

47. Offspring formed by sexual reproduction exhibits more variation than those formed by asexual reproduction because: **Tricky**
(a) sexual reproduction is a lengthy process.
(b) genetic material comes from two parents of different species.
(c) genetic material comes from two parents of same species.
(d) greater amount of DNA is involved in sexual reproduction.

48. Which of the labelled parts (1 and 2) in the transverse section of pea plant is/are diploid?

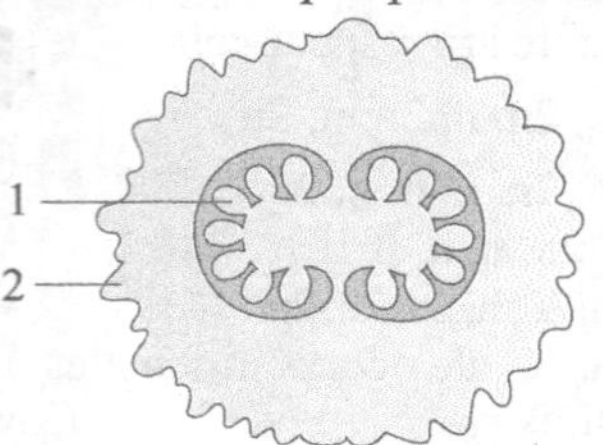

(a) 1 (b) 2
(c) Both (a) and (b) (d) None of these

49. Select the **incorrect** statements.
(a) Cucurbits and coconuts are monoecious plants.
(b) Papayas and date palms are dioecious plants.
(c) Leeches and tapeworms are bisexual animals.
(d) Sponges and coelenterates are unisexual animals.

50. Which of the following statements about animals that utilise external fertilisation is **incorrect**?
(a) They are divided equally between terrestrial and aquatic species.
(b) Many produce large numbers of gametes to ensure successful reproduction.
(c) The behaviours associated with mating are often highly synchronised.
(d) The probability of any one egg being fertilised and developing into an adult can be low.

51. The given figures P (plant *Chara*) and Q (Earthworm) have label 1, 2, 3 and 4 as their reproductive organs. Select the option which correctly identifies male reproductive organs of the two organisms.

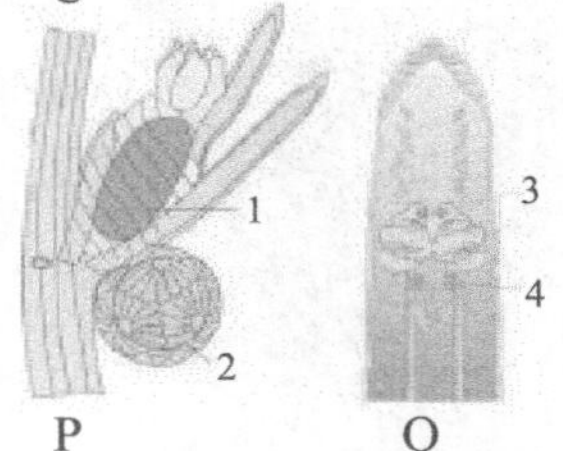

(a) 1 and 2 (b) 2 and 3
(c) 1 and 3 (d) 2 and 4

52. Match column-I with column- II and select the correct answer using the codes given below.

	Column-I		Column-II
A.	Gamete	I.	Result of fusion of male and female gametes.
B.	Budding	II.	Division of body into two equal halves.
C.	Fission	III.	Germ cells.
D.	Fertilisation	IV.	The fusion of male and female gametes.
E.	Zygote	V.	An unequal division of organisms in which individual arises as an outgrowth from the parent.

(a) A – I; B – II; C – III; D – IV; E – V
(b) A – III; B – V; C – II; D – IV; E – I
(c) A – III; B – I; C – V; D – II; E – IV
(d) A – V; B – IV; C – III; D – I; E – II

Answer KEYs

1	(b)	7	(b)	13	(a)	19	(b)	25	(a)	31	(a)	37	(a)	43	(c)	49	(d)
2	(d)	8	(b)	14	(b)	20	(b)	26	(b)	32	(d)	38	(b)	44	(a)	50	(a)
3	(c)	9	(d)	15	(d)	21	(a)	27	(a)	33	(b)	39	(c)	45	(a)	51	(b)
4	(c)	10	(c)	16	(c)	22	(b)	28	(d)	34	(b)	40	(b)	46	(d)	52	(b)
5	(a)	11	(d)	17	(d)	23	(d)	29	(d)	35	(b)	41	(b)	47	(c)		
6	(b)	12	(d)	18	(d)	24	(a)	30	(d)	36	(d)	42	(c)	48	(c)		

Hints & Solutions

1. (b) Buds do not germinate at very low temperature.
2. (d) Tubers are formed from underground branches of stem.
3. (c) Most of banana varieties are triploid and triploidy is associated with seedlessness.
4. (c) 5. (a)
6. (b) New plants in cane-sugar are formed from nodes, which are absent in this case.
7. (b) 8. (b) 9. (d) 10. (c)
11. (d) 12. (d)
13. (a) Onion - Bulb (Underground stem), Ginger - Rhizome, *Chlamydomonas* - Zoospore, Yeast - Ascospores
14. (b) 15. (d) 16. (c) 17. (d)
18. (d) 19. (b) 20. (b) 21. (a)
22. (b) Bulbils are modified floral buds, characteristic of family liliacae and many plants as *Cycas*.
23. (d) Grafting is performed between two plants of closely related varieties. The root system of one is grafted with the shoot system of other desired plant. In layering, stem branch chosen for propagation remains attached to the parent till it produces adventitious roots. Tissue culture or micro propagation is the latest method of obtaining a number of plantlets from plant tissue.
24. (a) Ramet is the synonym of clone. Plant apomixis and agamospermy are same. During bud grafting a cut, called scion is made, which is T-shaped and from this point bud formation takes place. Callus is the unorganised mass of tissue produced during tissue culture. After giving appropriate hormone it changes into the new plant. Infloresence gives rise to flower.
25. (a) Vegetative propagation by buds or notches occurs in potato
26. (b) A-Indicates male cockroach

    B- Plant body is thalloid and sexes are separate indicates *Marchantia.*

    C-Male and female gametes on same plant monoecious or bisexual flower.
27. (a) Conidia are non-motile gametes found singly or in chain form on the parent body, e.g., *Penicillum*

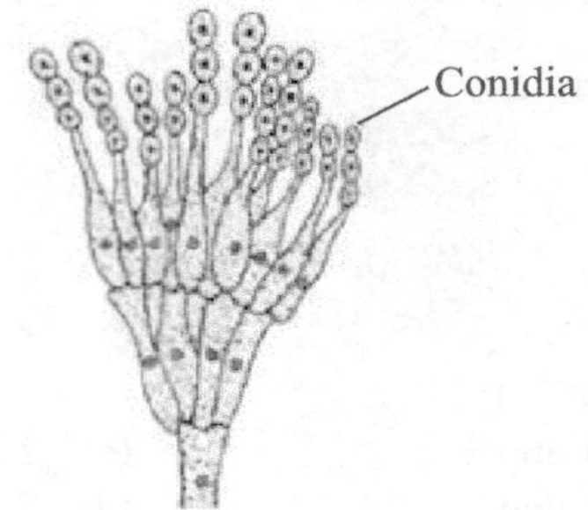

Conidia formation in *Pencillium*

28. (d)
29. (d) Roots are produced from stock. Stock is diploid so the number of chromosomes in root cells of stock is 48. But egg is haploid, so in egg, chromosome number will be 12 [as scion forms shoot system and its chromosome number is 24 (haploid)].
30. (d) Fungi reproduce by vegetative, asexual and sexual means. About 20% fungi propagate only by asexual means. Asexual reproduction takes place during favourable conditions by the formation of a variety of conidia or spores. The spores may be unicellular (e.g., *Aspergillus*) or multicellular (e.g., *Alternaria*). Bacteria also reproduce mainly by asexual method and therefore they have a dominant haploid phase in their life-cycle. Asexual reproduction in bacteria takes place by binary fission, budding, conidia, cysts and endospores.
31. (a) The birds and reptiles are not the only oviparous species. A group of mammals, the monotremes, are also egg layers.

32. (d) During spermatogenesis, a single male germ cell will undergo two meiotic divisions. This results in the production of 4 haploid sperm cells.

33. (b)

34. (b) In non-primate mammals such as cows, sheep, rats, deer, etc. oestrous cycle takes place.

35. (b) Runners, tubers and offsets are all the units of vegetative propagules in plants. These are all capable of giving rise to new offsprings. Since the formation of these structures does not involve two parents, the process involved is asexual.

36. (d)

37. (a) Homothallic and monoecious are used to denote bisexual condition. The example includes fungi and plants. Heterothallic and dioecious are used to denote unisexual condition.

38. (b)

39. (c) Some organisms are capable of asexual or sexual reproduction. Under favourable conditions, reproduction proceeds asexually. When conditions become more stressful reproduction switches to a sexual mode, because sexual reproduction produces individuals with new combinations of recombined chromosomes increasing diversity.

40. (b) In diploid organisms, specialised cells called meiocytes undergo meiosis. Blue-green algae (cyanobacteria) are unicellular where the most common method of multiplication is binary fission.

41. (b) The marked structure in the given figure is offset. Offsets are one internode long runners that occur in some aquatic plants. Breaking of offsets helps in propagation, e.g., *Eichhornia* (Water hyacinth), *Pistia* (Water lettuce).Vegetative propagation is a form of asexual reproduction in plants, in which multicellular structures become detached from the parent plant and develop into new individuals that are genetically identical to the parent plant. For example, liverworts and mosses form small clumps of tissue (called gemmae) that are dispersed by splashing raindrops to form new plants. Bulbs, corms, offsets, rhizomes, runners, suckers, and tubers are all important means of vegetative reproduction and propagation in cultivated plants.

42. (c)

A : Gemmule is a tough-coated dormant cluster of embryonic cells produced by a freshwater sponge for development in more favourable conditions.

B : Budding is a type of asexual reproduction in which a daughter individual is formed from a small projection, the bud, arising from the parent body, e.g., yeast.

C : Tuber is an underground stem structure which can take part in vegetative reproduction, e.g., Potato.

D : Offset is a type of sub aerial or creeping stem. These are one internode long runners that occur in some aquatic plants. Breaking of offsets helps in propagation, e.g., water hyacinth and water lettuce.

43. (c) Multiple fission occurs during unfavourable conditions to increase the chance of survival of daughter cells.

44. (a) When offspring is produced by a single parent with or without the involvement of gamete formation, the reproduction is called asexual. As a result, the offspring that are produced are not only similar to one another but are also exact copies of their parent. Only mitotic division occurs in asexual reproduction.

45. (a) Offspring's of oviparous animals are at greater risk as compared to offspring's of viviparous animals because proper embryonic care and protection is less in oviparous animals. Oviparous animals are those animals that lay eggs, with little or no other embryonic development within the mother. These eggs mature and hatch after being expelled from the body. Examples, most fish, amphibians, reptiles, all birds, etc. Whereas viviparous animals give birth to young lives. The fertilized eggs develop into a young one inside the body of the female organisms. Therefore, chances of the survival of the young ones in viviparous animals is more because of proper embryonic care and protection inside the mother body.

46. (d)

47. (c) Sexual reproduction is characterised by genetic recombination. Due to genetic recombination the progeny is different from the parents. In sexual reproduction, the genetic material comes from two parents of same species. But in asexual reproduction only one individual participate to produce offspring.

48. (c) Both 1 and 2 are diploid in the given figure of transverse section of pea plant. Part 1 represents seed and part 2 represents pericarp.

49. (d) Sponges and coelenterates are bisexual animals.

50. (a) Since external fertilisation can only take place in an aquatic habitat, there are no terrestrial animals that use it.

51. (b) In both the figures, the structure marked as 2 (in figure P) and C (in figure Q) are the male reproductive organs. 2 and 3 is respectively antheridium and testes sac. An antheridium (or antheridia) is a haploid structure (or organ) producing and containing male gametes (called antherozoids or sperms). It is present in the gametophyte phase of cryptogams like bryophytes and ferns, and also in the primitive vascular psilotophytes. Testis sac is the pouch containing the testes and their accessory organs.

52. (b)

A : Gamete is a mature haploid male or female germ cell which is able to unite with another of the opposite sex in sexual reproduction to form a zygote.

B : Budding is a form of asexual reproduction in which new individuals form, from outgrowths (buds) on the bodies of mature organisms. These outgrowths grow by means of mitotic cell division.

C : Fission is a form of asexual reproduction and cell division used by all prokaryotes (bacteria and archaebacteria) and some organelles within eukaryotic organisms (e.g., mitochondria). In this a unicellular organism divides into two or more independently maturing daughter cells.

D : Fertilisation involves the fusion of male and female gametes to form a zygote.

E : Zygote is a diploid cell resulting from the fusion of two haploid gametes; a fertilised ovum.

24 Sexual Reproduction in Flowering Plants

1. Male gametophyte of angiosperms/monocots is:
(a) Microsporangium (b) Nucellus
(c) Microspore (d) Stamen

2. The male gametophyte (microgametophyte) in case of monocots is:
(a) Tetrad (b) Nucellus
(c) Megaspore (d) Microspore

3. In a somatic cell, there are 16 chromosomes. What will be the number of chromosomes in the pollen mother cell of the plant? **Tricky**
(a) 16 (b) 8 (c) 24 (d) 32

4. Function of funicle includes:
(a) support to the body
(b) supplying nutrition to the body from the palcenta
(c) Both (a) and (b)
(d) None of these

5. If the haploid number in a flowering plant is 14. What shall be the number of chromosomes in integuments, antipodal cells, embryo, endosperm and nucellus respectively? **Critical Thinking**
(a) 14, 28, 7, 42, 21
(b) 7, 14, 42, 28, 14
(c) 28, 14, 28, 42, 28
(d) 42, 28, 14, 28, 14

6. As a pollen tube grows into the female organ, the nucleus that enters the synergid first is called the:
(a) Sperm nucleus
(b) Generative nucleus
(c) Tube nucleus
(d) Pollen nucleus

7. Central cell of embryo sac contains:
(a) a single haploid nucleus.
(b) two haploid polar nuclei.
(c) one diploid fusion or secondary nucleus.
(d) Either (b) or (c)

8. Identify 1, 2, 3, 4 and 5 structures shown in figure of a female gametophyte respectively.

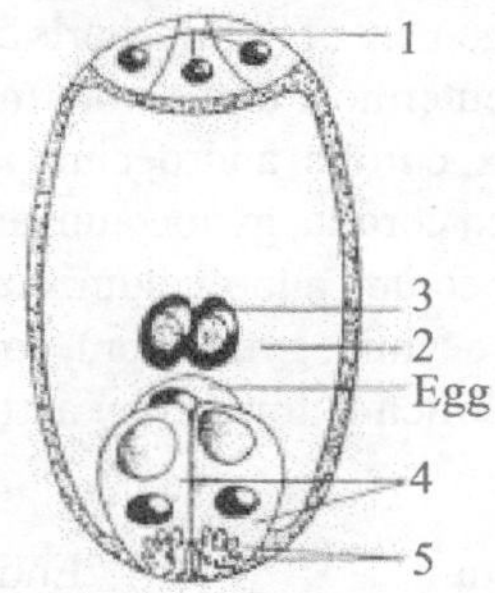

(a) Antipodal cells, Central cell, Polar nuclei, Synergids and Acrosome
(b) Antipodal cells, Central cell, Polar nuclei, Synergids and Filiform apparatus
(c) Synergids, Central cell, Polar nuclei, Antipodal cells and Filiform apparatus
(d) Synergids, Megaspore mother cell, Polar nuclei, Synergids, Filiform apparatus

9. In dicot embryo, the radicle is formed by:
(a) epibasal tier of embryo
(b) hypobasal tier of embryo
(c) hypophysis of suspensor
(d) terminal cell of suspensor

10. Dicot embryo consists of:
(a) radicle and plumule
(b) radicle, plumule, cotyledons and sometimes endosperm
(c) radicle, plumule, cotyledons and tegmen
(d) radicle, plumule, cotyledons, tegmen and testa

11. Generative cell was destroyed by laser but a normal pollen tube was still formed because:

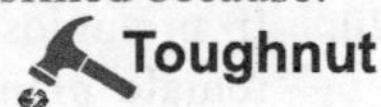

(a) vegetative cell is not damaged.
(b) contents of killed generative cell stimulate pollen growth.
(c) laser beam stimulates growth of pollen tube.
(d) the region of emergence of pollen tube is not harmed.

12. Female gametophyte of angiosperms is represented by:
(a) Ovule
(b) Megaspore mother cell
(c) Embryo sac
(d) Nucellus

13. Angiospermic plant having chromosome number of 12. What will be chromosome number in integuments and nucellus? **Tricky**
(a) 4 (b) 6 (c) 12 (d) 24

14. The arrangement of floral whorls on the thalamus from the outermost to the innermost is:
(a) Calyx, corolla, androecium and gynoecium
(b) Calyx, corolla, gynoecium and androecium
(c) Gynoecium, androecium, corolla and calyx
(d) Androecium, gynoecium, corolla and calyx

15. Exine of pollen is formed by activity of: **Tricky**
(a) Tapetum (b) Endothelium
(c) Middle layers (d) Endothecium

16. Male gametes or sperms are developed from generative cell by:
(a) Meiotic division (b) Mitotic division
(c) Amitotic division (d) None of these

17. Wall of a pollen sac consists of:
(a) endothecium and tapetum
(b) tapetum and middle layers
(c) endothecium, middle layers and tapetum
(d) epidermis, endothecium, middle layers and tapetum

18. In dicots, the most common pollen tetrad is: **Critical Thinking**
(a) Isobilateral (b) Tetrahedral
(c) Linear (d) Decussate

19. Anther are:
(a) externally tetralocular
(b) internally tetralocular
(c) externally and internally tetralocular
(d) internally bilobed

20. If an angiospermic male plant is diploid and female plant tetraploid, the ploidy level of endosperm will be: **Toughnut**
(a) Haploid (b) Triploid
(c) Tetraploid (d) Pentaploid

21. Aleurone cells are present in endosperm of:
(a) Cereals (b) Legumes
(c) Both (a) and (b) (d) None of these

22. Multinucleate condition is present in:
(a) Quiescent centre
(b) Maize
(c) Meristematic tissue
(d) Liquid endosperm of coconut

23. Match column-I with column- II and select the correct answer using the codes given below.

	Column-I		Column-II
A.	Zoophily	I.	Pollination by birds
B.	Ornithophily	II.	Pollination by insects
C.	Entomophily	III.	Pollination by bats
D.	Chiropterophily	IV.	Pollination by animals

(a) A – III; B – II; C – I; D – IV
(b) A – I; B – II; C – III; D – IV
(c) A – IV; B – I; C – II; D – III
(d) A – IV; B – II; C – I; D – III

24. Pea seeds are sown two feet deep in the soil. New plants will be: **Critical Thinking**
(a) Formed normally (b) Not formed
(c) Weaker (d) Healthier

25. Which one commonly develops haustoria in embryo sac?
(a) Synergids (b) Antipodal cells
(c) Oosphere (d) Central cell

26. An apomictic seed contains an embryo that is:
(a) produced when two sperm fertilise one egg.
(b) developed from one egg alone.
(c) the result of parental self-fertilisation.
(d) genetically identical to its parent.

27. Scutellum is present in the embryo of:
(a) Pea (b) *Ranunculus*
(c) *Triticum* (d) None of these

28. In ovule, meiosis occurs in:
(a) Archesporial cell (b) Megasporocyte
(c) Megaspore (d) Parietal cell

29. Seeds are adaptively important because:
(a) they maintain dormancy.
(b) they protect young plants during vulnerable stages.
(c) they store food for young plants, and facilitate disperal.
(d) All of the above

30. Number of prothallial cells present in the male gametophyte of angiosperms is: **Tricky**
(a) One (b) Two (c) Many (d) Zero

31. Growth of pollen tube is:
(a) Apical (b) Basal
(c) Intercalary (d) Diffused

32. Tapetum present in the microsporangial wall occurs between:
(a) Epidermis and endothecium
(b) Endothecium and middle layers
(c) Epidermis and middle layers
(d) Middle layers and sporogenous tissue

33. Pollen tube is covered by: **Tricky**
(a) Pectocellulose (b) Sporopollenin
(c) Cellulose (d) Lignocellulose

34. Polygonum type of embryo sac/typical female gametophyte of angiosperms is: **Critical Thinking**
(a) 7-celled, 7-nucleate
(b) 7-celled, 8-nucleate
(c) 8-celled, 7-nucleate
(d) 8-celled, 8-nucleate

35. Ovule is inverted with body fused to funicle, micropyle lying close to hilum and facing the placenta. It is:
(a) Hemitropous (b) Orthotropous
(c) Anatropous (d) Campylotropous

36. Bamboo reproduces by:
(a) Rhizomes (b) Tubers
(c) Corms (d) Stolons

37. When the pollen tube enters the ovule through the integuments, the phenomenon is known as:
(a) Isogamy (b) Porogamy
(c) Mesogamy (d) Chalazogamy

38. Select the correct order of endosperm types.

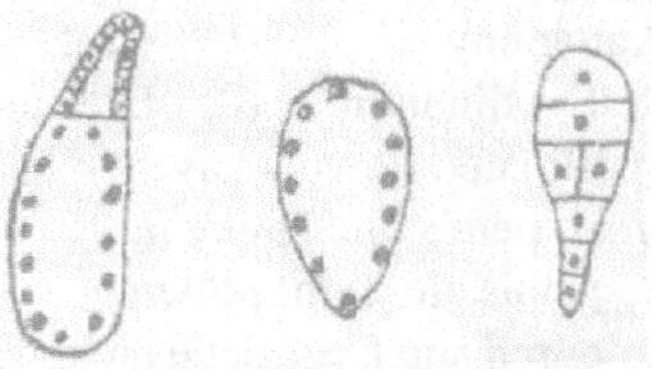

(a) Cellular, helobial, free nuclear
(b) Cellular, free nuclear, helobial
(c) Helobial, free nuclear, cellular
(d) Free nuclear, cellular, helobial

39. Which of the following statement(s) is/are false?
(i) Pollen grains represents immatured male gametophyte.
(ii) In angiosperms, partially developed male gametophytes are pollinated.
(iii) Generative cell is siphonogamous while vegetative cell is spermatogenous.
(iv) Formation and differentiation of pollen grains is called microsporogenesis.
(v) Pollen grains of some plants produce severe allergy and respiratory or bronchial diseases.
(vi) Pollen grains are poor in nutrients.
(a) (i) and (vi) (b) (iii) and (vi)
(c) (iv) and (v) (d) (v) and (vi)

40. For self pollination, flower must be:
(a) Unisexual (b) Bisexual
(c) Monosexual (d) Asexual

41. The phenomenon of two flowers, one having long stamens and short styles, and other having short stamens and long style:
(a) Allogamy (b) Heterostyly
(c) Dicliny (d) Herkogamy

42. Which of the following pair has haploid structures? **Tricky**
(a) Nucellus and antipodal cells.
(b) Antipodal cells and egg cell.
(c) Antipodal cells and megaspore mother cell.
(d) Nucellus and primary endosperm nucleus.

43. In angiosperms, female gametophyte is represented by:
(a) Synergids (b) Carpel
(c) Egg (d) Pollen grains

44. Intine of pollen grain is made of:
(a) Callose
(b) Pecto-cellulose
(c) Cellulose
(d) Fat-like sporopollenin

45. Abundant occurrence of fossilised pollen grains is due to resistant:
(a) Lignocellulose (b) Sporopollenin
(c) Pectocellulose (d) Pectolignin

46. Tapetal cells are: **Tricky**
(a) Haploid (b) Polyploid
(c) Diploid (d) Triploid

47. Embryo sac is surrounded by a wall of:
(a) specialised nucellar cells
(b) transfer cells
(c) membrane of megaspore
(d) membrane of egg

48. Vegetative nucleus occurs in:
(a) All flowering plants
(b) All seed plants
(c) All vascular plants
(d) All embryophytes.

49. Cross-pollination is preferred over self pollination because:
(a) it results in better offspring
(b) the new varieties are formed
(c) it is easy
(d) parthenogenesis can be induced

50. Which type of association is found in between entomophilous flower and pollinating agent?
(a) Mutualism (b) Commensalism
(c) Cooperation (d) Co-evolution

51. Unisexuality of flowers prevents: 

(a) geitonogamy but not xenogamy
(b) autogamy and geitonogamy
(c) autogamy but not geitonogamy
(d) both geitonogamy and xenogamy

52. The endosperm in angiosperm develops from:
(a) Zygote
(b) Secondary nucleus
(c) Chalazal polar nucleus
(d) Micropylar polar nucleus

53. Given below are the different types of ovules. Identify A to F. Critical Thinking

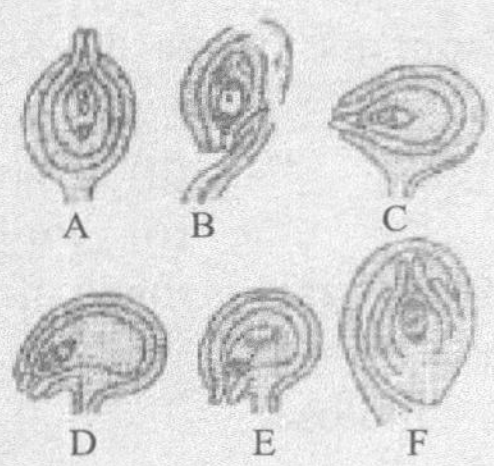

(a) A-Circinotropous, B-Amphitropous, C-Campylotropous, D-Hemitropous, E-Anatropous, F-Orthotropous
(b) A-Campylotropous, B-Anatropous, C-Hemitropous, D-Amphitropous, E-Circinotropous, F-Orthotropous
(c) A-Orthotropous, B-Anatropous, C-Hemitropous, D-Campylotropous, E-Amphitropous, F-Circinotropous
(d) A-Anatropous, B-Campylotropous, C-Hemitropous, D-Amphitropous, E-Circinotropous, F-Orthotropous

54. Which of the following statement is **true** about tapetum?
(a) It is the innermost layer of anther wall.
(b) Cells of tapetum are large in size and has more than one diploid nuclei or polyploid nucleus.
(c) It contributes in the formation of sporopollenin.
(d) All of the above

55. Hybrid seeds have to be produced every year because: Critical Thinking
(a) hybrid plants become sterile in coming years.
(b) they show more heterosis in coming years.
(c) hybrid vigour is not maintained beyond one generation as segregation of genes begins in the second generation.
(d) hybrid seed industry tends to increase cost of the seeds.

56. In a seed of maize, scutellum is considered as cotyledon because it:
(a) protects the embryo.
(b) contains food for the embryo.
(c) absorbs food materials and supplies them to the embryo.
(d) converts itself into a monocot leaf.

57. An example of a naturally occurring parthenocarpic fruit is:
(a) Guava (b) Mango
(c) Banana (d) Apple

58. The largest cell in a embryo sac is:
(a) Egg (b) Central cell
(c) Synergid (d) Antipodal cell

59. Meiosis of megaspore mother cell generally produces:
(a) Linear tetrad
(b) Tetrahedral tetrad
(c) Decussate tetrad
(d) Isobilateral tetrad

60. Xenogamy is also known as:
(a) Autogamy (b) Cross pollination
(c) Self pollination (d) Cleistogamy

61. Wind pollinated flowers have:
(a) small petals and sticky pollen.
(b) no petals and light pollen.
(c) coloured and large petals with large pollen.
(d) small petals and heavy pollen.

62. Function of suspensor of embryo is: 

(a) absorption of nourishment.
(b) push the embryo into nutritive endosperm region.
(c) formation of secondary embryos.
(d) All of the above

63. Wall of mature pollen grain consists of:
(a) Ektexine and endexine
(b) Tapetum and endothecium
(c) Exine and intine
(d) Foot and baculate layer

64. Which one of the following element is essential for the growth of pollen tube? **Tricky**
(a) Ca (b) B (c) Mg (d) Mo

65. Isogamy is:
(a) fusion of male and female gametes.
(b) fusion of physiologically and morphologically similar gametes.
(c) entry of pollen tube through integuments.
(d) None of the above

66. The insect pollinated pollen grain, tapetum helps in the development of:
(a) Sporopollenin (b) Cellulose
(c) Pectin (d) Pollen kit

67. At the time of anther dehiscence:
(a) middle layers develop fibrous thickenings.
(b) epidermis degenerates.
(c) endothecium develops fibrous thickenings.
(d) endothecium degenerates.

68. The given diagram refers to a T. S. of anther. Identify 1 to 5 respectively.

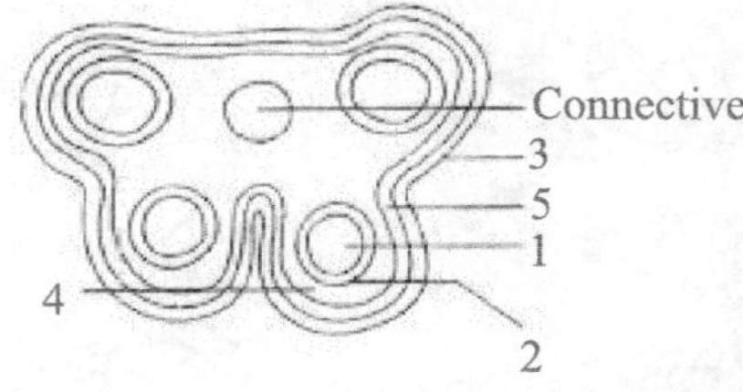

(a) Sporogenous tissue, tapetum, epidermis, middle layer, endothecium.
(b) Sporogenous tissue, epidermis, tapetum, middle layer, endothecium.
(c) Sporogenous tissue, epidermis, middle layer, tapetum, endothecium.
(d) Sporogenous tissue, tapetum, middle layer, epidermis, endothecium.

69. Give the of name the cell division type at A, B, C and D.

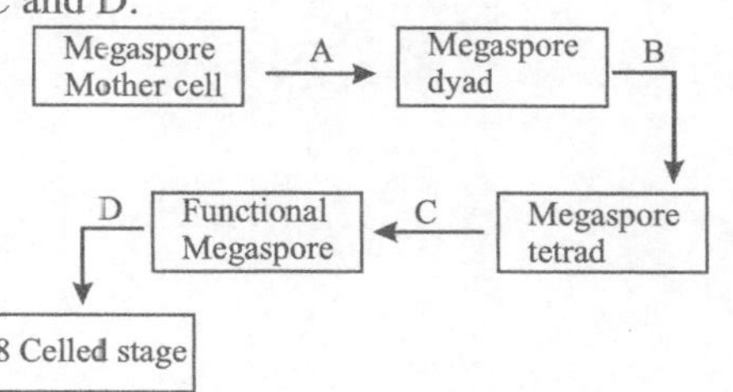

(a) A-Meiosis-I, B-Mitosis, C-Mitosis, D-Meiosis
(b) A-Meiosis-I, B-Meiosis-II, C-No division, D-Mitosis
(c) A-Mitosis, B-No division, C-Meiosis-II, D-Meiosis-I
(d) A-Mitosis, B-Mitosis, C-Meiosis-I, D-Meiosis-I

70. Apomixis is:
(a) development of plants in darkness.
(b) development of plants without fusion of gametes.
(c) inability to perceive stimulus for flowering.
(d) effect of low temperature on plant growth.

71. Which of the following floral parts forms pericarp after fertilisation?
(a) Nucellus (b) Outer integument
(c) Ovary wall (d) Inner integument

72. Plumule is produced from the:
(a) Proembryo (b) Hypophysis
(c) Apical octant (d) Micropylar octant

73. Flowers showing ornithophily, represent few characteristics like: **Tricky**
(a) blue flower with nectaries at base of corolla.
(b) red sweet scented flower with nectaries.
(c) bright red flower into thick inflorescence.
(d) white flowers with fragrance.

74. An angiospermic plant has to produce 88 viable ovules. How many meiotic divisions will be needed to produce equal number of female gametophytes by this plant? **Tricky**
(a) 88 (b) 22 (c) 44 (d) 132

75. In cross self fertilised plants, the flowers are first emasculated. Which organ of the plant is removed in this process? **Tricky**
(a) Ovary (b) Ovules
(c) Stigmas (d) Stamens or anthers

76. 3-celled stage of the mature male gametophyte is reached:
(a) before pollination (b) after pollination
(c) during fertilisation (d) after fertilisation

77. Endothecium layer of anther lobes is present:
(a) outside the epidermis
(b) inner to epidermis
(c) in the innermost layer
(d) in the middle region

78. How many pollen grains will be formed after meiotic division in ten microspore mother cells?

(a) 10 (b) 20 (c) 40 (d) 80

79. Pollination by water occurs in:
(a) *Ceratophyllum* (b) *Zostera*
(c) *Lemna* (d) All of these

80. Seed coat is derived from:
(a) Pericarp
(b) Epicarp
(c) Integuments of ovule
(d) Nucellus

81. Non-endospermic seeds are found in:
(a) Castor (b) Rice
(c) Wheat (d) Bean

82. The condition of maturation of stigma before anthers of the same flower is:
(a) Protandry (b) Herkogamy
(c) Protogyny (d) Prepotency

83. In a flowering plant, archesporium gives rise to:

(a) only tapetum and sporogenous cells
(b) only the wall of the sporangium
(c) both wall and the sporogenous cells
(d) wall and the tapetum

84. In the given diagram name the parts 1, 2, 3, 4 and 5.

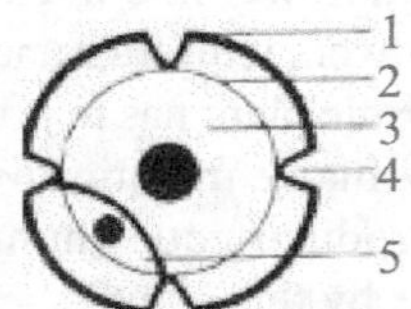

(a) 1-germpore, 2-generative cell, 3-intine, 4-exine, 5 - vegetative cell
(b) 1-germpore, 2-generative cell, 3-exine, 4-intine, 5 - vegetative cell
(c) 1-intine, 2-exine, 3-germpore, 4-generative cell, 5 - vegetative cell
(d) 1-exine, 2-intine, 3-vegetative cell, 4-germpore, 5 - generative cell

85. Which plant will lose its economic value, if its fruits are produced by induced parthenocarpy?
(a) Grape (b) Pomegranate
(c) Orange (d) Banana

86. Match column-I with column- II and select the correct answer using the codes given below.

Column-I	Column-II
A. Ovary	I. Groundnut, mustard
B. Ovule	II. Guava, orange, mango
C. Wall of ovary	III. Pericarp
D. Fleshy fruits	IV. Seed
E. Dry fruits	V. Fruit

(a) A – V; B – IV; C – III; D – II; E – I
(b) A – I; B – II; C – III; D – IV; E – V
(c) A – I; B – III; C – II; D – IV; E – V
(d) A – V; B – IV; C – I; D – II; E – III

87. The given diagram shows 2 plants of the same species. Identify the types of pollination indicated as P_1, P_2 and P_3 respectively.

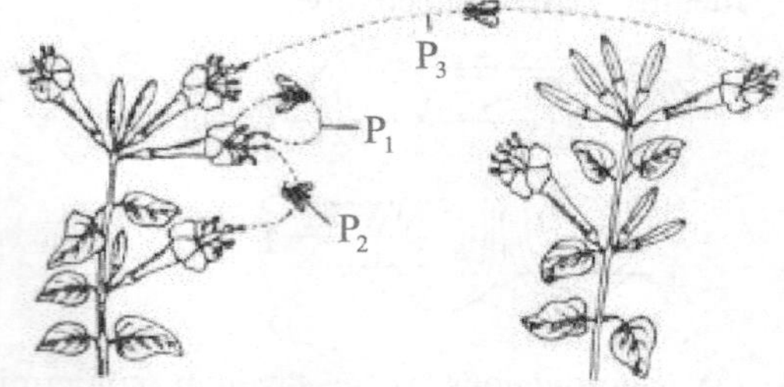

(a) Allogamy, Chasmogamy, Cleistogamy
(b) Autogamy, Xenogamy, Geitonogamy
(c) Autogamy, Geitonogamy, Xenogamy
(d) Geitonogamy, Allogamy and Autogamy

Answer KEYs

1	(c)	**10**	(b)	**19**	(b)	**28**	(b)	**37**	(c)	**46**	(b)	**55**	(c)	**64**	(b)	**73**	(a)	**82**	(c)
2	(d)	**11**	(a)	**20**	(d)	**29**	(d)	**38**	(c)	**47**	(c)	**56**	(c)	**65**	(b)	**74**	(a)	**83**	(c)
3	(a)	**12**	(c)	**21**	(a)	**30**	(d)	**39**	(b)	**48**	(a)	**57**	(c)	**66**	(d)	**75**	(d)	**84**	(d)
4	(c)	**13**	(c)	**22**	(d)	**31**	(a)	**40**	(b)	**49**	(a)	**58**	(b)	**67**	(c)	**76**	(b)	**85**	(b)
5	(c)	**14**	(a)	**23**	(c)	**32**	(d)	**41**	(b)	**50**	(a)	**59**	(a)	**68**	(a)	**77**	(b)	**86**	(a)
6	(c)	**15**	(a)	**24**	(b)	**33**	(a)	**42**	(b)	**51**	(c)	**60**	(b)	**69**	(b)	**78**	(c)	**87**	(c)
7	(d)	**16**	(b)	**25**	(b)	**34**	(b)	**43**	(d)	**52**	(b)	**61**	(b)	**70**	(b)	**79**	(d)		
8	(b)	**17**	(d)	**26**	(d)	**35**	(c)	**44**	(b)	**53**	(c)	**62**	(b)	**71**	(c)	**80**	(c)		
9	(c)	**18**	(b)	**27**	(c)	**36**	(a)	**45**	(b)	**54**	(d)	**63**	(c)	**72**	(c)	**81**	(d)		

Hints & Solutions

1. **(c)** Microspore is haploid, uninucleate, minute spore produced in large number as a result of meiosis in microspore mother cell inside the microsporangia. These are the first cell of gametophytic generation in angiosperms.

2. **(d)** Microspore is the beginning of male-gametophyte.

3. **(a)** The pollen mother cell is a diploid cell like the somatic cell.

4. **(c)**

5. **(c)** Integument is a diploid cell $2n = 2 \times 14 = 28$

Antipodal cells are haploid $n = 14$

Embryo is diploid $\therefore 2n = 2 \times 14 = 28$

Endosperm is triploid $\therefore 3n = 3 \times 14 = 42$

Nucellus is diploid $\therefore 2n = 2 \times 14 = 28$

6. **(c)** **7.** **(d)** **8.** **(b)** **9.** **(c)**

10. **(b)**

11. **(a)** Each microspore divide by mitotic division making a smaller generative cell and a larger vegetative cell or tube cell. If generative cell is damaged then the normal pollen tube will be formed because pollen tube is formed by vegetative cell, not by generative cell of microspore.

12. **(c)** Embryo sac is 7-celled structure. There is a large central cell with two polar nuclei, egg apparatus with egg cell and 2 synergids present at micropylar end and its chalazal end, 3 antipodal cells are present.

13. **(c)** Integument and nucellus develop from sporophytic tissue.

14. **(a)**

15. **(a)** Ubisch bodies provided by tapetum which help in thickening of exine.

16. **(b)** Generative cell is haploid and 2 male gametes are formed from it by mitotic division.

17. **(d)** **18.** **(b)**

19. **(b)** Anthers are internally tetralocular, i.e. they have two lobes. Each lobe contains two sporangia.

20. **(d)** The male gamete will be haploid (n). 2 polar nuclei will be diploid ($2n$). Endosperm formed by fusion of male gamete with two polar nuclei will be pentaploid.

$$\underset{n}{\text{Male gamete}} + \underset{(2n)+(2n)}{\text{2 polar nuclei}} \longrightarrow \underset{(5n)}{\text{Endosperm}}$$

21. **(a)** One or more specialised outer layers of endosperm of cereals are called aleurone layers.

22. (d) 23. (c)
24. (b) Seedling cannot emerge from this depth.
25. (b) 26. (d)
27. (c) Single cotyledon of grass embryo is called scutellum.
28. (b) 29. (d) 30. (d) 31. (a)
32. (d) 33. (a) 34. (b) 35. (c)
36. (a) 37. (c) 38. (c) 39. (b)
40. (b) 41. (b)
42. (b) (a) Nucellus-2n, antipodal cells-n
(b) Antipodal cells-n, egg cell-n
(c) Antipodal cells-n, megaspore mother cell-2n
(d) Nucellus-2n, primary endosperm nucleus-3n
Antipodal and egg cell are the product of meiotic division and rest are not.
43. (d) 44. (b) 45. (b) 46. (b)
47. (c) 48. (a)
49. (a) Crossing leads to hybrid vigour or superiority.
50. (a) Commensalism is interaction between two individuals in which one is benefitted while the other is unaffected. In mutualism, both individuals are beneficial for each other. The pollinating insect gets nectar from flower and in turn helps in pollination of flowers.
51. (c) 52. (b)
53. (c) **A-Orthotropous :** It is also called atropous. It is erect no bending is there, e.g., *Polygonum.*
B-Antropous : Completely inverted, e.g. *Helianthus.*
C-Hemitropous : The micropyle and chalaza line at 90° to funicle, e.g., *Ranunculus.*
D-Campylotropous : more curvature than hemitropous, e.g., *Capparis*, mustard.
E-Amphitropous : Horse shoe like, e.g., *Capsella.*
F-Circinotropous : Ovule straight micropyle upward due to unilateral growth of funicle it becomes inverted. e.g., *Opuntia.*
54. (d) 55. (c) 56. (c) 57. (c)
58. (b) 59. (a) 60. (b) 61. (b)
62. (b) 63. (c) 64. (b) 65. (b)
66. (d) 67. (c) 68. (a)
69. (b) After the meiosis-I and II, megaspore tetrad is formed where three out of four cells die from it, only one remain functioning called functional megaspore and in that three mitotic division give rise to 8 called stage.
70. (b) 71. (c) 72 (c)
73. (a) Transfer of pollen grains by birds is known as ornithophily. They usually have bright coloured flowers, tubular or cup shaped and having a large quantity of nectar.
74. (a)
75. (d) In cross self fertilised plants, the flowers are emasculated. The organ of the plant removed in this process is stamens or anthers. And this removal of anthers from this hermaphrodite flower is called emasculation.
76. (b) The fully matured 3-celled gametophyte is formed after pollination. It consists of two male gametes and a tube nucleus.
77. (b)
78. (c) Each microspore mother cell gives rise to 4 microspores which develop into pollen grains.
79. (d) 80. (c) 81. (d)
82. (c)
83. (c) In flowering plants, archesporium cells undergo periclinal (transverse) division to form outer primary parietal layer and inner sporogenous cells. Primary parietal wall after few more periclinal divisions form anther wall and sporogenous cells give rise to sporogenous tissue.
84. (d) Pollen grains contain the male gametophyte produced by meiosis of microspore mother cells that are located along the inner edge of the microsporangia. The outer part of the pollen is exine, which is composed of a complex polysaccharide, sporopollenin. Inner part is intine. The cell contains vegetative cell which develop into the pollen tube and germ pore and generative cell (degenerative) are also present.
85. (b) 86. (a) 87. (c)

25 Human Reproduction

1. Which of the following is a transporting tube leading from the bladder which brings urine outside the body *via* penis?

(a) Ureter (b) Epididymis
(c) Ejaculatory duct (d) Urethral meatus

2. Sperms produce an enzymatic substance for dissolving egg coverings. It is called
(a) Hyaluronic acid (b) Hyaluronidase
(c) Androgamone (d) Diastase

3. 10 oogonia yield 10 primary oocytes, then how many ova are produced on completion of oogenesis?

(a) 5 (b) 10 (c) 20 (d) 40

4. Ovulation or release of ovum occurs on day of menstrual cycle.
(a) 8-10 day
(b) 12-14 day
(c) 4-14 day
(d) Last two days

5. Given below is the diagrammatic sectional view of seminiferous tubule with their parts marked as 1, 2, 3, and 4. Select the option which shows the correct identification of the structure with its characteristics.

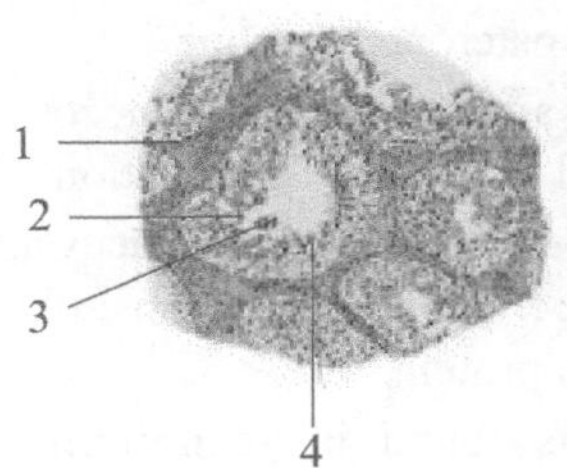

(a) 1: Spermatozoa, secretes testicular hormones that control spermatogenesis.
(b) 2: Spermatogonium, it is also called male germ cells which undergo meiotic division to from spermatozoa.
(c) 3: Interstitial cells, present in the interstitial spaces and store and transport the sperms from the testis to the outside through the urethra.
(d) 4: Sertoli cells, it maintains low temperature of the testis.

6. Select the option which correctly matches the endocrine gland with its hormone and its function.

	Endocrine gland	Hormone	Function
(a)	Ovary	FSH	stimulates follicular development and the secretion of oestrogens.
(b)	Placenta	Oestrogen	initiates secretion of the milk.
(c)	Corpus luteum	Oestrogen	essential for maintenance of endometrium
(d)	Leydig	Androgen	Initiates the cells for production of sperms.

7. Which one of the following is the most likely root cause for menstruation not taking place at regular cycling in human female?
(a) Maintenance of the hypertrophical endometrial lining.
(b) Maintenance of high concentration of sex hormones in the blood stream.
(c) Retention of well developed corpus luteum.
(d) Fertilisation of the ovum.

8. Which of the following statements about the mammalian blastocyst is **not true**?
   (a) The trophoblast gives rise to the embryo proper.
   (b) Maternal genes are expressed during cleavage.
   (c) The blastocyst implants in the mother's uterus.
   (d) Early mammalian development is slow.

9. Menstruation results in the discharge of:
   (a) the corpus luteum of the uterus.
   (b) surface cells from the vagina.
   (c) blood from the outer surface of the uterus.
   (d) the endometrial lining.

10. Which one of the following is the precise site of embryo implantation in a normal pregnancy?
   (a) Endometrium
   (b) Vagina
   (c) Oviduct
   (d) Cervix

11. Follicular atresia is:  Tricky
   (a) formation of corpus luteum.
   (b) degeneration of follicles till puberty.
   (c) formation of Graafian follicle.
   (d) increase in number of follicles.

12. The type of placenta found in human beings is:
   (a) Diffuse
   (b) Zonary
   (c) Cotyledonary
   (d) Discoidal

13. The role of the corpus luteum and the hormones it produces ends:
   (a) at the time of embryonic implantation in the uterus.
   (b) with the formation of the placenta.
   (c) during the second trimester of pregnancy.
   (d) just prior to parturition.

14. The given figure shows the male reproductive system. Some structures are marked as 1, 2, 3, and 4.
   Identify the structure whose removal will cause the sperm to be reacted with acidic urine in the urethra.
   (a) 1
   (b) 2
   (c) 3
   (d) 4

15. In the given columns, column-I contain structures of male reproductive system and column-II contains its feature. Select the correct match from the options given below.

	Column-I (Structure of male reproductive system)		Column-II (features)
A.	Seminiferous tubule	I.	Network of seminiferous tubule
B.	Rete testis	II.	Secondary sexual characters
C.	Leydig cells	III.	Meiosis and sperm formation occurs
D.	Prepuce	IV.	Place of implantation
		V.	Terminal skin of penis

   (a) A – I; B – II; C – III; D – V
   (b) A – III; B – I; C – II; D – V
   (c) A – III; B – I; C – IV; D – II
   (d) A – II; B – IV; C – III; D – V

16. Which of the following statements regarding sertoli cell is **correct**? Tricky
   (a) It is found in seminiferous tubule and secrete testosterone hormone.
   (b) It is a place where spermatozoa is concentrated and stored until ejaculation.
   (c) It secretes spermatozoa activating substances like fructose, citrate, inositol, prostaglandin and protein.
   (d) It is found in seminiferous tubule and function as nurse cells for differentiating spermatozoa.

17. Which one of the following statements is **incorrect** about menstruation?
   (a) During normal menstruation, about 40 ml blood is lost.
   (b) The menstrual fluid can easily clot.
   (c) At menopause in the female, there is especially abrupt increase in gonadotropic hormones.
   (d) The beginning of the cycle of menstruation is called menarche.

18. Which of the following best represents the normal path of a sperm cell as it makes its way from the point of entry into the female's reproductive tract to the place where fertilisation typically occurs?
   (a) Cervix, vagina, ovary, oviduct
   (b) Vagina, cervix, uterus, oviduct
   (c) Uterus, cervix, vagina, oviduct
   (d) Vagina, uterus, cervix, oviduct

19. "A" cells start division and enter in "B" stage of meiotic division and get temporarily "C" at this stage, called "D". Identify A, B, C and D.
   (a) A: Oogonia; B: Metaphase I; C: Arrested; D: Primary oocyte.
   (b) A: Oogonia; B: Anaphase I; C: Released; D: Secondary oocyte.
   (c) A: Oogonia; B: Prophase I; C: Arrested; D: Primary oocyte.
   (d) A: Oogonia; B: Telophase I; C: Released; D: Secondary oocyte.

20. Assume 2n =12 in a hypothetical species. In gametogenesis, the number of chromosomes in a spermatid would be ______ and in a first polar body would be ______.

   Toughnut

   (a) 3, 6 (b) 3, 3
   (c) 6, 3 (d) 6, 6

21. Which of the following groups of cells in the male gonad represent haploid cells?
   (a) Spermatogonial cells
   (b) Germinal epithelial cells
   (c) Secondary spermatocytes
   (d) Primary spermatocytes

22. The figure given below shows the structure of sperm. Identify the correct feature corresponding to the marked structure 1, 2, 3 and 4.

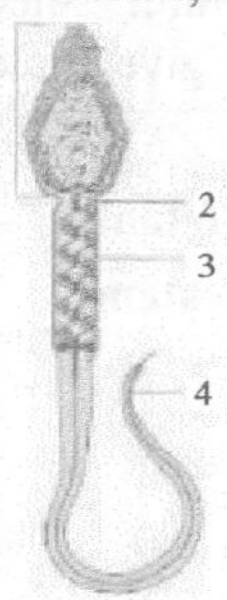

   (a) 1 – Head: Its anterior portion is covered by a structure filled with enzymes that help in the fusion of male and female gametes.
   (b) 2 – Middle piece: It contains a haploid nucleus.
   (c) 3 – Neck: It possesses few ribosomes which produces energy for the process of fertilisation.
   (d) 4 – Tail: It releases energy source for swimming of sperm.

23. Read the following statements (i) to (v) and answer the question.
   (i) Each testis has highly coiled 250 compartments called seminiferous tubules.
   (ii) Erection of the penis occurs due to presence of special tissues that facilitate insemination.
   (iii) Immunologically competent cells are also present in the interstitial spaces of seminiferous tubules.
   (iv) Testes lie outside the abdominal cavity in a thin pouch like skin called scrotum.
   (v) Bulbourethral gland is a single accessory gland.

   How many of the above statements are **incorrect**?
   (a) (i), (ii) and (iii)
   (b) (iii) and (v)
   (c) (i) and (v)
   (d) (ii), (iv) and (v)

24. In the given columns, column-I contain structures of female reproductive system and column-II contain its feature. Select the correct match from the option given below.

Column-I (Structures of female reproductive system)	Column-II (Features)
A. Ampulla	I. It undergoes cyclical changes during menstrual cycle.
B. Labia majora	II. It helps in collection of ovum after ovulation.
C. Oviduct	III. Wider part of fallopian tube where fusion of male and female gametes takes place.
D. Fimbriae	IV. Larger hairy folds which extend down from the mons pubis and surrounds the vaginal opening.
E. Endometrium	V. Also called fallopian tubes, which extend from the periphery of each ovary to the womb.

(a) A – I; B – II; C – III; D – V; E – IV
(b) A – III; B – I; C – II; D – V; E – IV
(c) A – III; B – IV; C – V; D – II; E – I
(d) A – II; B – IV; C – III; D – V; E – I

25. Milk secreted from the cells of alveoli of mammary lobes reaches nipple through lactiferous duct (L), mammary duct (M), mammary tubule (T) and mammary ampulla (A). Choose the option which shows the correct order.

(a) TMAL (b) MTLA
(c) MTAL (d) ATML

26. For approximately how long during the human female's menstrual cycle are progesterone concentrations high enough to maintain the uterus in a proper condition for pregnancy?

(a) All of the cycle
(b) None of the cycle
(c) During the first half of the cycle
(d) During the second half of the cycle

27. Shortest phase in the menstruation cycle of women is:

(a) Menses
(b) Luteal phase
(c) Ovulatory phase
(d) Follicular phase

28. Which gland releases a small amount of fluid just prior to ejaculation to decrease acidity in the urethra caused by urine?

Critical Thinking

(a) Prostate
(b) Glans penis
(c) Seminal vesicle
(d) Bulbourethral gland

29. The head of mature mammalian sperm is made of:

(a) An acrosome
(b) Elongated nucleus covered by acrosome
(c) Two centrioles and an axial filament
(d) Nucleus, acrosome, cytoplasm and mitochondrial sheath

30. A sac shaped like an upside down pear with a thick lining and muscles in the pelvic area where a fertilised egg or zygote comes to grow into a baby is called _______.

(a) Oviduct (b) Uterus
(c) Vagina (d) Vulva

31. Cleavage in the fertilised egg of humans:
(a) starts in uterus
(b) is meroblastic
(c) starts when egg is in fallopian tube
(d) is discoidal

32. Identical twins result when:
(a) the zygote divides and the daughter cells split and develop independently.
(b) two eggs with identical genetic material are fertilised, each by a different sperm.
(c) an embryo splits before cellular differentiation has occurred.
(d) Both (a) and (c)

33. Match column-I with column-II and select the correct answer using the codes given below.

	Column-I		Column-II
A.	Proliferative phase	I.	Testosterone
B.	Leydig cell	II.	Oestrogen
C.	Spermiogenesis	III.	Progesterone
D.	Secretory phase	IV.	Spermatid

(a) A – II; B – I; C – IV; D – III
(b) A – IV; B – II; C – III; D – I
(c) A – IV; B – III; C – II; D – I
(d) A – IV; B – III; C – I; D – II

34. Select the **correct** statements regarding oogenesis.
(i) It is initiated during the embryonic development stage when millions of oogonia are formed within each ovary.
(ii) Graafian follicle releases primary oocyte from the ovary by ovulation.
(iii) At puberty, only 60,000 – 80,000 primary follicles are left in each ovary.
(iv) Secondary oocyte within tertiary follicles grows in size and completes its second meiotic division.
(a) (i), (ii) and (iii)
(b) (i) and (iii)
(c) (ii) and (iv)
(d) All of the above

35. Given below is a diagrammatic sketch of a portion of human male reproductive system. Select the correct set of the names of the parts marked as 1, 2, 3, and 4 respectively.

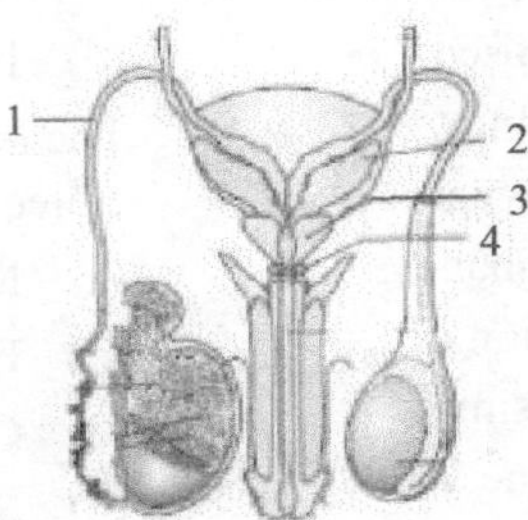

	1	2	3	4
(a)	Ureter	Seminal vesicle	Prostate	Bulbourethral gland
(b)	Ureter	Prostate	Seminal vesicle	Bulbourethral gland
(c)	Vas deferens	Seminal vesicle	Prostate	Bulbourethral gland
(d)	Vas deferens	Seminal vesicle	Bulbourethral gland	Prostate

36. In the given columns, column-I contain various phases of menstrual cycle and column-II contain its features. Select the correct match from the options given below.

Column-I (Phases of menstrual cycle)	Column-II (Features)
A. Menstrual phase	I. Breakdown of endometrial lining of uterus along with its blood vessels which form liquid that comes out of vagina.
B. Luteal phase	II. A temporary endocrine gland is formed and secretes a hormone which maintains endometrium and implantation of fertilised ovum and other events of pregnancy.
C. Follicular phase	III. Secretion of luteinising hormone at its maximum level and induces breakdown of mature follicle to release the female gamete.
D. Ovulatory phase	IV. Formation of mature Graafian follicle and regeneration of endometrium of uterus.

(a) A – IV; B – II; C – III; D – I
(b) A – III; B – I; C – II; D – IV
(c) A – III; B – I; C – IV; D – II
(d) A – I; B – II; C – IV; D – III

37. Keeping a normal, sexually mature woman in her mid-twenties on continuous high doses of progesterone will:
(a) cause her to ovulate repeatedly within each menstrual cycle.
(b) stimulate the production of mature follicles.
(c) prevent her from producing mature follicles and ovulating.
(d) cause the endometrial lining of her uterus to slough off.

38. Which of the following is the first change that occurs to the zygote after fertilisation?

(a) It divides to form a hollow ball of cells, called the blastocyst.
(b) It begins to secrete the hormones.
(c) It contacts the endometrial wall of the uterus and becomes buried inside it.
(d) It initiates the formation of a placenta.

39. The human embryo, with 8 to 16 blastomere is called:
(a) Morula
(b) Blastula
(c) Gastrula
(d) Foetus

40. In males, testes are contained in the scrotal sacs because:

Toughnut

(a) other organs do not make space for the testes in the abdominal cavity.
(b) testes in the abdomen will hamper maturation of sperms.
(c) it provides temperature that is slightly lower than body temperature required for formation of functional sperms.
(d) it facilitates ejaculation.

41. Ejaculation of human male contains about 200 – 300 million sperms, of which for normal fertility ____ % sperms must have normal shape and size and at least ____% must show energetic motility.

(a) 40, 60 (b) 50, 50
(c) 60, 40 (d) 30, 70

42. The given figure shows the diagrammatic sectional view of female reproductive system with few structures marked as 1, 2, 3, 4, 5 and 6.

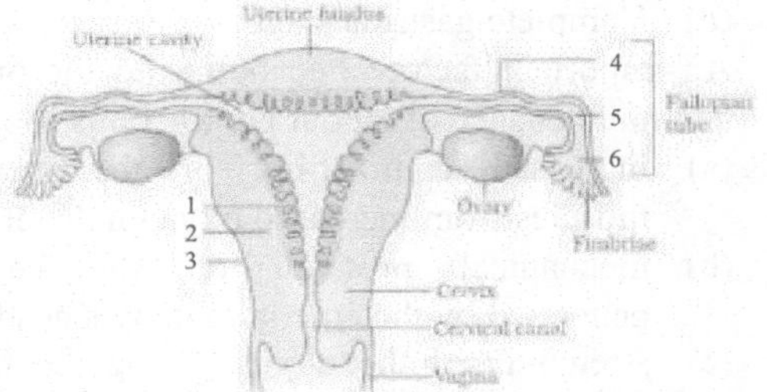

Which of the following options shows the correct labeling of A–F?

(a) 1 – Myometrium, 2 – Isthmus, 3 – Endometrium, 4 – Perimetrium, 5 – Ampulla, 6 – Infundibulum
(b) 1 – Infundibulum, 2 – Perimetrium, 3 – Endometrium, 4 – Myometrium, 5 – Ampulla, 6 – Isthmus
(c) 1 – Endometrium, 2 – Myometrium, 3 – Perimetrium, 4 – Isthmus, 5 – Ampulla, 6 – Infundibulum
(d) 1 – Perimetrium, 2 – Endometrium, 3 – Isthmus, 4 – Infundibulum, 5 – Ampulla, 6 – Myometrium

43. Match column-I with column-II and choose the correct option.

Column-I	Column-II
A. Hyaluronidase	I. Graafian follicle
B. Corpus luteum	II. Mammary gland
C. Colostrum	III. Progesterone
D. Antrum	IV. Acrosomal reaction

(a) A – II; B – I; C – IV; D – III
(b) A – IV; B – II; C – III; D – I
(c) A – IV; B – III; C – II; D – I
(d) A – IV; B – III; C – I; D – II

44. Hormones, secreted by human placenta are:

(i) Human chorionic gonadotropin (hCG)
(ii) Human placental lactogen (hPL)
(iii) Oestrogen and progesterone
(iv) Relaxin

(a) (i) and (ii)
(b) (i), (ii), (iii) and (iv)
(c) (iii) and (iv)
(d) (i), (ii) and (iii)

45. Polar body is produced during the formation of:

(a) Sperm
(b) Secondary oocyte
(c) Oogonium
(d) Spermatocytes

46. A couple have been trying to conceive a child for over a year without success. Recently, the man's doctor informed him that his seminal vesicles are defective. This means that:

(a) the man makes functional sperm but they are not delivered because he is unable to ejaculate.
(b) the man's sperm lack functional flagella
(c) the fluid needed to lubricate the man's urethra for easier sperm travel is defective.
(d) the man's semen lacks the fructose fuel needed for the sperm to swim.

47. Which of the following statement regarding human development is **correct**?

(a) Ovulation occurs at primary oocyte stage.
(b) Chorion develops from inner cell mass.
(c) Embryonic disc develops from trophoblast.
(d) Primitive streak is formed in the post part of the embryonic shield.

48. Which of the following is required for the increased production of oestrogen, progesterone, cortisol, prolactin and thyroxine etc. in the maternal blood?
(i) Metabolic changes in the mother.
(ii) Maintenance of pregnancy.
(iii) Supporting the foetal growth.
(iv) Destruction of Graafian follicle.
(a) (iii) and (iv)
(b) (i), (iv) and (v)
(c) (i), (ii) and (iii)
(d) All the four statements

49. All of the following statements concerning pregnancy are accurate except: **Tricky**
(a) the detection of human chorionic gonadotropin in the urine forms the basis for pregnancy tests.
(b) the cyclic release of pituitary gonadotropins and ovarian steroids is continued.
(c) the mammary gland tissue of the pregnant woman is stimulated to develop by placental hormones.
(d) the corpus luteum of pregnancy maintains the uterus until the placenta is well established.

50. Spermiogenesis/Spermatogenesis is the transformation of :
(a) spermatogonium into primary spermatocyte.
(b) spermatogonium into functional spermatozoa.
(c) primary spermatocytes into secondary spermatocytes.
(d) spermatids into spermatozoa.

51. Which of the following would you expect, if you compared the genetic makeup of an animal produced by parthenogenesis with that of its mother? **Tricky**
(a) About 100 percent genetic similarity
(b) About 50 percent genetic similarity
(c) No genetic similarity
(d) Parthenogenetic animals have no mother

52. By the end of how many weeks, major organ system are formed during the embryonic development? **Critical Thinking**
(a) 4 weeks (b) 8 weeks
(c) 12 weeks (d) 24 weeks

53. From which embryonic structure does vertebral column develop?
(a) Notochord
(b) Nerve cord
(c) Dermis
(d) Mesodermal somites

54. When a bird lays a fertilised egg, the embryo is in:
(a) Zygote stage
(b) Blastula stage
(c) Early gastrula stage
(d) Complete gastrula stage

55. An important factor contributing to the onset and maintenance of labour contractions is:
(a) increased contractility of the uterine muscles from increased oxytocin levels.
(b) mechanical pressure on the cervix generated by the emerging baby's head.
(c) stretching of the uterus during the final stages of foetal growth.
(d) All of the above

56. Which of the following organ is differentiated, first during development?
(a) Heart (b) Skin
(c) Brain (d) Neural tube

57. The figure given below shows the various events occurring during a menstrual cycle with few events marked as 1, 2, 3, 4 and 5. Which of the following options shows the correct events?

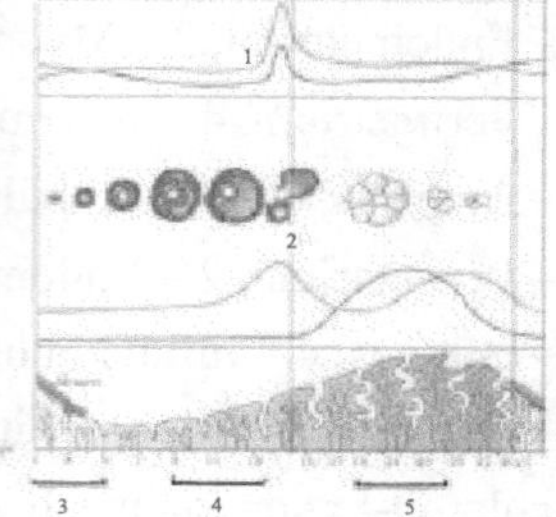

(a) 1 – LH, 2 – Ovulation, 3 – Menstruation, 4 – Proliferative phase, 5 – Luteal phase
(b) 1 – FSH, 2 – Implantation, 3 – Follicular phase, 4 – Menstruation phase, 5 – Luteal phase
(c) 1 – Estrogen, 2 – Parturition, 3 – Luteal phase, 4 – Follicular phase, 5 – Follicular phase
(d) 1 – Progesterone, 2 – Fertilisation, 3 – Menstruation phase, 4 – Secretory phase, 5 – Follicular phase

58. Which of the following human developmental stage becomes embedded in the uterine endometrium by a process called implantation and leads to pregnancy?

(a) (b)

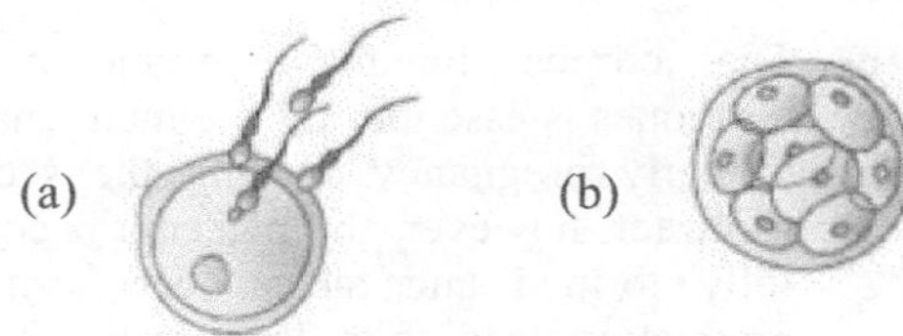

(c) (d)

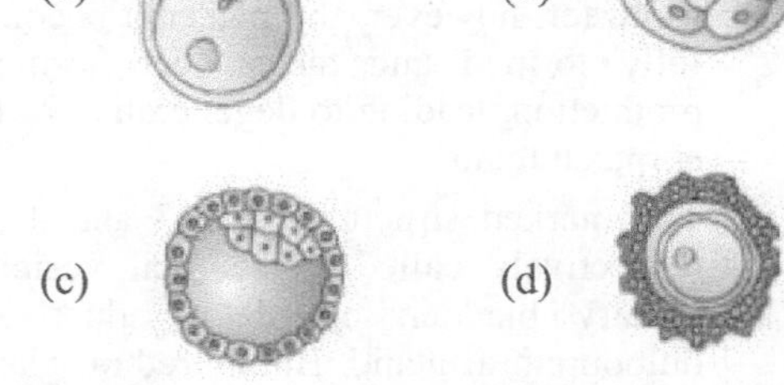

59. Which of the following statement regarding female reproductive system is (are) **correct**?

(i) Myometrium undergoes strong contraction at the time of delivery of baby.

(ii) Ovary is secondary female sex organ which produces female gamete and steroid hormones.

(iii) Ovarian stroma is divided into two zones: inner cortex and outer medulla.

(iv) Infundibulum possess finger like projections which help in collection of ovum after the release of secondary oocyte.

(v) A functional mammary gland is the characteristic of all the mammals (including male and female).

(a) (ii) and (iv) (b) (i), (ii), (iii) and (v)
(c) (iii), (iv) and (v) (d) All of the above

60. Select the correct match of terms given in column-I with their definition given in column-II.

	Column-I (Terms)		Column-II (Definition)
A.	Parturition	I.	Duration between pregnancy (of about 9 month) and birth.
B.	Ovulation	II.	Attachment of zygote to the endometrium.
C.	Gestation	III.	Childbirth
D.	Implantation	IV.	Stoppage of menstruation.
E.	Conception	V.	Release of egg from Graafian follicle.
		VI.	Process of milk secretion.
		VII.	Formation of zygote by fusion of the egg and sperm.

(a) A – I; B – II; C – VII; D – V; E – III
(b) A – III; B – I; C – IV; D – II; E – V
(c) A – III; B – V; C – I; D – II; E – VII
(d) A – III; B – V; C – I; D – IV, E – II

Answer KEYs

1	(d)	7	(d)	13	(c)	19	(c)	25	(a)	31	(c)	37	(c)	43	(c)	49	(b)	55	(d)
2	(b)	8	(a)	14	(d)	20	(d)	26	(d)	32	(d)	38	(a)	44	(b)	50	(d)	56	(c)
3	(b)	9	(d)	15	(b)	21	(c)	27	(c)	33	(a)	39	(a)	45	(b)	51	(a)	57	(a)
4	(b)	10	(a)	16	(d)	22	(a)	28	(d)	34	(b)	40	(c)	46	(d)	52	(c)	58	(c)
5	(b)	11	(b)	17	(b)	23	(c)	29	(b)	35	(c)	41	(c)	47	(d)	53	(a)	59	(a)
6	(d)	12	(d)	18	(b)	24	(c)	30	(b)	36	(d)	42	(c)	48	(c)	54	(c)	60	(c)

Hints & Solutions

1. **(d)** The urethra originates from the urinary bladder and extends through the penis to its external opening called urethral meatus.
2. **(b)** 3. **(b)** 4. **(b)**
5. **(b)** Spermatogonia are undifferentiated germ cells which originate in seminiferous tubules and divide into two primary spermatocytes (a kind of germ cell) in the production of spermatozoa.
6. **(d)** Leydig cells or interstitial cells, which are present in the intertubular spaces produces a group of hormones called androgens mainly testosterone. Androgens play a major stimulatory role in the process of spermatogenesis (formation of spermatozoa).
7. **(d)** Fertilisation of ovum is the most likely root cause for menstruation not taking place at regular cycling in human female. In order for fertilisation to occur, one of these sperm must attach itself to the ovum and penetrate its outer surface. The fertilised ovum will then continue travelling down the fallopian tube, taking several days until it reaches the uterus. When it arrives, it attaches itself to the lining of the uterus and continues to grow. If the ovum is not fertilised, it will die and the levels of progesterone and oestrogen hormone will drop. Without these hormones, the prepared uterus lining breaks up and is released through the vagina. This is known as a 'period'.
8. **(a)** In the mammalian blastocyst, the trophoblast forms the foetal part of the placenta. A disc-shaped portion of the inner cell mass surrounded by the trophoblast becomes the embryo.
9. **(d)** In menstrual cycles, the endometrium is shed from the uterus through the cervix and vagina.
10. **(a)** This is the capillary-rich inner wall of the uterus, where implantation is accomplished for the developing human embryo.
11. **(b)** 12. **(d)**
13. **(c)** The corpus luteum's production of hormones is essential for the maintenance of early pregnancy. During the second trimester, however, the placenta becomes fully formed and takes over hormone production, leading to degeneration of the corpus luteum.
14. **(d)** The marked structure 1, 2, 3 and 4 are respectively called as seminal vesicles, urinary bladder, ejaculatory duct and bulbourethral gland. Bulbourethral gland, also called Cowper's gland, is a pea sized gland that lies beneath the prostate gland and whose secretion neutralises the acidity of the urine in the urethra.
15. **(b)** Seminiferous tubules are located in the testes, which are the specific location of meiosis, and the subsequent formation of sperms. Rete testis is a network of small tubules found in the part of the testicle that carries sperm. Leydig cells or interstitial cells of Leydig, produce the hormone called testosterone which contributes to male secondary sexual characters. Prepuce is the foreskin, surrounding and protecting the head of the penis as well as clitoris.
16. **(d)**
17. **(b)** In human female, the periodic discharge of blood, mucus and cellular debris from uterine wall from non-pregnant women of sexual maturity is known as menstrual cycle. Few hours before the start of mensuration, the spiral arterioles constrict one by one resulting into the blanche mucosa. The mucosa shrinks and the death of the blood deprived tissues takes place.
18. **(b)** A sperm is ejected by the male into the vagina. From the vagina, the sperm move through the cervix into the uterus and finally the oviduct where fertilisation occurs.
19. **(c)**
20. **(d)** Haploid sex cells contain half as many chromosomes as diploid somatic cells.
21. **(c)**

22. (a) The mammalian sperm cell consists of a head, neck, middle piece and a tail. The head contains the elongated haploid nucleus with densely coiled chromatin fibres, surrounded anteriorly by an acrosome. Acrosome contains enzymes for penetrating the female egg. Neck contains centrioles (proximal centriole and distal centriole). Distal centriole gives rise to axial filament of the sperm which runs up to the end of the tail. The middle piece has a central filamentous core with many spirally arranged mitochondria which is used for ATP production. The tail or flagellum performs the beating movements that propel the spermatocyte.

23. (c) Each testes has about 250 compartments called testicular lobules. Each lobule contains one to three highly coiled seminiferous tubules. Bulbourethral glands are paired male accessory glands.

24. (c) 25. (a)

26. (d) High levels of progesterone are needed to maintain the uterus in the proper condition for pregnancy. The levels of progesterone are high only during the second half of the uterine cycle.

27. (c)

28. (d) Bulbourethral gland, also called as Cowper's gland, is one of two small exocrine glands in the male reproductive system. This gland releases a small amount of fluid just prior to ejaculation to decrease the acidity in the urethra. Bulbourethral gland are homologous to Bartholin's glands present in females.

29. (b)

30. (b) The uterus is a female reproductive organ located between the bladder and the rectum, in the pelvic area. The main purpose of the uterus is to nourish a foetus prior to birth. In menstruating females, the ovaries release eggs which travel *via* the fallopian tubes to the uterus.

31. (c)

32. (d) If an embryo splits before cell fate has been determined, identical twins may result.

33. (a)

34. (b) Graafian follicle releases secondary oocyte from the ovary by the process of ovulation. Primary oocyte within the tertiary follicle grows in size and completes its first meiotic division.

35. (c) 36. (d) 37. (c)

38. (a) The zygote divides mitotically to form 8, 16 daughter cells called blastomeres. This stage is called morula which continues to divide and transforms into blastocyst.

39. (a) 40. (c)

41. (c) Ejaculation of human male contains about 200 – 300 million sperms, of which for normal fertility 60 % sperms must have normal shape and size and at least 40 % must show energetic motility.

42. (c) In the given figure of female reproductive system, the marked structures (1 to 6) are the parts of uterus and fallopian tube. 1 to 6 are respectively endometrium, myometrium, perimetrium, isthmus, ampulla and infundibulum.

43. (c) 44. (b) 45. (b)

46. (d) Seminal vesicles produce the fructose sugars that power the sperm's mitochondria.

47. (d) Primitive streak extends maximum up to the middle of the disc.

48. (c) During pregnancy, the increased production of the levels of hormones like oestrogen, progesterone, cortisol, prolactin, thyroxine in the maternal blood are essential for supporting the foetal growth, metabolic changes in the mother and maintenance of pregnancy.

49. (b) The high levels of oestrogen and progesterone in the maternal circulation during pregnancy inhibit the cyclic release of pituitary gonadotropins and prevent the menstrual cycles. Pregnancy hormones include hCG, which maintains the corpus luteum of pregnancy and forms the basis for pregnancy diagnosis urine tests. Ovarian and placental oestrogen and progesterone, human placental lactogen, and pituitary prolactin stimulate development of the ducts and alveoli in the mammary glands.

50. (d)

51. (a) Species that exhibit parthenogenesis develops from unfertilised eggs produced by the mother. Therefore, the genetic make-up should be 100 percent same as the mother.

52. (c) By the end of 12 weeks (first trimester), most of the major organ systems are formed, for example, the limbs and external genital organs are well-developed.

53. (a) 54. (c)

55. (d) Labour contraction is brought on and intensified by all the given factors.

56. (c)

57. (a) 1 – LH, 2 – Ovulation, 3 – Menstruation, 4 – Proliferative, 5 – Luteal
LH is luteinising hormone secreted from the anterior pituitary gland. In females, an acute rise of LH (called LH surge) triggers ovulation and development of the corpus luteum. Ovulation is the process when a mature egg is released from the ovary, pushed down the fallopian tube, and is available for fertilisation. Menstruation is a woman's monthly bleeding. Menstrual blood flows from the uterus through the small opening in the cervix and passes out of the body through the vagina. Proliferative and luteal are phases of the menstrual cycle.

58. (c) Figure (c) refers to blastocysts stage. Blastocyst is embedded in the uterine endometrium by a process called implantation and leads to pregnancy. It possesses an inner cell mass (ICM) which subsequently forms the embryo. The outer layer of the blastocyst consists of cells collectively called the trophoblast. This layer surrounds the inner cell mass and a fluid-filled cavity known as the blastocoel. The trophoblast gives rise to the placenta.

59. (a) Statement (ii) and (iv) are correct. Ovary is the primary female sex organ which produces female gamete (ovum) and steroid hormones. Ovarian stroma is divided into peripheral cortex and inner medulla. Mammary gland is a paired structure that contains glandular tissue and variable amount of tissue. It is the characteristic of all female mammals.

60. (c) Parturition is the delivery of foetus or child birth. Ovulation is the release of egg from Graafian follicle on 14th day during menstrual cycle. Gestation is the duration of pregnancy and birth. Implantation is the attachment of the blastocysts to the endometrium layer of the uterus. Conception is the formation of zygote by fusion of the egg and sperm.

26 Reproductive Health

1. The process 'Amniocentesis' is the withdrawal of amniotic fluid during: **Critical Thinking**
   (a) Menopause (b) Lactation
   (c) Gestation (d) Pregnancy
2. The chemical method of contraception includes:
   (a) Jellies only
   (b) Creams and foams only
   (c) Oral contraceptives only
   (d) All of the above
3. Which one of the following is tested by the technique of amniocentesis? **Critical Thinking**
   (a) Biochemical abnormalities in the foetus.
   (b) Errors of metabolism in the foetus.
   (c) Chromosomal abnormalities in the foetus.
   (d) All of the above
4. Which of the following is a method for birth control?
   (a) IUDs (b) GIFT
   (c) HTF (d) IVF-ET
5. The accompanying diagram shows the uterine tubes of four women (P, Q, R and S).

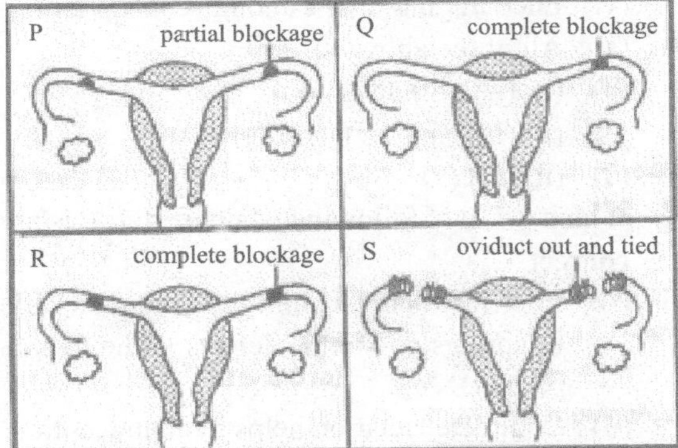

   In which of two women, fertilisation is impossible at present?
   (a) P and Q (b) Q and R
   (c) R and S (d) S and P
6. Identify the characteristics of IUDs.
   (i) Increase phagocytosis of sperms.
   (ii) Releases Cu ions that suppress sperm motility and fertilising capacity of sperms.
   (iii) Make the uterus unsuitable for implantation.
   (iv) Make the cervix hostile to sperms.
   (v) Prevent semen from entering the female reproductive tract.
   (a) (i), (ii), (iii), (iv) and (v)
   (b) (i), (ii), (iii) and (iv)
   (c) (i) and (ii)
   (d) (ii), (iii) and (iv)
7. Choose the **correct** statements given below, regarding contraception.
   (i) Medical Termination of Pregnancy (MTP) during first trimester is generally safe.
   (ii) Generally chances of conception are nil until mother breast-feeds the infant upto two years.
   (iii) Intrauterine devices like copper-T are effective contraceptives.
   (iv) Contraception pills may be taken upto one week after coitus to prevent conception.
   (a) (i) and (ii) (b) (ii) and (iii)
   (c) (iii) and (iv) (d) (i) and (iii)
8. The success of birth control programmes in controlling population growth is dependent on:
   (a) Use of contraceptives
   (b) Tubectomy
   (c) Vasectomy
   (d) Acceptability of the above by the people
9. World AIDS day is:
   (a) December 21 (b) December 1
   (c) November 1 (d) June 11
10. Hepatitis B is transmitted through:
    (a) Blood transfusion
    (b) Intimate physical contact
    (c) Sexual contact
    (d) All of the above

11. Emergency contraceptives are effective if used within:
(a) 72 hours of coitus
(b) 72 hours of ovulation
(c) 72 hours of menstruation
(d) 72 hours of implantation

12. Test-tube babies are produced in a process when:
(a) fertilising egg is removed from the body of the female and fertilised with the husband's sperm outside *in vitro* culture and then the zygote is transferred back to mother's uterus.
(b) external fertilisation and development *in vitro* culture till 32 cell stage before putting the embryo back into mother's uterus.
(c) complete development of a baby take place *in vitro*.
(d) development upto 32 cell stage of embryo takes place and its transplantation in the uterus of a surrogate mother occurs.

13. Intensely lactating mothers do not generally conceive due to the:
(a) suppression of gonadotropins.
(b) hypersecretion of gonadotropins.
(c) supression of gametic transport.
(d) supression of fertilisation.

14. Foetal sex can be determined by examining cells from the amniotic fluid by looking:
(a) Barr bodies (b) Autosomes
(c) Chiasmata (d) Kinetochore

15. Progesterone pill helps in preventing pregnancy by not allowing:
(a) Ova formation (b) Fertilisation
(c) Implantation (d) None of these

16. *In vitro* fertilisation is a technique that involves transfer of which one of the following into the fallopian tube? **Critical Thinking**
(a) Zygote only
(b) Embryo only, upto 8-cell stage.
(c) Either zygote or early embryo upto 8-cell stage.
(d) Embryo of 32-cell stage.

17. The given figure shows one of the elements releasing intrauterine device. Select the option which shows the **correct** identification of the device and its feature.

(a) CuT: Suppress sperm motility and its fertilising capacity.
(b) Cu7: Make uterus unsuitable for the attachment of blastocysts.
(c) Lippes loop: Protect the users from contracting AIDS and STDs.
(d) LNG - 20: Acts as spermicidal means and decrease the contraceptive efficiency.

18. Reproductive health in society can be improved by:
(i) Introduction of sex education in schools.
(ii) Increased medical assistance.
(iii) Awareness about contraception and STDs.
(iv) Equal opportunities to male and female child.
(v) Ban on aminocentesis.
(vi) Encouraging myths and misconceptions.
(a) All of the above
(b) (i), (ii), (iv) and (vi)
(c) (i), (ii), (iii), (iv) and (v)
(d) (ii) and (v)

19. Which of the following statement (s) is/are **correct**.
(i) Infertility is the inability to produce viable offsprings due to the defects in the female partner.
(ii) Complete lactation helps in contraception.
(iii) Spreading awareness can help in creating a reproductively healthy society.
(a) (ii) only (b) (i), (ii) and (iii)
(c) (ii) and (iii) (d) (i) and (iii)

20. Match column-I with column-II and select the correct answer using the codes given below.

Column-I	Column-II
A. Hepatitis B	I. Vitamin E
B. Saheli	II. 7'April, 1948
C. Normal functioning of reproductive organs	III. CDRI, Lucknow
D. World Health organisation	IV. Detection of antibody/ antigen
E. ELISA technique	V. Hepatitis B virus

(a) A – V; B – III; C – I; D – II; E – IV
(b) A – V; B – II; C – I; D – III; E – IV
(c) A – V; B – III; C – IV; D – II; E – I
(d) A – V; B – II; C – IV; D – III; E – I

21. Which of the following STDs are not completely curable?
(a) Chlamydiasis, gonorrhoea, trichomoniasis
(b) Chancroid, syphilis, genital warts
(c) AIDS, syphilis, hepatitis B
(d) AIDS, genital herpes, hepatitis B

22. Which of the following infections can also be transmitted by sharing of injection needles, surgical instruments, etc., with infected persons, transfusion of blood, or from an infected mother to the foetus too?
(a) Hepatitis B and HIV
(b) Genital herpes and HIV
(c) Syphilis and Hepatitis B
(d) Chlamydiasis and Trichomoniasis

23. 'Safe period', a means of birth control is based on:
(a) Ovulation which occurs on the 14th day of menstrual cycle.
(b) Ovum which survives for about 2 days.
(c) Sperms which remain alive for about 3 days.
(d) All of the above

24. Assisted reproductive technologies (ART) includes which of the following statement(s)?

Tricky

(a) Social awareness programmes to educate people about reproductive health and diseases.
(b) Research organisation working on to produce new and more effective contraceptives for birth control.
(c) A number of special techniques which assist infertile couples to have children.
(d) Both (b) and (c)

25. The stage transferred into the uterus after induced fertilisation of ova, in the laboratory is:

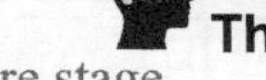

(a) Zygote
(b) Embryo at 4 blastomere stage
(c) Embryo at 2 blastomere stage
(d) Morula

26. The process, by which transfer of zygote or early embryo (up to 8 blastomeres) into fallopian tube takes places, is called ________.
(a) IVF and ET (b) ZIFT
(c) GIFT (d) IUT

27. Match the following figures (given as A, B, C, D and E) with their respective names given below as (i), (ii), (iii), (iv) and (v).
(i) Tubectomy (ii) Vasectomy
(iii) Implants (iv) Condoms
(v) Copper-T

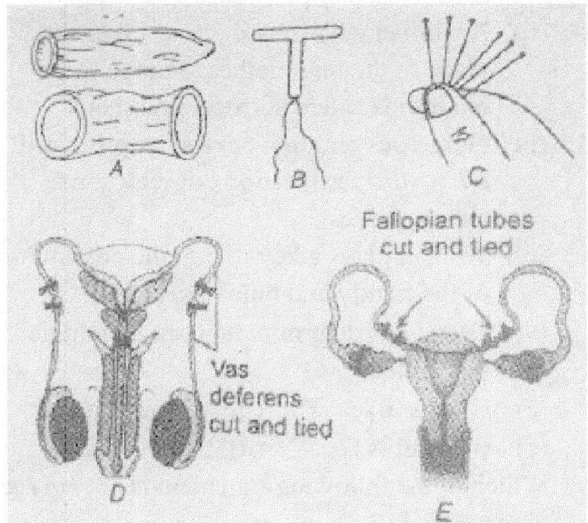

(a) A – (iv), B – (v), C – (iii), D – (ii), E – (i)
(b) A – (iv), B – (iii), C – (ii), D – (i), E – (v)
(c) A – (i), B – (ii), C – (iii), D – (iv), E – (v)
(d) A – (iii), B – (iv), C – (v), D – (i), E – (ii)

28. Which of the following statements are **correct** for an ideal contraceptive?
(i) It should be user-friendly.
(ii) It should be easily available.
(iii) It should be ineffective and reversible with least side effects.
(iv) It should be effective and reversible with least side effects.
(v) It should interfere with the sexual act of the user.
(a) All of these (b) (i), (ii) and (iii)
(c) (i), (ii) and (iv) (d) (i), (ii), (iv) and (v)

29. Select the correct match of the techniques given in column-I with its feature given in column-II.

Column-I		Column-II	
A.	IUT	I.	Copper-T
B.	Birth control	II.	Agent to kill spermatozoa
C.	Spermicides	III.	VD or RTI
D.	STD	IV.	To help infertile couple

(a) A – IV; B – III; C – I; D – II
(b) A – IV; B – III; C – II; D – I
(c) A – II; B – IV; C – I; D – III
(d) A – IV; B – I; C – II; D – III

30. Identify the **incorrect** statements and select the correct option.
(i) Birth control pills are likely to cause cardiovascular problems.
(ii) A woman who substitutes or takes the place of the real mother to nurse the embryo is called surrogate mother.
(iii) Numerous children have been produced by *in vitro* fertilisation but with some abnormalities.
(iv) Woman plays a key role in the continuity of the family and human species.
(v) Foetal sex determination test should not be banned.
(a) (i) and (ii) (b) (ii) and (iv)
(c) (iii) and (v) (d) None of these

31. Which of the following statement is not **correct** about syphilis?
(a) Primary stage of disease shows the appearance of chancre on genitalia and lips.
(b) In quaternary stage of disease, thick pus is discharged from urinogenital organ.
(c) Secondary stage of disease shows the appearance of multiple lesions on mucous membrane of lips, mouth and genitalia.
(d) Tertiary stage of disease shows the appearance of gumma on the skin.

32. Identify the **false** statement regarding the techniques of contraception.
(i) Tubectomy and vasectomy show many side effects.
(ii) Purpose of tubectomy is to prevent formation of ova.
(iii) The most important component of the oral contraceptive pills is progesterone.
(iv) Contraceptive oral pills help in birth control by preventing ovulation.
(a) (i), (ii) and (iii) (b) (i) and (ii)
(c) (iii), (iv) and (ii) (d) (iv) and (ii)

33. Non-gonococcal urethritis is caused by:
(a) *Neisseria gonorrhoeae*
(b) *Chlamydia trachomatis*
(c) *Treponema pallidum*
(d) *E. coli*

34. Recurrence of watery blisters on the genitalia and swollen lymph nodes in the groin are symptoms of: **Critical Thinking**
(a) Genital warts (b) Genital herpes
(c) Genital lesions (d) Genital hepatitis

35. In the given figure which marked number (1, 2, 3, 4) is tied and cut to block fertilisation.

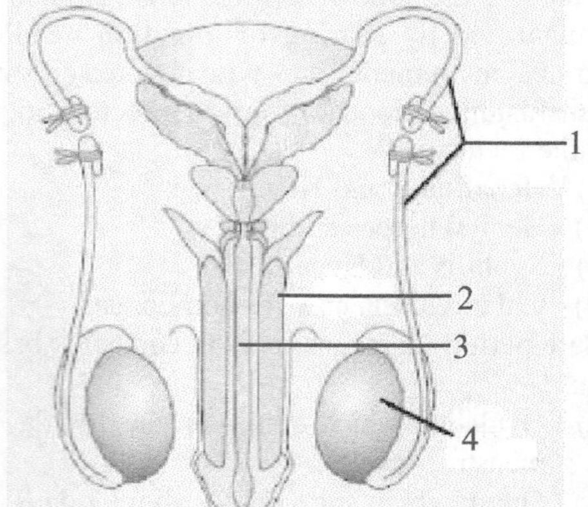

(a) 1 (b) 2 (c) 3 (d) 4

36. The most important component of the oral contraceptive pills is:
(a) Progesterone
(b) Growth hormone
(c) Thyroxine
(d) Luteinising hormone

37. The diaphragm, cervical cap and vaults are:
(a) Disposable contraceptive devices
(b) Reusable contraceptives
(c) IUDs
(d) Implants

38. Birth control pills check ovulation in female by inhibiting the secretion of:
(a) Follicle stimulating hormone (FSH)
(b) Luteinising hormone (LH)
(c) Both (a) and (b)
(d) None of the above

39. Which of the following ions plays an effective role in the activity of IUDs?
(a) Iron (b) Zinc
(c) Copper (d) Ammonium

40. Copper-T is a device that prevents:
(a) Implantation of blastocyst
(b) Ovulation
(c) Fertilisation
(d) Egg maturation

41. Contraceptive oral pills help in birth control by:
(a) killing the sperms in uterus.
(b) forming barriers between sperms and ova.
(c) preventing ovulation.
(d) killing the ova.

42. 'Saheli' is a new oral contraceptive, developed by:
(a) All Indian Institute of Medical Science
(b) Central Drug Research Institute
(c) Health Care Pvt. Ltd.
(d) Bharat Immunologicals & Biologicals Corp. Ltd.

43. Profuse, yellowish, greenish frothy smelling discharge from vagina is due to infection of:
(a) *Treponema pallidum*
(b) *Chlamydia*
(c) *Trichomonas vaginallis*
(d) *Neisseria*

44. Match column-I with column-II and select the correct answer using the codes given below.

Column–I (Method)	Column–II (Mode of Action)
A. The pill	I. Prevents sperms reaching cervix
B. Condom	II. Prevents implantation
C. Vasectomy	III. Prevents ovulation
D. Copper-T	IV. Vas deferens is removed or tied up.

(a) A – III; B – I; C – IV; D – II
(b) A – IV; B – I; C – II; D – III
(c) A – III; B – IV; C – I; D – II
(d) A – II; B – III; C – I; D – IV

45. Which of the following statement (s) is/are **correct**?
(i) According to the WHO, reproductive health is total well-being in the physical, social, emotional, behavioural aspects of reproduction.
(ii) According to the WHO, reproductive health is total well being in the physical, social and emotional aspects of reproduction.
(iii) A reproductively healthy society has people with physically and functionally normal reproductive organs.
(iv) Reproductively healthy societies have abnormal sex-related emotional and behavioural interactions.
(a) (i), (ii) and (iii) (b) (ii) and (iv)
(c) (i) and (iii) (d) (i) only

46. Which of the following statements are **correct** regarding "Lactational amenorrhoea"?
(i) It means absence of menstruation.
(ii) Ovulation does not occur during the lactational period.
(iii) Chances of failure of contraception are almost nil upto six months following parturition.
(iv) Side effects are almost nil.
(v) Contraceptive efficiency reduces after the period of intense lactation.
(vi) It is natural method of contraception.
(vii) It increases phagocytosis of sperms.
(a) (ii), (iii), (iv), (v) and (vi)
(b) All except (vii)
(c) (ii), (iii), (iv) and (v)
(d) All of these

47. Identify the **correct** match from the given option.
(a) Non Medicated IUDs - Multiload 375.
(b) Saheli - Contains a steroidal preparation.
(c) Lactational amenorrhea - Presence of menstruation.
(d) Diaphragms, cervical caps, vaults - Cover the cervix after the coitus.

48. Identify the **correct** statements regarding diaphragm as contraceptive device from the given options.

(i) They act as physical barrier for sperm entry.
(ii) They are placed to cover the cervical region during coitus.
(iii) They are introduced into the uterus to prevent implantation.
(iv) They act as spermicidal agents to increase their contraceptive effectiveness.

(a) (i), (ii) only (b) (i), (iv) only
(c) (ii), (iii) only (d) (iii), (iv) only

49. Read the following given statements and mark the option that are **correct** statements.

(i) MTP was legalised in 1971.
(ii) Inability to conceive or produce children even after 2 years of protected sexual cohabitation is called infertility.
(iii) Surgical method of contraception prevents gamete formation.
(iv) MTPs are relatively safe up to 3 weeks of pregnancy.

(a) (i) and (iv) (b) (ii) and (iii)
(c) (iii) and (iv) (d) (i) and (iii)

Answer KEYs

1	(d)	**6**	(b)	**11**	(a)	**16**	(c)	**21**	(d)	**26**	(b)	**31**	(b)	**36**	(a)	**41**	(c)	**46**	(b)
2	(d)	**7**	(d)	**12**	(b)	**17**	(a)	**22**	(a)	**27**	(a)	**32**	(b)	**37**	(b)	**42**	(b)	**47**	(b)
3	(c)	**8**	(d)	**13**	(a)	**18**	(c)	**23**	(d)	**28**	(c)	**33**	(b)	**38**	(c)	**43**	(c)	**48**	(a)
4	(a)	**9**	(b)	**14**	(a)	**19**	(c)	**24**	(c)	**29**	(d)	**34**	(b)	**39**	(c)	**44**	(a)	**49**	(a)
5	(c)	**10**	(d)	**15**	(a)	**20**	(a)	**25**	(d)	**30**	(c)	**35**	(a)	**40**	(a)	**45**	(c)		

Hints & Solutions

1. (d) Amniotic fluid is secreted by amnion of foetus during pregnancy.
2. (d) 3. (c)
4. (a) Intra utrine device (Copper-T) inhibit fertilisation and help in birth control.
5. (c) 6. (b)
7. (d) Contraceptive pills have to be taken daily for a period of 21 days starting preferably within the first five days of menstrual cycle. It helps in the prevention of conception. As long as the mother breast-feeds the child fully, chances of conception are almost nil only up to a maximum period of six months following parturition.
8. (d) 9. (b) 10. (d)
11. (a) Emergency contraceptives are effective if used within 72 hours of coitus.
12. (b) 13. (a) 14. (a) 15. (a)
16. (c)
17. (a) The given figure is that of copper CuT. CuT is a simple copper releasing IUD made of a flexible, "T" shaped piece of plastic wrapped with a thin copper containing wire. It helps the uterus and fallopian tubes to produce fluid that kills sperm. This fluid contains white blood cells, copper ions, enzymes, and prostaglandins. Copper ions prevent pregnancy by inhibiting the movement of sperm, because the copper-ion containing fluids are directly toxic to sperm. Even if an aggressive little spermatozoan fertilises an egg, the copper ion laden environment prevents implantation of the fertilised egg, and thus pregnancy.
18. (c) Reproductive health in society can be improved by creating awareness among people about various reproduction related aspects and providing facilities and support for building up a reproductively healthy society.
19. (c) Infertility is the inability to produce viable offspring due to the defects in female or male partner.
20. (a)
21. (d) AIDS, genital herpes and hepatitis B are sexually transmitted diseases which are not completely curable.
22. (a) Hepatitis B is an infectious disease caused by the hepatitis B virus (HBV) which affects the liver. HIV (human immunodeficiency virus) is a virus that attacks the immune system, the body's natural defense system. Without a strong immune system, the body has trouble fighting off disease. Both the diseases can also be transmitted by sharing of injection needles, surgical instruments, etc., with infected persons, transfusion of blood, or from an infected mother to the foetus too.
23. (d)
24. (c) Assisted reproductive technologies (ART) include a number of special techniques which assist infertile couples to have children. *In-vitro* fertilisation is one such technique.
25. (d) In Intra-Uterine Transfer (IUT), embryo with more than 8-blastomeres stage (morula) is used for transfer into the uterus.
26. (b) ZIFT (Zygote intra fallopian transfer) is the process in which the zygote or early embryos (with upto 8 blastomeres) is transferred into the fallopian tube.
27. (a) A-Condoms, B-Copper-T, C-Implants, D-Vasectomy, E-Tubectomy.
28. (c) An ideal contraceptive should be user-friendly, easily available, effective and reversible with no or least side-effects.
29. (d) 30. (c) 31. (b)
32. (b) There are little or no side effects of tubectomy and vasectomy. The purpose of tubectomy and vasectomy is to block the passage of ova and sperm respectively.

33. (b) 34. (b)

35. (a) The given figure is showing the procedure of vasectomy, in which a small part of the vas deferens is removed or tied up through a small incision on the scrotum.

36. (a) 37. (b) 38. (c)

39. (c) Copper ions play an important and effective role in the activity of IUDs. Copper IUDs primarily work by disrupting sperm mobility and damaging sperm so that they are prevented to take part in the process of fertilisation. The increased copper ions in the cervical mucus inhibit the sperm's motility and viability, preventing sperm from traveling through the cervical mucus or destroying it as passes through.

40. (a) Copper 'T' is an intrauterine device which prevents the fertilised egg becoming implanted in the wall of the womb.

41. (c)

42. (b) 'Saheli'–an oral contraceptive for the females was developed at Central Drug Research Institute (CDRI) in Lucknow, India.

43. (c)

44. (a)

A.	The pill	—	Prevents ovulation
B.	Condom	—	Prevents sperm reaching cervix
C.	Vasectomy	—	vas defers is removed or tied up.
D.	Copper-T	—	Prevent implantation

45. (c) Reproductive health is a crucial part of general health and a central feature of human development. It is a reflection of health during childhood, and crucial during adolescence and adulthood, Reproductive healthy societies have normal sex related emotional and behavioural interactions.

46. (b)

47. (b) Saheli is a non- hormonal birth control pill. It is different from other birth control pills because it doesn't contain any hormones. Instead of using the hormone oestrogen to prevent pregnancy, it contains a drug that blocks oestrogen.

Non Medicated IUDs - Lippes loop

Lactational amenorrhea - Absence of menstruation

Diaphragms, cervical caps, vaults - Cover the cervix during the coitus.

48. (a) The diaphragm is one of a barrier method of contraception. It is a shallow, dome-shaped cup with a flexible rim and made of silicon. It is inserted into the female reproductive tract to cover the cervix. To be effective in preventing pregnancy, diaphragms need to be used in combination with spermicide, which is a chemical that kills sperm.

49. (a)

27 Principles of Inheritance and Variation

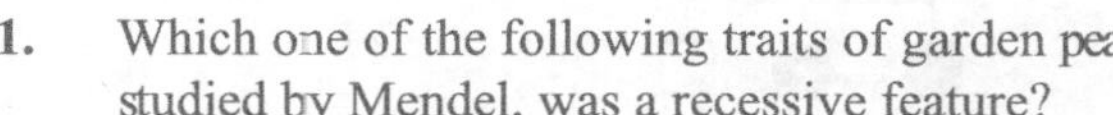

1. Which one of the following traits of garden pea studied by Mendel, was a recessive feature?
   (a) Round seed shape
   (b) Axial flower position
   (c) Green seed colour
   (d) Green pod colour

2. In a certain plant, red fruit (R) is dominant over yellow fruit (r) and tallness (T) is dominant over shortness (t). If a plant with RRTt genotype is crossed with a plant rrtt genotype, what will be the percentage of tall plants with red fruits in the progeny?

   Tricky

   (a) 50% (b) 100% (c) 75% (d) 25%

3. Suppose that in sheep, a dominant allele (B) produces black hair and a recessive allele (b) produces white hair. If you saw a black sheep, you would be able to identify:
   (a) Its phenotype for hair colour.
   (b) Its genotype for hair colour.
   (c) The genotypes for only one of its parents.
   (d) The genotypes for both of its parents.

4. Mendel's rules do not correctly predict patterns of inheritance for tightly linked genes or the inheritance of alleles that show incomplete dominance or epistasis. Does this mean that his hypothesis are **incorrect**?
   (a) Yes, because they are relevant to only a small number of organisms and traits.
   (b) Yes, because not all data support his hypothesis.
   (c) No, because he was not aware of meiosis or the chromosome theory of inheritance.
   (d) No, it just means that his hypothesis are limited to certain conditions.

5. In the F_2 generation of a dihybrid cross between yellow, round seeded and green, wrinkled seeded pea plants, 17 out of 254 pea seeds were green and wrinkled. Other seeds were: Yellow and round; green and round; yellow and wrinkled. These results indicate which of the following statement?
   (a) Crossing over has occurred.
   (b) Green and wrinkled are both recessive characters.
   (c) The alleles for green and wrinkled are linked.
   (d) The allele for green is recessive but not the allele for wrinkled.

6. The F_2 generation of a cross produced identical phenotypic and genotypic ratio. It is not an expected Mendelian result, and can be attributed to:
   (a) Independent assortment
   (b) Linkage
   (c) Incomplete dominance
   (d) None of the above

7. The total number of progeny obtained through dihybrid cross of Mendel is 1280 in F_2 generation. How many of them are recombinants?

   Tricky

   (a) 240 (b) 360
   (c) 480 (d) 720

8. Which of the following is not a hereditary disease?
   (a) Cystic fibrosis (b) Thalassemia
   (c) Haemophilia (d) Cretinism

9. A women with 47 chromosomes due to three copies of chromosome 21 is characterised by:
   (a) Superfemaleness
   (b) Triploidy
   (c) Turner's syndrome
   (d) Down's syndrome

10. A modified dihybrid mendelian ratio of 9 : 3 : 4 indicates:
    (a) Supplementary genes
    (b) Complementary genes
    (c) Lethal genes
    (d) Epistatic genes

11. In multiple allele system, a gamete possesses:

    (a) Two alleles (b) Three alleles
    (c) One allele (d) Several alleles

12. Which of the following parental phenotypes would produce offspring with blood group phenotypes in the expected ratio of 1 type A : 1 type B?

	Blood group of mother	Blood group of father
(a)	A	B
(b)	AB	AB
(c)	AB	B
(d)	AB	O

13. Which mendelian principle will not operate if two genes under study are close together? **Tricky**
(a) Paired unit factors
(b) Dominace
(c) Segregation
(d) Independent assortment

14. The person with Turner's syndrome has:
(a) 45 autosomes and X sex chromosome
(b) 44 autosomes and XYY sex chromosomes
(c) 45 autosomes and XYY sex chromosomes
(d) 44 autosomes and X sex chromosome

15. Mongolism, Edward's syndrome and Patau's syndrome are due to:
(a) Autosomal abnormalities
(b) Allosomal abnormalities
(c) Both (a) and (b)
(d) None of the above

16. Albinism and phenylketonuria are disorders due to:
(a) Recessive autosomal genes
(b) Dominant autosomal genes
(c) Recessive sex genes
(d) Dominant sex genes

17. In Klinefelter's syndrome, the sex chromosome compliment is:
(a) 22AA + XXY (b) 22AA + XO
(c) 22AA + XY (d) 22AA + XX

18. Which of the following is related to heart diseases?
(a) Turner's syndrome
(b) Edward's syndrome
(c) Cushing's syndrome
(d) Patau's syndrome

19. In *Drosophila* XXY represents a female but in human it is an abnormal male. It shows that:
(a) Y-chromosome is essential for male sex in human.
(b) Y-chromosome is essential for female sex determination in *Drosophila*.
(c) Y-chromosome is not essential for male sex determination in human.
(d) All of the above

20. Identify the type of inheritance shown in the given figure.

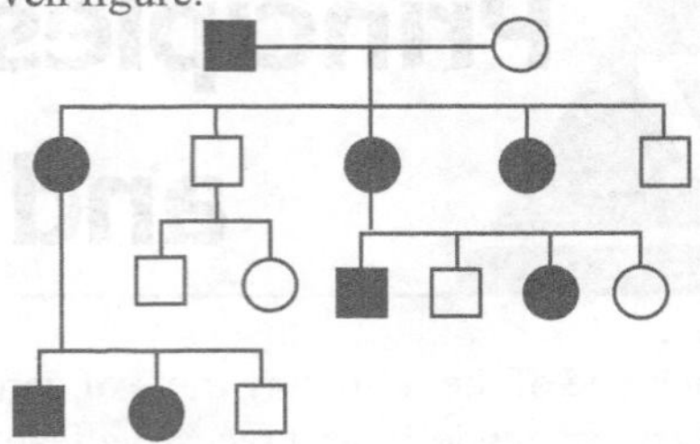

(a) Dominant X-linked
(b) Recessive X-linked
(c) Dominant Y-linked
(d) Recessive Y-linked

21. Given below is a highly simplified representation of the human sex chromosomes from a karyotype. The gene *a* and *b* could be of:

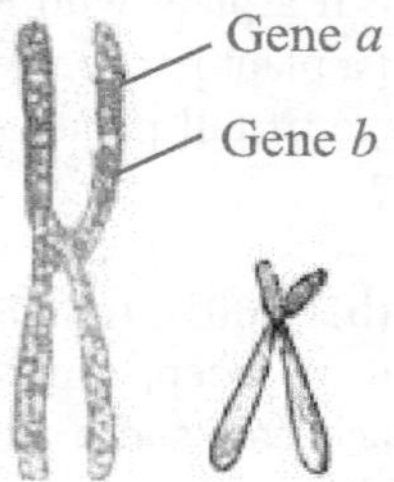

(a) colour blindness and body height.
(b) attached ear lobe and Rhesus blood group.
(c) haemophilia and red-green colour blindness.
(d) phenylketonuria and haemophilia.

22. Mary's father has haemophilia but her husband is also haemophilic. What would be the chance of her daughter having this disease?

(a) 0% (b) 50% (c) 100% (d) 25%

23. Match the terms given in column-I with their description given in column-II and choose the correct option.

Column-I	Column-II
A. ABO blood groups	I. Dihybrid cross
B. Law of segregation	II. Monohybrid cross
C. Law of Independent assortment	III. Base pairs substitution
D. Gene mutation	IV. Multiple allelism

(a) A – II; B – I; C – IV; D – III
(b) A – IV; B – I; C – II; D – III
(c) A – IV; B – II; C – I; D – III
(d) A – II; B – III; C – IV; D – I

24. Which of the following statements regarding crossing-over is **incorrect**? Critical Thinking
(a) Crossing-over disrupts the linkage between genes on the same chromosome.
(b) Crossing-over disrupts the linkage between genes on different chromosomes.
(c) Crossing-over produces new genetic combinations.
(d) Crossing-over produces non-parental chromosomes.

25. *Drosophila* flies with XXY genotype are females but in case of humans, such individuals are abnormal males (Klinefelter's syndrome). This indicates that:
(a) The Y chromosome has no role in sex determination.
(b) In *Drosophila*, the Y chromosome is essential for sex determination.
(c) The Y chromosome is male determining in humans.
(d) The Y chromosome is female determining in *Drosophila*.

26. Christmas disease is another name of:
(a) Sleeping sickness (b) Haemophilia
(c) Hepatitis B (d) Down's syndrome

27. Barr-body (seen in saliva test in Olympic games) is associated with:
(a) Y-chromosome (b) X-chromosome
(c) Autosome (d) Male sex only

28. The males of grasshoppers and bugs possess two sets of autosomes and:
(a) Only Y-chromosome
(b) Only X-chromosome
(c) X and Y-chromosome
(d) Neither X nor Y-chromosome

29. Which of the following match is **correct**?
(a) Sickel cell anaemia – X-chromosome
(b) Down Syndrome – 21 chromosome
(c) Parkinsonism – XY chromosome
(d) Haemophilia – Y chromosome

30. There are three genes a, b, c. Percentage of crossing over between a and b is 20%, b and c is 28% an a and c is 8%. What is the sequence of genes on chromosome? Tricky
(a) b, a, c (b) a, b, c
(c) a, c, b (d) None of these

31. Human skin colour is controlled by several gene pairs. Let us assume here that there are just three gene pairs on different chromosomes and that for each pair there are two alleles, an incompletely dominant one that codes for melanin deposition and an incompletely recessive one that codes for no melanin deposition. If a very dark skinned person marries a very light skinned woman, what will be the chance that their offspring will have very dark skin? Toughnut
(a) 1/64 (b) 1/4 (c) 5/8 (d) 9/64

32. Rh factor is responsible for which of the following disease?
(a) AIDS
(b) Turner's syndrome
(c) Erythroblastosis foetalis
(d) Sickle -cell anaemia

33. Repeated selfing produces:
(a) Heterozygosity
(b) Homozygosity
(c) Homozygosity in some and heterozygosity in other traits
(d) Pure hybrids

34. In genetics, the term test cross means:
(a) the crossing of F_1 individual with homozygous recessive.
(b) crossing an F_1 individual with either of the two parents.
(c) crossing F_1 individual with another F_1 individual.
(d) crossing F_1 individual with that of F_2.

35. In Mendelian dihybrid cross, when heterozygous round yellow are self crossed, round green offsprings are produced which are represented by genotype?
(a) RrYy, RrYY, RRYy
(b) RrYY, RRyy, rryy
(c) rrYy, rrYY
(d) RRyy, Rryy

36. Which of the following illustrates pleiotropy?
(a) In fruit flies, the genes for scarlet eyes and hairy body are located on the same chromosome.
(b) Matings between earless sheep and long-eared sheep always result in short-eared offspring.
(c) Wheat kernels can range from white to red in color, a trait controlled by several genes.
(d) The human cystic fibrosis gene causes many symptoms, from respiratory distress to digestive problems.

37. A pea plant with purple flowers is heterozygous for flower colour. Its genotype is Pp. The P and p alleles in the pea plant's cells are located:
(a) next to each other on the same chromosome.
(b) at the same location on homologous chromosomes.
(c) on the X and Y chromosomes.
(d) some distance apart on the same chromosome.

38. Consider a gene that has two alleles and shows complete dominance. When two heterozygotes for this gene breed, they have a 25% chance of producing a homozygous recessive offspring. The next time they breed, what are the chances that they will once again have a homozygous recessive progeny? **Toughnut**
(a) 0% (b) 25% (c) 75% (d) 100%

39. When dominant and recessive alleles express itself together it is called ________.
(a) codominance (b) dominance
(c) amphidominace (d) pseudodominance

40. The genes controlling the seven pea characters studied by Mendel are now known to be located on how many different chromosomes?
(a) Four (b) Seven
(c) Six (d) Five

41. Barr body is found in:
(a) Male somatic cells
(b) Male germinal cells
(c) Female somatic cells
(d) Female germinal cells

42. Which of the following chromosome arrangement is symbolised by $2n + 2$?
(a) Tetraploid (b) Nullisomic
(c) Trisomic (d) Tetrasomic

43. Match the column-I with column-II and choose the correct option.

Column-I	Column-II
A. Non-parental gene exchange	I. Crossing over
B. Non-sister chromatids	II. X and Y
C. Sex chromosome	III. Autosome-linked disease
D. Thalassaemia	IV. Recombination

(a) A – IV; B – I; C – II; D – III
(b) A – II; B – I; C – IV; D – III
(c) A – II; B – IV; C – III; D – I
(d) A – II; B – IV; C – I ; D – III

44. Human skin colour is the example of:
(i) Multiple gene inheritance.
(ii) Three separate genes control this trait.
(iii) Single gene controls this trait.
(iv) Two genes control this trait.
(v) A significant role played by environment.
Which of the following statements are correct?
(a) (i), (ii) and (iii)
(b) (i), (ii) and (v)
(c) (iii), (iv) and (v)
(d) (ii), (iii) and (iv)

45. The experiment shown in the figure has been carried out by Morgan to show the phenomenon of linkage and recombination. If in Cross I, genes are tightly linked and in Cross II, genes are loosely linked then what will be the percentage of recombinants produced in Cross I and Cross II respectively?

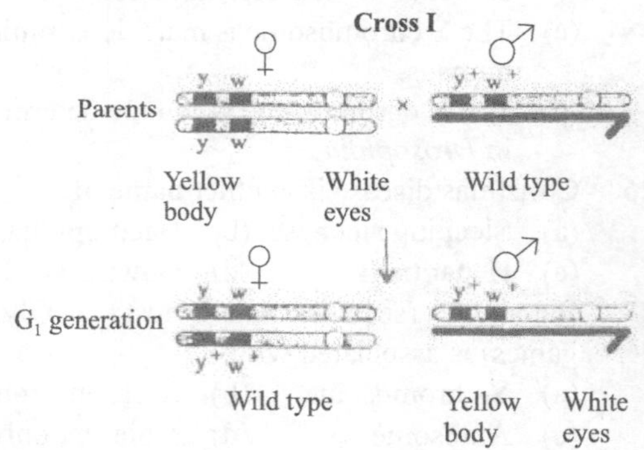

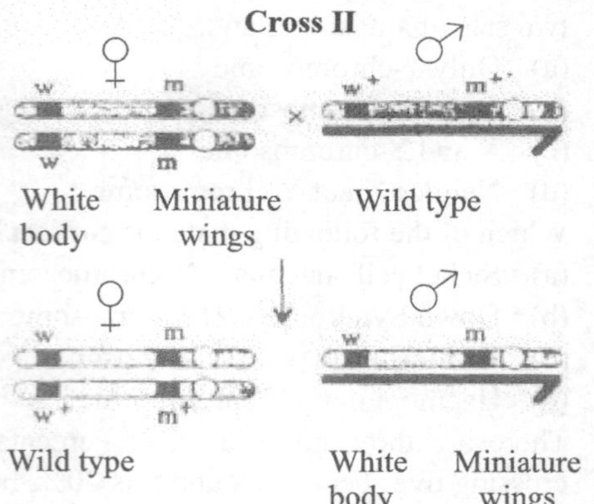

(a) 98.7% and 62.8%
(b) 1.3% and 37.2%
(c) 37.2% and 1.3%
(d) 62.8% and 98.7%

46. Match the symbols used in human pedigree analysis given in column-I with their name given in column-II and choose the correct option.

Column-I	Column-II
A.	I. Consanguineous mating
B.	II. Normal female
C.	III. Mating
D.	IV. Affected female
E.	V. Parents with male child affected with disease
	VI. Sex unspecified

(a) A – III; B – I; C – II; D – V; E – IV
(b) A – II; B – I; C – VI; D – III; E – IV
(c) A – III; B – IV; C – I; D – V; E – II
(d) A – III; B – I; C – VI; D – V; E – IV

47. Sickle cell anaemia is:
(a) caused by substitution of valine by glutamic acid in the beta globin chain of haemoglobin.
(b) caused by a change in a single base pair of DNA.
(c) characterised by elongated sickle like RBCs with a nucleus.
(d) an autosomal linked dominant trait.

48. The person affected by Turner's syndrome has:
(a) 45 autosomes and X sex chromosome
(b) 44 autosomes and XYY sex chromosomes
(c) 45 autosomes and XYY sex chromosomes
(d) 44 autosomes and X sex chromosome

49. In *Drosophila*, the sex is determined by:
(a) whether the egg is fertilised or develops parthenogenetically.
(b) the ratio of number of X-chromosomes to the sets of autosomes.
(c) X and Y chromosomes.
(d) the ratio of pairs of X-chromosomes to the pairs of autosomes.

50. Alkaptonuria is:
(a) Sex linked dominant disorder
(b) Sex linked recessive disorder
(c) Autosomal recessive disorder
(d) Autosomal dominant disorder

51. A normal woman is married to a man having hypertrichosis. They got one daughter and one son. What would be the possibility of their daughter to show hypertrichosis condition? **Tricky**

(a) 0% (b) 25% (c) 50% (d) 75%

52. A pregnant woman who has done amniocentesis test finds an extra Barr body in her embryo. The syndrome which is likely to be associated with the embryo is: **Critical Thinking**
(a) Edward's syndrome
(b) Down's syndrome
(c) Klinefelter's syndrome
(d) Patau's syndrome

53. The linkage map of X-chromosome of fruit fly has 66 units, with yellow body gene (y) at one end and bobbed hair (b) gene at the other end. The recombination frequency between these two genes (y and b) should be: **Critical Thinking**

(a) 100% (b) 66%
(c) > 50% (d) ≤ 50%

54. In Shepherd's Purse, the fruit shape is controlled by:
(a) Supplementary genes
(b) Complementary genes
(c) Duplicate genes
(d) Polymeric genes

55. A diseased man marries a normal woman. They have three daughters and five sons. All the daughters were diseased and sons were normal. The gene of this disease is: **Critical Thinking**
(a) Sex-linked dominant
(b) Sex-linked recessive
(c) Sex-linked character
(d) Autosomal dominant

56. One chromosome of a homologous pair carries the genes J and K. The other chromosome of the pair carries the genes j and k at corresponding loci. Crossing over results in exchange of chromosome segments and production of gametes with new combinations of genes. A recombinant type gamete resulting from this crossover might contain: **Critical Thinking**

(a) Genes J and K (b) Genes j and K
(c) Genes J and j (d) Genes j and k

57. Appearance of hidden character in some progeny of F_2 population indicates:
(a) Law of purity of gametes
(b) Law of independent assortment
(c) Law of dominance
(d) None of the above

58. Punnett square is used to know:
(a) outcome of a cross
(b) probable result of a cross
(c) types of gametes
(d) result of meiosis

59. In order to find out the different types of gametes produced by a pea plant having the genotype AaBb, it should be crossed to a plant with the genotype:

Critical Thinking

(a) AABB (b) AaBb
(c) aabb (d) aaBB

60. Universal donor blood group is:
(a) AB (b) O (c) A (d) B

61. Given pedigree shows that the trait is inherited as autosomal dominant. Trace the genotype of Mother and Father.

Tricky

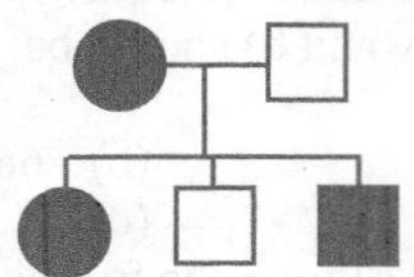

(a) Father AA, Mother aa
(b) Father AA, Mother Aa
(c) Father aa, Mother AA
(d) Father aa, Mother Aa

62. Match the column-I with column-II and choose the correct option.

	Column-I	Column-II
A.	Incomplete dominance	I. *Drosophila*
B.	Mendelian disorder	II. *Antirrhinum* sp.
C.	Transforming principle	III. Griffith
D.	Dihybrid cross	IV. Haemophilia

(a) A – I; B – IV; C – III; D – II
(b) A – IV; B – II; C – III; D – I
(c) A – II; B – III; C – IV; D – I
(d) A – II; B – IV; C – III; D – I

63. Which of the following statements are **incorrect**?
(i) A dominant allele determines the phenotype when paired with a recessive allele.
(ii) A recessive allele is weaker than a dominant allele.
(iii) A recessive allele do not shows its effects when paired with a dominant allele.
(iv) A dominant allele is always better for an organism.
(a) (ii), (i) and (iv) (b) (ii) and (iv)
(c) (i), (ii) and (iii) (d) (i), (iii) and (iv)

64. Which of the following statements are **correct**?
(i) Failure of segregation of chromatids during cell division results in aneuploidy.
(ii) Chromosomal disorders are mainly determined by alteration or mutation in a single gene.
(iii) Thalassemia and cystic fibrosis are Mendelian disorders.
(iv) Sickle cell anaemia is an X-linked trait.
(v) Haemophilia is an autosome linked recessive disease.
(a) (i) and (iii) (b) (i), (iii) and (iv)
(c) (iii) and (iv) (d) (ii) and (iii)

65. Harmful mutations does not get eliminated from gene pool because:
(a) they are recessive and carried by homozygous individuals.
(b) they are recessive and carried by heterozygous individuals.
(c) they are formed repeatedly.
(d) they show genetic drift.

66. Which of the following has normal sex chromosomes?
(a) Down syndrome
(b) Klinefelter syndrome
(c) Turner syndrome
(d) Super female

67. Haemophilia is mentioned as a trait carried by the mother and passed to her sons. What is the pattern of inheritance for this trait?

(a) Haemophilia is an allele carried on one of the mother's autosomal chromosomes.
(b) Haemophilia is an allele carried on the Y chromosome because more males have this genetic disorder than females.
(c) Haemophilia is an allele carried on the X chromosome and can be directly inherited by the son from the father or the mother.
(d) Haemophilia is carried on the X chromosome and can only be inherited by the son if the mother is a carrier.

68. Gynandromorphs are animals having:
(a) same sex in all cells of the body.
(b) different sexes in all cells of the body.
(c) same sex in different cells of the body.
(d) different sexes in different cells of the body.

69. Which of the following is an example of sex-linked inheritance?
(a) Anaemia (b) Cretinism
(c) Night-blindness (d) Colour-blindness

70. Which of the following is an example for sex-influenced inheritance?
(a) Baldness (b) Haemophilia
(c) Colour blindness (d) Down's syndrome

71. A dihybrid ratio of 1 : 4 : 6 : 4 : 1 is obtained instead of 9 : 3 : 3 : 1. This is an example of:
(a) Complementary genes
(b) Supplementary genes
(c) Polygenic inheritance
(d) Pleiotropic genes

72. One child is haemophilic (sex-linked trait), while its fraternal twin brother is normal. Which one of the following informations is most appropriate?

(a) The mother must have been heterozygous.
(b) The child is a monozygotic twin.
(c) The other child is a female and the father is haemophilic.
(d) The haemophilic child is a male.

73. Two organisms that are true-breeding for a certain genetic characteristic are mated and their offspring were analysed. Which of the following statements about this situation is **correct**?
(a) Both parents are homozygotes.
(b) The offspring are either all homozygotes or all heterozygotes.
(c) The offspring represent the F_1 generation and the gametes produced by the offspring will carry only one allele for this gene.
(d) All of the above

74. The law of segregation of characters is also called the law of purity of gametes because:
(a) gametes have only one of the two alleles for each character.
(b) gametes cannot be contaminated.
(c) Both (a) and (b)
(d) gametes are very different type of cells.

75. According to the law of independent assortment, in a dihybrid cross: Critical Thinking
(a) there are four genotypes in F_2.
(b) F_2 contains 16 phenotypes.
(c) there is a single individual which is homozygous recessive for both the characters.
(d) it is not possible to forecast the different phenotypes.

76. Match column-I with column-II and select the correct option from the codes given below.

Column-I		Column-II	
A.	Autosomal recessive trait	I.	Down's syndrome
B.	Sex-linked recessive trait	II.	Phenylketonuria
C.	Metabolic error linked to autosomal recessive trait	III.	Haemophilia
D.	Additional 21st chromosome	IV.	Sickle cell anaemia

(a) A – II; B – I; C – IV; D – III
(b) A – IV; B – I; C – II; D – III
(c) A – IV; B – III; C – II; D – I
(d) A – III; B – IV; C – I; D – II

77. Study the pedigree chart given below. Tricky

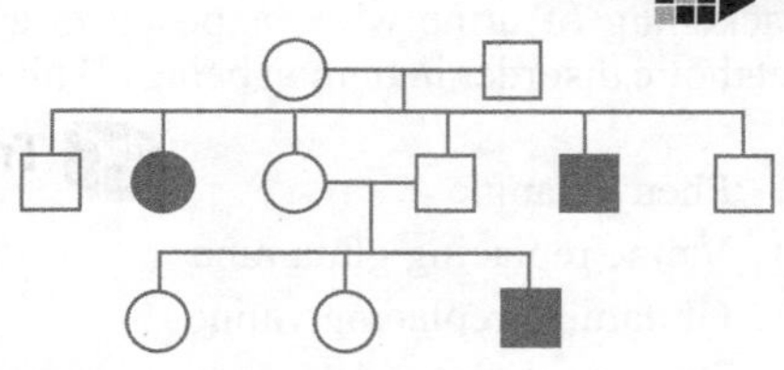

What does it show?
(a) This pedigree chart is wrong as this is not possible.
(b) Inheritance of a recessive sex-linked disease like haemophilia.
(c) Inheritance of a sex-linked inborn error of metabolism like phenylketonuria.
(d) Inheritance of a condition like phenylketonouria as an autosomal recessive trait.

78. Match column-I (Definition) with column-II (Terms) and select the correct option from the codes given below.

	Column-I (Definition)		Column-II (Terms)
A.	A single trait controlled by three or more than three alleles	I.	Pleiotropy
B.	A single trait controlled by three or more than three genes	II.	Multiple alleles
C.	A single gene exhibits multiple phenotypic expression	III.	Polygenic inheritance

(a) A – II; B – III; C – I
(b) A – III; B – II; C – I
(c) A – I; B – II; C – III
(d) A – II; B – I; C – III

79. A mutation is a:

(a) sudden temporary change in an organism's genetic material.
(b) change in phenotype followed by a change in genotype.
(c) change in hereditary material directed by a changing environment.
(d) change in genotype which may result in a new expression of a characteristic.

80. Barr body is found in the nucleus during:
(a) interphase in cell of female mammal.
(b) interphase in cell of male mammal.
(c) prophase in cell of female mammal.
(d) prophase in cell of male mammal.

81. Blackening of urine when exposed to air is a metabolic disorder in human beings. This is due to: Tricky
(a) Phenylalanine
(b) Valine replacing glutamine
(c) Glutamine replacing valine
(d) Homogentisic acid

82. In sickle-cell anaemia, shape of RBCs under oxygen tension becomes:
(a) Biconcave disc like
(b) Elongated and curved
(c) Circular
(d) Spherical

83. If a trait passes from father to all his daughters but none of his sons, the trait is: Critical Thinking
(a) Autosomal dominant
(b) Autosomal recessive
(c) Sex linked dominant
(d) Sex linked recessive

84. In fruit flies, dumpy wings are shorter and broader than normal wings. The allele for normal wings (D) is dominant to the allele for dumpy wings (d). Seventy five normal-winged and twenty five dumpy-winged flies. The genotype of parents would be. Tricky
(a) DD and DD (b) DD and Dd
(c) Dd and Dd (d) Dd and dd

85. How many pairs of contrasting character in pea plants were studied by Mendel in his experiments?
(a) Seven (b) Eight
(c) Five (d) Six

86. It is well known that Queen Victoria of England was a carrier for haemophilia. Since this is an X-linked disease, it can be predicted that:
(a) all of her sons would have that disease.
(b) all her daugthers would have been carriers.
(c) her father must definitely have had haemophilia.
(d) haemophilia would have occurred in more of her male than her female descendents.

87. Deuteranopia is a disease when there is no perception of which of the following colour?
(a) Red
(b) Blue
(c) White
(d) Green

88. A genetic disease in many Africans which reduces the oxygen uptake by RBC is known as:
(a) Sickle-cell anaemia
(b) Anaemia
(c) Pernicious anaemia
(d) Haemophilia

89. Match column-I with column-II and select the correct answer using the codes given below.

	Column-I		Column-II
A.	Turner syndrome	I.	Trisomy
B.	Linkage	II.	AA + XO
C.	Y-chromosome	III.	Morgan
D.	Down's syndrome	IV.	Testis determining factor

(a) A – II; B – I; C – IV; D – III
(b) A – IV; B – I; C – II; D – III
(c) A – IV; B – II; C – I; D – III
(d) A – II; B – III; C – IV; D – I

Answer KEYs

1	(c)	10	(a)	19	(a)	28	(b)	37	(b)	46	(d)	55	(a)	64	(a)	73	(d)	82	(b)
2	(a)	11	(c)	20	(a)	29	(b)	38	(b)	47	(b)	56	(b)	65	(b)	74	(a)	83	(c)
3	(a)	12	(d)	21	(c)	30	(a)	39	(a)	48	(d)	57	(a)	66	(a)	75	(c)	84	(c)
4	(d)	13	(d)	22	(d)	31	(a)	40	(a)	49	(b)	58	(b)	67	(d)	76	(c)	85	(a)
5	(b)	14	(d)	23	(c)	32	(c)	41	(c)	50	(c)	59	(c)	68	(d)	77	(d)	86	(d)
6	(c)	15	(a)	24	(b)	33	(b)	42	(d)	51	(a)	60	(b)	69	(d)	78	(a)	87	(d)
7	(c)	16	(a)	25	(c)	34	(a)	43	(a)	52	(c)	61	(d)	70	(a)	79	(d)	88	(a)
8	(d)	17	(a)	26	(b)	35	(d)	44	(b)	53	(b)	62	(d)	71	(c)	80	(a)	89	(d)
9	(d)	18	(d)	27	(b)	36	(d)	45	(b)	54	(c)	63	(b)	72	(a)	81	(d)		

Hints & Solutions

1. **(c)** Yellow seed colour is dominant over green seed colour.

2. **(a)** Parent : RRTt × rrtt
   Gametes : (RT), (Rt), (rt)

   Offspring : RrTt tall with red fruit; Rrtt short with red fruit

   Ratio = 1 : 1

3. **(a)** Black is the phenotype of the sheep. Without further information, you cannot identify the genotype of a black sheep because it could be either BB or Bb. The possible genotypes of the parents of a black sheep could be BB × BB, BB × Bb, Bb × bb, or Bb × Bb. Thus, there is no single genotype for either parent.

4. **(d)**

5. **(b)** The proportion of seeds that turn out to be green and/or wrinkled was very little, indicating that these two are both recessive characters, which are phenotypically expressed only if they are present in homozygous condition.

6. **(c)** Incomplete dominance is the phenomenon where dominant allele does not completely express itself. This phenomenon was first studied in flower colour of *Mirabilis jalapa* or four O' clock plant. The phenotypic as well as genotypic monohybrid ratio in F_2 generation in incomplete dominance is 1 : 2 : 1 *i.e.*, pure dominant : hybrid : pure recessive. F_1 generation expresses a phenotype which is intermediate between those of the parent, *e.g.*, pink flowers are obtained when red and white flowered plants are crossed.

7. **(c)** The total number of progeny obtained through dihybrid cross of Mendel is 1280 in F_2 generation, 480 will be recombinant. In a dihybrid cross, the ratio of pure is to recombinant is 10 : 6. So, when total progeny is 1280, 800 is pure and rest recombinant. The calculation is as:

   $$\text{Pure} = \frac{10}{16} \times 1280 = 800$$

   $$\text{Recombinant} = \frac{6}{16} \times 1280 = 480$$

8. **(d)** Cystic fibrosis : It is a common disorder of caucasian race in which thick and more salty mucus blocks the respiratory tract. The homozygous recessive condition produces the defective protein which regulates chloride transport channel.
   Cretinism : In this disorder, the physical growth, mental growth and sexual growth in children is retarded. Such a dwarf and sterile child is called a cretin. It is due to hyposecretion of thyroid hormones.
   Thalassemia : Due to defective production of α or β chains of haemoglobin, autosomal recessive.
   Haemophilia : Sex linked disorder due to defective recessive gene.

9. (d) Down's syndrome is caused by the presence of an extra chromosome number 21 and the offspring has 47 chromosomes.

10. (a) 11. (c)

12. (d) The mother would have genotype I^AI^B and the father would have genotype I^OI^O. Thus, the two genotypes possible for their offspring would be I^AI^O and I^BI^O, with equal probability. Hence, the expected phenotypes would be 1 type A : 1 type B.

13. (d) 14. (d) 15. (a) 16. (a) 17. (a)

18. (d) 19. (a) 20. (a) 21. (c)

22. (d)

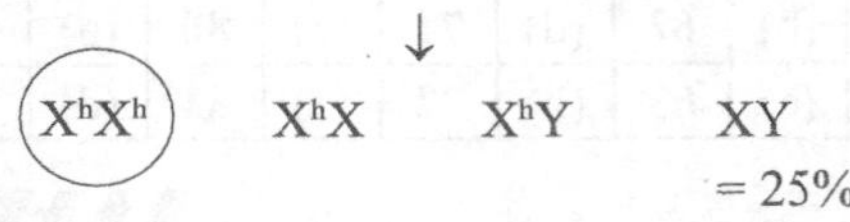

23. (c)

24. (b) Linked genes are found only on the same chromosome. Crossing-over can result in the breakage of the link between such genes.

25. (c) In human, the Y chromosome bears the androgenic factor or the male determining factor. Whereas, in *Drosophila,* the ratio of sex chromosomes to autosomes is the factor determining sex of the individual.

26. (b) 27. (b) 28. (b) 29. (b)

30. (a)

b ⟶ 20 units ⟵ a ⟷ c 8 units

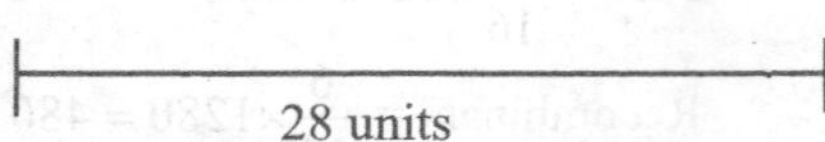

Percent crossing over between two genes is proportional to the distance between them.

31. (a) The phenotypic ratio is 1 very dark : 6 dark : 15 fairly dark : 20 intermediate : 15 fairly light : 6 light : 1 very light, hence very dark is $\frac{1}{64}$.

32. (c) 33. (b)

34. (a) A test cross is a cross involving the mating of F_1 individual with the homozygous recessive parent. It is done to find out homozygous and heterozygous individuals. The offsprings will be 100% dominant if the individual was homozygous dominant and the ratio will be 50% dominant and 50% recessive in case of hybrid or heterozygous individual. Here, AaBb, should be crossed with aabb. As aabb will produce only one type of gamete thus difference in traits if observed in their progenies will be due to the different gametes produced by hybrid dominant parent. This will indicate the genotype of the dominant parent.

35. (d)

36. (d) Pleiotropic genes have multiple effects on the phenotype. E.g. Cystic fibrosis.

37. (b) Genes coding for the same trait are located at the same locus on homologous chromosomes.

38. (b) Each conception is an independent event, so the probabilities remain the same at every time.

39. (a) According to principle of dominance, out of the two factors of alleles representing different traits of a character, only one expresses itself. But when both express themselves, it is codominance.

40. (a) The seven traits are now known to be present on 4 chromosomes. But they do not show linkage, because of large distances between them on the chromosome.

41. (c) 42. (d)

43. (a) Crossing over takes place only in non-sister chromatid. In sisters chromatids, the crossing over does not take place.

Non Parental gene exchange - Recombination

Non sister chromatid - Crossing over

Sex Chromosome - X and Y

Thalassaemia - Autosome-linked recessive disease.

44. (b) Human skin colour is the example of multiple gene inheritance. Human skin colour contributed by three separate genes. Environment also plays a significant role in determining human skin colour.

45. **(b)** **46.** **(d)**

47. **(b)** Sickle cell anaemia is an autosomal linked recessive trait that can be transmitted from parents to the offsprings when both the partners are carriers for the gene (or heterozygous). It is caused by a change in a single base pair of DNA. Sickle-cell anaemia is the name of a specific form of sickle-cell disease in which there is homozygosity for the mutation that causes HbS. Sickle-cell disease, or sickle-cell anaemia, is a life-long blood disorder characterised by red blood cells that assume an abnormal, rigid, sickle shape. Sickling decreases the cells flexibility and results in a risk of various complications.

48. **(d)** Turner's syndrome is caused by the absence of X-chromosomes in females. Persons suffering from turner's syndrome are monosomic for sex chromosomes *i.e.*, possess only one X and no Y-chromosome (XO). In other words, they have chromosome number $2n - 1 = 45$. They are phenotypic females but are sterile because they have under developed reproductive organs. They are dwarf about 4 feet 10 inches and are flat chested with wide spread nipples of mammary glands which never enlarge like those in normal woman. They develop as normal female in childhood but at adolescence their ovaries remain underdeveloped. They lack female hormone oestrogen. About one out of every 5,000 female births results in Turner's syndrome.

49. **(b)** Calvin Bridges demonstrated that in *Drosophila*, the sex is determined by ratio of the number of X chromosomes to the sets of autosomes.

50. **(c)** **51.** **(a)** **52.** **(c)**

53. **(b)** Mapping of genes on chromosomes is based on the assumption that genes are arranged on the chromosome and that the percentage of crossing over (recombination frequency) between two genes is an index of their distance apart. Distances between genes is expressed as map units, where, one map unit is defined as 1 per cent recombination.

54. **(c)**

55. **(a)** Sex linked disorders follow criss-cross inheritance pattern where affected father passes on the gene to their daughters. The daughters receive one of their X - chromosome from their fathers.

56. **(b)** Crossing over does not alter the traits that are coded for it only affects the variant that is coded for.

57. **(a)** **58.** **(b)**

59. **(c)** In order to find out the gamete or the genotype of an unknown individual, scientists perform a test cross. In test cross, the individual in question is crossed with the homozygous recessive parent.

60. **(b)** **61.** **(d)**

62. **(d)** Incomplete dominance (discovered by Correns) is the phenomenon where the dominant allele does not completely express itself. The inheritance of flower colour in *Antirrhinum sp.* is a good example of incomplete dominance. Haemophilia is a sex linked recessive disease. It is a type of Mendelian disorder since its inheritance follows the principles of Mendelian genetics and can be traced in a family by pedigree analysis. Transforming principle was given by Griffith. Dihybrid cross is a cross when two allelic pair are used for crossing. Several dihybrid crosses were carried out in *Drosophila* to study genes that were sex linked.

63. **(b)** A recessive allele is not weaker than the dominant allele. It (recessive allele) does not show its effect (in the presence of dominant allele) because of modified or different enzymes. A recessive allele makes its gene product even when paired with the dominant allele. It is not necessary that dominant allele is always better (in the case of dominant disease).

64. **(a)** Statements (i) and (iii) are correct. Sickle cell anaemia is an autosome linked recessive trait that can be transmitted from parents to the offspring when both the partners are carrier for the gene. Haemophilia is sex-linked recessive disease which shows its transmission from unaffected carrier female to some of the male progeny.

65. (b) 66. (a)

67. (d) Haemophilia is an X-linked trait, and can only be inherited by the son from his mother's X chromosome. The father contributes the Y chromosome to his son (not his X chromosome) and thus, cannot pass any of his X-linked alleles to his son.

68. (d) 69. (d) 70. (a) 71. (c)

72. (a) The mother must have been a haemophilic carrier (X^hX). One of the twins would have inherited the normal X chromosome and the other would have received the X chromosome carrying the gene for haemophilia.

73. (d) Given that both parents are true-breeding, the cross must be either AA × AA, AA × aa, or aa × aa. If you work out these crosses, you will see that all of the possible answers are true for each possible situation.

74. (a) The law of segregation states that when a pair of contrasting factors or genes or allelomorphs are brought together in a heterozygote (hybrid), the two members of the allelic pair remain together without being contaminated and when gametes are formed from the hybrid, the two separate out from each other and only one enters each gamete as seen in monohybrid and dihybrid cross. That is why the law of segregation is also described as law of purity of gametes.

75. (c)

76. (c) Sickle cell anaemia is an autosomal recessive trait. Haemophilia is a sex linked recessive trait. Metabolic error is linked to autosomal recessive: Phenylketonuria is an inherited disease due to faulty metabolism of phenylalanine. The affected individual lacks an enzyme that converts the amino acids phenylalanine into tyrosine. It is characterised by the presence of phenylketones in the urine and usually first noted by signs of mental retardation in infancy. Down's syndrome occurs due to presence of an additional 21st chromosomes.

77. (d) 78. (a)

79. (d) Mutation is the sudden inheritable discontinuous variation which appears in an organism due to permanent changes in their genotypes. The term mutation was coined by Hugo de Vries (1901).

80. (a) 81. (d) 82. (b) 83. (c)

84. (c) If both parents have normal wings (DD or Dd), there are three possible parent crosses: DD × Dd, Dd × Dd, or Dd × Dd. All of the progeny of DD × DD and of DD × Dd would have normal wings. Only the progeny of Dd × Dd would consist of 3/4 normal winged flies.

85. (a) 86. (a) 87. (d) 88. (a) 89. (d)

28 Molecular Basis of Inheritance

1. DNA polymerisation rate of DNA polymerase is:
   (a) 3.9×10^7 bp (b) 2000 bp/s
   (c) 4.6×10^6 bp (d) 5000 bp/s

2. If a nucleotide lacking a hydroxyl group at the 3' end is added to a PCR, what would be the outcome?
   Critical Thinking
   (a) No additional nucleotides would be added to a growing strand containing that nucleotide.
   (b) Strand elongation would proceed as normal.
   (c) Nucleotides would only be added at the 5' end.
   (d) *T. aquaticus* DNA polymerase would be denatured.

3. Which of the following would you expect to find in an inducible system? **Critical Thinking**
   (a) A repressor protein, which is bound to DNA in absence of any other factor.
   (b) A repressor protein, which is bound to DNA in the presence of a co-repressor.
   (c) An activator protein, which is bound to DNA in absence of any other factor.
   (d) An activator protein, which is bound to DNA only in the absence of an inhibitor.

4. Nonsense codon takes part in:
   (a) terminating message of gene controlled protein synthesis.
   (b) formation of unspecified amino acids.
   (c) conversion of sense DNA into non-sense one.
   (d) releasing tRNA from polypeptide chain.

5. Information flow or central dogma of modern biology is:
   (a) RNA → Proteins → DNA.
   (b) DNA → RNA → Proteins.
   (c) RNA → DNA → Proteins.
   (d) DNA → RNA → Proteins.

6. AUG initiation codon occurs over:
   (a) 3' end of mRNA
   (b) 5' end of mRNA
   (c) short arm of tRNA
   (d) long arm of tRNA

7. What effect would you expect if gene expression of the *lac* operon was completely repressed?
   (a) The cell would be more efficient without 'wasting' the energy required for the low level of *Lac* Z, *Lac*Y, and *Lac* A gene expression.
   (b) Allolactose would accumulate within the cell and become toxic.
   (c) Lactose would not be converted into the inducer and the operon could not be induced.
   (d) All of the above

8. Identify the correct labelling of A, B, C, D & E. **Tricky**

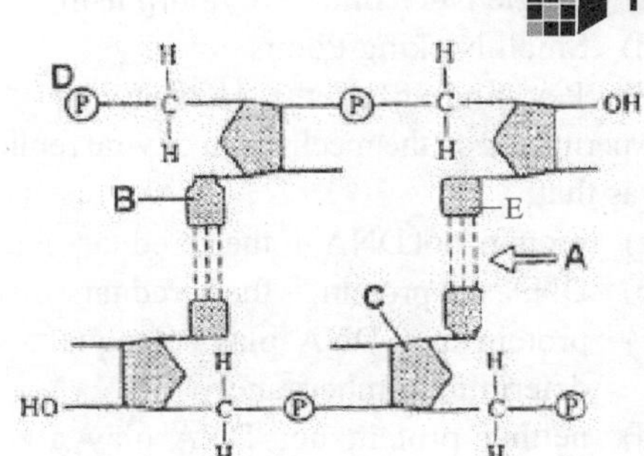

   (a) A–Hydrogen bonds, B–Pyrimidine, C–Hexose (deoxyribose) sugar, D–5' end, E–Purine base
   (b) A–Hydrogen bonds, B–Purine base, C–Hexose (deoxyribose) sugar, D–5' end, E–Pyrimidine
   (c) A–Hydrogen bonds, B–Pyrimidine, C–Pentose (deoxyribose) sugar, D–5' end, E–Purine base
   (d) A–Hydrogen bonds, B–Purine base, C–Pentose (deoxyribose) sugar, D– 5' end, E– Pyrimidine

9. Given diagram represents the schematic structure of a transcription unit with some parts labeled as A, B, C and D. Select the option which shows its correct labeling.

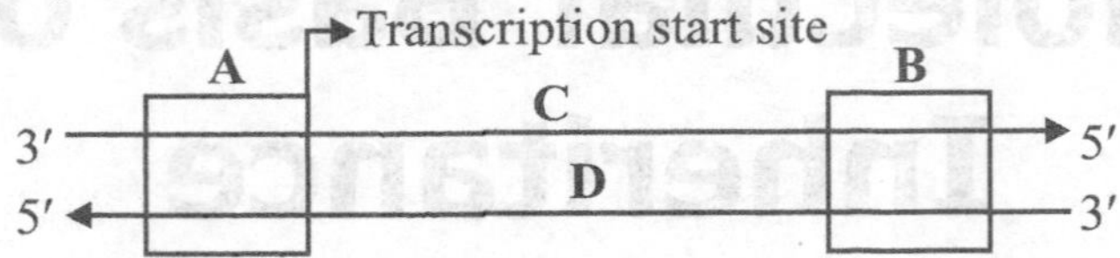

	A	B	C	D
(a)	Terminator	Promoter	Template strand	Coding strand
(b)	Promoter	Terminator	Coding strand	Template strand
(c)	Promoter	Terminator	Template strand	Coding strand
(d)	Terminator	Promoter	Coding strand	Template strand

10. Which enzymes will be produced in a cell in which there is a nonsense mutation in the lac Y gene?
(a) Lactose permease
(b) Transacetylase
(c) Lactose permease and transcetylase
(d) β- galactosidase

11. SNP which is pronounced as "snips" stands for:
(a) Small Nuclear Protein
(b) Single Nucleotide Particle
(c) Single Nucleotide Polymorphism
(d) Small Nicking Points

12. The key finding of the Hershey and Chase experiments on the mechanism of viral replication was that:
(a) protein, not DNA, is the hereditary material.
(b) DNA, not protein, is the hereditary material.
(c) protein and DNA play an equal role in determining inheritance.
(d) neither protein nor DNA play a role in determining inheritance.

13. Which of the following is/are **correct** matching(s)?

	Codon	Amino acid
(i)	$^{5'}AUG^{3'}$	Serine
(ii)	$^{5'}UAC^{3'}$	Tyrosine
(iii)	$^{5'}AUG^{3'}$	Methionine
(iv)	$^{5'}GUG^{3'}$	Valine

(a) (i) and (iii)
(b) All of these
(c) (ii), (iii) and (iv)
(d) (i), (ii) and (iii)

14. Select the two **correct** statements out of the four (i – iv) given below about *lac* operon.
(i) Glucose or galactose may bind with the repressor and inactivate it.
(ii) In the absence of lactose the repressor binds with the operator region.
(iii) The *z*-gene codes for permease.
(iv) This was elucidated by Francois Jacob and Jacque Monod.

(a) (ii) and (iii) (b) (i) and (iii)
(c) (ii) and (iv) (d) (i) and (ii)

15. Match column-I with column-II and select the correct answer using the codes given below.

Column-I	Column-II
A. Operator site	I. Binding site for RNA polymerase
B. Promoter site	II. Binding site for repressor molecule
C. Structural gene	III. Codes for enzyme protein
D. Regulator gene	IV. Codes for repressor molecules

(a) A – II; B – I; C – III; D – IV
(b) A – II; B – I; C – IV; D – III
(c) A – IV; B – III; C – I; D – II
(d) A – II; B – III; C – I; D – IV

16. A bacterium grown over medium having radioactive ^{35}S incorporates radioactivity in:
(a) Carbohydrates (b) Proteins
(c) DNA (d) RNA

17. If a double stranded DNA has 20% of cytosine, what will be the percentage of adenine in it?
(a) 20% (b) 40%
(c) 30% (d) 60%

18. Chargaff's rules of base pairing states that:
(a) the ratio of purines to pyrimidines is roughly equal in all tested organisms.
(b) the ratio of A to T is roughly equal in all tested organisms.
(c) the ratio of A + T and G + C is roughly equal in all tested organisms.
(d) Both (a) and (b)

19. Thirty percent of the bases in a sample of DNA extracted from eukaryotic cells is adenine. What percentage of cytosine is present in this DNA?
(a) 10% (b) 20%
(c) 30% (d) 40%

20. Lac operon is:
(a) Arabinose operon
(b) Repressible operon
(c) Inducible operon
(d) Overlapping genes

21. DNA packaging is the way DNA is folded into chromosomes that affects gene expression by:
(a) controlling access to DNA.
(b) positioning related genes near each other.
(c) protecting DNA from mutations.
(d) enhancing recombination of genes.

22. The most common way of gene expression is regulated in both prokaryotes and eukaryotes is through the:
(a) control of mRNA translation.
(b) breakdown of proteins formed by translation.
(c) prevention of DNA uncoiling prior to transcription.
(d) control of gene transcription.

23. Lactose operon produces enzymes:
(a) β-galactosidase, permease and glycogen synthetase.
(b) β-galactosidase, permease and transacetylase.
(c) Permease, glycogen synthetase and transacetylase.
(d) β-galactosidase, permease and phosphoglucose isomerase.

24. In negative operon:
(a) co-repressor binds with repressor
(b) co-repressor does not bind with repressor
(c) co-repressor binds with inducer
(d) cAMP have negative effect on *lac* operon

25. Which of the following is the first thing that happens when a signal molecule acts on a target cell?
(a) A transcription factor acts on the DNA.
(b) The signal molecule binds to RNA.
(c) A new protein is made in the target cell.
(d) The signal molecule binds to a receptor.

26. Antibiotic inhibiting interaction between tRNA and mRNA during protein synthesis in bacteria is:
(a) Tetracycline (b) Neomycin
(c) Erythromycin (d) Streptomycin

27. Which of the following statement forms the basis of DNA fingerprinting?
(a) The relative proportions of purines and pyrimidines in DNA.
(b) Satellite DNA occurring as highly repeated short DNA segments.
(c) The relative difference in the DNA occurrence in blood, skin and saliva.
(d) The relative amount of DNA in the ridges and grooves of the fingerprints.

28. Identify the statements pertaining to the regulatory genes (R), associated with the lac operon that are incorrect.

Critical Thinking

(i) mRNA is transcribed from the R gene whether lactose is present or not.
(ii) mRNA is transcribed from the R gene only when the lactose is present.
(iii) mRNA is transcribed from the R gene only when the lactose is not present.
(iv) Lactose inhibits the translation or R gene mRNA.
(a) (i) and (ii) (b) (ii) and (iii)
(c) (iii) and (iv) (d) (ii), (iii) and (iv)

29. Which one of the following group of codons is called as degenerate codons?
(a) UAA, UAG and UGA
(b) GUA, GUG, GCA, GCG and GAA
(c) UUC, UUG, CCU, CAA and CUG
(d) UUA, UUG, CUU, CUC, CUA and CUG

30. Which of the following characters is not associated with a leading strand?
(a) Its replication proceeds away from the fork.
(b) Its replication takes place in continuous manner.
(c) Its replication takes place in along 3'-5' direction of template strand.
(d) All of the above.

31. Number of amino acids in a polypeptide chain is 150. The number of mRNA bases required is:

(a) 150 (b) 450
(c) 100 (d) 50

32. In a mutational event, when adenine is replaced by guanine, it is a case of:

(a) Frame shift mutation
(b) Transcription
(c) Transition
(d) Transversion

33. The following ratio is generally constant for a given species: **Tricky**

(a) $\frac{A+G}{C+T}$ (b) $\frac{T+C}{G+A}$
(c) $\frac{A+T}{G+C}$ (d) $\frac{A+C}{T+G}$

34. In an eukaryote, a repressor protein may block gene expression by binding to a DNA site called:

(a) Operon (b) Histone
(c) Enhancer (d) Promoter

35. The given figure shows *lac* operon model and its functioning. Select the option which correctly labels A, B, X, Y and Z marked in the figure and also identify the label (L) which is primarily responsible for the hydrolysis of the disaccharide, lactose, into galactose & glucose.

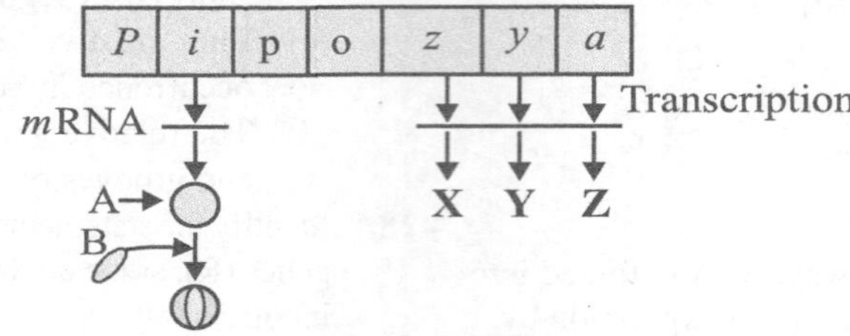

	A	B	X	Y	Z	L
(a)	Repressor	Inducer	β-Galactosidase	Permease	Transacetylase	X
(b)	Repressor	Inducer	Permease	β-Galactosidase	Transacetylase	Y
(c)	Inducer	Repressor	β-Galactosidase	Permease	Transacetylase	Z
(d)	Inducer	Repressor	β-Galactosidase	Transacetylase	Permease	B

36. Identify the labels A, B, C and D in the given structure of tRNA and select the correct option.

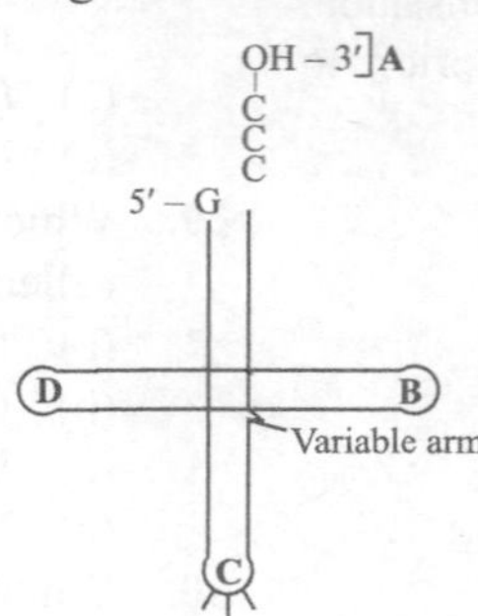

	A	B	C	D
(a)	Anticodon loop	TψC loop	AA binding site	DHU loop
(b)	AA binding site	TψC loop	Anticodon loop	DHU loop
(c)	AA binding site	DHU loop	Anticodon loop	TψC loop
(d)	AA binding site	DHU loop	TψC loop	Anticodon loop

37. Find out the **incorrect** statement.
   (a) Uracil is present in RNA at the place of thymine.
   (b) The complex of DNA and protein in chromosome is called chromatin.
   (c) Heterochromatin is the most highly condensed form of chromatin.
   (d) The process involved in the RNA formation on the DNA template is called replication.

38. DNA fingerprinting using Variable Number Tandem Repeats (VNTRs) is based on the observation that:
   (a) every individual has unique alleles at each VNTR locus.
   (b) the DNA of VNTR loci is more stable than that of loci which code for proteins.
   (c) VNTR sequences show little variability.
   (d) VNTR loci are highly polymorphic.

39. Satellite DNA:
   (a) is classified in many categories such as micro-satellites, minisatellites, etc. on the basis of base composition length of segments and number of repetitive units.
   (b) normally does not code for any protein.
   (c) shows polymorphism.
   (d) All of the above

40. Chargaff's rule is applicable to: Tricky
   (a) single stranded RNA
   (b) single stranded DNA and RNA
   (c) single stranded DNA
   (d) double stranded DNA

41. Match the enzymes given in column-I with its function given in column-II and select the correct option.

Column-I		Column-II	
A.	β-galactosidase	I.	Joining of DNA fragments
B.	Permease	II.	Peptide bond formation
C.	Ligase	III.	Hydrolysis of lactose
D.	Ribozyme	IV.	Increase permeability of galactosidase

   (a) A – II; B – I; C – IV; D – III
   (b) A – III; B – IV; C – I; D – II
   (c) A – II; B – IV; C – I; D – III
   (d) A – I; B – II; C – IV; D – III

42. Operon is a:
   (a) sequence of three nitrogen bases determining a single amino acid.
   (b) set of closely placed genes regulating a metabolic pathway in prokaryotes.
   (c) segment of DNA specifying a polypeptide.
   (d) gene responsible for switching on and switching off other genes.

43. Genes that are involved in turning on or off the transcription of a set of structural genes are called:
   (a) Operator genes (b) Redundant genes
   (c) Regulator genes (d) Polymorphic genes

44. If the gene encoding the *trp* repressor is mutated such that it can no longer bind to tryptophan, will transcription of the *trp* operon occur?

   (a) Yes, because the *trp* repressor can only bind the *trp* operon and block transcriptional initiation when it is bound to tryptophan.
   (b) No, because this mutation does not affect the part of the repressor that can bind the operator.
   (c) No, because the *trp* operon is repressed only when tryptophan levels are high.
   (d) Yes, because the *trp* operon can allosterically regulate the enzymes needed to synthesise the amino acid tryptophan.

45. Transcriptional regulation:
   (a) is highly efficient at completely preventing transcription.
   (b) allows the cell to only produce proteins that are needed at the time.
   (c) can be induced by a repressor protein.
   (d) All of the above

46. Post–transcriptional regulation includes:

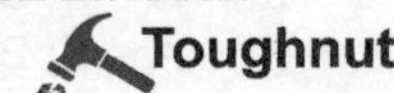

   (a) binding of repressor on silencer regions.
   (b) transport of messenger RNA into the cytoplasm.
   (c) decreasing messenger RNA stability in the cytoplasm.
   (d) Both (b) and (c)

47. Mutations which alter nucleotide sequence within a gene are called:
   (a) Frame shift mutations
   (b) Base pair substitutions
   (c) Both (a) and (b)
   (d) None of the above

48. What would happen if a mutation occurred in the DNA such that the second codon of a polypeptide, UGC, was changed to a UAG?
(a) Nothing. The ribosome would skip that codon and translation would continue.
(b) Translation would continue, but the reading frame of the ribosome would be shifted.
(c) Translation would stop at the second codon and no functional protein would be made.
(d) Translation would continue, but the second amino acid in the protein would be different.

49. Which statement is/are **correct**? 
(i) Nucleosome contains basic protein.
(ii) Tightly packed DNA of chromatin is called heterochromatin.
(iii) Loosely packed DNA of chromatin is called euchromatin.
(iv) H_2A, H_2B are linker proteins.
(v) DNA is basic in nature.
(a) (i), (ii) and (iii)
(b) (i) and (ii)
(c) (i), (ii), (iii), (iv) and (v)
(d) (iv) and (v)

50. Match column-I with column-II and select the correct answer using the codes given below.

Column-I	Column-II
A. Helicase	I. Joining of nucleotides
B. Gyrase	II. Opening of DNA
C. Primase	III. Unwinding of DNA
D. DNA polymerase III	IV. RNA priming

(a) A – II; B – I; C – III; D – IV
(b) A – II; B – I; C – IV; D – III
(c) A – IV; B – III; C – I; D – II
(d) A – II; B – III; C – IV; D – I

51. Which of the following statements about the process of DNA replication is **incorrect**?
(a) Many different enzymes are needed for the process to function properly.
(b) Mistakes can be corrected at multiple steps in the process.
(c) Uncorrected mistakes introduce mutations into the DNA base sequence.
(d) Mistakes in the copying process are very common occurrences.

52. Experiments by Avery, McLeod, and McCarty supported DNA as the genetic material by showing that:
(a) both protein and DNA samples provided the transforming factor.
(b) DNA was not complex enough to be the genetic material.
(c) only samples with DNA provided activity.
(d) even though DNA was molecularly simple, it provided adequate variation to act as the genetic material.

53. Enzyme required for removing RNA primer during DNA replication is:
(a) Primase
(b) Ligase
(c) DNA polymerase I
(d) DNA polymerase III

54. Termination of polypeptide chain is brought about by:
(a) UUG (b) UAA
(c) UUG (d) UCG

55. Wobble hypothesis establishes:
(a) peptide chain formation
(b) initiation of peptide chain
(c) termination of peptide chain
(d) economy in tRNA molecules.

56. Regulated unit of genetic material is termed as:
(a) Operon
(b) Regulator gene
(c) Operator gene
(d) Okazaki segment

57. Identify the **incorrect** statement. 
(a) In prokaryotes, the structural gene is polycistronic.
(b) In eukaryotes, structural genes have interrupted coding sequences.
(c) Eukaryotes have split gene arrangement.
(d) Intervening sequences appear in mature RNA.

58. Which of the following statements about RNA polymerase are **correct**?
(i) RNA polymerase I transcribes rRNAs.
(ii) RNA polymerase II transcribes snRNAs.
(iii) RNA polymerase III transcribes hnRNA.
(iv) RNA polymerase II transcribes hnRNAs.
(a) (i) and (ii) (b) (i) and (iii)
(c) (ii) and (iii) (d) (i) and (iv)

59. The lagging daughter strand of DNA is synthesised in what appears to be the "wrong" direction. This synthesis is accomplished by:

Tricky

(a) ligating (connecting) short Okazaki fragments that are synthesised in short spurts in the "right" direction.
(b) primase.
(c) using multiple primers and DNA polymerase I.
(d) Both (a) and (b)

60. Which of the following factor is required for the protein synthesis?
(i) Initiation codon
(ii) GTP and ATP
(iii) Peptidyl transferase
(iv) tRNA
(v) mRNA
(vi) Amino acid activating enzyme
(vii) rRNA
Choose the correct combination.
(a) (i), (ii) and (iii) (b) (iii), (iv) and (v)
(c) (v), (vi) and (vii) (d) All of these

61. During transcription, the DNA site at which RNA polymerase binds is called:
(a) Enhancer (b) Promoter
(c) Regulator (d) Receptor

62. If the nucleus of a frog egg is destroyed and replaced with the nucleus of an intestinal cell from a tadpole, the egg can develop into a normal tadpole. This demonstrates that:

(a) intestinal cells are fully differentiated.
(b) there is little functional difference between an egg cell and an intestinal cell.
(c) an intestinal cell possesses a full set of genes.
(d) intestinal cells are not differentiated.

63. Which of the following does not play a role in the genetic regulation of tryptophan synthesis in the bacterium *E. coli*?
(a) RNA polymerase (b) Homeotic gene
(c) Repressor protein (d) Operator

64. Which of the following is most abundant in human DNA?
(a) Regulatory genes (b) Non-coding DNA
(c) Transposons (d) Homeotic genes

65. The regulation of tryptophan synthesis in *E.coli* is an example of affecting gene expression through:
(a) translational control
(b) transcriptional control
(c) homeotic gene control
(d) breaking down mRNA molecules

66. Isolation and purification of specific DNA segment from a living organism was achieved by:
(a) Nirenberg
(b) Lederberg
(c) Zacharis
(d) Watson & Crick

67. Hershey and Chase used radioactive ^{35}S and ^{32}P in experiments to provide evidence that DNA was the genetic material. These experiments pointed to DNA because:
(a) progeny viruses retained ^{32}P but not ^{35}S.
(b) retention of ^{32}P in progeny viruses indicated that DNA was passed on.
(c) loss of ^{35}S in progeny viruses indicated that proteins were not passed.
(d) All of the above

68. Histones are rich in:
(a) Alanine and glycine
(b) Lysine and arginine
(c) Histidine and serine
(d) Cysteine and tyrosine

69. Which of the following mechanisms of gene regulation operates after mRNA transcription but before translation of mRNA into protein?

(a) mRNA splicing
(b) DNA packing
(c) Repressors and activators
(d) Protein degradation

70. The process of the elongation in transcription requires:
(a) Ca^{++} or Mn^{++}
(b) Na^{+} or Cl^{-}
(c) Na^{+} or K^{+}
(d) Mn^{++} or Mg^{+2}

71. DNA replication is:
(a) conservative and discontinuous.
(b) semi-conservative and semi-discontinuous.
(c) semi-conservative and discontinuous.
(d) conservative

72. DNA binding proteins:
(a) have distinct three-dimensional structures that allow them to bind to the DNA.
(b) are transcription factors.
(c) inhibit the loss of ends from the chromosome and help DNA condense in the nucleus.
(d) All of the above

73. A short sequence of bases on one strand of DNA is AGTCTACCGATAGT. If this sequence serves as a template for the formation of a new strand of DNA, what will be the corresponding base sequence in the new strand?
(a) AGTCTACCGATAGT
(b) TCAGATGGCTATCA
(c) TGATAGCCATCTGA
(d) GACATCGATTCGAT

74. DNA replication is semi-conservative as:
(a) only non-parent strand acts as template.
(b) both strands of new molecule are synthesised *de novo*.
(c) one of the strand in each new molecule is parental and the other is new.
(d) daughter strands are dispersive.

75. Chromatin structure must be altered in order for gene expression to occur because: **Tricky**
(a) condensed chromatin is replicated but not transcribed.
(b) condensed chromatin makes most DNA sequence inaccessible to the transcription complex.
(c) decondensed chromatin has more nucleosomes per DNA molecule.
(d) heterochromatin is actively transcribed and euchromatin is not transcribed.

76. Introns are DNA sequences that:
(a) code for functional domains in proteins.
(b) are removed from pre-mRNA by spliceosomes.
(c) allow one gene to make different gene products depending on which introns are removed during splicing
(d) Both (b) and (c)

77. Transcription:
(a) starts at initiator region and ends at stop region.
(b) starts at operator region and ends at telomeric end.
(c) starts at promoter region and ends at terminator region.
(d) starts at CAAT box and ends at TATA box.

78. Match column-I with column-II and choose the correct option.

	Column-I		Column-II
A.	Exon	I.	Coding sequence
B.	Intron	II.	Cistron
C.	Genetic code	III.	Triplet bases on mRNA
D.	DNA packaging	IV.	Nucleosome
		V.	Non-codding sequence

(a) A – I; B – III; C – II; D – V
(b) A – I; B – IV; C – II; D – III
(c) A – I; B – V; C – III; D – IV
(d) A – IV; B – I; C – V; D – III

79. Choose the sequence in which the following enzymes take part in DNA replication.
(i) Helicase (ii) Primase
(iii) SSB (iv) DNA polymerase
(v) DNA ligase
(a) (i) → (v) → (iv) → (iii) → (ii)
(b) (i) → (ii) → (iii) → (iv) → (v)
(c) (i) → (iii) → (ii) → (iv) → (v)
(d) (i) → (iv) → (iii) → (ii) → (v)

80. Code transfer for synthesis of polypeptide involves: **Critical Thinking**
(a) DNA, tRNA, rRNA and mRNA.
(b) mRNA, tRNA, rRNA and DNA.
(c) tRNA DNA, mRNA and rRNA.
(d) DNA, mRNA, tRNA and amino acids.

81. Eukaryotic chromosomes:
(a) are circular and contain origin and terminator sequences.
(b) are linear and have origins and telomeres.
(c) contain coding and non-coding sequences.
(d) Both (b) and (c)

82. DNA is a genetic material:
(i) due to its stability.
(ii) its ability to replicate.
(iii) its ability to mutate.
(iv) its ability to express itself.
Choose the correct combination.
(a) (i) and (ii)
(b) (i), (iii) and (iv)
(c) (iii) and (iv)
(d) (i), (ii), (iii) and (iv)

83. tRNA recognises ribosome by:
(a) TψC loop (b) DHU loop
(c) Anticodon (d) AA-site

84. Nucleotide arrangement in DNA can be seen by:
(a) X-ray crystallography
(b) Electron microscope
(c) Ultracentrifuge
(d) Light microscope

85. DNA synthesis can be measured by estimating incorporation of radio-labeled:
(a) Uracil (b) Ribose sugar
(c) Thymidine (d) Adenine

86. During conjugation:

(a) DNA from one bacterial cell is transferred to another bacterial cell using a bacteriophage.
(b) mutants that are auxotrophic for one nutrient can be converted to prototrophs when mixed with mutants that are auxotrophic for another nutrient.
(c) a pilus is synthesised and DNA is transferred from one bacterium across the conjugation tube to the recipient bacterium.
(d) Both (a) and (b)

87. Transposable genetic elements:

(a) always affect the cell adversely, because when they move, they inactivate genes.
(b) are retroviruses.
(c) provide a mechanism for moving genetic material from organelle genomes to the nuclear genome.
(d) always replicate their DNA when they want.

88. The scientists involved in discovery of DNA as chemical basis of heredity were:
(a) Hershey and Chase
(b) Griffith and Avery
(c) Avery, Mac Leod and McCarty
(d) Watson and Crick

89. Match column-I with column-II and select the correct combination from the given options.

Column-I	Column-II
A. Sigma factor	I. 5' – 3'
B. Capping	II. Initiation
C. Tailing	III. Termination
D. Coding strand	IV. 5' end
	V. 3' end

(a) A – III; B – V; C – IV; D – II
(b) A – II; B – IV; C – V; D – I
(c) A – II; B – IV; C – V; D – III
(d) A – III; B – V; C – IV; D – I

90. Select the **incorrect** statement(s) about genetic codon.
(i) Six codons do not code for any amino acid.
(ii) Codon is read in mRNA in a contiguous fashion.
(iii) Three codons function as stop codons.
(iv) The initiation codon AUG codes for methionine.
(a) (i) only (b) (ii) only
(c) (i), (ii) and (iv) (d) (i), (ii) and (iii)

91. Match column-I with column-II and select the correct answer using the codes given below.

Column-I	Column-II
A. Termination	I. Aminoacyl tRNA synthetase
B. Translation	II. Okazaki fragments
C. Transcription	III. GTP dependent release factor
D. DNA replication	IV. RNA polymerase

(a) A – II; B – I; C – III; D – IV
(b) A – III; B – I; C – IV; D – II
(c) A – IV; B – III; C – I; D – II
(d) A – II; B – III; C – I; D – IV

92. Co-repressor binds with:

(a) Promoter gene (b) Operator gene
(c) Aporepressor (d) Regulator gene

93. In lac operon, structural gene '*z*' synthesises:
(a) β-galactosidase
(b) Galactosidase permease
(c) Galactosidase transacetylase
(d) None of the above

94. *E.coli* cells with a mutated *z* gene of the *lac* operon cannot grow in medium containing only lactose as the source of energy because:

Critical Thinking

(a) the *lac* operon is constitutively active in these cells.
(b) they cannot synthesise functional beta-galactosidase.
(c) in the presence of glucose, *E.coli* cells do not utilize lactose.
(d) they cannot transport lactose from the medium into the cell.

95. Which one of the following triplet codes, is correctly matched with its specificity for an amino acid in protein synthesis or as 'start' or 'stop' codon?
(a) UAC – Tyrosine (b) UCG – Start
(c) UUU – Stop (d) UGU – Leucine

96. Which step of translation does not consume a high energy phosphate bond? **Critical Thinking**
(a) Translocation
(b) Amino acid activation
(c) Peptidyl-transferase reaction
(d) Aminoacyl tRNA binding to active ribosomal site

97. One gene one enzyme hypothesis was proposed by:
(a) Khorana and Nirenberg
(b) Beadle and Tatum
(c) Bateson and Punnet
(d) Bridges

98. In some viruses, RNA is present instead of DNA indicating that:
(a) their nucleic acid must combine with host DNA before replication.
(b) they cannot replicate.
(c) there is no hereditary information.
(d) RNA can act to transfer heredity.

99. What is the main function of tRNA in relation to protein synthesis?
(a) Initiates transcription
(b) Inhibits protein synthesis.
(c) Identifies amino acids and transport them to ribosomes.
(d) Proof reading.

100. In tertiary structure of DNA, what is a histone octamer? **Toughnut**
(a) A complex consisting of eight positively charged histone proteins (two of each H_2A, H_2B, H_3 and H_4) that aid in the packaging of DNA.
(b) A complex consisting of eight negatively charged histone proteins (two of each H_2A, H_2B, H_3 and H_4) that aid in the packaging of DNA.
(c) A complex consisting of nine positively charged histone proteins (H_1 and two of each H_2A, H_2B, H_3 and H_4) that aid in the packaging DNA.
(d) A complex consisting of nine negatively charged histone proteins (H_1 and two of each H_2A, H_2B, H_3 and H_4) that aid in the packaging of DNA.

101. Phosphodiester linkage is formed between:
(a) C-5 of one nucleotide and C-3 of the next
(b) C-1 of one nucleotide and C-2 of the next
(c) Purines of opposite strands
(d) Pyrimidines of opposite strands

102. From the experiments carried out by Avery, Macleod and McCarty by using various enzymes, which of the following results prominently proved that DNA is the transforming material?
(a) DNA of heat killed 'S' + R type + DNAase → virulent strain
(b) DNA of heat killed 'S' + R type + RNAase → virulent strain
(c) DNA of heat killed 'S' + R type → virulent strain
(d) DNA of heat killed 'S' + R type + Protease → virulent strain

103. A segment of DNA molecule contains 200 guanine and 200 thymine bases. What will be the total number of nucleotides in this segment of DNA?
(a) 400 (b) 200
(c) 800 (d) 100

104. In ABO system of blood group there are 4-types of:
(a) genotypes (b) phenotypes
(c) antibodies (d) alleles

105. During elongation of polypeptide chain, sigma factor is: **Tricky**
(a) functionless.
(b) retained for specific function.
(c) released for re-use.
(d) required during closing of chain.

106. Genetic code is:

(a) triplet, universal, ambiguous and degenerate.

(b) triplet, universal, non-ambiguous and non-degenerate.

(c) triplet, universal, non-ambiguous and degenerate.

(d) triplet, universal, ambiguous and non-degenerate.

107. Plasmid DNA contains genes that:

(a) confer drug resistance to the host cell.

(b) regulate conjugation.

(c) can confer resistance to heavy metals.

(d) All of the above

108. Who gave the Ist experimental proof of triplet code?

(a) George Gamow

(b) Nirenberg

(c) Kohenberg

(d) Khorana

109. Methyl guanosine triphosphate is added at 5' end of hn RNA in a process of:

(a) Tailing (b) Splicing

(c) Capping (d) None of these

110. The given figure represent one of the step in the process of transcription in bacteria. Identify the step and label A, B & C marked in the figure.

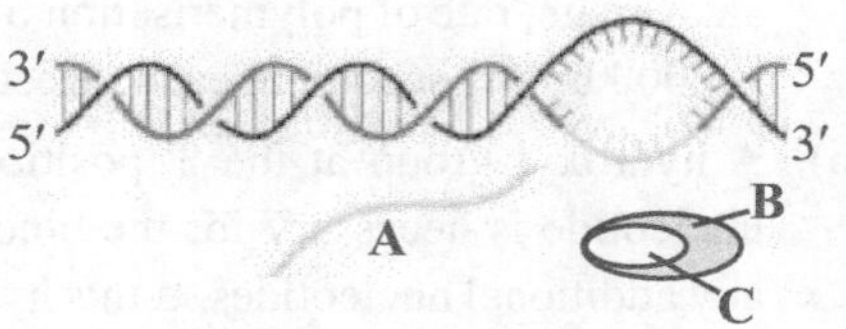

(a) Initiation; A – DNA, B – RNA, C – Promoter

(b) Termination; A – RNA, B – RNA polymerase, C – Rho factor

(c) Elongation; A – RNA, B – RNA polymerase, C – Sigma factor

(d) Elongation; A – DNA, B – DNA polymerase, C – RNA

Answer KEYs

1	(b)	12	(b)	23	(b)	34	(c)	45	(b)	56	(a)	67	(d)	78	(c)	89	(b)	100	(a)
2	(a)	13	(c)	24	(a)	35	(a)	46	(d)	57	(d)	68	(b)	79	(c)	90	(a)	101	(a)
3	(b)	14	(c)	25	(d)	36	(b)	47	(c)	58	(d)	69	(a)	80	(d)	91	(c)	102	(a)
4	(a)	15	(a)	26	(b)	37	(d)	48	(c)	59	(d)	70	(d)	81	(d)	92	(c)	103	(c)
5	(d)	16	(b)	27	(b)	38	(d)	49	(a)	60	(d)	71	(c)	82	(d)	93	(a)	104	(b)
6	(b)	17	(c)	28	(d)	39	(d)	50	(d)	61	(b)	72	(d)	83	(a)	94	(b)	105	(a)
7	(c)	18	(d)	29	(a)	40	(d)	51	(d)	62	(c)	73	(b)	84	(a)	95	(a)	106	(c)
8	(d)	19	(b)	30	(a)	41	(b)	52	(c)	63	(b)	74	(c)	85	(c)	96	(d)	107	(d)
9	(c)	20	(c)	31	(b)	42	(b)	53	(c)	64	(b)	75	(b)	86	(d)	97	(b)	108	(b)
10	(d)	21	(a)	32	(c)	43	(a)	54	(b)	65	(b)	76	(d)	87	(c)	98	(d)	109	(c)
11	(c)	22	(d)	33	(c)	44	(a)	55	(d)	66	(a)	77	(c)	88	(c)	99	(c)	110	(b)

Hints & Solutions

1. **(b)** 2000 base pair per second. DNA replication is rapid. The replication of DNA of E. coli with 4.6×10^6 bp requires 19 minutes. On an average, rate of polymerisation of bases is 2000 bp per second in each direction.

2. **(a)** A hydroxyl group at the 3' position of a nucleotide is necessary for the binding of any additional nucleotides. If this hydroxyl group were absent, no other nucleotides could be added to a growing strand.

3. **(b)** 4. **(a)** 5. **(d)** 6. **(b)**

7. **(c)** A low level of *Lac* z expression is required for conversion of lactose to the inducer, allolactose.

8. **(d)** 9. **(c)**

10. **(d)** A non-sense mutation is the one which stops polypeptide synthesis due to formation of a terminating or non-sense codon. e.g. ATT(UAA), ATC (UAG), ACT(UGA). The lactose or lac operon of *Esherichia coli* contains structural genes (*z, y, a*). If Y codes for termination of polypeptide chain then only the product of '*z*' gene transcribe to form β galactosidase.

11. **(c)** Single nucleotide polymorphism(SNP) is the most common type of genetic variation among people. Each SNP represents a difference in a single DNA building block, called a nucleotide. For example, a SNP may replace the nucleotide cytosine (C) with the nucleotide thymine (T) in a certain stretch of DNA.

12. **(b)** Hershey and Chase used radioactive labeling of sulphur and phosphorus to clearly demonstrate that DNA, and not protein, is the material that carries hereditary information.

13. **(c)**

14. **(c)** Jacob and Monod proposed the *lac* operon of *E. coli.* The *lac* operon contains a promoter, an operator, and three structural genes called *z, y,* and *a*, coding for the enzyme β galactosidase, permease and transacetylase respectively. The *lac* regulator gene, designated as *i* gene, codes for repressor. In the absence of the inducer, the repressor binds to the *lac* operator, preventing RNA polymerase from binding to the promoter and thus transcribing the structural gene.

15. **(a)**

16. **(b)** Few bacteriophages were grown in bacteria containing ^{35}S which was incorporated into the cysteine and methionine amino acids of proteins and thus these amino acids with ^{35}S formed the proteins of phage.

17. **(c)** The purines and pyrimidines are always in equal amounts as per chargaff's rule. So, if cytosine is 20%, the thymine will be 30% and thymine is equal to adenine. Then the percentage of adenine will be 30%.

18. **(d)** Chargaff found that the relative ratios of purine to pyrimidines were equal. Adenine and guanine are purines and cytosine and thymine are pyrimidines; therefore, ratios of adenine and thymine should be equal. Chargaff also found that there is no conserved ratio between specific pairs (*e.g.*, A + T and G + C).

19. **(b)** If 30 percent of DNA is adenine, then by Chargaff's rule 30 percent will be thymine. The remaining 40 percent of the DNA is cytosine and guanine. Since the ratio of cytosine to guanine must be equal, then each accounts for 20 percent of the bases.

20. **(c)**

21. (a) It is thought that RNA polymerase and other transcription factors cannot bind to tightly packed DNA.

22. (d) All of these are ways that gene expression can be regulated, but transcriptional control is clearly the most common mechanism.

23. (b)

24. (a) In negative (repressible) operon, the repressor co-repressor complex binds with the operator. The free repressor cannot bind to the operator.

25. (d) The first effect of any signal molecule must involve the binding of the molecule to a receptor.

26. (b)

27. (b) DNA fingerprinting is a test to identify and evaluate the genetic information called DNA (deoxyribonucleic acid) in a person's cells. DNA fingerprinting is a form of identification based on sequencing specific non-coding portions of DNA that are known to have a high degree of variability from person to person. These sections are known as tandem repeats. The test is used to determine whether a family relationship exists between two people, to identify organisms causing a disease, and to solve criminal cases.

28. (d) In lac opeon model, regulatory gene is constitutively working. Its translation product (repressor) is inactivated by the presence of an inducer (lactose).

29. (a) Degenerate codons (also called as non - sense codons or terminator codons) do not code for any amino acids. Three types of degenerate codons are UAG (amber), UAA (ochre) and UGA (opal).

30. (a) Leading Strand One strand with polarity 3'-5', forms its complementary strand continuously because 3' end of the latter is always open for elongation. It is called leading strand (continuous strand).

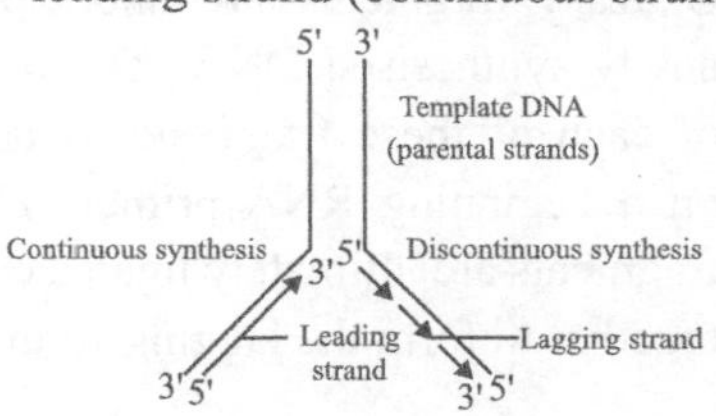

Newly synthesized strands (replication fork)

31. (b) An amino acid is coded by triplet of nitrogen bases known as codon.

32. (c) In transition substitution a purine is replaced by another purine base (A with G or vice versa). In transversion substitution a purine is replaced by a pyrimidine base or vice versa. Frameshift mutation is a type of mutation where the reading of codons is changed due to insertion or deletion of nucleotides. Transition is the formation of RNA over the template of DNA.

33. (c) According to Chargaff purines and pyrimidines are in equal amounts. Purine (adenine) is equimolar with pyrimidine (thymine) and purine (guanine) is equimolar with pyrimidine (cytosine). Base ratio is specific for species.

34. (c) Without the repressor, the genes that encode for enzymes that break down tetracycline could be expressed.

35. (a) In the given figure of lac operon model (proposed by Jacob and Monad), the labels A, B, X, Y and Z are respectively repressor, inducer, β- galactosidase (*z*), permease (*y*), transacetylase (*a*). *z*, *y* and a are three structural genes which produces three enzymes for the degradation of lactose to glucose and galactose. Label X (β- galactosidase) is primarily responsible for the hydrolysis of disaccharide lactose into galactose and glucose.

36. (b) tRNA or transfer RNA is a single stranded molecule and takes the shape of a clover leaf. In the process of transcription tRNA brings amino acid and reads the genetic code and acts as an adapter molecule. In the given structure of tRNA, the labels A, B, C and D are respectively AA binding site (amino acid binding site), TψC loop, anticodon loop (codon recognition site) and DHU loop (amino acid recognition site).

37. (d) The process involved in the RNA formation on the DNA template is called transcription.

38. (d) The technique of DNA fingerprinting was initially developed by Alec Jeffrey's. He used a satellite DNA as probe that shows high degree of polymorphisms. DNA fingerprinting using variable number tandem repeats is based on the observation that VNTR loci are highly polymorphic.

39. (d) Satellite DNA is a portion of DNA consisting of short, repeating sequences of nucleotide pairs near the region of the centromere. Normally it does not code for any protein but shows polymorphisms. It is classified in many categories like micro- or minisatellites based on the composition, length of segments and number of repetitive units.

40. (d) Chargaff's rule is applicable to double stranded DNA because according to the rule, percentage of adenine is equal to the percentage of thymine and percentage of guanine is equal to the percentage of cytosine.

41. (b) A-III, B-IV, C-I, D-II

β-galactosidase	–	Hydrolysis of lactose
Permease	–	Increases permeability to β-galactosidase
Ligase	–	Joining of DNA fragments
Ribozyme	–	Peptide bond formation

42. (b) Operons are segments of genetic material (DNA) which functions as regulated unit or units that can be switched on or switched off. It is a sequence of closely placed genes regulating a metabolic pathway in prokaryotes.

43. (a) Operator gene allows the functioning of the operon.

44. (a) If the repressor can no longer bind to tryptophan, then it cannot bind the operator and transcription of the *trp* operon will always be on, whether tryptophan levels in the cell are high or low.

45. (b) 46. (d)

47. (c) In frame shift mutation the reading of the frame of the base sequence shifts laterally either in forward direction due to addition of one or more nucleotides or in backward direction due to deletion of one or more nucleotides. Whereas in base pair substitution a base pair is replaced by another base which results in change of nucleotide sequence.

48. (c) UAG is a stop codon and translation would terminate at that site.

49. (a) H_2A, H_2B, H_3 and H_4 are called the core histone proteins, H_1, is called the linker histone protein because it joins the two histone octamers. DNA is acidic in nature.

50. (d)

51. (d) Even though DNA replication typically occurs millions of times during the life of a multicellular organism, it is remarkably error-free. Those errors that do occur are usually corrected with a high degree of reliability.

52. (c) These researchers were able to isolate nearly pure DNA samples. It was only these samples that provided transformation activity.

53. (c) 54. (b) 55. (d) 56. (a)

57. (d) Introns or intervening sequences do not appear in mature or processed RNA.

58. (d) RNA pol I transcribes rRNAs, whereas the RNA pol III is responsible for transcription of tRNA, 5srRNA and snRNAs. RNA pol II transcribes hnRNA.

59. (d) Okazaki fragments are short pieces of newly synthesised DNA. The production of each of these fragments is dependent on a beginning RNA primer. The small fragments are ultimately ligated connected together to form the lagging strand.

60. (d) Machinery for Protein Synthesis

(i) Ribosome protein synthesis occurs over the ribosome. They are also called protein factories. There are two kinds of ribosomes found in eukaryotes and prokaryotes respectively called prokaryotic ribosome (70S) and eukaryotic ribosome (80S).

There are three reactive sites in ribosome

P-site (Peptidyl transfer or donor site)

A-site (Amino acyl or acceptor site)

E-site (Exit site)

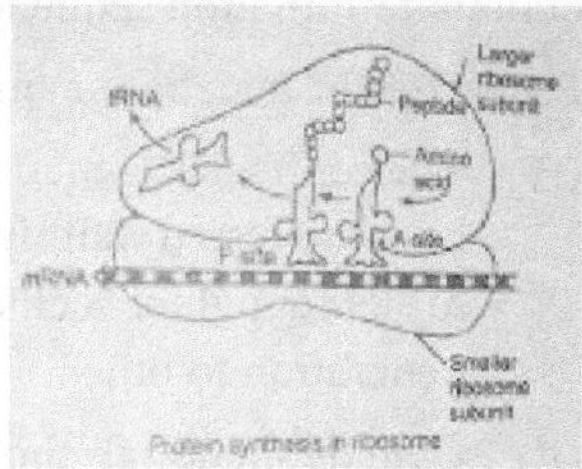

Protein synthesis in ribosome

(ii) Amino Acid- There are about 20 amino acids and amides which constitutes the building blocks of protein. They occurs in cellular pool.

(iii) mRNA- It brings coded information from DNA and takes part in its translation by bringing amino acids in a particular sequence.

(iv) tRNA - They pick up particular amino acids in the process called charging. The charged tRNAs take codons corresponding to their anticodons.

(v) Aminoacyl tRNA synthetase enzyme helps in combining amino acid to its particular tRNA. The enzyme is specific for each amino acid and also called as amino acid activating enzyme.

61. (b) Regulator is a gene which forms a biochemical for suppressing the activity of operator gene. Promoter is the gene which provides the point of attachment to RNA polymerase required for transcription of structural genes.

62. (c) This illustrates that all of the genetic information needed by the different types of tissues making up a frog are retained by an intestinal cell.

63. (b) Homeotic genes are used by eukaryotes to regulate the interaction among groups of genes in long-term, complex processes such as development.

64. (b) Over 95 percent of the human genome is made up of the various kinds of non-coding DNA.

65. (b) The presence or absence of tryptophan determines whether the genes that code the necessary enzymes in tryptophan synthesis will even be transcribed.

66. (a)

67. (d) Only DNA incorporates radioactive P. Radioactive S is incorporated into proteins. The fact that progeny virus retained radioactive P and not radioactive S indicates that DNA is heritable and protein is not.

68. (b) Histones are positively charged, basic proteins, enriched in the amino acids arginine and lysine.

69. (a) RNA processing occurs after transcription and before translation.

70. (d)

71. (c) DNA replication is the process in which a double-stranded DNA molecule is copied to produce two identical DNA molecules. In DNA replication each new strand is half the original parent strand hence called semiconservative and one strand is synthesised continuously and other discontinuous hence called discontinuous.

72. (d) DNA binding proteins must have a three structure to bind DNA. Some DNA binding proteins are transcription factors, some are telomerases, which bind the ends of chromosomes, and some are nucleosomes which land the DNA and help pack it into nucleus.

73. (b) The complementary base will pair with each base on the template to form the specific sequence shown.

74. (c) Replication is the formation of exact carbon copy or replica. According to semi -conservative method of DNA replication, the two strands of DNA molecule separate and the complementary strand is synthesised from the medium. After the completion of replication, each DNA molecule would have one parental and one newly synthesised strand.

75. (b) Nucleosomes condense the DNA. Heterochromatin is transcriptionally inactive, while euchromatin is transcriptionally active.

76. (d) Introns are removed by spliceosomes, and differential splicing can produce different gene products as seen in the tropomyosin example. The DNA loops containing introns, seen in hybridization experiments were done with DNA and mature messenger RNA, not pre-mRNA.

77. (c) Formation of mRNA from DNA is called as transcription. The segment of DNA involved in transcriptions is cistron, which have a promoter region where initiation is started and terminator region where transcription ends. Enzyme involved in transcription is RNA polymerase-II.

78. (c) **Exon** : coding (functional) part of DNA or RNA.

Intron : non-coding (non-functional) part of DNA or RNA.

Genetic code : Triplet bases on mRNA.

DNA packaging : Nucleosomal model.

79. (c) Steps of DNA Replication in the initiation step, several key factors are recruited to an origin of replication. This origin of replication in unwound and the partially unwound strands form a 'replication bubble'. with one replication fork on either end. Each group of enzymes at the replication fork moves away from the origin, unwinding and replicating the original DNA as they proceed. The factors involved are collectively called the pre-replication complex. It consists of the following A helicase, which unwinds and splits the DNA ahead of the fork. Thereafter, Single-strand Binding Proteins (SSB) swifty bind to the separated DNA, thus preventing the strands from reunting. A primase, which generates an RNA primer to be used in DNA replication.

80. (d)

81. (d) Prokaryotic chromosomes are circular and contain termination sequences.

82. (d) Properties of genetic material:

(i) The replicated genetic material must be transferred faithfully from cells to its daughter cells or from one generation to other.

(ii) Genetic material should be able to express itself.

(iii) There are occasional changes or mutation in the structure and functioning of its genes, which are of permanent nature and inheritable.

(iv) It should be stable.

(v) It should be same in both quantity and quantity in all cells of an individual.

All the property mentioned above are fulfilled by DNA, so DNA is the genetic material.

83. (a) **84. (a)** **85. (c)**

86. (d) Option (a) describes transduction; option (b) describes how auxotrophic mutants can be rescued by mating with other bacteria.

87. (c) Transposable elements of not always have an adverse effect on genes; sometimes their insertion can create new genes. Retrotransposons are like retroviruses, but lack the genes for the viral proteins. DNA transposons do not replicate their DNA before they move to a different site on the chromosome.

88. (c) The scientists involved in the discovery of DNA as chemical basis of heredity were Avery, Mac Leod and McCarty. They expanded the work of Griffith on the process of transformation.

89. (b) Sigma factor is associated with the initiation of transcription. Sigma factor conferrs the specificity of RNA synthesis at the promoter region. Capping involves the addition of unusual nucleotide at the 5' end of hn RNA. Tailing involves the addition of adenylate residues at 3' end in a template independently. Coding strand (also called leading strand) is a strand synthesized by an enzyme in continuous piece in 5' - 3' direction.

90. (a) 3 codons do not code for any amino acid. Such codons are called non-sense codons or terminator codon. Eg UAG, UAA & UGA.

91. (c) **92. (c)** **93. (a)**

94. (b) Operons are segments of genetic material which function as regulated unit or units that can be switched on and switched off. An operon consists of one to several structural genes. (Three in *lac* operon).

These are genes which produce mRNAs for forming polypeptides / proteins / enzymes. Z (produces enzyme β-galactosidase for splitting lactose into glucose and galactose). Y (produces enzyme galactoside permease required in entry of lactose). A (produces enzyme thiogalactoside trans-acetylase).

The three structural genes of the operon produce a single polycistronic mRNA.

95. (a) AUG is initiating codon. UCG codes for serine, UUU codes for phenylalanine, UGU codes for cysteine.

96. (d) During the activation of amino acids, in the presence of Mg^{2+} and ATP an amino acid gets attached to a specific enzyme aminoacyl t-RNA synthetase. Pyrophosphate is released which breaks up to release energy. During translocation, in the presence of the enzyme translocase and energy from GTP the ribosome moves in such a way that the peptidyl bearing t-RNA of A - site comes to lie on the P-site, exposing a new codon at A - site. In the peptidyl transferase reaction energy is provided by GTP.

97. (b) **98. (d)**

99. (c) tRNA (or transfer RNA) is a single stranded RNA molecule which brings amino acid and reads the genetic code in the process of transcription. It helps decode a messenger RNA (mRNA) sequence into a protein. It functions at specific sites in the ribosome during translation, which is a process that synthesizes a protein from a mRNA molecule.

100. (a) A histone octamer is a complex of eight positively charged histone proteins (two of each H_2A, H_2B, H_3 and H_4) that aid in the packaging of DNA. Negatively charged DNA wraps around these histone octamers to form the nucleosome. The DNA is held there by ionic bonds. Linker histone H1 binds to each nucleosome where the DNA enters and exits and this draws a string of nucleosomes closer together to form the 10 nm fibre. The nucleosomes in chromatin are seen as beads-on string structure when viewed under electron microscope.

101. (a)

102. (a) Heat-killed S-type bacteria caused a transformation of the R-type bacteria into S-type bacteria but he was not able to understand the cause of this bacterial transformation. The occurrence of living 19 S-type virulent bacteria is possible only by their formation from R-type non-virulent bacteria, which pick up the trait of virulence from dead bacteria. This phenomenon is called Griffith effect or transformation. Griffith proposed that a chemical substance released by heat killed bacteria changed the R-bacteria into S-bacteria which was a permanent genetic change.

103. (c) In a given DNA segment, 200 guanine bases will pair with 200 cytosine bases. Similarly, 200 thymine bases will pair with 200 adenine bases. Thus, total number of nucleotides in the given segment of DNA is 800.

104. (b)

105. (a) The function of sigma factor is to confer the specificity of RNA synthesis at the promoter site. But during elongation of polypeptide chain, sigma factor is functionless.

106. (c) Genetic code is the depiction of codon by which the information in RNA is decoded in a polypeptide chain. The information is transferred in the form of triplet of bases coding for one amino acid. It is triplet, universal, non-ambiguous and degenerate in nature.

107. (d) All of these genes can be found on plasmid molecules.

108. (b) **109. (c)**

110. (b) In the given figure, the step shown is termination of transcription in bacteria. The label A, B and C are respectively RNA, RNA polymerase and rho factor. RNA polymerase is an enzyme that synthesizes the formation of RNA from a DNA template during transcription. Rho factor is a termination factor which releases RNA from the DNA template.

29 Evolution

1. The Miller-Urey abiotic synthesis experiment (and other subsequent, similar experiments) shows that:

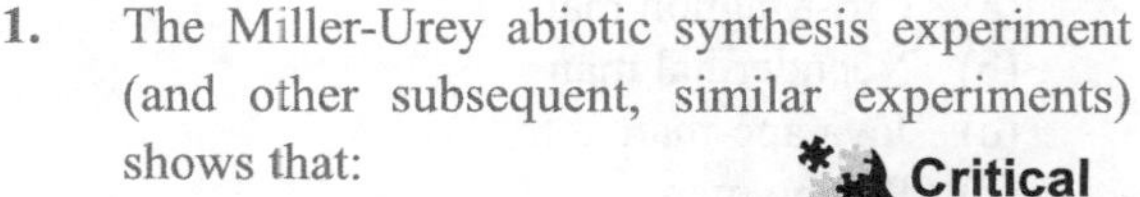

Critical Thinking

(a) simple organic molecules can form spontaneously under conditions like those thought to prevail early in the earth's history.
(b) the earliest life forms introduced large amounts of oxygen to the atmosphere.
(c) life can be created in a test tube.
(d) long chains of DNA can form under abiotic conditions.

2. The atmosphere of early earth probably contained no O_2 until the emergence of organisms that:
(a) were chemoautotrophic.
(b) were oxygen-respiring.
(c) used hydrogen sulfide as a hydrogen source for photosynthesis.
(d) used water as a hydrogen source for photosynthesis.

3. Which of the following has been basic material for the origin of life?
(a) Carbohydrates (b) Proteins
(c) Nucleic acids (d) Nucleoproteins

4. Miller synthesised simple amino acids from water vapours and mixtures containing:
(a) H_2, O_2, N_2 (1 : 2 : 1)
(b) H_2, O_2, N_2 (2 : 1 : 2)
(c) CH_4, NH_3, H_2 (2 : 1 : 2)
(d) CH_4, NH_3, H_2 (1 : 2 : 1)

5. The most primitive cell-like chemical aggregates capable of growth and division were:
(a) Chemoautotrophs (b) Eobionts
(c) Prokaryotes (d) Microspheres

6. Under certain conditions scientists have obtained cell-like structures. These are known as:
(a) Prebiotic soup (b) Coacervates
(c) Protists (d) Microbes

7. A population of organisms has a gene for which there are two alleles, D and d. The allele frequency of D = 0.8. If this population satisfies all five of the Hardy-Weinberg conditions, what are the genotype frequencies that are expected in the next generation?

Tricky

(a) DD = 0.04; Dd = 0.32; dd = 0.64
(b) DD = 0.64; Dd = 0.32; dd = 0.64
(c) DD = 0.04; Dd = 0.64; dd = 0.32
(d) DD = 0.64; Dd = 0.32; dd = 0.04

8. What was the earliest hominid to have an enlarged brain (relative to body size)?
(a) *Neanderthal* (b) *Homo habilis*
(c) *Australopithecus* (d) *Homo erectus*

9. Which characteristic of living monkeys suggests that the earliest anthropoids were monkey-like?

Critical Thinking

(a) All monkeys lack tails, like prosimians.
(b) All monkeys are arboreal, like prosimians.
(c) Like prosimians, monkeys rely more on smell than on eyesight for finding food and avoiding predators.
(d) Of all the anthropoids, monkeys have brains that are most similar in size (relative to body size) to those of prosimians.

10. The earliest primates were most similar to modern:
(a) Arboreal prosimians
(b) Arboreal anthropoids
(c) Ground-dwelling anthropoids
(d) Ground-dwelling prosimians

11. *Australopithecus* was the first of our ancestors known to:
(a) move out of Africa and spread throughout Eurasia.
(b) have definitely diverged from the apes.
(c) leave written records.
(d) have a brain size equal to or larger than ours.

12. Following is the diagrammatic representation of the operation of natural selection of different traits. Which of the following options correctly identifies all the three graphs A, B and C.

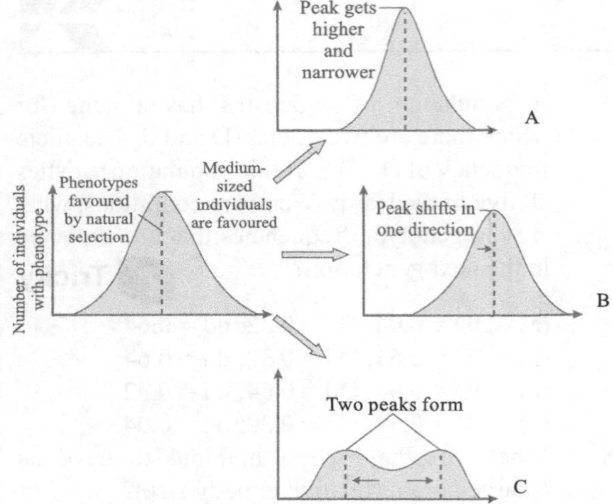

	A	B	C
(a)	Directional	Stabilising	Disruptive
(b)	Stabilising	Directional	Disruptive
(c)	Disruptive	Stabilising	Directional
(d)	Directional	Disruptive	Stabilising

13. Match column-I with column-II and select the correct option.

	Column-I		Column-II
A.	Saltation	I.	Darwin
B.	Formation of life was preceded	II.	Louis Pasteur
C.	Reproductive fitness	III.	de Vries
D.	Life comes from pre-existing life	IV.	Oparin and haldane

(a) A – III ; B – IV ; C – I ; D – II
(b) A – IV ; B – III ; C – II ; D – I
(c) A – II ; B – III ; C – I ; D – IV
(d) A – I ; B – IV ; C – III ; D – II

14. Which of the following is/are **incorrect** about *Neanderthal* man?
(i) Had large brain around 900 c.c.
(ii) Lived in near east and central Asia between 100000-40000 years back.
(iii) Used hides to protect their body and buried their dead.
(iv) Had no religious feeling.
(a) All of these (b) None of these
(c) (i) and (iv) (d) (ii) and (iii)

15. Coloured rock-paintings were presumably first made by:
(a) Cro-Magnon man
(b) Neanderthal man
(c) Java ape-man
(d) Peking man

16. Man differs from apes in having:
(a) Legs longer than arms.
(b) Protruding jaws.
(c) Grasping hands.
(d) Poorly developed forehead.

17. Man who first built huts and burried the dead with ceremonies are:
(a) Java man (b) *Neanderthal*
(c) Cro-Magnon (d) Peking man

18. The presence of which one of the following was absolutely essential for the development of the first life forms? **Critical Thinking**
(a) Heat produced by a planet whose surface was still warm.
(b) A reducing atmosphere.
(c) Volcanoes spewing forth various gases.
(d) UV light.

19. Both prokaryotic and eukaryotic fossils are found in:
(a) Azoic (b) Coenozic
(c) Proterozoic (d) Archaeozoic

20. *Desulfovibrio* is a bacterium that breaks down organic matter which it must have and uses sulphate not oxygen as an electron acceptor. As a result, it produces hydrogen sulphide (H_2S), accounting for the 'rotten egg' smell of swamp muck. Oxygen is a deadly poison to *Desulfovibrio*. We would call *Desulfovibrio* a(n): **Critical Thinking**
(a) facultatively anaerobic chemoautotroph.
(b) obligately anaerobic chemoheterotroph.
(c) facultatively aerobic chemoheterotroph.
(d) obligately anaerobic chemoautotroph.

21. The first ribozyme was found in: 

    (a) nuclear gene for a DNA replicating enzyme.
    (b) mitochondrial gene for a respiratory enzyme.
    (c) mRNA for a mitochondrial enzyme.
    (d) an intron within a pre-rRNA molecule.
22. *Homo habilis* was the first of our ancestors to:
    (a) have a brain size equal to (or larger than) ours.
    (b) leave written records.
    (c) use tools.
    (d) have definitely diverged from the apes.
23. The similarity of mitochondrial DNA among different human races suggests that the races:
    (a) all arose from a single ancestral population.
    (b) are closely related to chimpanzees.
    (c) share no recent common ancestor.
    (d) arose from distinct ancestral populations.
24. The finches of Galapagos islands provide an evidence in favour of:
    (a) evolution due to mutation.
    (b) retrogressive evolution.
    (c) biogeographical evolution.
    (d) special creation.
25. The shape of the heals of Darwins finches, industrial melanism, or the changes in horse teeth are all examples of:
    (a) Artificial selection
    (b) Natural selection
    (c) Convergent evolution
    (d) Homologous structures
26. Choose the **incorrect** statement regarding Hardy-Weinberg principle.
    (a) Allele frequencies in a population are stable and constant from generation to generation.
    (b) Sum total of all the allelic frequencies in a population is 1.
    (c) Variation due to genetic drift results in changed frequency of genes and alleles in future generations.
    (d) Genetic recombination helps in maintaining Hardy-Weinberg equilibrium.
27. Which of the following pairs of terms is not related?
    (a) Evolution - Natural selection
    (b) Vestigial structure - Appendix
    (c) Analogous structures - Butterfly wings
    (d) Adaptive radiations - Vertebrates hearts or brains
28. What was the most significant trend in the evolution of modern man (*Homo sapiens*) from his ancestors? 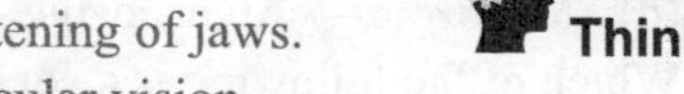

    (a) Shortening of jaws.
    (b) Binocular vision.
    (c) Increasing cranial capacity.
    (d) Upright posture.
29. Trilobite is an extinct:
    (a) Mammal (b) Reptile
    (c) Mollusc (d) Arthropod
30. The tendency of population to remain in genetic equilibrium may be disturbed by: Critical Thinking
    (a) Random mating
    (b) Lack of immigration
    (c) Lack of mutations
    (d) Lack of random mating
31. Which one of the following supports the 'multi-regional' hypothesis of the origin of modern humans from archaic *Homo sapiens*?
    (a) The high variability of nuclear DNA in some modern Africans.
    (b) The uniformity in mtDNA among modern human populations.
    (c) The oldest known fossils of modern *Homo sapiens* have been found in Africa.
    (d) The existence of distinctive human races.
32. In human evolution, which of the following occurred before the switch to walking upright? 

    (a) Tool use
    (b) Mastery of fire
    (c) Spoken language
    (d) Enhanced grasping ability
33. Which is thought to be the first hominid to use fire?
    (a) *Homo erectus*
    (b) *Homo habilis*
    (c) *Australopithecus africanus*
    (d) The Neanderthals

34. The ultimate source of all genetic variation is:
(a) Natural selection
(b) Genetic drift
(c) Sexual recombination
(d) Mutation

35. Select the correct order of geological time scale of earth.
(a) Palaeozoic → Archaeozoic → Coenozoic
(b) Archaeozoic → Palaeozoic → Proterozoic
(c) Palaeozoic → Mesozoic → Coenozoic
(d) Mesozoic → Archaeozoic → Proterozoic

36. Which of the following is a **correct**?
(a) *Homo erectus* lived in east and central Asia and used hides to protect their bodies.
(b) Agriculture came around 18000 years back.
(c) The skull of modern human resembles more closely to baby chimpanzee than to adult chimpanzee.
(d) All of the above

37. The diversity within the wild bird species in the diagram below can best be explained by which process? Tricky

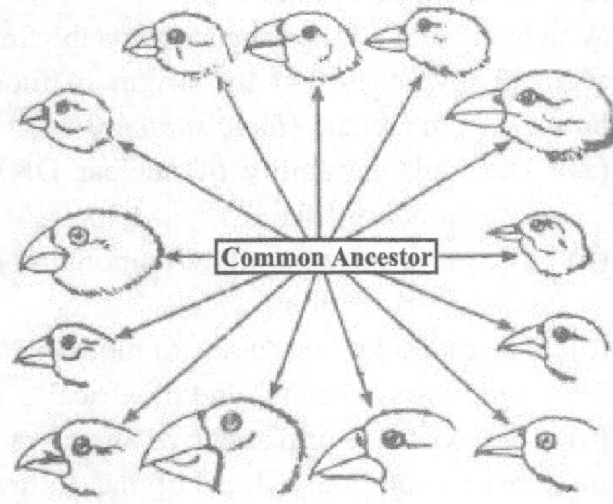

(a) Natural selection
(b) Ecological succession
(c) Adaptive radiation
(d) Both (a) and (c)

38. Coacervates belongs to which of the following categories?
(a) Cyanobacteria
(b) Protozoans
(c) Molecular aggregate with lipid bilayer.
(d) Molecular aggregate which lack definite lipid membrane.

39. Which one of the following amino acids was not found to be synthesised in Miller's experiment?

(a) Alanine (b) Glycine
(c) Aspartic acid (d) Glutamic acid

40. How has the free O_2 been released into the earth's atmosphere as there was no free O_2 at the time of origin of life? Tricky
(a) By catabolic activity of animals
(b) By photosynthetic activity
(c) By heating of the earth
(d) None of the above

41. The first formed organism (riboorganism) used only........for catalysing reactions.
(a) DNA (b) Amino acids
(c) Fatty acids (d) RNA

42. Read the following three statements (i - iii) and mark the right option.
(i) The thorns in *Bougainvillea* and tendrils in cucurbits represent divergent evolution.
(ii) The similarity in the eyes of *Octopus* and monkeys is the result of convergent evolution.
(iii) The potato and sweet potato are the examples of homology.
(a) (i) and (ii) (b) (ii) and (iii)
(c) (i) and (iii) (d) Only (i) is

43. Identify the **incorrect** statements regarding origin of life:
(i) Swan-neck flask experiment was performed by Louis Pasteur.
(ii) Louis Pasteur is famous for germ theory of disease.
(iii) Louis Pasteur disapproved spontaneous theory forever.
(iv) Cosmozoic theory of origin of life was proposed by Richter.
(v) Theory of Catasrophism was given by Georges Cuvier.
(a) (i), (ii) and (iv)
(b) (i), (ii), and (v)
(c) (i), (ii), (iii), (iv) and (v)
(d) None of the above

44. Match the evolution concepts given in column-I with their description given in column-II and select the correct answer.

Column-I	Column-II
A. Mutation	I. Change in population's allele frequencies due to chance alone.
B. Gene flow	II. Differences in survival and reproduction among variant individuals.
C. Natural selection	III. Immigration, emigration change allele frequencies.
D. Genetic drift	IV. Source of new alleles.

(a) A – I ; B – II ; C – III ; D – IV
(b) A – IV ; B – II ; C – III ; D – I
(c) A – III ; B – I ; C – IV ; D – II
(d) A – IV ; B – III ; C – II ; D – I

45. Sita saw an old Tarzan movie on television. The movie supposedly took place in Africa, but Trina easily spotted that it was not really filmed there. Which of the following could have tipped her off?

Critical Thinking

(a) Chimps like Cheetah do not live in Africa.
(b) The monkeys in the jungle all had prehensile tails.
(c) There were no lemurs shown in the forest.
(d) Tarzan wrestled with a gorilla.

46. Evolutionary convergence is characterised by:
(a) development of dissimilar characteristics in closely related groups.
(b) development of common set of characteristics in groups of different ancestory.
(c) development of characteristics by random mating.
(d) replacement of common characteristics in different groups.

47. The given figure shows an example of:

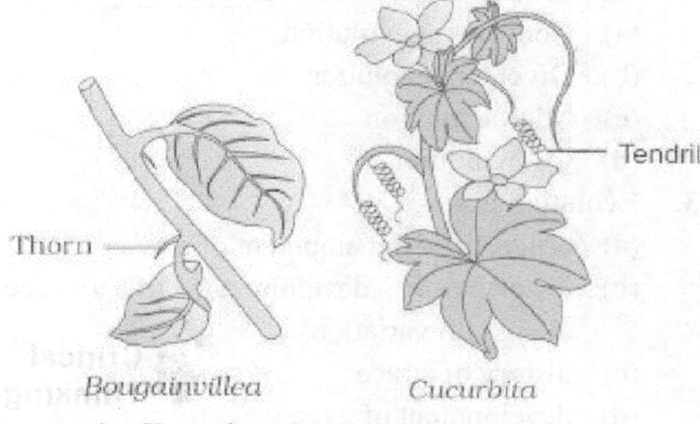

(a) Homologous organs
(b) Convergent evolution
(c) Divergent evolution
(d) Both (a) and (c)

48. A new species is formed when:
(a) an individual with a new genotype is formed due to exchange of chromosome segments during crossing-over in gametogenesis.
(b) genotypic changes accumulate in a population resulting in its reproductive isolation.
(c) variants with new phenotypes are produced due to new combinations of genes during reproduction.
(d) homologous chromosomes exchange segments during crossing-over in gametogenesis.

49. Ancestor of man who first stood erect was:
(a) Australopithecus (b) Java ape man
(c) Peking man (d) Cro-Magnon

50. Cro-Magnon man was:
(a) Carnivorous (b) Omnivorous
(c) Herbivorous (d) Frugivorous

51. Diversity in the type of beaks of finches adapted to different feeding habits on the Galapagos Islands, as observed by Darwin. This provides evidence for:
(a) Intraspecific variations
(b) Intraspecific competition
(c) Interspecific competition
(d) Origin of Species by natural selection

52. When two species of different genealogy come to resemble each other as a result of adaptation, the phenomenon is termed as ________.
(a) Convergent evolution
(b) Divergent evolution
(c) Microevolution
(d) Co-evolution

53. Evolution is a:
(a) progressive development of a race.
(b) history and development of a race alongwith variations.
(c) history of a race.
(d) development of a race.

Critical Thinking

54. One of the important consequences of geographical isolation is:
(a) random creation of new species.
(b) no change in the isolated fauna.
(c) preventing speciation.
(d) speciation through reproductive isolation.

55. A geneticist mixed together many different kinds of fruit flies–some with long wings, some with short wings, some with red eyes, some with brown eyes, and so on. He allowed the flies to feed, mate randomly, and reproduce by the millions. After many generations, most of the flies in the population had medium wings and red eyes, and most of the extreme types had disappeared. This experiment appears to demonstrate:

Critical Thinking

(a) Stabilising selection
(b) Geographic variation
(c) Diversifying selection
(d) Genetic drift

56. Most biologists think that RNA was the first genetic material because:

Critical Thinking

(a) amino acids were produced in Stanley Miller's apparatus.
(b) DNA is the universal genetic material of eukaryotes.
(c) existence of ribozymes suggests RNA was used to catalyse chemical reations.
(d) RNA is simpler than DNA.

57. In primordial atmosphere, when the living organism evolved, the respiration process was:
(a) anaerobic as it released more energy.
(b) aerobic as it was more complex.
(c) aerobic as it released more energy.
(d) anaerobic because early atmosphere contained little or no oxygen.

58. Reproductive isolation can occur in a number of ways, but the net effect is always the same in that:
(a) hybridisation must take place before new species can form.
(b) it causes geographic isolation between organisms.
(c) the rate of speciation slows down.
(d) few or no genes flow between populations.

59. Reproductive isolation is to speciation as selection is to:
(a) Adaptive evolution
(b) Hybridisation
(c) Ecological isolation
(d) Polyploidy

60. What is observed in evolution?
(a) A sudden change occurring in a population.
(b) Progeny with modifications.
(c) It is a discontinuous process.
(d) All of the above

61. The evolution occurring at a genetic level is called:
(a) Macroevolution
(b) Megaevolution
(c) Inorganic evolution
(d) Microevolution

62. Identify the **correct** conclusion regarding the given statements.

Critical Thinking

(i) There are patterns in the fossil record that suggest other species have diverged from a single ancestor species.
(ii) There are biogeographic patterns in the distribution of species, for instance distinct bird species on an island tend to resemble one another, suggesting a common ancestor.

(iii) There are common stages in the early embryological development of organisms representing several distinct vertebrate groups.

(iv) Anatomical structures, such as forelimbs, in different groups appear to be modified versions of structures that might have been present in a common ancestor.

(a) These are the example of a macro evolutionary process.
(b) They are the evidences for Lamarck's theory of inheritance.
(c) They are the evidences for Darwin's theory of common descent.
(d) They are the examples of natural selection which occurs within populations, where artificial selection does not occur.

63. Select the **correct** statement(s).
(i) Adaptative ability is inherited.
(ii) Adaptative ability has a genetic basis.
(iii) Fitness is the end result of the ability to adapt and get selected by nature.
(a) (i), (ii) and (iii)
(b) None of these
(c) Only (i) and (iii)
(d) Only (ii)

64. Match the column-I with column-II and select the correct option.

	Column-I (Name of the Scientist)		Column-II (Contributions)
A.	Charles Darwin	I.	Mutation theory
B.	Lamarck	II.	Germ plasm theory
C.	Hugo De Vries	III.	*Philosophie Zoologique*
D.	Ernst Haeckel	IV.	The Origin of species
E.	August Weismann	V.	Biogenetic law
		VI.	Essay on population

(a) A – IV; B – III; C – I; D – V; E – II
(b) A – IV; B – III; C – V; D – I; E – VI
(c) A – IV; B – VI; C – V; D – III; E – I
(d) A – II; B – III; C – I; D – V; E – II

65. The diagram (A and B) given below shows the skull of two different mammals. Which of the following accurately describes the differences between these skulls?

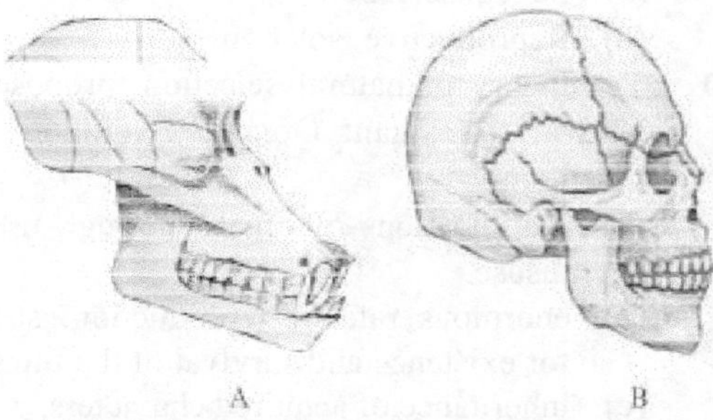

(a) Skull A has more teeth than skull B.
(b) Skull A has more brain capacity than skull B.
(c) Skull A is of a human and skull B is of an ape.
(d) Skull A is of an ape and skull B is of human.

66. More recent prebiotic simulations in the lab mimic conditions found in ____________.
(a) deep sea hydrothermal vents
(b) shallow ponds
(c) underside of sea ice
(d) All of the above.

67. Consider a population with a total of 500 alleles for a specific gene. How many individuals are in this population? **Critical Thinking**

(a) 1000 (b) 500
(c) 250 (d) 125

68. If a guppy species found in a stream is placed in a lake, its colour will evolve rapidly from colourful to drab over as few as 10 generations because of the:
(a) lack of food.
(b) presence of additional predators in the lake.
(c) temperature of the air.
(d) taste of the water.

69. The smallest unit that can evolve, is a:

(a) Species (b) Genotype
(c) Gene (d) Population

70. Most modern breeds of domestic dogs are evolved by:
(a) Natural selection
(b) Artificial selection
(c) Sexual selection
(d) Reproductive isolation

71. The theory of natural selection, proposed by Darwin to explain Organic Evolution was based on:
(a) modifications of organs through use and disuse.
(b) enormous rate of reproduction, struggle for existence and survival of the fittest.
(c) inheritance of acquired characters.
(d) appearance of sudden large variations, inheritance and survival of those having these variations.

72. Suppose that a new fossil hominid is discovered. What criterion can be used to determine where in the human evolutionary lineage this new fossil belongs?

Tricky

(a) Brain size (b) Tooth structure
(c) Foot structure (d) All of these

73. The given figure provides an example of:

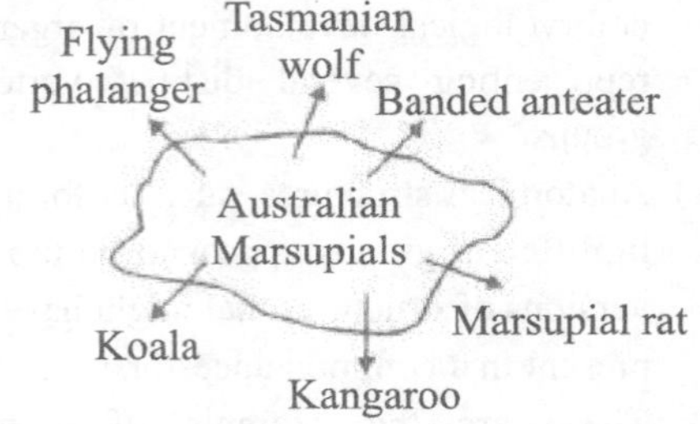

(a) Convergent evolution
(b) Parallel evolution
(c) Recapitulation
(d) Divergent evolution

74. Which of the following is **correct** statement ?

(a) Life originated in mesozoic era.
(b) Earth was formed about 4.6 million years ago.
(c) Ist living organism in primitive ocean was coacervate.
(d) Homologous organs show divergent evolution.

Answer KEYs

1	(a)	9	(d)	17	(b)	25	(b)	33	(a)	41	(d)	49	(a)	57	(d)	65	(d)	73	(d)
2	(d)	10	(a)	18	(b)	26	(d)	34	(d)	42	(a)	50	(a)	58	(d)	66	(a)	74	(d)
3	(c)	11	(b)	19	(d)	27	(d)	35	(c)	43	(d)	51	(d)	59	(a)	67	(c)		
4	(d)	12	(b)	20	(b)	28	(c)	36	(c)	44	(d)	52	(a)	60	(b)	68	(b)		
5	(d)	13	(a)	21	(d)	29	(d)	37	(d)	45	(b)	53	(b)	61	(d)	69	(d)		
6	(b)	14	(c)	22	(c)	30	(d)	38	(d)	46	(b)	54	(d)	62	(c)	70	(b)		
7	(d)	15	(a)	23	(a)	31	(d)	39	(d)	47	(d)	55	(a)	63	(a)	71	(b)		
8	(b)	16	(a)	24	(c)	32	(d)	40	(b)	48	(b)	56	(c)	64	(a)	72	(d)		

Hints & Solutions

1. (a) In 1953, Stanley Miller and Harold Urey created, in the laboratory, conditions comparable to those of early Earth, with water vapour, hydrogen, methane, and ammonia. The Miller-Urey apparatus produced a variety of amino acids and other organic compounds found in living organisms today.
2. (d) Around 2.5 billion years ago, during the Precambrian, oxygen began to accumulate in the atmosphere due to the action of photosynthetic cyanobacteria.
3. (c) 4. (d) 5. (d) 6. (b)
7. (d) We are told that the allele frequency for D = 0.8; therefore, the frequency of d = 0.2. These values are the p and q that we need to calculate the genotype frequencies in the next generation. Using the Hardy-Weinberg equation, p^2 (DD) = 0.64, 2pq (Dd) = 0.32, and q^2 (dd) = 0.04.
8. (b) The brain was intermediate in size between that of *Australopithecus* and that of *Homo sapiens. Homo habilis* had come in the age between that of *Australopithecus* and *Homo sapiens*.
9. (d) The brains of monkeys are larger than those of prosimians, but they are smaller than the brains of apes and hominids.
10. (a) The promisians probably resemble early arboreal primates.
11. (b) Discovered in 1924 by Raymond Dart, *Australopithecus* was a hominid that walked fully erect and had human like hands and teeth.
12. (b) The graph A, B and C shows stabilising, directional and disruptive traits of natural selection. In stabilizing selection, the median phenotype is selected during natural selection and which does not tilt the bell curve in any way. Instead, it makes the peak of the bell curve even higher than what would be considered normal. Directional selection of natural selection derives its name from the shape of the approximate bell curve that is produced when all individual's traits are plotted. Instead of the bell curve falling directly in the middle of the axes on which they are plotted, it tilts either to the left or the right by varying degrees. Hence, it has moved in one direction or the other. In disruptive selection instead of the bell curve having one peak in the middle, it has two peaks with a valley in the middle of them.
13. (a) Saltation theory was given by de Vries who was a Dutch botanist working with primroses. He believed that a new species had suddenly sprung into existence as a mutation. He theorized that new species "saltated" (leaped), that is continually sprung into existence. Oparin and Haldane give origin of life, according to which life evolved in the oceans during a period when the atmosphere was reducing - containing H_2, H_2O, NH_3, CH_4, and CO_2, but no free O_2. Reproductive fitness was given by Darwin. Louis Pasteur was a French chemist and microbiologist renowned for his discoveries of the principles of vaccination, microbial fermentation and pasteurisation. According to him, life comes from pre-existing life.
14. (c) 15. (a) 16. (a) 17. (b)
18. (b) Reducing atmosphere would have enhanced the joining of simple molecules to form more-complex ones.
19. (d)
20. (b) Obligate anaerobes are poisoned by oxygen. Chemoheterotrophs must consume organic molecules for both energy and carbon.
21. (d)
22. (c) Simple handmade stone tools are sometimes found with the fossils of *Homo habilis*.
23. (a) Analysis of mitochondrial DNA from an exceptionally well-preserved Neanderthal fossil in Europe indicates that the individual was very different from modern Europeans, supporting the monogenesis model.
24. (c) Biogeographical evolution is a process in which gene pool of a population gradually changes in response to environmental pressures, natural selection and genetic mutations.

25. (b) Natural selection is the Darwin's most famous theory of evolution. Natural selection is the gradual process by which transmissible biological traits become either more or less common in a population as a function of the effect of inherited traits on the differential reproductive success of organisms interacting with their environment. It is a key mechanism of evolution. The shape of heal of Darwin finches, industrial melanism or the changes in horse teeth are all examples of natural selection.

26. (d) Genetic recombination leads to variation results in changed frequency of genes and alleles in future generation.

27. (d) Adaptive radiations - Darwin finches and marsupials.
Vertebrate's hearts and brains are the examples of homologous organ.

28. (c) Binocular vision, smaller jaw and upright posture are the main adaptations that lead to evolution of man from its ancestors. But during human evolution major and most significant changes occurred in the cranial capacity of man. In living modern man it is about 1450 cc compared to 500 cc in *Australopithecus*. Increased cranial capacity accommodates larger brain and forms the basis of social, cultural and educational evolution of modern man.

29. (d) Trilobites are well-known fossil group of extinct-marine arthropods that form the class Trilobite. The first appearance of trilobites in the fossil record defines the base of the Atdabanian stage of the early Cambrain period (526 million years ago), and they flourished throughout the lower palaeozoic era before beginning a drawn-out decline to extinction when, during the Devonian, all trilobite orders with the sole exception of proetids, died out.

30. (d) The tendency of population to remain in genetic equilibrium may be disturbed by lack of random mating. Random mating is a mating between individuals where the choice of partner is not influenced by the genotypes. The mating of individuals in a population such that the union of individuals with the trait under study occurs according to the product rule of probability.

31. (d) In fact, this observation led to the proposal of the 'multi-regional' hypothesis.

32. (d) Enhanced grasping ability was a very early primate adaptation for life in the trees, and so it pre-dates upright walking in the hominids by tens of millions of years.

33. (a) This appears to be the first hominid group to use fire.

34. (d) Changes in DNA are the ultimate source of all variation.

35. (c) The correct geological time scale of earth is as follows:

Precambrian (Era/Age of early life)
↓
Palaeozoic (Age of fishes)
↓
Mesozoic (Age of reptiles dinosaurs and cycads)
↓
Coenozoic (Age of mammals)

A geological time scale is a diagram that deals with the history of earths geology, noting major events like the formation of the earth, the first life forms and mass extinctions. The first geological time scale was proposed in 1913 by the British geologist **Arthur Holmes** (1890-1965).
The history of the earth has been subdivided into eras, eras into periods and periods into major divisions.

36. (c) Neanderthal man lived in east and central Asia and used hides to protect their bodies. Agriculture came around 10000 years back and human settlement started.

37. (d)

38. (d) Eobionts are also called protocell or protobiont. There are two types of eobionts– coacervates and microsphere. Coacervates lack membrane, no one claims coacervates are alive, but they do exhibit some life like characters. They are able to grow and divide. Microsphere is a non-living collection of organic molecule with double layered outer boundary. The terrm was given by Sidney Fox (1958 - 1964).

39. (d) 40. (b)

41. **(d)** It was reported in 1989 by Nobel Laureaute, Thomas Cech that RNA molecules can act as their own enzymes in certain microorganisms.
42. **(a)** The potato (stem modification) and sweet potato (root modification) are the examples of analogy. Analogy is the similarity of function and superficial resemblance of structures that have different origins.
43. **(d)** Theories of origin of life and their creators or supporter are as follows:
    (i) Theory of Special Creation - The greatest supporter of this theory was father Suarez.
    (ii) Theory of Spontaneous Creation - This concept was held by early Greek philosophers like Thales, Plato Aristotle.
    (iii) Cosmozoic Theory or Theory of Panspermia - This theory was proposed by Richter (1828).
    (iv) Theory of Eternity of Life - This theory was proposed by Prayer in 1880.
    (v) Theory of Catastrophism - This theory was given by Georges Cuvier (1769-1832)
    (vi) Modern Theory - Oparin (1938) and Haldane (1929) gave similar views regarding the origin of life called chemical or naturalistic theory.
44. **(d)**
45. **(b)** Only certain new world monkeys have a prehensile tail.
46. **(b)**
47. **(d)** The given figure of *Bougainvillea* and *Cucurbita* shows an example of homologous organ or divergent evolution. Homologous organ are those organs which are dissimilar in shape, size and function but their origin, basic plan and development are similar. Other examples in animals are fore limbs of frog, reptile, birds and mammals. Such differences are due to divergent evolution or adaptation for varied conditions.
48. **(b)** **49. (a)** **50. (a)**
51. **(d)** Nature selects those set of characters that are best adapted to the environment. This has resulted in the great diversity seen in the population of finches.
52. **(a)** Convergent evolution is the process whereby organisms which are not closely related independently evolve similar traits as a result of capability to adapt to similar environments or ecological niches. It is the independent evolution of similar features in species of different lineages. Convergent evolution creates analogous structures that have similar form or function, but that were not present in the last common ancestor of those groups.
53. **(b)** The term evolution was given by Herbert Spencer that is "descent with modifications". Evolution helps us to understand the history of life.
54. **(d)** Speciation through reproductive isolation is one of the important consequences of geographical isolation. Geographical isolation is a group of plants, animals or other living individuals being separated from mixing genes within their same species. Geographical isolation may ultimately lead to the populations becoming separate species by adaptive radiation.
55. **(a)** Stabilising selection favours individuals about the mean.
56. **(c)**
57. **(d)** Primordial conditions on the earth were high temperature, volcanic storms, reducing atmosphere containing CH_4, NH_3, H_2, etc. There was no oxygen on the primitive earth. The first living organisms developed among the organic molecule and in oxygen free atmosphere. They presumably obtained energy by the fermentation of organic compounds and were almost certainly heterotrophs, requiring readymade organic materials as food. They were also anaerobic *i.e.* capable of respiration in the absence of oxygen.
58. **(d)** Reproductive isolation recalls that gene flow does not occur, which sets the stage for possible divergence of traits that eventually can lead to speciation.
59. **(a)** Reproductive isolation is necessary for speciation to occur. Likewise, natural selection must take place for adaptive evolution to occur.
60. **(b)** Evolution is a continuous process is which progeny with modification are produced.

61. (d) Microevolution is a series of changes within a species due to gene mutation and accumulation of variations.
62. (c) The given statements are the evidences for Darwin's theory of common descent. A group of organisms share common descent, if they have a common ancestor. There is strong quantitative support for the theory that all living organisms on Earth are descended from a common ancestor. Charles Darwin proposed the theory of universal common descent through an evolutionary process in the 'Book Origin of Species'.
63. (a) 64. (a)
65. (d) The diagrams of two skulls of two different mammals show that skull A is of an ape and skull B is of human.
66. (a) More recent prebiotic simulations in the lab mimic conditions found in deep sea hydrothermal vents.
67. (c) Since each diploid individual has two alleles, the number of individuals is one-half of the total number of alleles in the population.
68. (b) The guppies that are found in the stream can be colourful because there are few predators. In a lake, there are more predators and colourful individuals. Therefore, the colour of the transplanted fish will change from bright to drab relatively rapid.
69. (d) This is the least inclusive level that exhibit variation.
70. (b) 71. (b)
72. (d) All of these hominid characteristics have been influenced by evolution, so they serve as legitimate bases for comparison of the new fossil to existing fossils.
73. (d) 74. (d)

30 Human Health and Disease

1. Infectious proteins are present in:
   (a) Satellite viruses (b) Gemini viruses
   (c) Prions (d) Viroids
2. Robert Koch's postulates to establish a micro-organism as pathogen can be obtained, after his work on:
   (a) Hydrophobia (b) Anthrax
   (c) Small pox (d) Diphtheria
3. RBC counting is low in anaemia and:
   (a) Typhoid (b) Leukemia
   (c) Influenza (d) Myxoedema
4. A child is suffering from Kwashiorkor, and if this child is compared with other marasmus children then what additional symptoms are present in Kwashiorkor child? **Toughnut**
   (a) Decrease in body weight
   (b) Wasted muscles
   (c) Oedema
   (d) Impaired physical growth
5. Chronic anaemia and multiple sclerosis are:
   (a) Allergic diseases
   (b) Hormonal diseases
   (c) Autoimmunity diseases
   (d) Hereditary diseases
6. Vaccines are:
   (a) treated bacteria or viruses or one of their proteins.
   (b) MHC (major histocompatibility complex) proteins.
   (c) curative medicines.
   (d) monoclonal antibodies.
7. For which of the following disease, there is preventive vaccine? **Tricky**
   (a) AIDS (b) Hepatitis B
   (c) Syphilis (d) Gonorrhoea
8. Which one of the following statements is **correct** about T-lymphocytes in mammals?
   (a) These are produced in thyroid.
   (b) There are three main types — cytotoxic T-cells, helper T-cells and suppressor T-cells.
   (c) These are originated in lymphoid tissues.
   (d) They scavenge damaged cells and cellular debris.
9. Active immunity is:
   (a) borrowed from an active disease case.
   (b) developed in direct response to a disease agent.
   (c) the product of borrowed antibodies.
   (d) passive immunity that is activated.
10. In the cell-mediated immune response, T-lymphocytes divide and secrete:
    (a) Antigens (b) Plasmogens
    (c) Collagens (d) Cytokines
11. B-lymphocytes are primarily involved in:
    (a) Humoral immunity
    (b) Autoimmune disorders
    (c) Graft rejection
    (d) Cell-mediated immunity
12. Passive immunity is defined as immunity:
    (a) Inherited from the parents.
    (b) Achieved through vaccination.
    (c) Acquired through first exposure to the disease.
    (d) Achieved through the sera of other animals enriched in antibodies.
13. If a person shows production of interferons in his body, the chances are that he has got an infection of:
    (a) Typhoid (b) Measles
    (c) Tetanus (d) Malaria

14. Passive immunity is achieved from injection of:
(a) Antibodies (b) Vaccines
(c) Antigens (d) Histamines

15. Antiserum is rich in:
(a) Steroids (b) Antibodies
(c) Antigens (d) RBC

16. Identify the disease against which human beings have innate immunity?
(a) Distemper (b) Influenza
(c) Small-pox (d) Rabies

17. Immunoglobulin is present in sebum of:
(a) IgM (b) IgG
(c) IgA (d) IgE

18. Lymphoid tissue is found in:
(a) Thymus (b) Tonsils
(c) Lymph nodes (d) All of these

19. A non-infectious unnatural and unusual reaction to a substance or condition is:
(a) Toxin (b) Allergy
(c) Infection (d) Immunity

20. A person is injected with globulin against hepatitis. This is: **Critical Thinking**
(a) Naturally acquired active immunity
(b) Naturally acquired passive immnity
(c) Artificially acquired active immunity
(d) Artificially acquired passive immunity

21. The antibody dependent cytotoxicity is seen in: **Tricky**
(a) Local anaphylaxis
(b) Generalise anaphylaxis
(c) Immune complex reaction
(d) Non-compatable-blood transfusion

22. What is released in body during disease which results in fever?
(a) Pyrogens (b) Antibodies
(c) Interferons (d) Interleukins

23. E. Jenner prepared the vaccine for smallpox virus using:
(a) Attenuated cowpox virus
(b) Attenuated smallpox virus
(c) Large dose of smallpox virus
(d) Small dose of smallpox virus

24. Monoclonal antibodies are produced from hybrid cells, called hybridomas. The cells employed to obtain these hybridoma cells, are:
(a) B-lymphocytes and myeloma cells
(b) Lymphoma cells and bone marrow cells
(c) T-lymphocytes and myeloma cells
(d) B-lymphocytes and carcinoma cells

25. Types of antibodies involved in asthma are:
(a) IgA (b) IgM
(c) IgE (d) IgD

26. The class of antibodies which appear first after the infection, is:
(a) IgG (b) IgM
(c) IgA (d) IgD

27. All allergens are non-infectious foreign substances that:
(a) increase the secretion of IgA
(b) increase the secretion of IgE
(c) increase the secretion of IgG
(d) increase the secretion of IgM

28. Mast cells secrete a protein which causes the dilation of blood vassels. The protein is:
(a) Pyrogens
(b) Histamine
(c) Interferon
(d) None of the above

29. Which one of the following conditions though harmful in itself, is also a potential saviour from a mosquito borne infectious disease? **Tricky**
(a) Leukemia
(b) Thalassemia
(c) Sickle cell anaemia
(d) Pernicious anaemia

30. Identify the odd one out. **Critical Thinking**
(a) Rabies, Influenza, AIDS
(b) Amoebiasis, Giardiasis, Trypanosomiasis
(c) Taeniasis, Ascariasis, Elephantiasis
(d) Cancer, Tuberculosis, Tetanus

31. 'Encephalitis' in man is a viral disease and is transmitted by:

(a) *Anopheles* (b) *Culex*

(c) *Aedes* (d) *Culex* and *Aedes*

32. *Aedes aegypti* is a vector of:

(a) Both dengue and yellow fever

(b) Dengue fever

(c) Yellow fever

(d) Japanese encephalitis

33. An immune disease in which the body destroys the ill-functioning of thyroid gland, is called:

(a) Cretinism

(b) Myxoedema

(c) Simmond's disease

(d) Hashimoto's disease

34. Food poisoning is caused by:

(a) *Entamoeba histolytica*

(b) *Escherichia coli*

(c) *Clostridium botulinum*

(d) *Corynebacterium diphtheriae*

35. Toxin produced by tetanus affects:

(a) Jaw bones

(b) Voluntary muscles

(c) Involuntary muscles

(d) Both voluntary and involuntary muscles

36. Amoebiasis is caused by:

(a) *Plasmodium vivax*

(b) *Entamoeba gingivalis*

(c) *Trypanosoma gambiense*

(d) *Entamoeba histolytica*

37. The infective stage of *Entamoeba histolytica* is:

(a) Binucleate form

(b) Tetranucleate form

(c) Minuta form

(d) Sporozoite stage

38. Parkinson's disease is caused due to deficiency of:

Critical Thinking

(a) GABA (b) Endorphins

(c) Dopamine (d) Acetylcholine

39. Radiation is health hazard because it causes:

(a) Haemophilia (b) Leukemia

(c) Pneumonia (d) Leucopenia

40. If you keep the sanitary system around yourself then the disease which will not most probably break out is:

(a) Cholera (b) Malaria

(c) Beri-beri (d) Scurvy

41. The interval between infection and appearance of a disease is known as:

Critical Thinking

(a) Inoculation

(b) Penetration

(c) Infection period

(d) Incubation period

42. Which of the following is transmitted through the serum in man?

(a) Hepatitis B

(b) Cirrhosis

(c) Tuberculosis

(d) Venereal disease

43. Which one of the following pair of diseases is viral?

(a) Cholera and TB

(b) AIDS and Syphilis

(c) Mumps and Rabies

(d) Tetanus and Typhoid

44. Which of the following diseases is not transmitted by houseflies?

(a) Typhoid (b) Yellow fever

(c) Dysentery (d) Cholera

45. There is a patient having a disease in which a semisolid material oozes out and forms a tough membrane over it in air passage. The disease is:

Critical Thinking

(a) Diphtheria (b) Pertussis

(c) Tetanus (d) TB

46. The diseases mumps, cholera and TB are the type of ______.

(a) Bacterial diseases

(b) Endemic diseases

(c) Viral diseases

(d) Communicable diseases

47. DPT stands for: **Tricky**

(a) Diphtheria, Pertussis and Typhoid
(b) Diarrhoea, Plague and Tetanus
(c) Diarrhoea, Whooping cough and Tuberculosis
(d) Diphtheria, Pertussis and Tetanus

48. Deposition of cholesterol in the arteries causes:

(a) Angina pectoris
(b) Arteriosclerosis
(c) Heart failure
(d) None of the above

49. Leprosy is also famous as:

(a) Koch's disease
(b) Hensen's disease
(c) Pertussis
(d) Cholera

50. Gaucher's disease is associated with:

(a) Abnormal protein metabolism
(b) Abnormal carbohydrate metabolism
(c) Abnormal fat metabolism
(d) Malnutrition

51. Which of the following is not a water borne disease?

(a) Asthma (b) Cholera
(c) Amoebiasis (d) All of these

52. The disease marasmus in children is caused due to the deficiency of:

(a) Carbohydrates (b) Proteins
(c) Fats (d) All of these

53. Damage to thymus in a child may lead to:

(a) a reduction in haemoglobin content of blood.
(b) a reduction in stem cell production.
(c) loss of antibody mediated immunity.
(d) loss of cell mediated immunity.

54. Cirrhosis of the liver is caused by:

(a) Cocaine (b) LSD
(c) Alcohol (d) Morphine

55. The vector of 'break bone fever' is:

(a) *Culex* (b) *Aedes*
(c) *Anopheles* (d) None of these

56. A triple antigen (vaccine) is: **Tricky**

(a) BCG (b) DPT
(c) ATS (d) TAB

57. Lung cancer is caused by:

(a) Asbestos
(b) Cement factory
(c) Bauxite mining
(d) Calcium fluoride

58. AIDS is due to:

(a) Reduction in number of helper-T cells
(b) Lack of interferon
(c) Reduction in number of killer-T cells
(d) Autoimmunity

59. Which type of cancer is found in lymph nodes and spleen?

(a) Sarcoma (b) Carcinoma
(c) Adenoma (d) Leukemia

60. Radiotherapy is used for:

(a) detecting bone fracture.
(b) detecting cardiac trouble.
(c) getting whole body's photograph.
(d) treating cancer by controlled exposure of X-rays.

61. Which of the following is used in the treatment of thyroid cancer? **Tricky**

(a) I_{131} (b) U_{258}
(c) Ra_{224} (d) C_{14}

62. 'Western blot test' is done for the confirmation of:

(a) Filaria (b) AIDS
(c) Anaemia (d) Malaria

63. Co is used for the treatment of:

(a) Cancer (b) Thyroid
(c) AIDS (d) Hepatitis

64. Cancer cells are:

(a) Hela cells
(b) CD_4 cell
(c) Memory cell
(d) Plasma cell

65. At which stage of HIV infection does one usually show symptoms of AIDS?

Critical Thinking

(a) Within 15 days of sexual contact with an infected person.
(b) When the infecting retrovirus enters host cells.
(c) When viral DNA is produced by reverse transcriptase.
(d) When HIV replicates rapidly in helper T-lymphocytes and damages large number of these cells.

66. Hodgkin's disease is the cancer of:
(a) Bone marrow
(b) Lymphoid tissue
(c) Mammary glands
(d) Liver

67. Carcinogen present in the cigarette-smoke is:
(a) Nitrosamine (b) Aflatoxin
(c) Nicotine (d) None of these

68. LSD is obtained from:
(a) *Cannabis sativa*
(b) *Erythroxylon coca*
(c) *Claviceps purpurea*
(d) *Papaver somniferum*

69. Reserpine alkaloid is a drug extracted from the roots of:
(a) *Ferula foetida*
(b) *Digitalis purpurea*
(c) *Rauwolfia serpentina*
(d) *Atropa belladona*

70. The drug "Belladona" is obtained from:
(a) *Rauwolfia* (b) *Atropa*
(c) *Capsicum* (d) *Solanum*

71. Which of the following is a narcotic drug?
(a) Heroin (b) Codeine
(c) Morphine (d) All of these

72. Hashish, marijuana and LSD are:
(a) Toxins
(b) Stimulants
(c) Hallucinogens
(d) None of the above

73. Smoke of tobacco usually contains:
(i) Alcohol (ii) Nicotine
(iii) Phenol (iv) Tar
(a) (i) and (ii)
(b) (i), (ii) and (iii)
(c) (ii), (iii) and (iv)
(d) All of the above

74. From which part of *Atropa belladona* plant, belladona drug is obtained?

Critical Thinking

(a) Stems
(b) Flowers
(c) Leaves
(d) Leaves and roots

75. Drugs which induce dreamy state of consciousness are called:
(a) Sedative (b) Barbiturate
(c) Stimulant (d) Hallucinogen

76. Ergot, a drug is derived from fungus:
(a) *Aspergillus* (b) *Phytophthora*
(c) *Claviceps* (d) *Peronospora*

77. A person, showing unpredictable moods, outbursts of emotion, quarrelsome behaviour and conflicts with other, is suffering from:
(a) Borderline personality disorder (BPD)
(b) Mood disorders
(c) Addictive disorders
(d) Schizophrenia

78. Atropine, an alkaloid, is obtained from:
(a) *Datura stramonium*
(b) *Atropa belladona*
(c) *Hyoscyamus niger*
(d) *Withania somnifera*

79. Use of anti-histamines and steroids give a quick relief from:
(a) Headache (b) Allergy
(c) Nausea (d) Cough

80. The drug "Heroin" is obtained from the plant of the which of the following family?
(a) Leguminosae
(b) Papaveraceae
(c) Liliaceae
(d) Solanaceae

81. Which of the following statement is **incorrect**?

(a) Rheumatoid arthritis is an autoimmune disease.

(b) The use of drugs like antihistamine, adrenaline, and steroids quickly reduces the symptoms of bacterial infection.

(c) Several genes (called cellular oncogenes) have been identified in normal cells which when activated under certain conditions, could lead to oncogenic transformation of the cells.

(d) The vaccine also generates memory – B and T cells that recognise the pathogen quickly on subsequent exposure and overwhelm the invaders with a massive production of antibodies.

82. Which of the following statement is **correct**?

(a) Injecting microbes during immunisation induces passive immunity.

(b) Cell-mediated immune response is responsible for graft rejection.

(c) Colostrum during initial days of lactation provides active immunity to infant.

(d) None of the above

83. Which of the following statement(s) is/are **correct**?

(i) Antigen is usually a foreign body which causes the formation of antibody.

(ii) Antibody is a protein molecule.

(iii) Antigen can be either protein or polysaccharide molecule.

(iv) Antibody joins an antigen to destroy the latter.

(a) (i) and (ii) (b) (i) and (iii)

(c) All are incorrect (d) All are correct

84. Which of the following statements are **incorrect**?

(i) Antigen-binding site of an antibody is found between two light peptide chains.

(ii) The pathogen of elephantiasis is transmitted to a healthy person through the bite of same mosquito species which also cause malarial disease.

(iii) Lymphoid tissues are also located within the lining of the major tracts called mucosal-associated lymphoid tissue.

(iv) Increase intake of coffee or tea can cause indigestion, insomnia and disturb renal functions.

(v) Anaphylactic shock is an autoimmune reaction which involves all the tissues of the body that results in drastic fall in blood pressure.

(a) (i), (iii) and (v)

(b) (ii) and (iv)

(c) (i), (ii) and (v)

(d) (i), (ii) and (iv)

85. Which of the following statement is **correct** regarding the number of helper-T cells and the viral concentration in the blood?

(a) As the number of HIV increases, so do the numbers of helper-T cells.

(b) As the number of HIV increases, the number of helper-T cells decreases.

(c) As the concentration of HIV decreases, the number of helper-T cells increases.

(d) There is no relationship between the concentration of HIV and the number of helper-T cells.

86. Which of the following statements are **correct**?

(i) Abstinence from drugs of dependence causes withdrawal symptoms but not craving.

(ii) Chikungunya is confirmed by Widal test.

(iii) Rheumatoid arthritis, which affects many people in our society, is an auto-immune disease.

(iv) AIDS was first reported in 1981 and is caused by a member of a group of viruses called retroviruses.

(v) Benign tumors are normally considered with metastasis.

(vi) Most powerful stimulant is cocaine.

(a) (i), (ii) and (v)

(b) (i), (iii), (iv) and (vi)

(c) (iii), (v) and (vi)

(d) (iii), (iv) and (i)

Answer KEYs

1	(c)	10	(d)	19	(b)	28	(b)	37	(b)	46	(d)	55	(b)	64	(a)	73	(c)	82	(b)
2	(b)	11	(a)	20	(c)	29	(c)	38	(c)	47	(d)	56	(b)	65	(d)	74	(d)	83	(d)
3	(b)	12	(d)	21	(d)	30	(d)	39	(b)	48	(b)	57	(a)	66	(b)	75	(d)	84	(c)
4	(a)	13	(b)	22	(a)	31	(b)	40	(a)	49	(b)	58	(a)	67	(d)	76	(c)	85	(c)
5	(c)	14	(a)	23	(a)	32	(a)	41	(d)	50	(c)	59	(a)	68	(c)	77	(a)	86	(b)
6	(a)	15	(b)	24	(a)	33	(d)	42	(a)	51	(a)	60	(d)	69	(c)	78	(b)		
7	(b)	16	(a)	25	(c)	34	(c)	43	(c)	52	(d)	61	(a)	70	(b)	79	(b)		
8	(c)	17	(c)	26	(b)	35	(b)	44	(b)	53	(d)	62	(b)	71	(d)	80	(b)		
9	(b)	18	(d)	27	(b)	36	(d)	45	(a)	54	(c)	63	(a)	72	(c)	81	(b)		

Hints & Solutions

1. **(c)** **2.** **(b)**

3. **(b)** In leukemia the number of WBCs (especially granulocytes) increases but the number of RBCs decreases.

4. **(a)** **5.** **(c)**

6. **(a)** Vaccine contains dead, attenuated form or antigen of a pathogen which can be injected to provide immunity towards that pathogen. Monoclonal antibodies are homogenous immunological reagents of defined specificity, so that these can be utilised for diagnosis and screening with certainty.

7. **(b)** Hepatitis is a viral disease; two subgroups are hepatitis type A virus (HAV) and hepatitis type B virus (HBV). HBV causes serum hepatitis, it is most frequently transmitted by blood, products of blood or blood contaminated instruments. Since only adults are susceptible to HBV, but vaccines are available. While children and young adults are susceptible to HAV, no vaccines are available.

8. **(c)** The lymphocytes which differentiate in the thymus are known as the T-lymphocytes. These are responsible for the cellular immune response.

9. **(b)** Active immunity is the immunity which result from the production of antibodies by the immune system in response to the presence of an antigen.

10. **(d)** T-lymphocytes perform several immune functions including the release of cytokines. Cytokines (such as interleukin or interferon) has a specific effect on the interactions between cells, or communications between cells or on the behaviour of cell.

11. **(a)** B-lymphocytes are involved in humoral immunity by producing antibodies.

12. **(d)** **13.** **(b)** **14.** **(a)** **15.** **(b)**

16. **(a)** **17.** **(c)** **18.** **(d)** **19.** **(b)**

20. **(c)** **21.** **(d)** **22.** **(a)** **23.** **(a)**

24. **(a)**

25. **(c)** IgE antibodies are involved in allergic reactions.

26. **(b)** IgM is a pentamer and has maximum (10) binding sites. IgM is the largest antibody and is the first antibody to appear in response to initial exposure to an antigen.

27. (b) Allergens are non-infectious foreign substances that causes allergic reaction. The common allergens are dust, pollen, mould spores, fabrics, lipsticks,nail paints, fur, heat beacteria, etc. Allergy involves mainly the secretion of IgE antibodies and histamines.The allergy causes marked dilation of all the peripheral blood vessels.

28. (b) Mast cells are large cells with densely granular cytoplasm that are found in connective tissues. Mast cell granules contain mediators of inflammation, such as histamine, serotonin, and various chemotactic factors (substances that attract white blood cells to the area of infection). Histamine released from ruptured mast cells causes marked dilation of all the arteries so that a large amount of fluid is passed from the blood to the tissues and there is a drastic fall in blood pressure.

29. (c) A sickle cell anaemia affected person is more resistant to mosquito born infectious disease because the sickle - cell shaped RBCs are hostile to the protozoa *Plasmodium*.

30. (d) Rabies, influenza and AIDS are viral diseases. Amoebiasis, ascariasis and trypanosomiasis are caused by protozoa. Taeniasis, ascariasis and elephantiasis are the diseases caused by helminths but cancer, tuberculosis and tetanus are not related diseases. Tuberculosis and tetanus are bacterial diseases while cancer is not.

31. (b) 32. (a) 33. (d) 34. (c)

35. (b) 36. (d) 37. (b) 38. (c)
39. (b) 40. (a) 41. (d) 42. (a)
43. (c) 44. (b) 45. (a) 46. (d)
47. (d) 48. (b) 49. (b) 50. (c)
51. (a) 52. (d) 53. (d) 54. (c)
55. (b)

56. (b) DTP vaccine is for three disease- Diphtheria, Pertussis and Tetanus. BCG is for tuberculosis, ATS for tetanus and TAB for typhoid.

57. (a) 58. (a) 59. (a) 60. (d)
61. (a) 62. (b) 63. (a) 64. (a)
65. (d)

66. (b) Cancer of lymphatic tissue is commonly known as lymphoma.

67. (d) Carcinogen in tobacco smoke is N-Nitrosodimethyline.

68. (c) 69. (c) 70. (b) 71. (d)
72. (c) 73. (c) 74. (d) 75. (d)
76. (c) 77. (a) 78. (b) 79 (b)
80. (b) 81. (b)

82. (b) Injecting microbes during immunisation induces active immunity. Colostrum secreted from the mother during the initial days of lactation has abundant antibodies (IgA) to protect the foetus. This protection provides passive immunity.

83. (d) 84. (c)

85. (c) In general, as the concentration of HIV increases, the number of helper T cells destroyed which increases the number of cell count to fall.

86. (b)

31 Strategies for Enhancement in Food Production

1. The world's highly priced wool yielding 'Pashmina' breed is:
   (a) Goat
   (b) Sheep
   (c) Goat-sheep cross
   (d) Kashmir sheep-Afghan sheep cross

2. The most common species for bee-keeping in India is:
   (a) *Apis florae* (b) *Apis mellifera*
   (c) *Apis dorsata* (d) *Apis indica*

3. Of the world's top five crops in terms of annual production, which of the following statement is **correct**?
   (a) All five belong to the family Poaceae (Gramineae).
   (b) Four belong to Poaceae (Gramineae) and one to Leguminosae.
   (c) Four belong to Poaceae (Gramineae) and one to Solanaceae.
   (d) Three belong to Poaceae (Gramineae), one to Leguminosae and one to Solanaceae.

4. Which of the following is a viral disease of poultry birds?
   (a) Anthrax (b) Ranikhet
   (c) Coccidiosis (d) None of these

5. Famous Angora wool is obtained from:
   (a) Goat (b) Sheep
   (c) Rabbit (d) Arctic fox

6. Chemically, the silk is:
   (a) Cellulose (b) Resin
   (c) Protein (d) Lipid

7. Blue revolution is related to which of the following field?
   (a) Dairy products (b) Fisheries
   (c) Egg production (d) Agriculture

8. Worker Bees are:
   (a) Fertile males (b) Fertile females
   (c) Sterile females (d) Sterile males

9. Rinderpest in cattle and Ranikhet in poultry are:
   (a) Both bacterial diseases
   (b) Both viral diseases
   (c) One protozoan and other viral disease
   (d) One viral and other bacterial disease

10. Crop plants grown in monoculture are:
   (a) characterised by poor root system.
   (b) highly prone to pests.
   (c) low in yields.
   (d) free from intraspecific competition.

Tricky

11. In tissue culture, roots can be induced by:

   (a) lower concentration of cytokinin and higher concentration of auxins.
   (b) only cytokinin and no auxins.
   (c) no cytokinin and only auxins.
   (d) higher concentration of cytokinin and lower concentration of auxins.

12. The tissue culture technique used to grow nonviable inter-specific hybrids is:

Critical Thinking

   (a) Androgenic haploids
   (b) Shoot tip culture
   (c) Somatic embryogenesis
   (d) Embryo rescue

13. Transgenic plants are the ones:
   (a) produced by a somatic embryo in artificial medium.
   (b) generated by introducing foreign DNA in to a cell and regenerating a plant from that cell.
   (c) produced after protoplast fusion in artificial medium.
   (d) grown in artificial medium after hybridisation in the field.

14. A collection of all the alleles of all the genes of a crop plant is called:
(a) Germplasm collection
(b) Protoplasm collection
(c) Herbarium
(d) Somaclonal collection

15. Which of the following plants yields oil and fibre both?
(a) *Cocos nucifera*
(b) *Brassica campestris*
(c) *Eucalyptus*
(d) *Euphorbia hirta*

16. Which chemical is used in protoplast fusion?
(a) DMSO (b) Liquid N_2
(c) Pectinase (d) PEG

17. Match column-I with column-II and select the correct answer using the codes given below.
Tricky

	Column-I		Column-II
A.	Many people have deficiencies as they cannot buy fruits & vegetables.	I.	Single cell proteins
B.	Crops with higher vitamins, proteins and fats.	II.	Micropropagation
C.	Growing microbes as the alternative source of proteins.	III.	Somaclones
D.	Capacity to generate plant from a single cell or explant.	IV.	Hidden hunger
E.	Production of thousand plants through tissue culture.	V.	Biofortification
F.	Genetically identical plants.	VI.	Totipotency

(a) A – IV; B – V; C – VI; D – I; E – II; F – III
(b) A – IV; B – V; C – VI; D – I; E – III; F – II
(c) A – IV; B – V; C – I; D – VI; E – II; F – III
(d) A – VI; B – V; C – I; D – IV; E – II; F – III

18. Which of the following statement(s) is are/ correct?
(i) The maintenance of hives of honeybees for the production is called apiculture.
(ii) A group of animals related by descent and similar in most characters are called a breed.
(iii) The agricultural practice of breeding and raising livestock is called animal husbandry.
(iv) The ability of plant cells to regenerate into complete plant is called somaclonal variation.
(a) (i) and (ii) (b) (ii) and (iii)
(c) (iii) and (iv) (d) (i), (ii) and (iii)

19. A man made allopolyploid cereal crop is:
(a) *Hordeum vulgare*
(b) *Raphano brassica*
(c) *Triticale*
(d) *Zea mays*

20. Heterosis requires:
(a) Selection (b) Crossing
(c) Transformation (d) Mutation

21. Which one is the oldest breeding method?
(a) Hybridisation (b) Selection
(c) Mutation breeding (d) Introduction

22. Which one of the following is an exotic Indian fish?
(a) *Catla catla*
(b) *Heteropneustes fossilis*
(c) *Cyprinus carpio*
(d) *Labeo rohita*

23. The yellow colouration of cow's milk is due to the:
(a) Presence of carotene
(b) Absence of casein
(c) Absence of casein
(d) Presence of lactose

24. If chicken and eggs cooked properly (above 100°C), then the chances of catching bird flu are:
(a) Very high (b) High
(c) Moderate (d) Nil

25. Two bacteria found to be very useful in genetic engineering experiment are:

(a) *Nitrosomonas and Klebsiella*

(b) *Escherichia and Agrobacterium*

(c) *Nitrobacter and Azotobacter*

(d) *Rhizobium and Diplococcus*

26. The scientific process by which crop plants are enriched with certain desirable nutrients is called:

(a) Crop protection

(b) Breeding

(c) Bio-fortification

(d) Bio-remediation

27. The tumor inducing capacity of *Agrobacterium tumefaciens* is located in large extra-chromosomal plasmids called:

(a) R_i plasmid (b) Lambda phage

(c) pBR 322 (d) T_i plasmid

28. The bacterial disease which is found in chickens is?

Critical Thinking

(a) Rickets (b) Ranikhet

(c) Coccidiosis (d) Fowl cholera

29. Which is not a variety of sheep?

(a) Lohi (b) Beetal

(c) Nellore (d) Rampuri

30. Ranikhet is the disease of:

(a) Poultry (b) Fish

(c) Pigs (d) Honeybee

31. Coconut milk is used in tissue culture which consists of:

Tricky

(a) Cytokinin

(b) Auxin

(c) Gibberellin

(d) Ethylene

32. India's wheat yield revolution in the 1960s was possible primarily due to:

(a) hybrid seeds.

(b) increased chlorophyll content.

(c) mutations resulting in plant height reduction.

(d) quantitative trait mutations.

33. Match column-I with column-II and select the correct answer using the codes given below.

Column-I	Column-II
A. Usage of bisexual flowers as female parents.	I. Clonal selection
B. Incorporation of several desirable characters into a single variety.	II. Pure line selection
C. Exploiting hybrid vigour for many generations.	III. Emasculation
D. Improving local varieties of self-pollinated crops.	IV. Hybridisation
	V. Polyploidy breeding

(a) A – IV; B – V; C – III; D – I

(b) A – II; B – III; C – IV; D – V

(c) A – III; B – IV; C – I; D – II

(d) A – I; B – V; C – II; D – IV

34. Which of the following statements about bee colony is **incorrect**?

(a) Domesticated species are *Apis indica* and *A. mellifera.*

(b) Sex differentiation is due to haplodiploidy.

(c) They are female dominant societies.

(d) Fertilised eggs develop into sterile females only.

35. Which of the following disease(s) is/are poultry disease?

(i) Aspergillosis

(ii) Ranikhet disease

(iii) Gill rot

(iv) Black rot

(a) (i) and (ii) (b) (iii) and (iv)

(c) Only (iii) (d) Only (iv)

36. Which of the following statement(s) is/are **incorrect**?

(a) In 2000, maize hybrids that had twice the amounts of the amino acids, lysine and proline, compared to existing maize hybrids were developed.

(b) The Indian Agricultural Research Institute, New Delhi has released several vegetable crops that are rich in vitamins and minerals.

(c) Some of the diseases caused by fungi are black rot of crucifers, tobacco mosaic, turnip mosaic etc.

(d) Both (a) and (c)

37. Lathyrism which is caused by consumption of Khesari dal is a disease characterised by:

(a) reproductive failure, susceptibility to diabetes mellitus and skeletal abnormalities.

(b) retardation of body growth, precocious puberty and renal dysfunction.

(c) mental retardation, delay in the onset of puberty and cardiovascular abnormalities.

(d) gross skeletal deformation and thinning of collagen fibres and fibrils.

38. The alkaloid from *Colchicum automnale* of Liliaceae induces:

Critical Thinking

(a) Sterility

(b) Dormancy

(c) Cell division

(d) Polyploidy

39. Haploid plants can be obtained by culturing:

(a) Young leaves (b) Endosperm

(c) Pollen grains (d) Root tips

40. The process of removing stamens from the flower bud during hybridisation is called:

(a) Crossing (b) Selfing

(c) Emasculation (d) Capping

41. The *Triticum aestivum* wheat is:

(a) haploid (7 chromosome)

(b) diploid (14 chromosome)

(c) tetraploid (30 chromosome)

(d) hexaploid (42 chromosome)

42. In order to obtain virus-free plants through tissue culture, the best method is:

(a) Embryo rescue

(b) Anther culture

(c) Meristem culture

(d) Protoplast culture

43. Which among the following is the real product of the honey bee?

(a) Pollen (b) Bee wax

(c) Honey (d) Propolis

44. Natural silk contains:

(a) Potassium (b) Phosphorus

(c) Nitrogen (d) Magnesium

45. In India, which of the following species of honey bee is reared in artificial hives?

(a) *Apis indica* (b) *Apis dorsata*

(c) *Apis florae* (d) None of these

46. Gelatin, a very important raw material for preparation of photographic emulsion, is a byproduct of:

(a) Chicken (b) Forest

(c) Fish (d) Cattle

47. 'Inland fishery' refers to:

(a) culturing fish in fresh water.

(b) trapping and capturing fishes from sea coast.

(c) deep sea fishing.

(d) extraction of oil from fishes.

48. The reason for vegetatively reproducing crop plants for maintaining hybrid vigour is that:

(a) they can be easily propagated.

(b) they have a longer life span.

(c) they are more resistant to diseases.

(d) once a desire hybrid is produced, no chances of losing it.

49. In crop plants, clonal selection is used in:

(a) Non-flowering species

(b) Low seed setting species

(c) Apomictic species

(d) All of the above

50. Sharbati sonora variety of wheat was obtained by: Toughnut

(a) X-ray treatment
(b) crossing with wild varieties of wheat.
(c) hybridisation between wild grasses.
(d) irradiation of sonora-64 with gamma rays.

51. Which one of the following is the most suitable medium for culture of *Drosophilam elanogaster*? Tricky

(a) Agar agar (b) Ripe banana
(c) Cow dung (d) Moist bread

52. Tussar silk is extracted from:

(a) *Bombyx mori* (b) *Apis indica*
(c) *Antheraea paphia* (d) *Apis dorsata*

53. Following table shows the steps and the aims in the breeding of a new genetic variety of crop with few steps or aims marked as A, B, C and D. Identify correct explanation of A, B, C and D from the given option. Toughnut

	Step	Aim
I.	A	To evaluate and selection as parents
II.	Pure line creation	Collection of pre-existing genetic variability
III.	B	To combine desired characters of different plants.
IV.	Screening the superior recombinants	C
V.	Release and commercialisation of new cultivars	D

(a) A–Germplasm collection, B–Cross hybridisation, C–To select and test the hybrid for desired character combination for success of breeding objective, D–Obtained hybrid is evaluated in the crop fields and is compared with local crop cultivar.
(b) A–Germplasm collection, B–Self pollination, C–To select and test the hybrid for desired character combination for success of breeding objective, D–Obtained hybrid is evaluated in the crop fields and is compared with foreign crop cultivar.
(c) A–Mutation, B–Cross hybridisation, C–To select and test the hybrid for desired character combination for success of breeding objective, D–Obtained hybrid is evaluated in the crop fields and is compared with local crop cultivar.
(d) A–Selfpollination, B–Germplasm collection, C–Mutation, D–Cross-hybridisation.

54. Which one of the following statement is **incorrect**?

(a) Pectinase and cellulase dissolve the cell wall.
(b) Some cyanobacteria form symbiotic association with the fern *Azolla*.
(c) Regeneration of cell wall in somatic hybridisation is induced by PEG.
(d) Plants obtained through pollen culture are always haploids.

55. In the given methodology which has been used for cattle, sheep, buffaloes etc., few steps are marked as 1, 2 and 3. Identify 1, 2 & 3. Critical Thinking

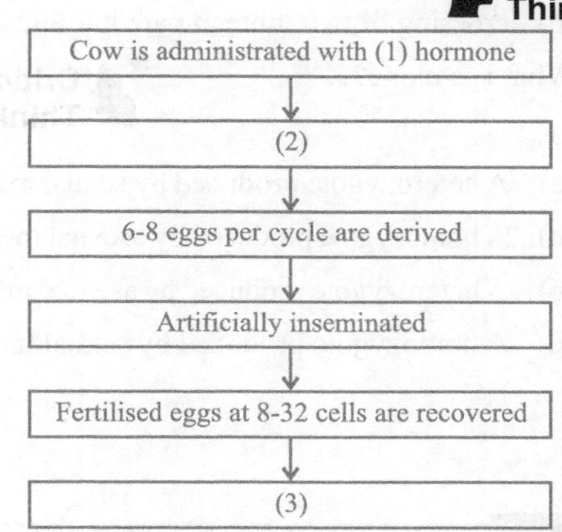

(a) 1-FSH, 2-Superovulation due to induced follicular maturation, 3-Transfer to surrogate mother
(b) 1-LH, 2-Superovulation due to induced follicular maturation, 3-Transfer to surrogate mother
(c) 1-Progesterone, 2-Superovulation due to induced follicular maturation, 3-Transfer to surrogate mother
(d) 1-Oestrogen, 2-Superovulation due to induced follicular maturation, 3-Transfer to surrogate mother

56. The new varieties of plants are produced by:
(a) Selection and hybridisation
(b) Mutation and selection
(c) Introduction and mutation
(d) Selection and introduction

57. The dwarf varieties of wheat brought from Mexico into India were

Critical Thinking

(a) Sonora-64 and Sonalika
(b) Sonora-64 and Lerma Rojo-64
(c) Sharbati Sonora and Pusa Lerma
(d) Sonalika

58. In maize, hybrid vigour is exploited by:

Tricky

(a) harvesting seeds from the most productive plants.
(b) inducing mutations.
(c) bombarding the protoplast with DNA.
(d) crossing of two inbreed parental lines.

59. What is a clone?

(a) A heterozygote produced by sexual means.
(b) A homozygote produced by asexual means.
(c) A heterozygote produced by asexual means.
(d) A homozygote produced by sexual means.

60. In crop improvement programme, haploids are of great importance because they:

(a) are useful in studies on meiosis.
(b) require only about half the amount of chemical fertilisers as compare to diploids.
(c) give homozygous lines following diploidisation.
(d) grow better under adverse conditions.

61. Match column-I with column-II and select the correct answer using the codes given below.

	Column-I		Column-II
A.	Hybrid variety	I.	X-ray
B.	Mutation	II.	Allopolyploidy
C.	Pure line	III.	F_1 generation
D.	*Triticale*	IV.	Selection in self pollinated crops
		V.	Genetic engineering

(a) A – III; B – II; C – I; D – IV
(b) A – III; B – I; C – IV; D – II
(c) A – I; B – III; C – IV; D – II
(d) A – II; B – I; C – III; D – IV

62. The larva of *Bombyx mori* is:
(a) Cocoon (b) Nymph
(c) Caterpillar (d) Trochophore

Answer KEYs

1	(a)	8	(c)	15	(a)	22	(c)	29	(b)	36	(d)	43	(b)	50	(d)	57	(b)
2	(d)	9	(b)	16	(d)	23	(a)	30	(a)	37	(d)	44	(c)	51	(b)	58	(d)
3	(c)	10	(b)	17	(c)	24	(d)	31	(a)	38	(d)	45	(a)	52	(c)	59	(b)
4	(b)	11	(a)	18	(d)	25	(b)	32	(c)	39	(c)	46	(d)	53	(a)	60	(c)
5	(c)	12	(d)	19	(c)	26	(c)	33	(c)	40	(c)	47	(a)	54	(c)	61	(b)
6	(c)	13	(b)	20	(b)	27	(d)	34	(d)	41	(d)	48	(d)	55	(b)	62	(c)
7	(b)	14	(a)	21	(b)	28	(d)	35	(a)	42	(c)	49	(d)	56	(a)		

Hints & Solutions

1. (a) The world's highly priced wool yielding Pashmina breed is the fur of Kashmiri and Tibetan goats.

   Pashmina is an almost generic name for accessories made from a type of mohair that is obtained from a special breed of goat indigenous to high altitudes of the Himalayan Range Belt of Asia.

   The name comes from Pashmineh, made from Persian pashm

   (= "wool"). The special goat's fleece has been used for thousands of years to make high-quality shawls that also bear the same name.

2. (d)

3. (c) The top five crops are : Potato (which belongs to the family Solanaceae) and Wheat, Rice, Corn, Barley (which belongs to the family Poaceae).

4. (b) 5. (c)

6. (c) The protein of the silk is 'Fibroin'.

7. (b) The revolution related to dairy products and egg production is called white revolution and silver revolution respectively.

8. (c) 9. (b) 10. (b) 11. (a)

12. (d) 13. (b) 14. (a) 15. (a)

16. (d) 17. (c) 18. (d)

19. (c) Allopolyploid means a mixture of two different genetic forms. Intergeneric hybridisation of cereal crops, *i.e., Secale cereale* and *Triticum* forms *Triticale* (man made).

20. (b) 21. (b)

22. (c) *Catla catla* and *Labeo rohita* are the two Indian major carps whereas *Heteropneustes* is a catfish. *Cyprinus* is the exotic breed.

23. (a) The yellow colour of cow milk is due to presence of carotene, which is a precursor for the yellow colour in cow's milk and is in the form of Vitamin-A.

24. (d) The chances of catching bird flu from properly cooked chicken and egg can be nil. The major causes of diseases in the poultry birds are over crowding dampness, insufficient light, unhygienic environmental conditions and dirty air.

25. (b) 26. (c) 27. (d) 28. (d)

29. (b) 30. (a)

31. (a) Coconut milk is used in tissue culture which consist of cytokinin.

32. (c) Green revolution was possible due to hybridisation. Increased chlorophyll content has no effect on yield of crops. Hybrid plants were dwarf because of selection. Quantitative trait mutations are not involved in hybridisation.

    In 1963, ICAR introduced many selections from CIMMYT including those developed by Norman Borlaug using Norin-10 as the source of dwarfing genes (CIMMYT-Centro International de Mejoramiento de Maiz Y Trigo).

33. (c) 34. (d) 35. (a)

36. (d) In 2000, maize hybrids that had twice the amount of the amino acids, lysine and tryptophan, compared to existing maize hybrids were developed. Some of the diseases caused by fungi are brown rust of wheat, red rot of sugarcane, late blight of potato, etc. Black rot of crucifers is caused by bacteria and tobacco mosaic and turnip mosaic are caused by viruses.

37. (d) Lathyrism which is caused by consumption of Khesari dal is a disease characterised by gross skeletal deformation and thinning of collagen fibres and fibrils.

38. (d) 39. (c) 40. (c) 41. (d)

42. (c) Meristem culture is done for the development of virus free plants. Meristematic tissue cell can be taken either from shoot or root tip.

43. (b) 44. (c) 45. (a) 46. (d)

47. (a) 48. (d) 49. (d)

50. (d) Sharbati Sonora is mutant yielding variety of wheat developed by irradiation with gamma ray from Sonora 64.

51. (b) The most suitable culture medium for *Drosophila melanogaster* is ripe banana. *Drosophila melanogaster* is commonly called as fruitfly and is often used in genetic and developmental biology researches. Moist bread is a culture medium for the fungus *Rhizopus* while Agar-agar is used as a tissue culture medium.

52. (c) 53. (a) 54. (c) 55. (b)

56. (a) Selection is picking up only those plants for reproduction which have desired qualities. Hybridisation is crossing of two or more types of plants for bringing their traits together in the progeny. Introduction is taking a plant or a new variety from an area where it grows naturally to a region where it does not occur before.

57. (b) Sonora-64 and Lerma Rojo-64 is dwarf variety of wheat, which were brought to India, modified through mutations and integrated to Indian Agriculture.

58. (d)

59. (b) In biology and agriculture, clone is defined as any organism whose genetic information is identical to that of a parent organism from which it was created. Natural reproductive processes producing clones include parthenogenesis and apomixis.

60. (c) 61. (b) 62. (c)

32 Microbes in Human Welfare

1. Vitamin B_2 is obtained from: Tricky
   (a) *Penicillium* (b) *Acetobacter*
   (c) *Aspergillus* (d) *Ashbya gossypii*
2. Methanogenic bacteria present in:
   (a) Anaerobic sludge
   (b) Rumen (a part of stomach) of cattle
   (c) Both (a) and (b)
   (d) None of these
3. Lichen that yields antibiotic is:
   (a) Ampicillin (b) Oxacillin
   (c) Both (a) and (b) (d) Tetracycline
4. Acetic acid is produced with the help of:
   (a) *Albugo* species
   (b) *Acetobacter* species
   (c) *Aspergillus* species
   (d) *Lactobacillus* species
5. Antibiotics are mostly got from:
   (a) Fungi
   (b) Virus
   (c) Bacteria
   (d) Cyanobacteria
6. Which antibiotic inhibits peptide bond formation?

   (a) Streptomycin (b) Tetracyclin
   (c) Chloramphenicol (d) Neomycin
7. Penicillin inhibits bacterial multiplication because it:
   (a) checks RNA synthesis.
   (b) checks DNA synthesis.
   (c) destroys chromatin formation.
   (d) inhibits cell wall formation.
8. Pasteurisation is heating at:
   (a) 120°C for 60 minutes
   (b) 60°-63°C for 30 minutes
   (c) 70°C for 60 minutes
   (d) 80°C for 30 minutes
9. Which of the following is not an antibiotic?
   (a) Griseofulvin (b) Cephalosporin
   (c) Citric acid (d) Streptomycin
10. Probiotics are:

   (a) live microbial food supplement.
   (b) cancer inducing microbes.
   (c) new kind of food allergens.
   (d) safe antibiotics.
11. Streptokinase, used as a 'clot buster' is obtained from: Tricky
   (a) *Streptococcus* (b) *Staphylococcus*
   (c) *Lactobacillus* (d) *Saccharomyces*
12. Secondary sludge uses:
   (a) gravitational sedimentation
   (b) primary clarifier
   (c) non-biodegradable materials
   (d) micro-organism to consume waste organic
13. Indicator of water pollution is:
   (a) *E. coli* (b) *Chlorella*
   (c) *Beggiatoa* (d) *Ulothrix*
14. Germinating barley is employed for preparation of:
   (a) Wine (b) Beer
   (c) Lactic acid (d) Cheese
15. Most widely used enzyme in detergents is:
   (a) Amylases (b) Lipases
   (c) Proteases (d) Glucoisomerases
16. Passage of effluents into oxidation tank is for:
   (a) Primary treatment
   (b) Secondary treatment
   (c) Tertiary treatment
   (d) Both (a) and (b)
17. Enzymes are immobilised through:
   (a) covalently attaching to solid support
   (b) cross-linking
   (c) entrapping in gel
   (d) All of the above
18. Cheese and yogurt are products of:
   (a) Pasteurisation (b) Fermentation
   (c) Dehydration (d) Distillation
19. Citric acid is obtained from:
   (a) *Aspergillus niger*
   (b) *Rhizobium nigricans*
   (c) *Penicillium citrinum*
   (d) *Lactobacillus bulgaricus*

20. Baculoviruses are excellent pathogens for:

(a) species-specific narrow spectrum pesticidal applications.
(b) species-specific broad spectrum pesticidal applications.
(c) species-specific narrow spectrum insecticidal applications.
(d) species-specific broad spectrum insecticidal applications.

21. Human insulin is being commercially produced from a transgenic species of:
(a) *Rhizobium* (b) *Saccharomyces*
(b) *Escherichia* (d) *Mycobacterium*

22. The amount of oxygen required by the microbes in the decomposition of organic matter is called:
(a) Chemical oxygen demand
(b) Biochemical oxygen demand
(c) Total oxygen demand
(d) Dissolved oxygen

23. Enzyme immobilisation is:

(a) conversion of an active enzyme into inactive form.
(b) providing enzyme with protective covering.
(c) changing a soluble enzyme into insoluble state.
(d) changing pH so that enzyme is not able to carry out its function.

24. In sewage treatment process, which part of decomposer bacteria is recycled into starting of the process is called:
(a) Cyclic treatment
(b) Primary treatment
(c) Tertiary treatment
(d) Activated sludge treatment

25. A lake with an inflow of domestic sewage rich in organic waste may result in:
(a) drying of the lake very soon due to algal bloom.
(b) an increased production of fish due to lot of nutrients.
(c) death of fish due to lack of oxygen.
(d) increased population of aquatic food web organisation.

26. Sewage purification is done by:
(a) Microbes (b) Fertilisers
(c) Antibiotics (d) Antiseptics

27. Cider is prepared by:

Tricky

(a) action of *Lactobacillus* on milk.
(b) action of *Saccharomyces* on apple juice.
(c) action of *Saccharomyces* on cane sugar.
(d) action of *Bacillus* on ramie.

28. Activated sludge has the ability to settle quickly this enables it to:
(a) be rapidly pumped back from sedimentation to aeration tank.
(b) absorb pathogenic bacteria while sinking to the bottom of the settling-tank.
(c) anaerobically digested and be discarded.
(d) absorb colloidal organic matter.

29. *Jatropha* is a:
(a) Biodiesel crop (b) Biopetro crop
(c) Fibre crop (d) Food crop

30. Consider the following statements about, secondary sewage treatment. Which of the statements given are **correct**?
(i) In secondary treatment useful aerobic microbes grow rapidly and form flocs of bacterial and fungal filaments in a mesh-like structure.
(ii) The growing microbes consume organic matter reducing the BOD after which the effluent is passed into settling tank.
(iii) In settling tank, the bacterial flocs settle and the sediment is called activated sludge.
(iv) A small part of the sludge is used as an inoculum. The remaining part is passed into anaerobic sludge digesters.
(v) On anaerobic digestion of bacteria and fungi mixture of gases such as, carbon dioxide, nitrogen and carbon monoxide are produced.
(a) (i), (ii), (iii) and (iv)
(b) (i), (iii), (iv) and (v)
(c) (ii), (iii), (iv) and (v)
(d) (i), (ii), (iii), (iv) and (v)

31. Name the blank spaces A, B, C and D given in the following table.

Type of Microbe	Scientific Name	Commercial Product
Bacterium	A	Clot buster enzyme
B	*Aspergillus niger*	Citric acid
Fungi	*Trichoderma polysporum*	C
Bacterium	D	Butyric acid

(a) A – *Streptococcus*, B – Fungus, C – Cyclosporin-A, D – *Clostridium butylicum*.
(b) A – *Clostridium butylicum*, B – *Streptococcus*, C – Fungus, D – Cyclosporin-A
(c) A – Cyclosporin-A, B - *Clostridium butylicum*, C – *Streptococcus*, D – Fungus
(d) A – Fungus, B – Cyclosporin-A, C – *Clostridium butylicum*, D - *Streptococcus*

32. Following is the sewage treatment. In which of the following options, correct word for all the four numbers (1,2,3 and 4) are indicated. *Critical Thinking

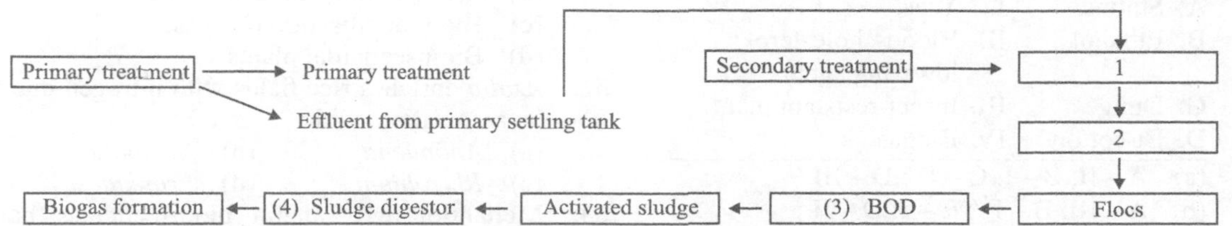

(a) 1 – Large aeration tanks; 2 – Chemical agitation; 3 – High; 4 – Anaerobic
(b) 1 – Large aeration tanks; 2 – Mechanical agitation; 3 – Low; 4 – Anaerobic
(c) 1 – Large aeration tanks; 2 – Chemical agitation; 3 – Low; 4 – Aerobic
(d) 1 – Large aeration tanks; 2 – Mechanical agitation; 3 – High; 4 – Anaerobic

33. Pyrethrin is a common ingredient of:
(a) Mosquito coils (b) Fly sprays
(c) Mosquito mats (d) All of these

34. Thurioside is an: *Critical Thinking
(a) Insecticide (b) Fungicide
(c) Antibiotic (d) Weedicide

35. The free-living fungus *Trichoderma* can be used for:
(a) killing insects
(b) biological control of plant diseases
(c) controlling butterfly caterpillars
(d) producing antibiotics

36. A genetically engineered micro-organism used successfully in bioremediation of oil spills is a species of:
(a) *Pseudomonas* (b) *Trichoderma*
(c) *Xanthomonas* (d) *Bacillus*

37. Viral pesticide used against gypsy moths is:
(a) NPV (b) *Nosema locustae*
(c) *Fusarium* (d) *Beauveria bassiana*

38. Identify the microorganism which is commercially exploited as the source of bioinsecticides?
(a) *Chrysanthemum cinerariifolium*
(b) *Phytophthora palmivora*
(c) *Bacillus thuringiensis*
(d) Both (a) and (c)

39. Methane content of biogas is:
(a) 24.6% (b) 55.8% (c) 8% (d) 4%

40. For biogas production besides dung, an extensive use of which weed is recommended in our country?
(a) *Mangifera indica*
(b) *Hydrilla*
(c) *Eichhornia crassipes*
(d) *Solanum*

41. IPM (Integrated Pest Management) involves: *Critical Thinking
(a) Tissue culture
(b) Biological control
(c) Biofertilisers
(d) Confusion technique

42. Organic farming includes:
(a) use of fertilisers and pesticides of biological origin.
(b) IPM (integrated pest management programme).
(c) locally developed pest resistant varieties.
(d) All of the above

43. Which of the following is the pair of biofertilisers? *Critical Thinking
(a) *Azolla* and blue green algae
(b) *Nostoc* and legume
(c) *Rhizobium* and grasses
(d) *Salmonella* & *E. coli*

44. Which of the following plants are used as green manure in crop fields and in sandy soils?

(a) *Dichanthium annulatum* and *Azolla pinnata*
(b) *Crotalaria juncea* and *Alhagi maurorum*
(c) *Calotropis procera* and *Phyllanthus niruri*
(d) *Saccharum munja* and *Lantana camara*

45. Consider the following four statements (i)-(iv) and select the option which includes all the correct ones only.
(i) Single cell *Spirulina* can produce large quantities of food rich in protein, minerals, vitamins etc.
(ii) Body weight wise, the micro-organism *Methylophilus methylotrophus* may be able to produce several times more proteins than the cows per day.
(iii) Common button mushrooms are a very rich source of vitamin C.
(iv) A rice variety has been developed which is very rich in calcium.
(a) (iii) and (iv)
(b) (i), (iii) and (iv)
(c) (ii), (iii) and (iv)
(d) (i) and (ii)

46. Match the column-I with column-II and choose the right option.

Column-I	Column-II
A. Statins	I. Yeast
B. Ethanol	II. Blood-cholesterol lowering agent
C. Dung	III. Insect-resistant plant
D. Bt-cotton	IV. Biogas

(a) A – II; B – I; C – IV; D – III
(b) A – III; B – IV; C – I; D – II
(c) A – I; B – II; C – III; D – IV
(d) A – IV; B – II; C – I; D – III

47. Which of the following statements is **incorrect** for stirred tank fermentaion?
(a) Buffer needed to control pH.
(b) Batch and feed possible.
(c) Control dissolved oxygen.
(d) Easy in process sampling.

48. Sewage or municipal sewer pipes should not be directly left into rivers, streams and other water bodies because:
(i) It contains human excreta and other organic waste.
(ii) It contains a number of pathogenic microbes.
Which of the statement given above is/are **correct**?
(a) Only (i) (b) Only (ii)
(c) Both (a) and (b) (d) None of the above

49. Which of the following statements regarding Baculoviruses as bio-control agents is/are **correct**?
(i) Baculoviruses are pathogens that attack insects and other arthropods.
(ii) Most of these biocontrol agents belong to the genus *Nucleopolyhedrovirus*.
(iii) They do not harm plants mammals, birds, fish and other non-target insects.
(iv) Baculoviruses are helpful in Integrated Pest Management (IPM) programme, in which beneficial insects are conserved.
Choose the correct option.
(a) (i), (ii) and (iii) (b) (i), (ii) and (iv)
(c) (ii), (iii) and (iv) (d) All of these

50. Which of the following is common *to* *Azospirillum*, *Azotobacter*, *Anabaena*, *Nostoc* and *Oscillatoria*?
(a) Prokaryotes (b) N_2-fixers
(c) Both (a) and (b) (d) Eukaryotes

51. *Bacillus thuringiensis* (Bt) strains have been used as:
(a) Biofertilisers
(b) Biometallurgical techniques
(c) Biomineralisation processes
(d) Bioinsecticidal plants

52. *Azolla* enriches rice fields with nitrogen due to its association with:
(a) *Anabaena* (b) *Nostoc*
(c) *Rhizobium* (d) *Frankia*

53. Meliantiol and salanin bioinsecticides have been extracted from: **Critical Thinking**
(a) *Chrysanthemum cinerariifolium*
(b) *Derris elliptica*
(c) *Azadirichta indica*
(d) *Eichhornia*

54. The bioactive molecule Cyclosporine A is used in:
(a) Whooping cough
(b) Diphtheria
(c) Leprosy
(d) Organ-transplant patients

55. The bacterium that commonly lives in animal and human intestines is:
(a) *Bacillus anthracis* (b) *Vibrio cholerae*
(c) *Escherichia coli* (d) *Corynebacterium*

56. Cochineal insects have proved very useful for: **Tricky**
(a) Weeds control
(b) *Parthenium* control
(c) *Cactus* prevention
(d) *Eichhornia* prevention

57. Consider the following statements about organic farming. Which of the given statements are **correct**?
(i) It utilises genetically modified crops like *Bt* cotton.
(ii) It uses only naturally produced inputs like compost.
(iii) It does not use pesticides and urea.
(iv) It produces vegetables rich in vitamins and minerals.
(a) (ii), (iii) and (iv) (b) (iii) and (iv) only
(c) (ii) and (iii) only (d) (i) and (ii) only

58. Most of eubacterial antibiotics are obtained from: **Critical Thinking**
(a) *Rhizobium* (b) *Bacillus*
(c) *Streptomyces* (d) *Cephalosporium*

Answer KEYs

1	(d)	7	(d)	13	(a)	19	(a)	25	(c)	31	(a)	37	(a)	43	(a)	49	(d)	55	(c)
2	(c)	8	(b)	14	(b)	20	(c)	26	(a)	32	(b)	38	(c)	44	(b)	50	(c)	56	(c)
3	(c)	9	(c)	15	(c)	21	(c)	27	(b)	33	(d)	39	(b)	45	(d)	51	(d)	57	(c)
4	(b)	10	(a)	16	(b)	22	(b)	28	(a)	34	(a)	40	(c)	46	(a)	52	(a)	58	(b)
5	(c)	11	(a)	17	(d)	23	(b)	29	(a)	35	(b)	41	(b)	47	(c)	53	(c)		
6	(c)	12	(d)	18	(b)	24	(d)	30	(a)	36	(a)	42	(d)	48	(c)	54	(d)		

Hints & Solutions

1. **(d)** Ashbya gossypii ad Yeast are good source of Vitamin B_2.

2. **(c)**

3. **(c)** Ampicillin and oxacillin are modified antibiotics and once known as semisynthetic antibiotics.

4. **(b)** The bacterium, *Acetobacter aceti* is used for the production of acetic acid in industries. It is mainly used in pharmaceutical industries, colouring agent, insecticides and plastics.

5. **(c)** 6. **(c)** 7. **(d)** 8. **(b)**

9. **(c)** 10. **(a)**

11. **(a)** Streptokinase is produced by *Streptococcus* and are used as a 'clot buster' to remove clots from the blood vessels of patients who have myocardial infarction.

12. **(d)** 13. **(a)** 14. **(b)** 15. **(c)**

16. **(b)** 17. **(d)** 18. **(b)** 19. **(a)**

20. **(c)** Baculoviruses are suitable for species-specific, narrow spectrum insecticidal applications.

21. **(c)** 22. **(b)** 23. **(b)** 24. **(d)**

25. **(c)** A river with sewage, increases the organic waste and decreases the dissolved oxygen.

26. **(a)** Sewage is a waste material (like human urine and faeces) that is carried away from homes and other buildings in a system of pipes. Sewage treatment is the process of removing contaminants from wastewater which is done by heterotrophic microbes naturally present in sewage.

27. **(b)**

28. **(a)** Activated sludge should have the ability to settle quickly so that it can be rapidly pumped back from sedimentation to aeration tank.

29. **(a)** *Jatropha* is a genus of flowering plants in the spurge family, euphorbiaceae. Currently the oil from *Jatropha curcas* seeds is used for making biodiesel fuel in Phillippines and in Brazil.

30. **(a)** All statements are correct except (v). Methane, hydrogen sulphide and CO_2 are the main gases produced in the digesters by the anaerobic digestion by heterotrophic bateria and fungi.

31. **(a)** 32. **(b)**

33. **(d)** Pyrethrin are obtained from the dry inflorescence of *Chrysanthemum cinerariifolium*. Pyrethrin is also used for sprays, mosquito coils, mats and aerosols.

34. **(a)** Thurioside is toxic to several insects and produced by bacterium *Bacillus thuringiensis*.

35. (b) *Trichoderma* is a free-living saprophytic fungi that most commonly lives on dead organic matter in the soil and rhizosphere (root ecosystem). It inhibits pathogens through the release of gliotoxin, viridin, gliovirin and trichodermin like substances.

36. (a)

37. (a) NPV i.e. *Nucleopolyhedrovirus* is a Baculovirus which is used to control gypsy moths, aphids and beetles.

38. (c) A bacterial species namely *Bacillus thuringiensis* is known to kill a wide range of insects such as beetles, mosquitoes, ants etc. It is a bioinsecticide.

39. (b) 40. (c) 41. (b) 42. (d)

43. (a) 44. (b)

45. (d) *Spirulina* is SCP rich in protein, vitamins & minerals. 250 gram biomass of *Methylophilus methylotrophus* produces 25tone protein/day while cow of 250 Kg. produces only 200 gm. protein/day.

46. (a)

47. (c) Controlled dissolved oxygen is not involved in stirred tank fermentation.

48. (c) Sewage or municipal waste should not be directly passed into rivers, streams and other water bodies because it does not only contain human excreta and other organic waste but a number of pathogenic microbes. It is made less pollution by passing it through Sewage Treatment Plants (STPs).

49. (d) *Baculovirus heliothis* is known to infect the larval stages of many harmful insects beetles, wasps and ants. A number of baculovirus, which are used as biopesticides belongs to the genus *Nucleopohyledrovirus*. These biological weapons are not only effective as potential biological control of harmful insects, but are also harmless to non-target organisms (plants, birds, mammals, non-targets insects etc). They are important in organic farming because of their specific action on harmful insects without causing any damage to beneficial insects as well as to the environment.

Baculoviruses are helpful in Integrated Pest Management (IPM) Programme, in which beneficial insects are conserved.

50. (c) 51. (d)

52. (a) *Azolla pinnata* contains symbiont *Anabaena* in its leaf cavities.

53. (c) 54. (d) 55. (c) 56. (c)

57. (c) Statement (ii) and (iii) are the correct statements regarding the organic farming. Organic farming comprises the production of crop and livestock that involves much more than selecting not to use pesticides, fertilizers, genetically modified organisms, antibiotics and growth hormones. It supports the use of crop rotations and cover crops, and encourages balanced host/predator relationships.

58. (b) *Bacillus subtilis* is used to extract antibiotics subtilin or bacitracin.

33 Biotechnology: Principles & Processes

1. Restriction endonucleases:
   (a) are enzymes that process pre-RNA.
   (b) are enzymes that degrade DNA.
   (c) protect bacterial cells from viral infection.
   (d) None of the above.
2. A gene is said to be cloned if:
   (a) the DNA sequence of the gene is known.
   (b) the function of the gene is known.
   (c) there is a DNA probe for the gene.
   (d) the gene has been isolated and copied.
3. Which of the following is a critically important tool used in experiments involving DNA hybridisation?
   (a) DNA sequencing machines
   (b) Ligases
   (c) DNA probes
   (d) Vectors
4. Introduction of one or more genes into an organism which normally does not possess them or their deletion by using artificial means (not by breeding) comes under: Critical Thinking
   (a) Molecular biology
   (b) Cytogenetics
   (c) Genetic hybridisation
   (d) Genetic engineering
5. A genetic marker is a:

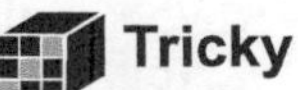

   (a) place where a restriction enzyme cuts DNA.
   (b) chart that traces the family history of a genetic trait.
   (c) nucleotide sequence near a particular gene.
   (d) radioactive probe used to find a gene.
6. A nucleic acid probe might be used to:
   (a) insert genes into a host cell.
   (b) make DNA for gene cloning.
   (c) splice pieces of DNA.
   (d) identify complementary segments present in nucleic acid sequence.
7. DNA fragments with sticky ends are not allowed to undergo self ligation by:
   (a) Unwindase
   (b) Single strand binding proteins
   (c) Gyrase
   (d) Alkaline phosphatase
8. In genetic engineering, a chimera is:

   Tricky

   (a) an enzyme that links DNA molecules.
   (b) a pasmid that contains foreign DNA.
   (c) a virus that infects bacteria.
   (d) a fungi.
9. Restriction enzymes are used in genetic engineering because:
   (a) they can join different DNA fragments.
   (b) they can cleave DNA at a specific target.
   (c) they are nucleases that cut DNA at variable sites.
   (d) they are proteolytic enzymes which can degrade harmful enzymes.
10. Which of the following is known as specific molecular scissors?
    (a) Ligase
    (b) Helicase
    (c) Restriction endonuclease
    (d) DNA polymerase
11. The RTF region enables the plasmid to:

    (a) be transmitted to other bacteria by conjugation.
    (b) undergo transformation.
    (c) replicate in the host cell.
    (d) code for enzymes that inactivate specific drugs.
12. Which one of the following option is **correct** for 1, 2, 3 and 4?

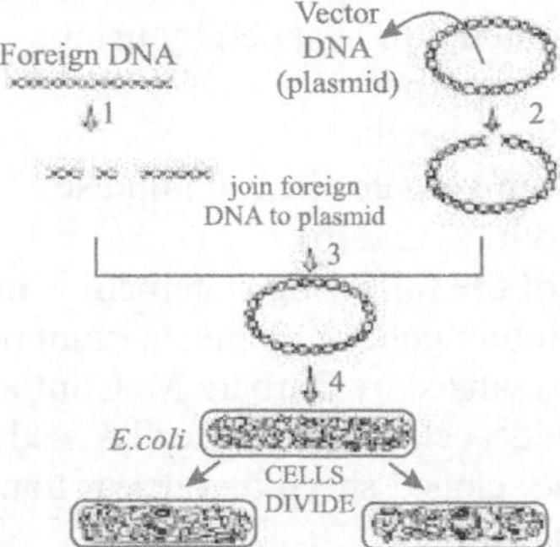

    (a) 1-Exonuclease; 2-Endonuclease; 3-DNA ligase; 4-Transformation
    (b) 1-Endonuclease; 2-Exonuclease; 3-DNA ligase; 4-Transformation
    (c) 1-Exonuclease; 2-Endonuclease; 3- Hydrolase; 4-Transduction
    (d) 1-Restriction endonuclease; 2-Restriction endonuclease; 3-DNA ligase; 4-Transformation

13. Match column-I with column-II and choose the correct option.

Column - I	Column - II
A. *EcoRI*	I. *Bacillus amyloliquefaciens*
B. *Bam* HI	II. *Haemophilus influenzae*
C. *Hind* III	III. *Escherichia coli*
D. pBR 322	IV. Artificial plasmid

(a) A – III; B – I; C – II; D – IV
(b) A – II; B – I; C – IV; D – III
(c) A – IV; B – I; C – III; D – II
(d) A – I; B – IV; C – II; D – III

14. Temperature required for various steps of polymerase chain reaction (i), (ii) and (iii) are:
(i) Denaturation
(ii) Annealing
(iii) Extension

Critical Thinking

(a) (i) – 94°C, (ii) – 60°C, (iii) – 72°C
(b) (i) – 40°C, (ii) – 72°C, (iii) – 94°C
(c) (i) – 72°C, (ii) – 94°C, (iii) – 40°C
(d) (i) – 90°C, (ii) – 72°C, (iii) – 40°C

15. A T_4 bacteriophage has a gene for the enzyme, lysozyme. The function of this enzyme is to digest:
(a) Cell wall (b) Cell membrane
(c) Golgi (d) Plasmid

16. Most sensitive technique to detect malignant cell in non-Hodgkin's lymphoma is:

Tricky

(a) Polymerase chain reaction
(b) Gene therapy
(c) Stem cell therapy
(d) None of the above

17. In microinjection method, the DNA is incorporated into host cell by using:
(a) Micro-pipettes
(b) Micro-needles
(c) High voltage electric impulse
(d) Both (a) and (b)

18. Which of the following statement is **incorrect**?
(a) Mobile genetic elements, tranposons were visualised by Barbara McClintock.
(b) Udder cell and somatic cell is used to produce the cloned sheep by nuclear transplanation method.
(c) In pedigree analysis, a person immediately affected by and action is called propositus.
(d) DNA ligases are used to cleave a DNA molecule.

19. Which one of the following pairs of terms/ names means one and the same thing?
(a) Gene pool of an organism-genome
(b) Codon-gene
(c) Cistron-triplet
(d) DNA fingerprinting-DNA profiling

20. Match column-I with column-II and select the correct answer using the codes given below.

Column-I	Column-II
A. Southern blotting	I. Running of DNA fragments on gel.
B. Gel electrophoresis	II. Cutting of DNA into fragments.
C. Cleavage	III. Transfer of DNA fragment from gel to nitrocellulose disc.
D. DNA probing	IV. Searching for desired DNA fragments.

(a) A – IV; B – I; C – II; D – III
(b) A – III; B – I; C – II; D – IV
(c) A – IV; B – I; C – III; D – II
(d) A – I; B – IV; C – II; D – III

21. Polyethylene glycol method is used for:
(a) biodiesel production.
(b) seedless fruit production.
(c) energy production from sewage.
(d) gene transfer without a vector.

22. Plasmids are suitable vectors for gene cloning because:
(a) these are small circular DNA molecules which can integrate with host chromosomal DNA.
(b) these are small circular DNA molecules with their own replication origin site.
(c) these can shuttle between prokaryotic and eukaryotic cells.
(d) these often carry antibiotic resistance genes.

23. Problems in obtaining large amounts of proteins encoded by recombinant genes can often be overcome by using:
(a) BACs (b) Expression vectors
(c) YACs (d) All of these

24. Which of the following is not a genetic vector?
(a) Plasmid (b) Phage
(c) Cosmid (d) Virusoid

25. The introduction of TDA into plants involves:

(a) exposing the plants to cold for a brief period.
(b) allowing the plant roots to stand in water.
(c) infection of the plant by *Agrobacterium tumefaciens*.
(d) None of these

26. Due to chloramphenicol resistance gene, one is able to select a transformed cell in the presence of chloramphenicol. The chloramphenicol resistance gene in this case is called:
(a) Origin of replication
(b) Cloning sites
(c) Selectable marker
(d) None of the above

27. Electrophoresis is used to:
(a) separate fragments of DNA.
(b) clone genes.
(c) cut DNA into fragments.
(d) match a gene with its function.

28. A biologist intends to use a polymerase chain reaction to perform a genetic task. The biologist probably is trying to:
(a) discover new genes.
(b) clone a gene.
(c) cut DNA into many small fragments.
(d) isolate DNA from a living cell.

29. Consider the given figure and identify the correct option .

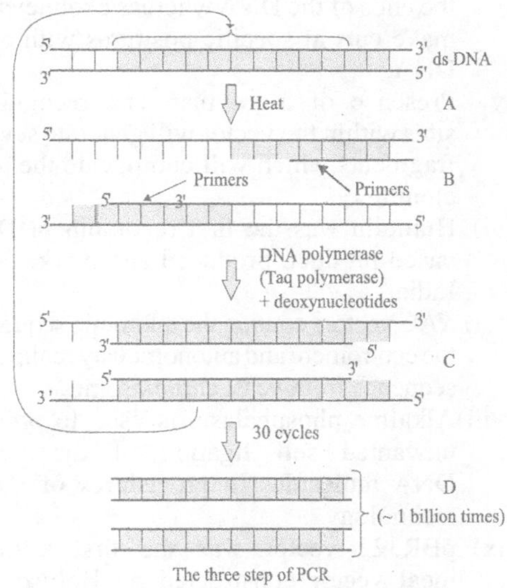

The three step of PCR

(a) A - Denaturation, B - Annealing, C - Extension, D - Amplified
(b) A - Annealing, B - Denaturation, C - Extension, D - Amplified
(c) A - Denaturation, B - Annealing, C - Amplified, D - Extension
(d) A - Annealing, B - Denaturation, C - Amplified, D - Extension

30. Which one of the following pairs is not **correctly** matched?
(a) Plasmid – Small piece of extra chromosomal DNA in bacteria
(b) Interferon – An enzyme that interferes with DNA replication
(c) Cosmid – A vector for carrying large DNA fragments into host cells
(d) Myeloma – Antibody - producing tumour cells

31. Which of the following is/are related to biotechnology?
(i) *In vitro* fertilisation.
(ii) Synthesis of a gene.
(iii) Correcting a defective gene.
(iv) Developing a DNA vaccine.
(a) (i) and (ii) (b) (ii) and (iii)
(c) (iii) and (iv) (d) (i), (ii), (iii) and (iv)

32. Match column-I with column-II and choose the correct option.

Column-I	Column-II
A. Recombinant DNA	I. Sea weeds
B. Ethidium bromide	II. DNA staining
C. Gel electrophoresis	III. Plasmid DNA that has incorporated into human DNA
D. Agarose	IV. Process by which DNA fragments are separated based on their size

(a) A – III; B – II; C – IV; D – I
(b) A – III; B – II; C – I; D – IV
(c) A – II; B – I; C – IV; D – III
(d) A – III; B – IV; C – I; D – II

33. DNA fragments are separated using gel electrophoresis:
(a) because DNA is pulled through the gel toward the negative end of the field.
(b) because larger DNA fragments move faster through the gel than smaller DNA fragments.
(c) to identify and isolate DNA fragments.
(d) to synthesise DNA for cloning.

34. Complementary base pairing is important for:
(a) ligation reactions with blunt-end DNA molecules.
(b) hybridisation between DNA and transcription factors.
(c) restriction endonucleases to cut cell walls.
(d) synthesising cDNA molecules from mRNA templates.

35. In genetic engineering, genes can be inserted from one organism into another or back into the original organism by using which of the following techniques?
(a) Polymerase chain reaction
(b) Gene gun
(c) DNA hybridisation
(d) Gel electrophoresis

36. A restriction fragment containing a specific gene of interest can be identified by gel electrophoresis followed by transferring the DNA to a membrane as solid support matrix using a procedure called:
(a) Gene amplification
(b) Southern blotting
(c) Identification of a gene
(d) Mc Clintock

37. Restriction endonuclease - *Hin*d II always cuts DNA molecules at a particular point by recognising a specific sequence of:

(a) six base pairs (b) five base pairs
(c) four base pairs (d) seven base pairs

38. Which of the following is/are used in recombinant DNA technology?
(i) Agarose gel
(ii) Ethidium bromide
(iii) Plasmid vector
(iv) Restriction endonuclease
(a) (i) and (ii) (b) (ii) and (iii)
(c) (iii) and (iv) (d) (i), (ii), (iii) and (iv)

39. Which of the following statements about restriction enzymes is **incorrect**?
(a) They work on DNA extracted from all types of organisms.
(b) They are used to glue short segments of DNA.
(c) They come in many varieties, each with its own DNA target sequence.
(d) They are highly specific for their DNA target sequences.

40. Which of the following statements represent/s 'cloning'?

Critical Thinking

(a) all of the cells are derived from one cell and are genetically identical.
(b) a gene from one organism has been inserted into a vector and successfully introduced into a host cell.
(c) all of the cells in a particular organism are identical.
(d) All of the above

41. Recombinant DNA can be transferred into host cell by:

Toughnut

(a) growing the host cell in growth medium containing ampicillin.
(b) coating the DNA with carbohydrates so that the cells will engulf the DNA.
(c) treating cells with calcium ions or electrical pulses to increase cell permeability.
(d) injecting proteins into host cells to make them more permeable.

42. Identify the **correct** statements.

(i) The first recombinant DNA was constructed by using a piece of DNA from a plasmid carrying antibiotic resistance gene in the bacterium *Salmonella typhimurium* and linked it to the plasmid of *E. coli*.
(ii) Cohen and Boyer are known as father of genetic engineering.
(iii) When cut by the same restriction enzyme, the resultant DNA fragments have the same kind of sticky ends and these can be joined together using DNA ligases.
(iv) Endonucleases remove nucleotides from the ends of the DNA whereas exonucleases make cuts at specific positions within the DNA.
(v) Presence of more than one recognition sites within the vector will generate several fragments which will complicate the gene cloning.
(vi) Humulin was the first recombinant DNA based product, produced and marketed in India.
(vii) YAC vectors contain the telomeric sequence, the centromere and autonomously replicating sequence from yeast chromosomes.
(viii) Alkaline phosphatase, is used to prevent unwanted self ligation of the vector DNA molecules in procedures of rDNA technology.
(ix) pBR322 vector was the first artificial ideal vector constructed by Boliver and Rodriguez.
(x) Plasmid DNA is coated with histone proteins and can act as genetic factor.
(a) (ii), (iv), (vi) and (x)
(b) (ii), (iv), (v) and (x)
(c) (i), (iii), (v), (vii), (viii) and (ix)
(d) (i), (ii), (iv), (vi) and (x)

Answer KEYs

1	(b)	6	(d)	11	(a)	16	(a)	21	(d)	26	(c)	31	(d)	36	(b)	41	(c)
2	(d)	7	(d)	12	(d)	17	(d)	22	(b)	27	(a)	32	(a)	37	(a)	42	(c)
3	(c)	8	(b)	13	(a)	18	(d)	23	(d)	28	(b)	33	(c)	38	(d)		
4	(d)	9	(b)	14	(a)	19	(d)	24	(d)	29	(a)	34	(d)	39	(b)		
5	(c)	10	(c)	15	(a)	20	(b)	25	(c)	30	(b)	35	(b)	40	(d)		

Hints & Solutions

1. **(b)**
2. **(d)** Cloned gene is one that has been isolated and repeatedly duplicated.
3. **(c)** DNA probes are specially constructed sequences that are complementary to the target sequence in the sample being tested. The probes bond (hybridise) with the target sequence wherever it occurs in the DNA sample.
4. **(d)** Genetic engineering is the science of gene manipulation.

   Genetic engineering, genetic modification (GM) and gene splicing are terms for the process of manipulating genes, generally implying that the process is outside the organism's natural reproductive process. It involves the isolation, manipulation, transfer, and reintroduction of DNA into cells or model organisms usually to express a protein. The aim is to introduce new characteristics or attributes physiologically or physically, such as making a crop resistant to a herbicide, introducing a novel trait, or producing a new protein or enzyme, along with altering the organism to produce more of certain traits.
5. **(c)** 6. **(d)** 7. **(d)** 8. **(b)**

9 **(b)**

10. **(c)** Restriction endonuclease are called "molecular scissors or biological scissors". They recognise and cut double stranded DNA at specific points and are called palindromic sequences. Palindromic sequence are the ones which read same on both the strand in $5' \rightarrow 3'$ direction. Same is true about $3' \rightarrow 5'$ direction.

    5′ —— G′AATTC —— 3′
    3′ —— CTTAAG —— 5′
11. **(a)** Resistance Transfer Factor (RTF) is an old name for a plamid that codes for antibiotic resistance.
12. **(d)**
13. **(a)** A : *Eco*RI is an endonuclease enzyme isolated from strains of *E. coli*, and is part of the restriction modification system.

    B : *Bam* HI (from *Bacillus amyloliquefaciens*) is a type II restriction endonuclease having the capacity for recognising short sequences of DNA and specifically cleaving them at a target site.

    C : *Hind* III is a type II site-specific deoxyribonuclease restriction enzyme isolated from *Haemophilus influenzae* that cleaves the DNA palindromic sequence AAGCTT in the presence of the cofactor Mg^{2+} *via* hydrolysis.

    D : pBR 322 is a plasmid and was one of the first widely used *E. coli* cloning vectors.
14. **(a)** (i) Denaturation - 94°C, (ii) Annealing - 60°C, (iii) Extesnsion - 72°C

15. **(a)** Lysozyme digest the bacterial cell wall and serves as a strong chemical defense against bacterial infections.

16. **(a)** Most sensitive technique or detect malignant cell in non- Hodgkin's lymphoma is polymerase chain reaction. In recent times, PCR is being used in the detection of HIV (Virus of AIDS).

17. **(d)** In microinjection method, foreign DNA is directly injected into the nucleus of animal cell or plant cell by using micro needles or micro pipettes.

18. **(d)** Restriction endonuclease cleaves DNA molecules only at specific nucleotide sequence called restriction sites. DNA ligase enzymes is used to joins bits of DNA.

19. **(d)** The dark bands on X-rays film represent the DNA fingerprints or the DNA profiles.

20. **(b)** **21.** **(d)** **22.** **(b)** **23.** **(d)**

24. **(d)**

25. **(c)** T_i plasmid (tumor inducing) from the soil bacterium *Agrobacterium tumefaciens* is effectively used as vector for gene transfer to plant cells. The part of T_i palsmid transferred into plant cell DNA, is called the T-DNA. The T-DNA with desired DNA spliced into it, is inserted into the chromosomes of the host plant where it produces copies of itself, by migrating from one chromosomal position to another at random. Such plant cells are then cultured, induced to multiply and differentiate to form plantlet, transferred into soil, the plantlets grow into mature plants, carrying the foreign gene, expressed throughout the new plant.

26. **(c)** Due to chloramphenicol resistant gene, one is able to select a transformed cell in the presence of chloramphenicol. The chloramphenicol resistant gene in this case is called selectable marker.

27. **(a)** Electrophoresis separates DNA fragments according to size.

28. **(b)** PCR is a way to dramatically increase the numbers of a particular sequence of DNA (or an entire gene). This technique is most useful when cloning DNA, since prior knowledge of the DNA sequence must first be obtained.

29. **(a)**

30. **(b)** Interferons are antiviral proteins produced by recombiant DNA technology in *E.coli*.

31. **(d)** **32.** **(a)**

33. **(c)** DNA fragments migrate towards the positive end of the electrical field with the smallest fragments, migrating the fastest.

34. **(d)**

35. **(b)** Besides the gene gun, other methods of genetic transfer between cells or organisms include the use of vectors such as viruses and plasmids.

36. **(b)**

37. **(a)** Restriction endonuclease-*Hin*d II, always cuts DNA molecules at a particular point by recognising a specific sequence of six base pairs. This specific base sequence is known as the recognition sequence for *Hin*d II.

38. **(d)**

39. **(b)** Ligases are the enzymes used to glue DNA fragments.

40. **(d)** If all of the cells are derived from one cell, they are all genetically identical and thus clones of one another. If all of the cells of an organism are identical (derived from a cell isolated from another organism as seen in plant and mammalian clones), then that organism is called a clone. Cloning also refers to creating a recombinant molecule by inserting a DNA into a plasmid vector and successfully introducing that recombinant molecule into the cell.

41. **(c)** Treating cells with ampicillin does not make them more permeable to DNA. Coating DNA with lipids (not carbohydrates) is used to introduce recombinant DNA molecules into host cells, injecting DNA (not protein) is a method used to place recombinant DNA into best cells.

42. **(c)**

34 Biotechnology and its Applications

1. Maximum utilisation of the techniques of biotechnology has been made in the field of:
   (a) Agriculture (b) Medicines
   (c) Industries (d) Biogas production
2. Golden rice is a transgenic crop of the future with the following improved trait:

   (a) High lysine (essential amino acid) content
   (b) Insect resistance
   (c) High protein content
   (d) High vitamin-A content
3. The bacteria generally used for genetic engineering is:
   (a) *Agrobacterium* (b) *Bacillus*
   (c) *Pseudomonas* (d) *Clostridium*
4. Polyploidy can result from: Tricky
   (a) Double fertilisation (b) Polyspermy
   (c) Diploid gametes (d) All of these
5. Strains of *Bacillus thuringiensis* are used in producing:
   (a) Bioinsecticidal plants
   (b) Biomineralisation
   (c) Biometallurgical techniques
   (d) Biofertilisers
6. Which is not a transgenic plant?
   (a) Soyabean (b) Maize
   (c) Golden rice (d) Cucumber
7. The pesticide most persistent in the soil is: Critical Thinking
   (a) DDT
   (b) BHC
   (c) Dieldrin
   (d) Baygon
8. The crops having *cry* genes need:
   (a) no insecticide
   (b) mild quantity of insecticide
   (c) large amount of insecticide
   (d) 5 kg insecticide / hectare
9. cry II Ab and cry I Ab produce toxins that control:

   (a) cotton bollworms and corn borer respectively
   (b) cotton borer and cotton bollworms respectively.
   (c) tobacco budworms and nematodes respectively.
   (d) nematodes and tobacco budworms respectively.
10. Which of the following *Bt* crops is being grown in India by the farmers?
   (a) Cotton (b) Brinjal
   (c) Soyabean (d) Maize
11. A GM crop is transgenic. It can be harmful because it may cause:
   (a) allergies and toxicity.
   (b) incorporation of antibiotic resistance in human beings.
   (c) disturbance in metabolism due to enzyme for antibiotic resistance.
   (d) All of the above
12. *Flavr savr* variety of Tomato is improved variety developed through incorporation of a transgene. *Flavr savr* variety of tomato remains fresh for long because it: Toughnut
   (a) has little polygalacturonase
   (b) abundant polygalacturonase
   (c) has gene for antibiotic resistance
   (d) lacks amylase
13. Which of the following is/are **correct**?

(i)	Biowar	–	Biowar is the use of biological weapons against humans and or their crops and animals.
(ii)	Bioethics	–	Bioethics is the unauthorised use of bioresources and traditional knowledge related to bioresources for commercial benefits.
(iii)	Biopatent	–	Exploitation of bioresources of other nations without proper authorisation.

   (a) Only (ii) (b) Only (i)
   (c) (i) and (ii) (d) (i) and (iii)

14. Match column-I with column-II and choose the correct option.

Column-I		Column-II	
A.	*Escherichia coli*	I.	*nif* gene
B.	*Bacillus thuringiensis*	II.	Interferon
C.	*Rhizobium meliloti*	III.	*Bt* toxin
D.	*Pseudomonas putida*	IV.	Bioremediation

(a) A – II; B – I; C – IV; D – III
(b) A – II; B – I; C – III; D – IV
(c) A – II; B – III; C – I; D – IV
(d) A – IV; B – I; C – II; D – III

15. Which one of the following statement is correct?
(a) The proteins encoded by the genes *cry* I *Ac* and *cry* II *Ab* control cotton bollworms.
(b) Protein encoded by *cry* I *Ab* controls corn borer.
(c) Both (a) and (b)
(d) Proteins encoded by *cry* I *Ac* and *cry* I *Ab* control flies.

16. Modifications by germ line gene therapy are heritable as: **Critical Thinking**
(a) the functional gene is incorporated into their genome.
(b) the functional gene is incorporated into one of the gene.
(c) the functional gene is incorporated into somatic cells.
(d) All of the above

17. Biotechnology deals with industrial scale production of biopharmaceuticals and biological products using genetically modified:
(a) microbes only
(b) plants and animals only
(c) fungi only
(d) All of the above

18. Demerit of bovine and porcine insulin is: **Toughnut**
(a) Hypercalcemia
(b) Expensive
(c) Allergic reactions
(d) Mutations in human genome

19. ....and................are present at N and C terminal of β chain of threonine. **Critical Thinking**
(a) Phenylalanine and alanine
(b) Glycine and asparagine
(c) Leucine and isoleucine
(d) Methionine and tryptophan

20. In the treatment of SCID, cDNA is introduced into a patient's cells using.............as a vector. **Critical Thinking**
(a) *E. coli*
(b) *Retrovirus*
(c) *Bacillus thuringiensis*
(d) *Agrobacterium*

21. Select the correct set of the names for 1, 2, 3 and 4.

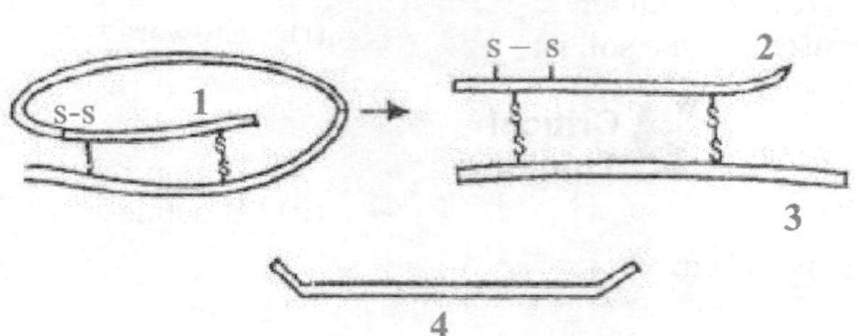

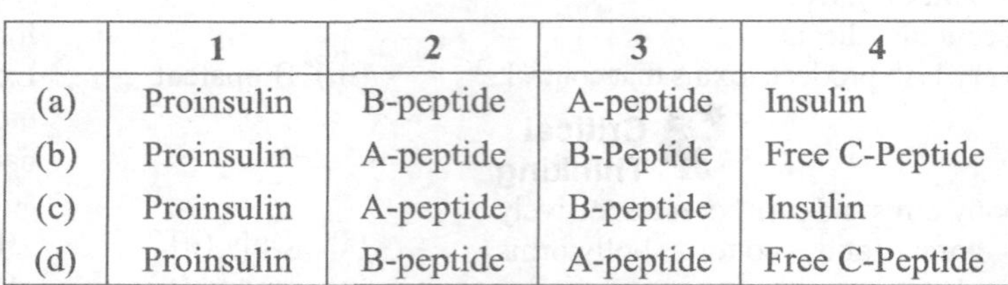

	1	2	3	4
(a)	Proinsulin	B-peptide	A-peptide	Insulin
(b)	Proinsulin	A-peptide	B-Peptide	Free C-Peptide
(c)	Proinsulin	A-peptide	B-peptide	Insulin
(d)	Proinsulin	B-peptide	A-peptide	Free C-Peptide

22. Match column-I with column-II and choose the correct option.

	Column-I		Column-II
A.	Golden rice	I.	Increased shelf life
B.	*Flavr Savr* tomato	II.	hGH
C.	Mouse	III.	Vitamin A
D.	Transgenic pig	IV.	Organ transplantation

(a) A – III; B – I; C – II; D – IV
(b) A – II; B – I; C – III; D – IV
(c) A – II; B – III; C – I; D – IV
(d) A – IV; B – I; C – II; D – III

23. Which of the following statement is **incorrect**?
(i) Recombinant DNA technology is used to improve crop plants by increasing their productivity, by making them more nutritious and by developing disease resistant.
(ii) *Bt* cotton is resistant to bollworm infestation.
(iii) *Bacillus thuringiensis* form cry protein during any phase of their growth.
(iv) *Bacillus thuringiensis* is not harmed by self cry protein because of its occurrence as protoxin (inactive).
(v) Protoxin cry protein is changed into active cry protein in the stomach of insects due to alkaline pH in stomach.
(a) (i), (ii) and (iv)
(b) (i) and (iv)
(c) Only (iii)
(d) All of the above

24. Multicellular organisms that carry a specific genetic change in each cell because of an intervention at the embryo level is a:
(a) Transversion (b) Transition
(c) Transgenic (d) Transformant

25. One of the following are transgenic organisms.
(a) Holly sheep and Tomato
(b) Dolly sheep and Subabul
(c) Molly sheep and Banana
(d) *Bt* Cotton and Tomato (*Flavr Savr*)

26. Bioaugmentation is:
(a) the addition of commercially prepared catabolic activated bacterial strains to degrade wastes.
(b) the production of fertilisers by using bacteria and virus.
(c) the metals are deposited as insoluble oxides and sulphides by activity of bacteria.
(d) the removal of pests.

27. Which was the first product contributed to human kind by science of biotechnology in early 1970's?
(a) Somatotropin (b) Insulin
(c) Cytokines (d) Erythropoietin

28. Most widely used bioweapon is:
(a) *Bacillus subtilis*
(b) *Pseudomonas putida*
(c) *Bacillus anthracis*
(d) None of the above

29. In a criminal trial in which DNA fingerprinting has been used, a genetic "match" between the suspect and evidence left at the crime scene provides:

(a) definitive proof that the suspect is guilty.
(b) a high probability that the suspect is guilty.
(c) a low probability that the suspect is guilty.
(d) definitive proof that the suspect is innocent.

30. *Yersinia pestis* is used as biological weapon. It causes a disease:
(a) Small pox (b) Plague
(c) Tularemia (d) Anthrax

31. In 1921, two Canadian scientists Frederick G. Banting and Charles. H. Best sucessfully purified insulin from a: Tricky
(a) Pig's pancreas (b) Camel's pancreas
(c) Dog's pancreas (d) Cattle's pancreas

32. Which of the following biotechnological product can enhance the immune system?
(a) IL-2 (b) GM-CSF
(c) TNF (d) All of these

33. Protein engineering and drug designing requires CAMM (computer aided molecular modelling) because of:
(a) study of three dimensional structures of receptors and functional molecules.
(b) requirement of DNA recombinant technology.
(c) specificity of proteins.
(d) All of the above

34. Which of the following statements is **incorrect**?
(i) Insulin was originally extracted from pancreas of slaughtered pigs and cattle.
(ii) Animal insulin is difficult to obtain.
(iii) Animal insulin is identical to human insulin.
(iv) Non human insulin caused some patients to develop allergy.
(v) Recombinant DNA allowed scientists to insert a human insulin gene into a bacterial expression vector.
(vi) Recombinant insulin is actually obtained from *E.coli* bacterial cell.
(a) (i) and (ii) (b) (iii) and (iv)
(c) Only (iii) (d) Only (vi)

35. Match column-I with column-II and choose the correct option.

	Column-I		Column-II
A.	Gene therapy	I.	*Rhizobium*
B.	Biofertiliser	II.	*Cry* gene
C.	*Bt* cotton	III.	SCID
D.	Humulin	IV.	Diabetes

(a) A – II; B – I; C – IV; D – III
(b) A – III; B – I; C – II; D – IV
(c) A – II; B – III; C – I; D – IV
(d) A – IV; B – I; C – II; D – III

36. Match column-I with column-II and choose the correct option.

	Column-I		Column-II
A.	*Bacillus thuringiensis*	I.	Production of chitinases
B.	*Rhizobium meliloti*	II.	Scavenging of oil spill
C.	*Pseudomonas putida*	III.	Incorporation of *nif* gene
D.	*Trichoderma*	IV.	Production of *Bt* toxin
		V.	Production of human insulin

(a) A – II; B – IV; C – I; D – V; E – III
(b) A – II; B – IV; C – V; D – I; E – III
(c) A – IV; B – III; C – V; D – II; E – I
(d) A – V; B – IV; C – III; D – I; E – II

37. Which of the following GM crop is produced by antisense RNA technology where the expression of native gene is blocked? Tricky
(a) *Bt* cotton
(b) Golden rice
(c) Flavr Savr
(d) Golden Mustard

38. The main use of recombinant DNA technology are:
(a) production of transgenic humans.
(b) the creation of cells capable of synthesising economically important molecules.
(c) the efficient reduction of useful proteins.
(d) Both (b) and (c)

39. ....infects the roots of tobacco plants which reduces the production of tobacco?
(a) Nematode (*Meloidogyne incognita*)
(b) Coleopterans (beetles)
(c) Lepidopterans (army worm)
(d) Dipterans (mosquitoes)

40. Golden rice was created by transforming rice with two beta-carotene biosynthesis genes, namely:

(a) Psy and cry I genes (b) LCY-E
(c) CHY-1 (d) CHY-2

41. Humulin is:
(a) A form of chitin
(b) A new digestive enzyme
(c) A powerful antibiotic
(d) Human insulin

42. A genetically engineered micro-organism used successfully in bioremediation of oil spills is a species of:
(a) *Pseudomonas* (b) *Trichoderma*
(c) *Xanthomonas* (d) *Bacillus*

43. When a patient with defective ADA was treated, which of the following steps was performed for gene therapy?
(i) Lymphocytes were obtained from the patients.
(ii) Lymphocytes are transferred to culture dishes.
(iii) Lymphocytes were transfected with normal ADA genes.
(iv) The transfected cells are returned to the patients.
(a) All of the above
(b) (iii) and (iv)
(c) Only (iv)
(d) SCID cannot be treated

44. Following are the steps of southern blot procedure.
(i) Autoradiography
(ii) Hybridisation with radioactive nucleic acid (probe).
(iii) Blotting
(iv) DNA fragments are treated to make them single stranded.
(v) Electrophoresis
(vi) Clearing of DNA by restriction endonuclease.
(vii) Isolation of DNA from sample
The correct sequence is:
(a) (i), (ii), (iii), (iv), (v), (vi), (vii)
(b) (vii), (vi), (v), (iv), (iii), (ii), (i)
(c) (i), (ii), (vi), (vii), (iii), (iv), (v)
(d) (vii), (vi), (i), (ii), (v), (iii), (iv)

45. Match column-I with column-II and choose the correct option.

	Column-I		Column-II
A.	Virus free plant	I.	*Rhizobium*
B.	Biofertiliser	II.	Cry gene
C.	*Bt* cotton	III.	Shoot apex
D.	Humulin	IV.	Diabetes

(a) A – II; B – I; C – IV; D – III
(b) A – III; B – I; C – II; D – IV
(c) A – II; B – III; C – I; D – IV
(d) A – IV; B – I; C – II; D – III

46. DNA vaccines are: Critical Thinking
(a) mixture of hormones
(b) recombinant vaccines
(c) synthetic vaccines
(d) pure DNA or RNA

47. The name of the drug used in cancer treatment produced by biotechnology is:
(a) Interferon (b) hGH
(c) TSH (d) Insulin

48. By the use of biotechnology in which of the following bacteria production of B_2 vitamins has been increased to about 20,000 times? Tricky
(a) *Ashbya gossypii*
(b) *E. coli*
(c) *Pseudomonas denitrificans*
(d) *Propionibacterium shermanii*

49. Recombinant DNA Technology prepared vaccine is called:
(a) First generation vaccine
(b) Second generation vaccine
(c) Third generation vaccine
(d) Synthesised vaccine

50. Human insulin is being commercially produced from a transgenic species of:
(a) *Rhizobium* (b) *Saccharomyces*
(c) *Escherichia* (d) *Mycobacterium*

51. Somatostatin or human growth hormone (HGH) for the first time was genetically isolated by:
(a) Banting and Best (b) Ross
(c) Pasteur (d) Goeddel

52. Hirudin gene inserted in *Brassica napus* was: Tricky
(a) synthesised chemically.
(b) obtained from leech.
(c) obtained from *Bacillus thuringiensis*.
(d) *Brassica campestris*.

53. Given below are certain characteristics of thurioside protein. Which of the statement given are **correct**? Tricky
(i) It is activated by alkaline pH of the gut of the insect pest ultimately killing it.
(ii) It binds with the epithelial cells of the midgut of the insect pest ultimately killing it.
(iii) It does not kill the carrier bacterium which is itself resistant to this toxin.
(a) (i) and (ii)
(b) (i) and (iii)
(c) (ii) and (iii)
(d) (i), (ii) and (iii)

54. Following are the steps in one type of gene therapy.
(i) Inject engineered cells into patients bone marrow.
(ii) Viral DNA carring the normal allele inserts into chromosome.
(iii) Let retrovirus infect bone marrow cells that have removed from patient and cultured.
(iv) Insert RNA version of normal allele into retrovirus.
The correct sequence is:
(a) (i), (ii), (iii) and (iv)
(b) (iv), (iii), (ii) and (i)
(c) (i), (ii), (iv) and (iii)
(d) (iv), (iii), (i) and (ii)

55. Select the correct statement. 

(a) Genetic engineering works only on animals and has not yet been successfully used on plants.
(b) There are no risks associated with DNA technology.
(c) The first step in PCR is heating which is used to separate both the strands of target DNA.
(d) DNA from one organism will not bond to DNA from another animal.

56. Consider the following statements.

(i) Transgenic animals are more sensitive to toxic substance than non-transgenic animals.
(ii) Useful biological products can be produced by introducing portion of DNA into transgenic animals which codes for a particular product.
(iii) Brazzein is protein produced by a west African plant, *Pentadiplandra brazzeana* which is approximately 2000 times as sweet as sugar.

Which of the statements are given above is **correct**?

(a) (i), (ii) and (iii)
(b) (i) and (ii)
(c) (i) and (iii)
(d) None of these

57. Consider the following statements.

(i) In earlier days insulin extracted from pancreas of slaughtered cattle and pigs, was more efficient than genetically engineered insulin.
(ii) PCR technique is being used for the detection of HIV and genetic mutations, in suspected cancer patients.
(iii) Cystic fibrosis, heamophilila, cancer, Parkinson's etc., are treated by gene therapy.

Which of the statements given above are **correct**?

(a) (i) and (ii)
(b) (i) and (iii)
(c) (ii) and (iii)
(d) (i), (ii) and (iii)

Answer KEYs

1	(b)	7	(c)	13	(b)	19	(a)	25	(d)	31	(c)	37	(c)	43	(a)	49	(b)	55	(c)
2	(d)	8	(a)	14	(c)	20	(b)	26	(a)	32	(d)	38	(d)	44	(b)	50	(c)	56	(a)
3	(a)	9	(a)	15	(c)	21	(b)	27	(a)	33	(a)	39	(a)	45	(b)	51	(d)	57	(c)
4	(c)	10	(a)	16	(a)	22	(a)	28	(c)	34	(c)	40	(a)	46	(d)	52	(a)		
5	(a)	11	(d)	17	(d)	23	(c)	29	(b)	35	(b)	41	(d)	47	(a)	53	(a)		
6	(d)	12	(a)	18	(c)	24	(c)	30	(b)	36	(c)	42	(a)	48	(a)	54	(b)		

Hints & Solutions

1. (b) Maximum utilisation of the techniques of biotechnology has been made in the field of medicines.
2. (d) 3. (a) 4. (c) 5. (a) 6. (d)
7. (c) 8. (a) 9. (a) 10. (a)
11. (d) The transgene may get transferred through pollen to their wild relatives, which may make those weeds more persistent and damaging. The transgenic crop themselves may become persistent weeds. Such crops can also damage the environment in some yet unknown manner. The enzyme produced by the antibiotic resistance gene could cause allergies, as it is a foreign gene product. The bacteria present in the human alimentary canal could take up the antibiotic-resistance gene from the food and become resistant to that particular antibiotic.
12. (a) In *Flavr Savr* variety of tomato, expression of gene for production of polygalacturonase was blocked. Enzyme polygalacturonase degrades pectin.
13. (b) 14. (c)
15. (c) Statement (a) and (b) are correct. (d) Proteins encoded by *cry 1 Ac* and *cry 1 Ab* controls cotton bollworms and *cry1Ab* controls corn borer.
16. (a)
17. (d) Biotechnology is the exploitation of biological processes for industrial and other purposes, especially the genetic manipulation of microorganisms (like microbes, fungi, plants and animals) for the production of antibiotics, hormones, etc.
18. (c) Insulin obtained from cattle and pigs slightly differ from the human insulin in their amino acid sequence. Also the quantity of insulin produced from pigs and cattle does not cater the needs of growing number of diabetic patients. Also on injecting of insulin into the patients, produces sensitivity reaction and side effects occasionally. These factors led reseachers to look for some alternative source of human insulin. The search for new source was soon fulfilled by recombinant DNA technology.
19. (a)
20. (b) A functional ADA (cDNA) can be introduced into cells of the patients receiving gene therapy by using vector constituted by *retrovirus*. The SCID patient has a defective gene for the enzyme Adensosine Deaminase (ADA). He/she lacks functional T-lymphocytes which are extracted from the patient's bone marrow and a normal functional copy of human gene coding for ADA is introduced into these lymphocytes with the help of *retrovirus*.
21. (b) 22. (a) 23. (c)
24. (c) A multicellular organism that carres a specific genetic change in each cell because of an intervention at the fertilised egg stage is a transgenic.
25. (d) 26. (a) 27. (a) 28. (c)
29. (b) Although DNA fingerprinting is a powerful technique, a genetic match can demonstrate only a high probability of guilt since there is a 1/100,000 to 1 / 1,000,000,000 chance that two individuals will share the same genetic profile.
30. (b) 31. (c) 32. (d) 33. (a) 34. (c)
35. (b) A: Gene therapy is the first clinical therapy which is given to a 4 year old girl with adenosine deaminase deficiency (caused due to deletion of the gene for adenosine deaminase).

    B: Biofertiliser is a substance which contains living microorganisms that, promotes growth by increasing the supply or availability of primary nutrients to the host plant. Bio-fertilisers such as Rhizobium, Azotobacter, Azospirilium and blue green algae (BGA) have been in use for a long time.

    C: *Bt* cotton is a genetically modified variety of cotton producing an insecticide. *Bt* cotton was created through the addition of genes encoding toxin crystals in the *Cry* group of endotoxin. When insects attack and eat the cotton plant the *Cry* toxins are dissolved due to the high pH level of the insects stomach.

D: Humulin is used for a preparation of insulin produced by genetic engineering and structurally identical to insulin made by the human pancreas. It is used to treat diabetes.

36. (c) 37. (c) 38. (d)

39. (a) A nematode, *Meloidogyne incognita* infects the roots of tobacco plants, which reduce the production of tobacco.

40. (a) Transgenic golden rice was created by transforming rice with the gene Psy (Phytoene synthase) from daffodil (*Narcissus pseudonarcissus*) and cry1 from the soil bacterium *Erwinia uredovora.*

41. (d) The first commercial example of enzyme modification of a protein for human use is the conversion of pig insulin to human insulin called "humulin".

42. (a) 43. (a) 44. (b) 45. (b)

46. (d) DNA vaccines are used as injection of pure DNA or RNA into skeleton muscles leading to expression of DNA in the muscle cell.

47. (a) Interferons are group of signalling proteins released by host cells in response to presence of tumor cells.

48. (a) By the use of bacteria Ashbya gossypii the production of B_2 vitamine has been increased by 20,000 times while production of B_{12} vitamines has been increased by 50,000 times.

49. (b) Recombinant DNA Technology prepared vaccine is called second generation vaccine.

50. (c) 51. (d)

52. (a) Hirudin is a protein that prevents blood clotting. The gene encoding hirudin was chemically synthesised. This gene was then transferred into *Brassica napus*, where hirudin accumulates in seeds. Hirudin is purified and used as medicine. In this case, the transgene product itself is the product of interest.

53. (a) *Bacillus thuringiensis* forms the protein crystals, which contains a toxic insecticidal protein. *Bt* toxins are initially inactive protoxins but after ingestion by the insects their inactive toxin becomes active due to the alkaline pH of the gut, which solubilises the crystals. The activated toxin binds to the surface of the midgut, epithelial cells swelling and lysis, further leading to death of the insect.

54. (b)

55. (c) Statement (c) is correct. The first step in PCR (polymerase chain reaction) is heating which is used to separate both the strands of target DNA. PCR is a laboratory technique used to make multiple copies of a segment of DNA. PCR is very precise and can be used to amplify, or copy, a specific DNA target from a mixture of DNA molecules.

56. (a)

(i) Transgenic animals are made to carry genes which makes them more sensitive to toxic substances than non-transgenic animals.

(ii) Useful biological products can be produced by introducing into transgenic animals the portion of DNA (or genes) which codes for a particular product, e.g., human protein (α - 1 -antitrypsin) is used to treat emphysema.

(iii) Brazzein is protein produced by a west African plant, *Pentadiplandra brazzeana*, which is approximatlely 2000 times as sweet as sugar. It is used as a low calorie sweetener. Local people have been using the super sweet berries of this plant for centuries. But the protein brazzein was patented in USA.

57. (c) Earlier, insulin was extracted from the pancreas of slaughtered cattle and pigs but some patients began developing allergies. The injection of this insulin into patients occasionally produces sensitivity reactions and side effects.

35 Organisms and Population

1. Sequence of humification and mineralisation is:
(a) Dead organic matter → Litter → Duff → Humus → Minerals
(b) Humus → Minerals → Litter → Duff
(c) Minerals → Humus → Litter → Duff
(d) Dead organic matter → Duff → Litter → Minerals → Humus

2. Permafrost is the most distinctive feature of:
(a) Taiga
(b) Temperate grasslands
(c) Tundra
(d) Pacific northwest

3. Desert regions are characterised by ___ centimeters of rainfall per year.
(a) less than 5 (b) less than 15
(c) less than 25 (d) over 50

4. For plants inhabiting a tropical rain forest, the critical competition, is for– **Tricky**
(a) Soil nutrients (b) Water
(c) Space (d) Light

5. Biome is:
(a) sum of ecosystems in a geographical area.
(b) sum of ecosystems of the whole earth.
(c) biotic component of an ecosystem.
(d) biotic potential of a population.

6. Which of the following communities is more vulnerable to invasion by outside animals and plants?
(a) Tropical evergreen forests
(b) Oceanic island communities
(c) Mangroves
(d) Temperate forests

7. The sun loving plants are referred to as:
(a) Halophytes (b) Heliophytes
(c) Heterotrophs (d) Sciophytes

8. What is best pH of soil for cultivation? **Tricky**
(a) 3.4 – 5.4 (b) 4.5 – 5.5
(c) 5.5 – 6.5 (d) .5 – 7.5

9. Mechanical tissue is best developed in:
(a) Hydrophytes (b) Halophytes
(c) Xerophytes (d) Mesophytes

10. Adaptation to low temperature and freezing in animals occurs due to the production of–
(a) Antifreeze proteins
(b) Chaperonins
(c) Proline
(d) Analine

11. Which of the following is not a reason why population growth declines as population size approaches the carrying capacity? **Tricky**
(a) Climate becomes unfavorable.
(b) Competition for resources increases.
(c) Predation rates increase.
(d) Disease rates increase.

12. Explosive growth in human population size can be attributed largely to: **Critical Thinking**
(a) declining death rates due to improved medical care.
(b) proliferation of fast-food outlets, leading to easier access to nutrients.
(c) increasing birth rates due to better health and fertility.
(d) declining death rates due to declining incidence of war.

13. No population can grow indefinitely. The ultimate size of any population is limited by:
(a) the carrying capacity of its environment.
(b) its birth rate.
(c) its intrinsic growth rate.
(d) reproductive isolation.

14. In an equilibrium population (at its carrying capacity), thousands of eggs and hundreds of tadpoles are produced by a single pair of frogs. About how many offsprings will live to reproduce the next year? **Toughnut**
(a) 0 (b) 2
(c) 10 to 20 (d) More than 100

15. Which of the following statement is **incorrect**?
(a) Habitat includes both biotic and abiotic factors.
(b) Abiotic and biotic components interact constantly with each other.
(c) Abiotic components alone can characterise the habitat of an organism.
(d) Major abiotic factors includes temperature, water, light and soil.

16. For a population that is stable in size, the following age distribution indicates that:

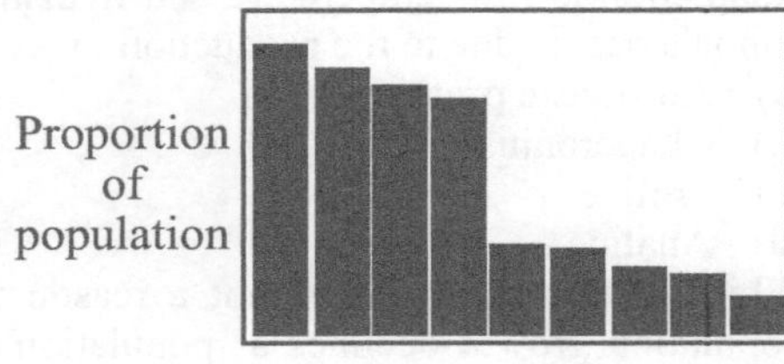

(a) the population's birth and death rates are both high.
(b) the population's birth and death rates are both low.
(c) the population's birth rate is low but its death rate is high.
(d) the population's birth rate is high but its death rate is low.

17. Which statement does not reflect a reason why grasses predominate in savannah?
(a) There are few native herbivores in this biome.
(b) Grasses grow well in fine, sandy soils.
(c) The above ground portions of grass plants die during dry seasons.
(d) The deep roots of the grasses can survive many months of drought.

18. Competitive exclusion principle was put forward by G.F. Gause on the basis of experimental observation on:
(a) *Chlorella* (b) *Paramoecium*
(c) *Microcystis* (d) Cuckoo bird

19. Which of the following type of interactions occur in predation and parasitism?
(a) (+, +) (b) (+, 0)
(c) (+, –) (d) (–, –)

20. If b = 65 and d = 45, N = 100 then find out $\frac{dN}{dt}$.

Critical Thinking

(a) 2000 (b) 1000
(c) 200 (d) 100

21. Organisms occupying similar ecological niche but different geographical areas of distribution are called as:
(a) Edge species
(b) Ecological equivalents
(c) Ecoclines
(d) Inquilines

22. In the models that describe population growth, *r* stands for:
(a) population density
(b) a time interval
(c) total number of individuals in the population.
(d) growth rate

23. Human population growth __________.

Tricky

(a) has an S-shaped curve.
(b) is currently in a logistic phase.
(c) is currently exponential.
(d) has reached carrying capacity.

24. The maximum expansion rate of a population under the best conditions is referred to as ______. There are abundant resources that last forever and no limits on growth.
(a) biotic potential (b) cohort
(c) mortality rate (d) biome

25. *Casuarina equisetifolia* is a:
(a) Mesophytes (b) Xerophyte
(c) Halophyte (d) Forest epiphyte

26. The chief advantage of encystment to *Amoeba proteus* is:

Critical Thinking

(a) an ability to live for a time without ingesting food.
(b) the cessation of most metabolic activity so that available energy may be used for reproduction.
(c) the opportunity to get rid of accumulated waste products.
(d) the ability to survive during unfavourable physical conditions.

27. Photosynthetically Active Radiation (PAR) is:

(a) 5% of incident solar radiation
(b) 50% of incident solar radiation
(c) 0.5% of incident solar radiation
(d) 99% of incident solar radiation

28. In an age pyramid, the number of individuals of reproductive age is lesser than pre-reproductive but higher than post reproductive ones. The population is:
(a) Growing
(b) Declining
(c) Stable
(d) Can not be predicted

29. In biotic community, which of the following can be called protective device?
(a) Mimicry
(b) Symbiosis
(c) Competition
(d) Parasitism

30. Bell-shaped polygonal pyramid indicates:
(a) high percentage of young individuals.
(b) moderate percentage of young individuals.
(c) low percentage of young individuals.
(d) low percentage of old individuals.

31. Mark the correct identification from the following picture.

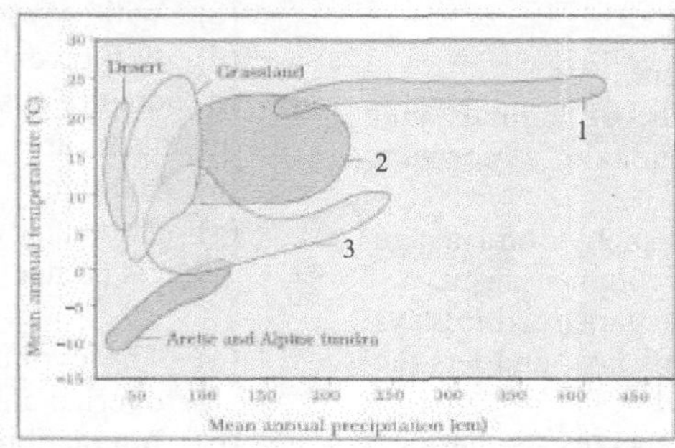

	1	2	3
(a)	Tropical forest	Temperate forest	Coniferous forest
(b)	Temperate forest	Tropical forest	Coniferous forest
(c)	Temperate forest	Coniferous forest	Tropical forest
(d)	Coniferous forest	Tropical forest	Temperate forest

32. What does the shape of the given age pyramids (A to C) reflect about the growth status of populations?

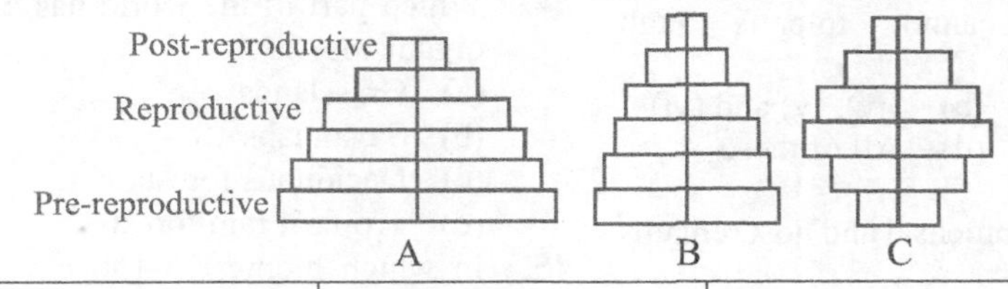

	A	B	C
(a)	Declining	Stable	Expanding
(b)	Stable	Expanding	Declining
(c)	Expanding	Stable	Declining
(d)	Declining	Expanding	Stable

33. Which of the following statement(s) is/are **incorrect**?
(i) Light is essential for life to exist on the earth.
(ii) Many species of small plant under the canopy to tall trees in forest show optimal use of available light due to having large sized antenna and higher number of thylakoids.
(iii) UV-rays is not harmful for many organisms.
(iv) Photoperiodic requirement is not essential for many plant for flowering.
(v) Red algae can live in deeper water of sea because of having pigment, phycoerythrin.
(a) Only (i)
(b) Only (ii)
(c) Only (iii)
(d) (iv) and (v)

34. The given, examples are associated with which type of adaptation?
(i) Basking by desert lizards in sun.
(ii) Hiding in burrows by some animals.
(iii) Wearing of woollen clothes.
(iv) Thermal gaping
(a) Scansorial adaptation
(b) Cursorial adaptation
(c) Biochemical adaptation
(d) Behavioural adaptation

35. Select the **incorrect** statement.
(a) Overwhelming majority of animals and nearly all plants maintain a constant internal temperature.
(b) An orchid growing as an epiphyte on a mango branch is an example of commensalism.
(c) In brood parasitism, the parasitic bird lays its eggs in the nest of its host and lets the host to incubate them.
(d) In amensalism, one species is harmed whereas the other is unaffected.

36. Which of the following statement(s) is/are **incorrect**?
(i) Plants have no thermoregulation.
(ii) 99% animals and almost all plants are conformers.
(iii) Heat loss or heat gain is a function of surface area.
(iv) Thermoregulation is energetically least expensive process for many organisms like shrews and humming birds.
(v) 99% animals are thermoregulators.
(vi) Archaebacteria cannot tolerate high temperature.
(a) (i) and (ii) (b) (iv), (v) and (vi)
(c) None of these (d) All of these

37. In commensalism:
(a) population of commensal and host remains unaffected.
(b) population of commensal may increase while that of host remains unaffected.
(c) population of both commensal and host increases.
(d) population of commensal increases while the population of host gradually decreases.

38. The relationship between hermit crab and sea anemone is:
(a) Symbiosis (b) Mutualism
(c) Parasitism (d) Commensalism

39. During succession, inhibition:
(a) may prevent the achievement of a climax community.
(b) is evidence for the equilibrium theory of succession.
(c) may interfere with the successful colonisation of other species.
(d) may involve changes in soil pH or accelerated accumulation of humus.

40. A small, isolated population:

(a) will do extremely well if protected.
(b) is unlikely to survive, even within a reserve.
(c) will grow, but extremely slowly.
(d) will given the right conditions.

41. In aquatic system, actively moving animal organisms are:
(a) Benthic (b) Nektons
(c) Zooplanktons (d) Phytoplanktons

42. What is competitive exclusion?

(a) The evolution of traits that reduce niche overlap and competition.
(b) Interactions that allow species to occupy their fundamental niche.
(c) The degree to which the niches of two species overlap.
(d) The claim that species with the same niche cannot coexist.

43. The combination of conditions that a species will occupy in the absence of competitors is:
(a) The fundamental niche
(b) The realised niche
(c) Competitive exclusion
(d) Mimicry

Toughnut

44. Which part of the world has a high density of organisms?
(a) Grasslands
(b) Savannahs
(c) Deciduous forests
(d) Tropical rain forests

45. In which biome, 75-150 cm. annual rain fall takes place?
(a) Grassland
(b) Mediterranean scrub forest
(c) Deciduous forest
(d) North coniferous forest

46. A group of individuals of a plant inhabiting a particular area form a:
(a) Population (b) Community
(c) Ecosystem (d) Species

47. Sunken stomata can be seen in–

(a) *Hydrilla* (b) *Nymphaea*
(c) *Acacia* (d) *Rhizophora*

48. Refer to the graph below, which shows the changes in population size over time.

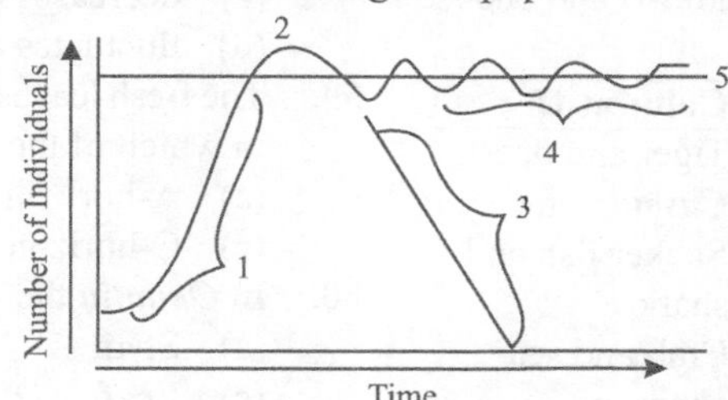

Which part of this graph describes or refers to the exponential growth phase of a new population?

(a) 1 (b) 5 (c) 2 (d) 3

49. In laboratory experiments, two species of the protist *Paramoecium* were grown alone and in the presence of the other species. The following graphs show growth of species 1 (left) and species 2 (right), both alone and when in mixed culture.

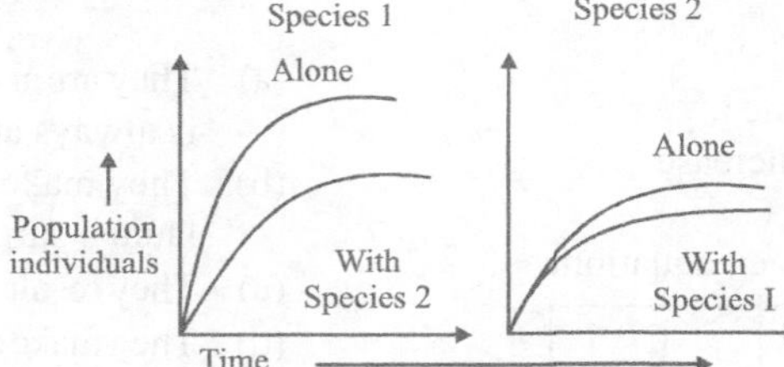

Interpretation of these graphs shows that:

(a) competitive exclusion occurred in the these experiments.
(b) both species are affected by interspecific competition but species 1 is less affected.
(c) both species are affected by interspecific competition but species 2 is less affected.
(d) both species are affected equally by interspecific competition.

50. Heat loss or heat gain is a function of surface area.

(i) Small animals in colder areas have large surface area.
(ii) They expend less amount of energy and do not loose body heat.
(iii) Small animals are rarely found in cold regions.

Which of the above statements are **correcct** regarding the function of surface area?

(a) Only (i) (b) (ii) and (iii)
(c) Only (iii) (d) (i) and (iii)

51. Which statement explains the concept of Allen's rule?

(a) Aquatic mammals have blubber as insulator.
(b) Mammals of colder climate have shorter ears and limbs.
(c) Mammals of humid and warmer region have more melanin in their skin.
(d) The bear undergoes hibernation during winter.

52. Consider the following statements (A) - (D) each with one or two blanks.

A. Lichens represent an intimate ____(i)____ relationship between a fungus and ____(ii)____.
B. The ____(iii)____ are associations between fungi and the roots of higher plants.
C. Plants need the help of ____(iv)____ for pollinating their flowers and dispersing their seeds.
D. The ____(v)____ pollinates the fig inflorescence while searching for suitable egg - laying sites.

Which one of the following options, gives the correct fill ups for the respective blank numbers from (i) - (v) in the statements?

(a) (i) Parasitic; (ii) - Cyanobacteria; (iii) - Mycorrhizae; (iv) - Wind; (v) - Bee
(b) (i) Mutualistic; (ii) - Cyanobacteria; (iii) - Mycorrhizae; (iv) - Animals; (v) Wasp
(c) (i) Parasitic; (ii) - Cyanobacteria; (iii) - Mycorrhizae; (iv) - Insect; (v) Bumble bees
(d) (i) Mutualistic; (ii) - Cyanobacteria; (iii) - Lichen; (iv) - Wind; (v) Wasp

53. Match the column-I with column-II and choose the correct option.

Column-I		Column-II	
A.	Mutualism	I.	Tiger and deer
B.	Commensalism	II.	*Cuscuta* on *Cissus*
C.	Parasitism	III.	Sucker fish and shark
D.	Predation	IV.	Crab and sea anemone

(a) A – I; B – II; C – III; D – IV
(b) A – III; B – IV; C – II; D – I
(c) A – II; B – III; C – I; D – IV
(d) A – IV; B – II; C – III; D – I

54. $\frac{dN}{dt} = rN\left(\frac{K-N}{K}\right)$

A – Population density at time t.
B – Intrinsic rate of nature increase.
C – Carrying capacity.
Identify A, B and C from given equation.

	A	B	C
(a)	N	K	r
(b)	N	r	K
(c)	K	N	r
(d)	K	r	N

55. During summer stratification, the upper layer of water is called: Tricky

(a) Thermocline (b) Hypolimnion
(c) Epilimnion (d) Ectolimnion

56. Pneumatophores are special features of plants growing in:
(a) Dry and hot conditions
(b) Water
(c) Mangrove forests
(d) Oligotrophic soils

57. When birth rate equals death rate:

(a) a population grows rapidly.
(b) the size of a population remains constant.
(c) density-dependent limiting factors do not affect the population.
(d) a population is in danger of extinction.

58. At carrying capacity, a population _____. Tricky
(a) increases rapidly
(b) decreases slowly
(c) decreases rapidly
(d) fluctuates around a median number

59. The fresh leaves and recognisable litter are found in which of the following horizon of soil?
(a) A-horizon (b) B-horizon
(c) C-horizon (d) O-horizon

60. In *Opuntia* the spine is the modification of:
(a) Stem (b) Root
(c) Leaf (d) None of these

61. Rainfall is absent but precipitation occurs in:
(a) Arctic desert (b) Tropical desert
(c) Chaparral (d) Savannah

62. Why are inducible defenses advantageous?

(a) They are always present - thus, an individual is always able to defend itself.
(b) They make it impossible for a consumer to launch surprise attacks.
(c) They result from a coevolutionary arms race.
(d) They make efficient use of resources, because they are produced only when needed.

63. Leaf cutter ants keep fungal gardens. The ants provide food for the fungus but also feed on the fungus. This is an example of
(a) Competition (b) Predation
(c) Parasitism (d) Mutualism

64. Which of the following statement is **not correct** about K-selection?
(a) Having long life span
(b) Developed Interspecific competition
(c) Very stable population
(d) Having short generation time

65. A population of 600 that experiences 65 births and 5 deaths during a one-year period. If the population maintains the current growth pattern, a plot of its growth would resemble: Toughnut
(a) Exponential growth
(b) Fluctuating growth
(c) K-selected growth
(d) Logistic growth

66. In the formula G = rN (K – N)/K, the factor rN tends to cause the population to _____.
(a) remain stable at the carrying capacity
(b) grow at a slower rate than the (K – N/K) factor
(c) grow increasingly rapidly
(d) decrease in size

67. Dense evergreen vegetation of broad sclerophyllous leaves and shrubs with fire resistant resinons plants is known as:
(a) Chaparral vegetation
(b) Savannah vegetation
(c) Steppe grassland
(d) Tundra vegetation

68. The shallow water zone around the edge of the lake is called:
(a) Littoral zone
(b) Limnetic zone
(c) Profundal zone
(d) Benthic zone

69. Genetically adapted local population of a species (that has a wide range of distribution) are called:
(a) Phenotypes
(b) Ecotype
(c) Sciophytes
(d) Mesophytes

70. Which one of the following is example of species that breed in captivity and produce infertile hybrids?
(a) Mallard (a duck) and the pin tail duck
(b) Male donkey and female horse
(c) Polar bear and the Alaskan brown bear
(d) Platy and sword tail fishes

71. Which one of the following is the interspecific hybrid variety?
(a) Mule (b) Sugarcane
(c) Pony (d) All of these

72. Which type of plants are found in Taiga & Tundra?
(a) Halophytes (b) Xerophytes
(c) Epiphytes (d) Psychrophytes

73. Succulents occur in:
(a) Deserts
(b) Tundra
(c) Temperate deciduous forests
(d) Tropical rain forests

74. You are growing a bacterial population in a laboratory flask with a flow-through system that continuously replaces old culture medium with fresh medium without removing any bacteria. Which of the following graphs best predicts the growth of this population over time?

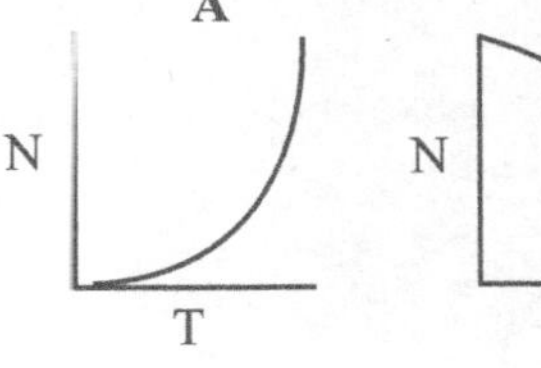

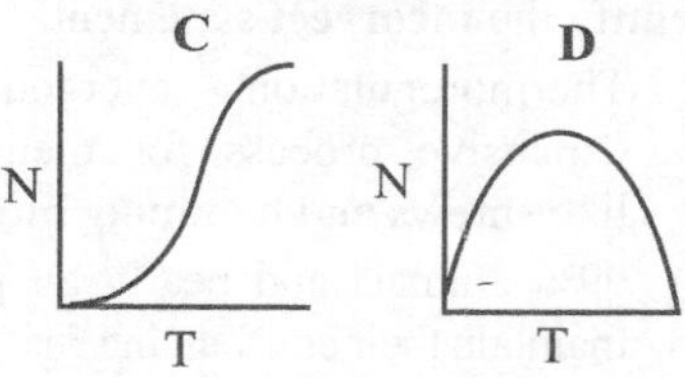

(a) A (b) D
(c) B (d) C

75. Which of the following factors (given below) determines the vegetation of any area?
(i) pH of soil.
(ii) Mineral compositon of the soil.
(iii) Water holding capacity of soil.
(iv) Precipitation.
(a) (i) and (ii)
(b) (ii) and (iii)
(c) (i), (ii), and (iii)
(d) (i), (ii), (iii) and (iv)

76. Match the column-I with column-II and choose the correct option.

Column-I Population		Column-II Example
A. Predation	I.	*Cuscuta* and Hedge plants
B. Commensalism	II.	*Balanus* and *Chthamalus*
C. Parasitism	III.	Cactus and Moth
D. Competition	IV.	Orchid and Mango

(a) A – III, B – IV, C – I, D – II
(b) A – IV, B – III, C – II, D – I
(c) A – I, B – III, C – II, D – IV
(d) A – III, B – IV, C – II, D – I

77. Plants have developed various means of protecting themselves from predators. One of the methods is by producing toxic chemicals. Identify such chemicals from the list given below.
(i) Nicotine
(ii) Caffeine
(iii) Quinine
(iv) Strychnine
(v) Opium
(a) Only (i)
(b) (i), (ii), (iii) and (iv)
(c) (i), (ii) and (iii)
(d) All of the above

78. Identify the **incorrect** statement.

(i) Thermoregulation energetically least expensive process for many organisms like shrews and humming birds.

(ii) 99% animals and nearly all plants cannot maintain their constant internal environment.

(iii) During the course of evolution, the costs and heights of maintaining a constant internal environment are discarded.

(iv) In aquatic animals, the osmotic concentration of the body fluids change with that of the ambient water osmotic concentration.

(a) (i) and (ii)
(b) (iii) and (iv)
(c) (i) and (iii)
(d) (ii) and (iii)

79. Mark the **correct** statement.

(a) In case of bear, going into hibernation during winter, is an example of 'escape in time'.
(b) To prevent desiccation, some snails and fishes enter into diapause.
(c) Under unfavourable conditions many zooplankton species go into aestivation.
(d) All of the above

80. Which of the following statement(s) is/are **correct**?

(i) Organisms living in oceans, lakes and rivers do not face any water-related problems.

(ii) Euryhaline can tolerate a wide range of salinities.

(iii) Stenohaline are restricted to a narrow range of salinities.

(iv) No fresh water animals cannot live for long in sea water but sea animals can live in fresh water for long time because of osmotic balance.

(v) The salt concentration is less than 5% in inland water, 30-50% in sea and >100% in some hypersaline lagoons.

(a) All are correct except (i)
(b) None of the above
(c) Only (iv)
(d) (i), (iii) and (iv)

81. A country with a high rate of population growth took measures to reduce it. The figure below shows age-sex pyramids of populations A and B twenty years apart.

Select the correct interpretation about them.

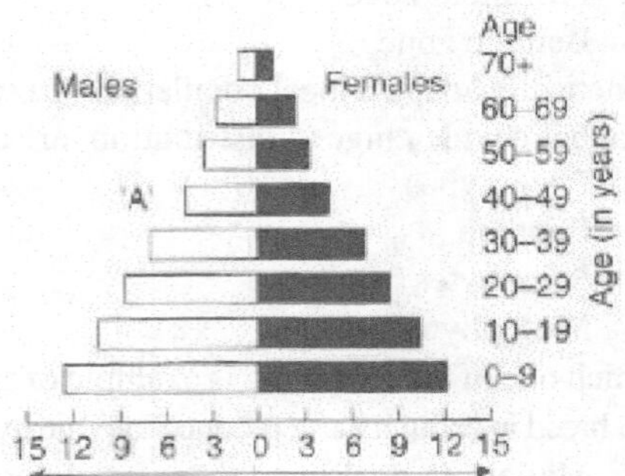

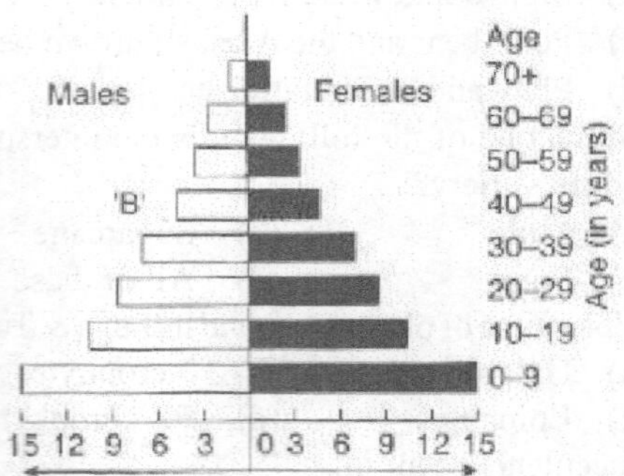

(a) "B" is earlier pyramid and shows stabilised growth rate.
(b) "B" is more recent showing that population is very young.
(c) "A" is the earlier pyramid and no change has occurred in the growth rate.
(d) "A" is more recent and shows slight reduction in the growth rate.

82. Study the population growth curves shown in the given diagram.

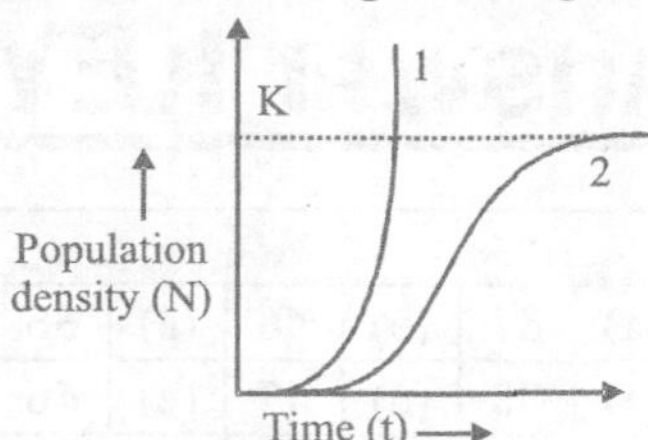

Which options is the best for curve (1) and (2)?

	Type of (1) curve	Type of (2) curve	Equation for curve (1)	Equation for curve (2)
(a)	Logistic curve	Exponential curve	$\frac{dN}{dt} = rN$	$\frac{dN}{dt} = rN\left(\frac{K-N}{K}\right)$
(b)	Exponential curve	Logistic curve	$\frac{dN}{dt} = rN\left(\frac{K-N}{K}\right)$	$\frac{dN}{dt} = rN$
(c)	Logistic curve	Exponential curve	$\frac{dN}{dt} = rN\left(\frac{K-N}{K}\right)$	$\frac{dN}{dt} = rN$
(d)	Exponential curve	Logistic curve	$\frac{dN}{dt} = rN$	$\frac{dN}{dt} = rN\left(\frac{K-N}{K}\right)$

83. In the given graph arrow indicates specifically:

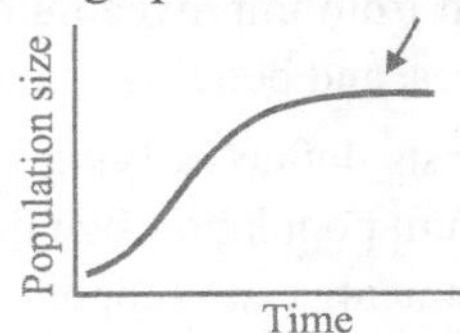

(a) r
(b) K–N
(c) K
(d) Density independent effect

84. An oak tree produces thousands of acorns, but very few grow into mature oak trees. The oak tree exhibits a ______ survivorship curve.

(a) type I (b) type II
(c) type III (d) type I or II

85. Savannahs are:
(a) Tropical rain forests
(b) Desert scrubs
(c) Grassland with scattered trees
(d) Dense forests with closed canopy

86. Most molluscs and cry fishes are:
(a) Thermoconformers
(b) Osmoconformers
(c) Oxyregulators
(d) All of the above

87. Which of the following statement(s) is/are correct?
(i) In a population growth rate curve, the lag phase represents the period when a population is adjusting to a new environment.
(ii) In a population growth rate curve the steepest part of growth phase represents the period when a population is growing exponentially.
(iii) In the stationary phase of a population growth rate curve, the population is in dynamic equilibrium.
(iv) Humans have changed the carrying capacity of the earth.
(a) All of these
(b) (i), (ii) and (iii)
(c) (ii) and (iii)
(d) (i) and (iv)

Answer KEYs

1	(a)	**10**	(a)	**19**	(c)	**28**	(a)	**37**	(b)	**46**	(a)	**55**	(c)	**64**	(d)	**73**	(a)	**82**	(d)
2	(c)	**11**	(a)	**20**	(a)	**29**	(a)	**38**	(d)	**47**	(c)	**56**	(c)	**65**	(a)	**74**	(d)	**83**	(c)
3	(c)	**12**	(a)	**21**	(b)	**30**	(b)	**39**	(c)	**48**	(c)	**57**	(b)	**66**	(c)	**75**	(c)	**84**	(c)
4	(d)	**13**	(a)	**22**	(d)	**31**	(a)	**40**	(b)	**49**	(c)	**58**	(d)	**67**	(a)	**76**	(a)	**85**	(c)
5	(a)	**14**	(b)	**23**	(c)	**32**	(c)	**41**	(b)	**50**	(d)	**59**	(d)	**68**	(a)	**77**	(d)	**86**	(d)
6	(b)	**15**	(c)	**24**	(a)	**33**	(c)	**42**	(d)	**51**	(b)	**60**	(c)	**69**	(b)	**78**	(c)	**87**	(a)
7	(b)	**16**	(a)	**25**	(b)	**34**	(d)	**43**	(a)	**52**	(b)	**61**	(a)	**70**	(b)	**79**	(a)		
8	(c)	**17**	(a)	**26**	(d)	**35**	(a)	**44**	(d)	**53**	(b)	**62**	(d)	**71**	(d)	**80**	(a)		
9	(c)	**18**	(b)	**27**	(b)	**36**	(b)	**45**	(c)	**54**	(b)	**63**	(d)	**72**	(d)	**81**	(d)		

Hints & Solutions

1. **(a)**
2. **(c)** Permafrost, a permanently frozen subsoil, is a distinctive characteristic of the tundra.
3. **(c)** Deserts have less than 25 centimeters of rainfall per year.
4. **(d)** Competition for light is a critical survival factor among plants living in tropical rain forests
5. **(a)** A biome is a major ecological community of organisms that extends over a large geographical area characterised by a dominant type of vegetation.
6. **(b)** This is so because oceanic islands are normally cut off from other habitats.
7. **(b)** The plants which grow best in full sunlight are called sun plants or heliophytes.
8. **(c)** Most cultivated soils have a pH range of 4.5 to 7.0. However, most plants grow best in soils with a neutral or slightly acidic reaction (pH).
9. **(c)** Mechanical tissues are best developed in xerophytic plants because these plants live in xeric habitats which require stronger cells, tissues, organs and then whole body.
10. **(a)** **11.** **(a)**
12. **(a)** Human death rates have decreased due to declining infant mortality rates that has resulted from improved nutrition, medical advances, and better sanitation.
13. **(a)** Ecologists define carrying capacity as the maximum population size that a particular environment can support with no net increase or decrease over a relatively long period of time.
14. **(b)** Ecologists define carrying capacity as the maximum population size that a particular environment can support with no net increase or decrease over a relatively long period of time. A single pair of frogs would be expected to only replace itself.
15. **(c)** Abiotic components are physical factors which affect the structure, behaviour and life history of organisms. These components (such as soil, moisture, range of temperature, and availability of light) along with biotic factors (such as the availability of food and the presence of predators) can characterise the habitat of an organism.

16. **(a)** When a population's birth rate and death rate are both high, the age distribution is dominated by young individuals. A similar age distribution would occur in a population with a high birth rate and a low death rate, here such a population would not be stable in size.

17. **(a)** Grasslands are typically maintained by grazing herbivores and by periodic fires. Overgrazing, may occur if agricultural livestock are introduced which can lead to destruction of the grassland.

18. **(b)**

19. **(c)** Predation occurs when members of one species eat the other species, while parasitism occurs when one member of a species parasite on the other body. In this case one member of a species are benefitted (+) while the other member of a species are harmed (–).

20. **(a)** For a closed population

$\Delta N = B - D$

$$\frac{dN}{dt} = B - D$$

$\Delta \quad B = bN; \qquad D = dN$

$$\frac{dN}{dt} = (b-d)N = (65-45)100$$

$= 2000$

(b and d are instantaneous birth and death rates)

21. **(b)** Examples of ecological equivalents are owl and cat; rabbit and rat.

22. **(d)** r is the intrinsic rate of increase of an organism.

23. **(c)** The exponential growth model virtually describes the population explosion of humans.

24. **(a)** The mortality rate is a percentage of individuals that began a specific year alive but have passed away before the conclusion of the same year. Biotic potential of a species is influenced by the following length of reproductive life, how often reproduction occurs, how many progeny are created at each birth occurrence, what is the reproductive age, and how many live to actually reproduce.

25. **(b)** It is a non-succulent xerophyte. It can withstand atmospheric as well as soil droughts.

26. **(d)** In the encystment state the *Amoeba* can withstand extremes of heat and cold as well as desiccation and their power of resistance is considereably increased.

27. **(b)** **28.** **(a)**

29. **(a)** Mimicry is the superficial protective device in a biotic community. It is resemblance of one organism to another which has evolved as a means of protection.

30. **(b)** **31.** **(a)** **32.** **(c)** **33.** **(c)**

34. **(d)** The given examples show behavioural adaptations. Behavioural adaptation is the process by which an organism or a species changes its pattern of action to better suit its environment.

35. **(a)** Statement (a) is incorrect. An overwhelming majority (around 99 percent) of animals and nearly all plants cannot maintain a constant internal temperature.

36. **(b)** **37.** **(b)**

38. **(d)** Commensalism is a relationship between organism and its host in which the host neither benefits nor suffers from the association, *e.g.*, hermit crab and sea anemone.

39. **(c)** **40.** **(b)**

41. **(b)** Nektons or neustons are actively swimming animals. Fish turtle, mammals and some marine birds are the example of nektons.

42. **(d)** **43.** **(a)**

44. **(d)** Tropical rain forests occur in equatorial and subequatorial regions. The forests receive all the external inputs for optimum plant growth. Due to abundant plant growth, a large number of animals live in tropical rain forests. In grasslands, fire occur periodically which prevent tree growth. In savannah periods of drought are common. In deciduous forests, broad leaved, hard wood deciduous trees are found predominantly.

45. **(c)** **46.** **(a)** **47.** **(c)**

48. **(c)** A population living under ideal conditions will experience exponential population growth.

49. (c) 50. (d)

51. (b) Allen's rule states that the limbs, ears, and other appendages of the animals living in cold climates tend to be shorter than in animals of the same species living in warm climates. Shorter and more compact body parts have less surface area than elongated ones and thus radiate less body heat.

52. (b) 53. (b) 54. (b) 55. (c)

56. (c)

57. (b) When birth rate is equal to death rate, individuals are being added to the population at the same rate at which they are being removed from the population (note that this fails to consider migration).

58. (d) Ecologists define carrying capacity as the maximum population size that a particular environment can support with no net increase or decrease over a relatively long period of time. Many populations oscillate about their carrying capacity or overshoot it at least once before attaining a relatively stable size.

59. (d) 60. (c) 61. (a) 62. (d)

63. (d) 64. (d)

65. (a) If the reproductive rate is greater than zero (r>0), then the population is growing exponentially.

66. (c) The difference between birth rate and death rate is symbolised as r. N is the population size. So rN is the population growth rate.

67. (a) Chaparral is broad leaved scrub forest (sclerophyllous or thick leaved shrubs and small trees) in which several plants are fire resistant.

68. (a) 69. (b) 70. (b) 71. (d)

72. (d) 73. (a)

74. (d) The logistic model of population growth produces a sigmoid (S-shaped) growth curve when the population size (N) is plotted over time. New individuals are added to the population most rapidly at intermediate population sizes when there is not only a breeding population of substantial size but also lots of available space and other resources in the environment. The population growth rate slows dramatically as the population size approaches carrying capacity (K).

75. (c) Weather changes have little impact on flora and fauna of place because it is the short term property of the atmosphere and it changes from place to place. pH, mineral, water holding capacity of soil determine flora and fauna of any area. Precipitation also plays a key role in determining the vegetation of an area.

76. (a) Predation is a relation between two organisms in which one organism captures and feeds on other. Commensalism is a relation between two organisms in which one benefits and the other derives neither benefit nor harms. Parasitism is a relation between organisms in which one lives as a parasite on another and harm the host. Competition is process in which the fitness and survival ability of one species is significantly lower in the presence of another species.

77. (d) A wide variety of chemical substances that we extract from plants on a commercial scale (nicotine, caffeine, quinine, strychnine, opium etc); are actually produced by them as defence against grazers and browers.

78. (c) Thermoregulation is energetically expensive for many organisms particularly true for small animals like shrews and humming birds. During the course of evolution, the costs benefits of maintaining a constant internal environment are taken into consideration.

79. (a) 80. (a)

81. (d) 'A' is more recent and shows slight reduction in growth rate.

82. (d) 83. (c)

84. (c) Production of large numbers of offspring with few growing to maturity describes

85. (c)

86. (d) Most molluscs and cry fishes are oxyregulators but with the exception of birds and mammals, they are thermoconformers and osmoconformers.

87. (a)

36 Ecosystem

1. The largest ecosystem of the world is:
   (a) Forests (b) Grasslands
   (c) Great lakes (d) Oceans
2. The driving force for an ecosystem is:
   (a) Biomass
   (b) Producers
   (c) Carbohydrates in producers
   (d) Solar energy
3. Which most often limits the primary productivity of the ecosystem? **Tricky**
   (a) Solar radiation/light
   (b) Oxygen
   (c) Consumers
   (d) Nitrogen
4. Benthos of pond ecosystem are:
   (a) Producers
   (b) Primary consumers
   (c) Secondary consumers
   (d) Tertiary consumers
5. A recently discovered ecosystem is: **Tricky**
   (a) Crater (b) Tundra
   (c) Floating iceberg (d) Vent
6. An incomplete ecosystem is:
   (a) Grassland (b) River
   (c) Wetland (d) Cave
7. How much portion of the PAR is captured by the plants?
   (a) 5 – 10% (b) 7 – 10%
   (c) 8 – 10% (d) 2 – 10%
8. In terrestrial ecosystem, maximum energy is present in:
   (a) T_3 (b) T_2
   (c) T_1 (d) T_4
9. Plankton, Nekton and Benthos are not the components of one of ecosystems which is:
   (a) Oceans (b) Fresh water rivers
   (c) Grassland (d) Pond or lake
10. Which of the following is the most stable ecosystem?
    (a) Ocean (b) Desert
    (c) Forest (d) Mountain
11. The process by which a lake ecosystem is altered by eutrophication involves several stages, each causing the next. Which of the following stages would occur second in the chain of causation? **Critical Thinking**
    (a) Algal blooms occur.
    (b) Oxygen levels drop in deeper water.
    (c) Phosphorus input from sewage and agricultural runoff increases.
    (d) Respiratory demand from decomposers increases.
12. Phosphorus usually becomes a limiting factor at certain times of the year in:
    (a) Aquatic ecosystem
    (b) Grassland ecosystem
    (c) Forest ecosystem
    (d) All of the above
13. The pyramid of number in grassland ecosystem is:
    (a) Linear (b) Inverted
    (c) Upright (d) Negative
14. A pasture of 40 hectares has a biomass of 40 kg/hectare, or roughly 1,000,000 calories of energy. Cattle grazing on this pasture are sold for beef. About how many calories of energy from the pasture reach human consumers of the beef? **Critical Thinking**
    (a) 10,000 (b) 1000
    (c) 100,000 (d) 1,000,000
15. Which hypothesis regarding what limits food-chain length has the most support?
    (a) Energy transfer hypothesis
    (b) Stability hypothesis
    (c) Environmental complexity hypothesis
    (d) All of the above
16. Which among the following is likely to have the highest levels of DDT depositions in its body?
    (a) Eel (b) Crab
    (c) Sea gull (d) Phytoplankton
17. Energy flow through an ecosystem differs from nutrient flow in that: **Critical Thinking**
    (a) energy flow is linear.
    (b) nutrient flow is linear.
    (c) energy flow is cyclical.
    (d) energy builds as it flows through the ecosystem.

18. Which of the given below pyramids represents the variation in biomass at different trophic levels in pond ecosystem?

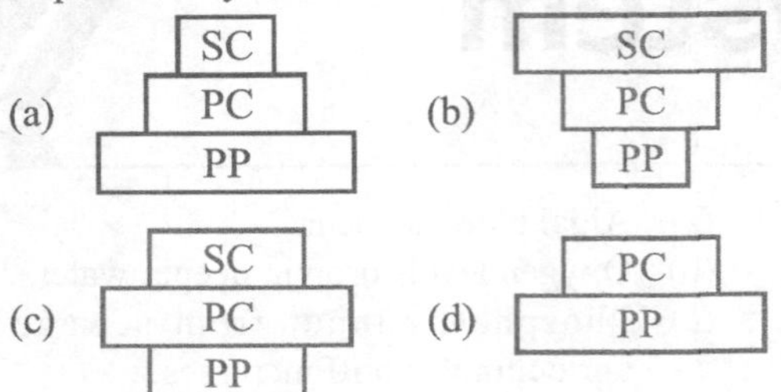

19. Which one of the following statement is correct about decomposition?
(a) Decomposition rate is slower if detritus is rich in lignin and chitin, and quicker if detritus is rich in nitrogen and water - soluble substances like sugars.
(b) Decomposition rate is slower if detritus is rich in nitrogen and water - soluble substances like sugars, and quicker if detritus is rich in lignin and chitin.
(c) Decomposition rate is slower if detritus is rich in cellulose, and quicker if detritus is rich in phosphorus.
(d) Decomposition rate is quicker if detritus is rich in lignin, and lower if detritus is rich in sulphur.

20. Which of the following statement are **correct**?
(i) Deserts and deep lakes show least productivity.
(ii) 10g by mass is passed to the next trophic level in the food chain.
(iii) Most productive ecosystem is coral reef.
(iv) Sugarcane is the most productive crop.
Choose the correct option.
(a) (i) and (iv) (b) (i) and (iii)
(c) (ii) and (iii) (d) (i), (iii) and (iv)

21. Because of limitations on resources, organisms must ______ materials in order to survive.
(a) Reuse (b) Reinvent
(c) Reintroduce (d) Recycle

22. The term ecosystem was coined by:
(a) E.P. Odum (b) A.G. Tansley
(c) E. Haeckel (d) E. Warming

23. Which of the following is not one of the three aspects studied in biogeochemical cycling?
(a) The nature and size of element reservoir.
(b) The rate of movement between reservoirs.
(c) How different biogeochemical cycles interact.
(d) How new species create their own biogeochemical cycles.

24. One way cycle is:
(a) CO_2 cycle (b) H_2O cycle
(c) Free energy cycle (d) O_2 cycle

25. Productivity at the second trophic level is always:

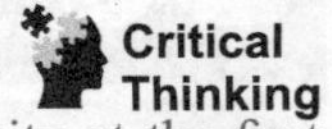

(a) greater than the productivity at the first trophic level always.
(b) less than the productivity at the first trophic level.
(c) equal to the productivity at the first trophic level.
(d) extremely variable compared to the productivity at the first trophic level.

26. The free floating organisms, of an open sea and shore, are collectively called:
(a) Planktons (b) Nektons
(c) Benthoic (d) Sea anemone

27. Decomposers such as fungi and bacteria are vital to the ecosystem because they:
(a) are the primary organic recyclers.
(b) are the terminal energy consumers.
(c) keep populations in check.
(d) Both (a) and (b)

28. In a food web, each successive trophic level has: Tricky
(a) increased total energy
(b) less total energy content
(c) more total energy content
(d) non-estimated energy content

29. Which of the following organism types would probably rank the highest in terms of their expected efficiency of energy transfer between trophic levels? Critical Thinking
(a) Herbivorous mammal
(b) Carnivorous mammal
(c) Invertebrate herbivore
(d) Invertebrate carnivore

30. A complex of several species and communities in various stages of succession maintained under similar climatic conditions is known as:
(a) Biome
(b) Environmental complex
(c) Biological spectrum
(d) Biological clocks

31. Which of the following features would be characteristic of oceans but not of freshwater ecosystems? Critical Thinking
(a) Receives material from land mostly via ground water.
(b) Seasonal mixing of materials.
(c) Elements buried in bottom sediments for long periods of time.
(d) Bottom waters frequently lacking oxygen.

32. The occupational address of a species is often called as its:

(a) Biome (b) Niche
(c) Environment (d) Habitat

33. The two vegetations of an ecosystem are separated by: Tricky

(a) Ecesis (b) Ecoline
(b) Ecophane (d) Ecosystem

34. The lentic ecosystem includes which of following water?

(a) Rain (b) Running
(c) Standing (d) Gravitational

35. Which of the following could not be considered an ecosystem? Tricky

(a) A small pond
(b) All the fish in a coral reef
(c) Earth
(d) A pile of dung in a pasture

36. Energy flow and energy transformations in living systems strictly confirm to the: Critical Thinking

(a) Law of limiting factors
(b) Liebig's law of minimum
(c) Laws of thermodynamics
(d) Shelford's law of tolerance

37. Each trophic level has a certain mass of living material at a particular time is known as:

(a) Catabolism
(b) Standing crop
(c) Humification
(d) Primary productivity

38. Which of the following is **incorrect** for a pyramid of productivity?

(a) Only about 10% of the energy in one trophic level is passed into the next level.
(b) Because of the loss of energy at each trophic level, most food chains are limited to three to five steps.
(c) The pyramid of productivity of some aquatic ecosystems is inverted because of the large zooplankton primary - consumer level.
(d) Eating grain-fed beef is an inefficient means of obtaining the energy trapped by photosynthesis.

39. Which of the following statement(s) is/are **correct**?

(i) An important characteristic of all communities is that their composition and structure constantly change in response to the changing environmental conditions.
(ii) The climax community remains stable as long as the environment remains changed.
(iii) The entire sequence of communities that successively change in a given area are called sere(s).

(a) (i), (ii) and (iii)
(b) (i) and (ii)
(c) (ii) and (iii)
(d) (i) and (iii)

40. Given below is one of the types of ecological pyramids. This type represents: Tricky

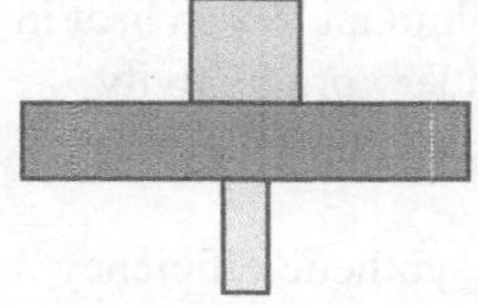

(a) Pyramid of numbers in a grassland
(b) Pyramid of biomass in a fallow land
(c) Pyramid of biomass in a lake
(d) Energy pyramid in a spring

41. Which is not true regarding wetlands?

(a) Natural ground water recharging systems
(b) Used as pasture land
(c) Purify runoff water
(d) Act as sedimentation traps

42. In which life zone of ocean, the producers in the food chains are rarely found?

(a) Littoral (b) Limnetic
(c) Profundal (d) Intertidal

43. A community which starts succession in a habitat is:

(a) Pioneer community (b) Seral community
(c) Biotic community (d) Ecosere

44. The flow of materials and energy in an ecosystem is respectively: Critical Thinking

(a) Cyclic only
(b) Cyclic and linear
(c) Linear and cyclic
(d) Linear only

45. Which food chain produces greater available energy? **Tricky**

(a) Diatoms → fish → large fish → birds → man
(b) Green plant → rabbit → man
(c) Green plant → grass hopper → frog → snake → peacock
(d) Green plant → rabbit → wolf → man

46. The zone of effective light penetration in oceanic biome is called as:

(a) Epipelagic zone (b) Bathypelagic zone
(c) Abyssal zone (d) Benthic zone

47. What will happen if decomposers are removed from the ecosystem?

(a) Energy cycle is stopped
(b) Mineral cycle is stopped
(c) Consumers cannot absorb solar energy
(d) Rate of decomposition of mineral increases

48. The amount of energy captured by photosynthesis, minus the amount lost as heat in respiration, is:

(a) Secondary productivity
(b) Consumption efficiency
(c) NPP
(d) Photosynthetic efficiency

49. Choose the **incorrect** statements regarding ecological pyramids.

(i) The relationship between organisms at different trophic levels is expressed in terms of number, biomass and energy.
(ii) Any calculations of energy content, biomass or number has to include one group of organism at that trophic level.
(iii) In most ecosystems, all the pyramids of number, biomass and energy are upright.
(iv) The pyramid of biomass in sea is generally inverted.
(v) Pyramid of energy is always inverted and can never be upright.

(a) (i), (iii) and (iv) (b) Only (i)
(c) (ii) and (v) (d) (i) and (v)

50. Match column-I with column-II and select the correct answer using the codes given below.

Column-I	Column-II
A. Primary succession	I. Autotrophs
B. Climax community	II. Community that has completed succession
C. Consumer	III. Colonisation of a new environment
D. Producer	IV. Animals

(a) A – III; B – II; C – IV; D – I
(b) A – III; B – I; C – IV; D – II
(c) A – I; B – III; C – II; D – IV
(d) A – II; B – III; C – IV; D – I

51. Which of the following statement(s) is/are **correct** about biogeochemical cycle?

(i) Gaseous exchanges of phosphorus between organism and environment are negligible.
(ii) All biogeochemical cycles include both organisms and non-living components.
(iii) Most elements remain longest in the living portion of their cycle.
(iv) The chemical elements used by organisms in large quantities cycle back and forth between organisms and environment.

(a) (i) and (iii) (b) (i), (ii) and (iv)
(c) (i), (ii) and (iii) (d) All of these

52. Two food chains are shown.

(i) Tree → aphid → insectivorous bird → prey feed on bird
(ii) Phytoplankton → zooplankton → plankton-feeding fish → carnivorous fish

Which diagram is a pyramid of energy representing both food chains?

(a)
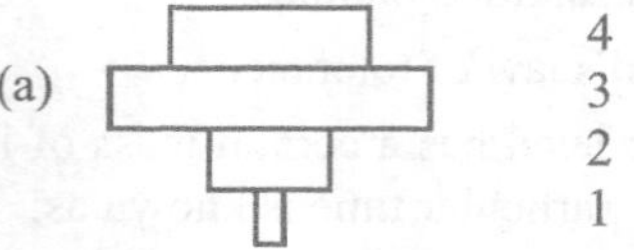

(b)
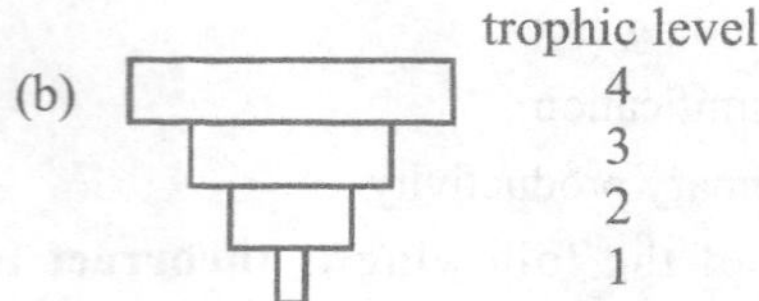

(c)
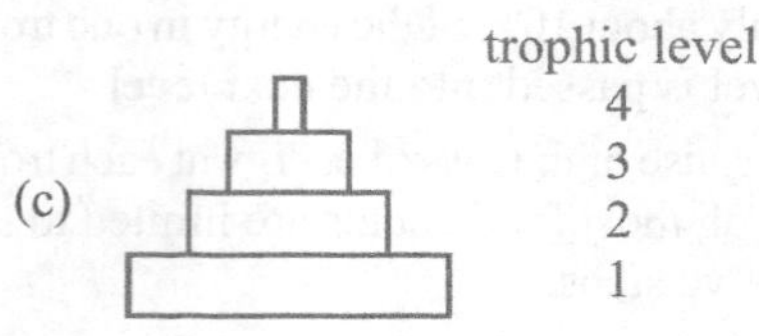

(d)
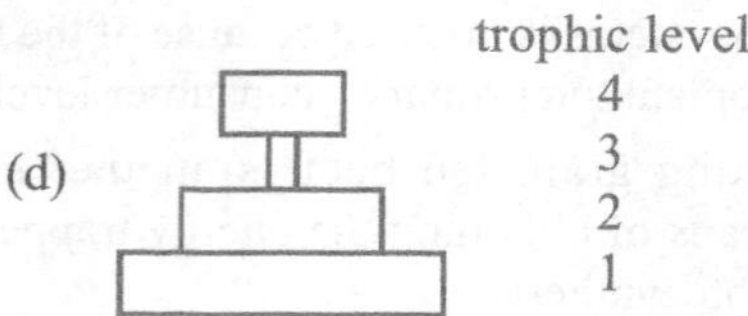

53. When the number of organisms at successive levels are plotted, they give the shape of a pyramid. This is called the pyramid of:
(a) Number (b) Energy
(c) Biomass (d) None of these

54. The flow of DDT through the food chain, culminating in the deaths of cats in Borneo, represents the: **Critical Thinking**
(a) number of links in an ecosystem.
(b) need to understand how resources are passed through an ecosystem.
(c) need for stringent testing of anything that might affect the environment.
(d) All of the above

55. If an ecosystem is composed of only three trophic levels then how much energy will be conserved at the III trophic level: **Tricky**
(a) 30% (b) 20%
(c) 10% (d) 1%

56. The animals which occupy the same trophic level are: **Critical Thinking**
(a) Lion & bees
(b) Deer & bees
(c) Snakes & earthworm
(d) Crow & cow

57. In an aquatic ecosystem, maximum biomagnification is seen among:
(a) Fishes
(b) Phytoplanktons
(c) Zooplanktons
(d) Macroscopic plant

58. Pick up the correct food chain.
(a) Grass → Chamelion → Insect → Bird
(b) Grass → Fox → Rabbit → Bird
(c) Phytoplankton → Zooplankton → Fish
(d) Fallen leaves → Bacteria → Insect larvae

59. Which of the following statements is not a major concern about our altering of the carbon cycle?
(a) The Greenland and Antarctic ice caps are expected to melt if global warming continues.
(b) High-sulphur fuels are used by power plants because they are less expensive than low-sulphur fuels.
(c) The increase in atmospheric CO_2 exceeds the ability of the oceans to absorb the increase.
(d) CO_2 is a gas that traps infrared radiation.

60. Which of the following statements concerning acid precipitation is **incorrect**?
(a) Acids that enter the atmosphere primarily affect ecosystems located less than 100 kilometers from the source of the pollution.
(b) The effects of acidification of lakes appear to be quickly reversible if their pH can be returned to normal.
(c) While regulation of emission sources has raised, the pH of precipitation in parts of the eastern United States, precipitation in many parts of the West is becoming more acidic.
(d) Sulfuric acid and nitric acid from smokestack and automobile emissions are the major causes of acid precipitation.

61. Which of the following ecological pyramid is never inverted?
(a) Pyramid of number in forest ecosystem.
(b) Pyramid of biomass in pond ecosystem.
(c) Pyramid of energy in parasitic food chain.
(d) Pyramid of biomass in parasitic food chain.

62. Which of following is/are trend (s) in ecological succession?
(i) An increase in complexity of species .
(ii) An increase in productivity.
(iii) An increase in community stability and species diversity.
(iv) A decrease in nonliving organic materials.
(a) (i) and (iii) (b) (i) and (iv)
(c) (i), (ii) and (iii) (d) (i), (ii), (iii) and (iv)

63. The following diagram shows a mountain with a sea breeze blowing as indicated by the arrow. Circle the letter for the area with air that would be both relatively warm and dry.

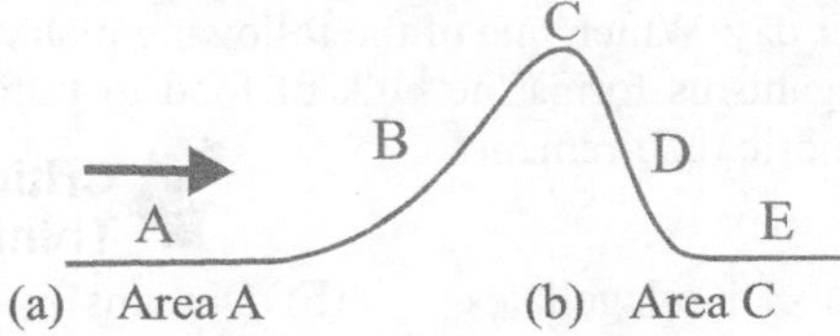

(a) Area A (b) Area C
(c) Area D (d) Area E

64. The given figure best represents.

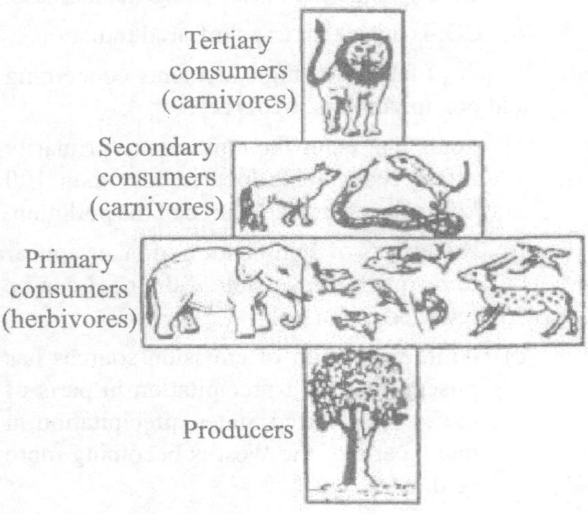

(a) Pyramid of number in parasitic food chain
(b) Pyramid of biomass in forest ecosystem
(c) Pyramid of number in grassland ecosystem
(d) Pyramid of number in forest ecosystem

65. Which type of pyramid shows best efficiency of ecosystem?
(a) Pyramid of number
(b) Pyramid of biomass
(c) Pyramid of volume
(d) Pyramid of energy

66. Tip of ecological pyramid is occupied by:
(a) Producers (b) Herbivores
(c) Carnivores (d) Omnivores

67. A crocodile has just eaten a lamb. A plover accommodatingly picks the crocodile's teeth. By this act, the plover, in ecological terms, is a:

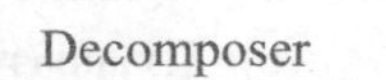

(a) Decomposer
(b) Primary consumer
(c) Secondary consumer
(d) Tertiary consumer

68. Blue whale, which inhabits Antarctica, has a requirement of more than a million calories per day. Which one of the following planktonic organisms forms the bulk of food to fulfill its caloric requirement?

(a) Dinoflagellates (b) Diatoms
(c) Green algae (d) Krill

69. Competition for water, minerals, light and space is most severe between two:
(a) Closely related species occupying the same niche.
(b) Closely related species occupying different niche.
(c) Unrelated species occupying the same niche.
(d) Species occupying different overlapping ecosystems.

70. Consider the following statements.
(i) An ecosystem is a functional unit of nature and comprises abiotic and biotic components.
(ii) Abiotic components are organic materials.
(iii) Biotic components are producers, consumers and decomposers.
(iv) Energy flow is bidirectional.
Which two statements are **correct**?
(a) (i) and (ii) (b) (ii) and (iv)
(c) (iii) and (iv) (d) (i) and (iii)

71. Which of the following is not a hypothesis for why food-chain length is limited? Critical Thinking
(a) Food-chain length is limited by one organism's ability to consume another.
(b) Food-chain length is limited because they are easily disrupted by environmental perturbations.
(c) Food-chain length is limited by productivity.
(d) Food-chain length is a function of an ecosystem's physical structure.

72. Which of the following pair is **not** correctly matched?
(a) Autotrophs – Fungi
(b) Primary consumers – Zooplankton
(c) Secondary consumers – Fishes
(d) Decomposers – Flagellates

73. Food chains differ from food webs in that:

(i) food chains are single sequence of who eats whom in a community.
(ii) food chains better represent the entire community.
(iii) food webs represent the complex interaction among food chains.
(iv) food chain is the flow of energy in a population.
(a) (i) and (iii) (b) (i) and (iv)
(c) (i), (ii) and (iii) (d) (i), (ii), (iii) and (iv)

74. Decomposition is largely an / a

(a) carbon – requiring process

(b) oxygen – requiring process

(c) nitrogen – requiring process

(d) sulphur – requiring process

75. Phosphorus is a sedimentary nutrient. This means it:

(a) cycles very slowly.

(b) never enters the atmosphere.

(c) settles as sediment in the ocean.

(d) All of the above

Answer KEYs

1	(d)	9	(c)	17	(a)	25	(b)	33	(b)	41	(b)	49	(c)	57	(a)	65	(d)	73	(a)
2	(c)	10	(a)	18	(c)	26	(a)	34	(c)	42	(c)	50	(a)	58	(c)	66	(c)	74	(b)
3	(a)	11	(a)	19	(a)	27	(d)	35	(b)	43	(a)	51	(b)	59	(b)	67	(c)	75	(d)
4	(b)	12	(a)	20	(d)	28	(b)	36	(c)	44	(b)	52	(c)	60	(a)	68	(d)		
5	(d)	13	(c)	21	(d)	29	(a)	37	(b)	45	(b)	53	(a)	61	(c)	69	(a)		
6	(d)	14	(c)	22	(b)	30	(a)	38	(c)	46	(a)	54	(c)	62	(c)	70	(d)		
7	(d)	15	(c)	23	(d)	31	(c)	39	(d)	47	(b)	55	(d)	63	(d)	71	(a)		
8	(c)	16	(c)	24	(c)	32	(b)	40	(c)	48	(c)	56	(b)	64	(d)	72	(a)		

Hints & Solutions

1. (d) 2. (c) 3. (a) 4. (b)
5. (d) 6. (d) 7. (d) 8. (c)
9. (c) 10. (a)

11. (a) The correct sequence is :
    1. Phosphorus input from sewage and agricultural runoff increases.
    2. Algal blooms occur.
    3. Respiratory demand from decomposers increases.
    4. Oxygen levels drop in deeper water.

12. (a)

13. (c) The number of organisms in grassland ecosystem, decreases progressively from the first trophic level (producers) to the last trophic level. Thus, the pyramid of number in grassland ecosystem, becomes upright in shape.

14. (c) Only about 10% of the energy is transferred between trophic levels; thus, 100,000 calories are in the cows and available to human consumers.

15. (c)

16. (c) DDT is a non-biodegradable pollutant (either do not degrade or degrade very slowly in nature). They are often biologically magnified with their subsequent movement in food chain. Its concentration in sea gull is highest.

17. (a) Energy cannot be recycled.

18. (c)

19. (a) Decomposition is an oxygen - requiring process. The rate of decomposition is controlled by chemical composition of detritus and climatic factors.

20. (d) (i) Deserts have the lowest primary productivity as the soil is deficient in moisture.
    (ii) Some plants have more efficiency to trap sunlight (sugar cane), so they accumulate more primary productivity.
    (iii) Productivity is maximum in the coral reefs because they grow in areas having good light, enough warm water and abundant nutrients.

21. (d) Nutrient cycling is necessary for life on earth.

22. (b) 23. (d) 24. (c) 25. (b)

26. (a) Planktons are free-floating animals (zooplankton) and plants (phytoplanktons) living near the shore of sea or lake which float or drift almost passively.

27. (d) Decomposers are vital to the health of the ecosystem because they prevent nutrients from being tied up in one place for extended periods of time.

28. (b) In food web, each successive trophic level has less total energy content as per law of Lindeman. Some energy is lost in transfer from one to another level.

29. (a) Because they spend so much energy maintaining constant high body temperatures, mammals (and birds) have very low production efficiencies. Herbivores are less efficient than carnivores because plant tissues take more energy to digest than animal tissues do.

30. (a) The biosphere is distinguishable into biomes and biomes into ecosystems.

31. (c) Most elements remain in bottom sediments in the ocean for millions of years, until elevated above sea level by movements of Earth's crust.

32. (b) Niche is occupation or profession of an organism.

33. (b) The two vegetations of an ecosystem are separated by ecoline or vegetational gradient. Whittaker (1970) recognised five kinds of ecoline correlated with climatic conditions in different geographic areas. All living commuities are grouped into 20 biomes.

34. (c) Lentic ecosystem is a type of limnetic ecosystem where the water is stationary (standing water) as in a pond or lake ecosystem where water does not flow.

35. (b) An ecosystem includes the environment and the biological community living there. A coral reef ecosystem would include much more than just the fish.

36. (c) In ecosystem energy transfers from one form to another and some energy loss in the form of heat, it follows the IInd law of thermodynamics.
37. (b)
38. (c) In an aquatic ecosystem such as pond, the producers (macrophytes, phytoplankton) are small organisms. Their biomass is least, and this value gradually shows an increase towards the apex of the pyramid, thus making the pyramid inverted in shape.
39. (d) The climax community remains stable as long as the environment remains unchanged.
40. (c) In aquatic ecosystem such as pond, as the producers (macrophytes, phytoplankton) are small organisms. Their biomass is least, and this value gradually shows an increase towards the apex of the pyramid, thus making the pyramid inverted in shape.
41. (b) 42. (c) 43. (a)
44. (b) The materials in an ecosystem are recovered but not the energy.
45. (b) Shorter the food chain, more is the availability of food to consumer.
46. (a) 47. (b) 48. (c)
49. (c) Any calculations of energy content, biomass, or numbers has to include all organisms at that trophic level. Pyramid of energy is always upright, can never be inverted, because when energy flows from a particular trophic level to the next trophic level, some energy is always lost as heat at each step.
50. (a)
51. (b) Regarding biogeochemical cycles, statement (i), (ii), and (iv) are correct. Biogeochemical cycle is the flow of chemical elements and compounds between living organisms and the physical environment. Chemicals absorbed or ingested by organisms are passed through the food chain and returned to the soil, air, and water by such mechanisms as respiration, excretion, and decomposition. As an element moves through this cycle, it often forms compounds with other elements as a result of metabolic processes in living tissues and of natural reactions in the atmosphere, hydrosphere, or lithosphere.
52. (c)
53. (a) In the pyramid of number, the number of individual organisms at each trophic level is shown. The pyramid may be upright or inverted.
54. (c) Quite often humans innocently introduce something into the environment that has devastating effects. It pays to be cautious.
55. (d) 56. (b)
57. (a) The term biomagnification refers to increase in the concentration of a persistant chemical by the organism at successive trophic levels in a food chain. The maximum biomagnification is among fishes in an aquatic ecosystem.
58. (c)
59. (b) The use of high-sulfur coal has more to do with acid precipitation than it does with global warming.
60. (a) Acid precipitation is a regional, not a local, environmental problem. Acids may travel hundreds of kilometers before they settle to Earth in precipitation or as dry particles.
61. (c) 62. (c)
63. (d) The air would have lost most of its moisture while rising on the windward side of the mountain, and in descending to area E, it would have warmed again.
64. (d) 65. (d) 66. (c)
67. (c) In the feeding relationship of complex natural communites, organisms whose food is obtained from plants by the same number of steps are said to belong to same trophic level. Here Lamb is a herbivore (primary consumer) and Crocodile a secondary consumer (primary carnivore). Since Plover is feeding on flesh of Lamb stuck in between the teeth of Crocodile, it is also referred to as secondary consumer.
68. (d) Krill (*Euphausia superba*) is the only answer to fulfil the caloric requirement of whale, because it is rich in protein and fat. 1 lb of krill yields at least 460 calories. Its production in the total area of Antarctica is approximately 1350 million tons and feeding requirement for the total population of whale (half a million consuming 3 tones

of krill per day for six months) is around 270 million tons of krill.

69. (a)

70. (d) Statements (i) and (iii) are correct.

(ii) Abiotic factors are non-living chemical and physical parts of the environment that affect living organisms and the functioning of ecosystems. Abiotic factors can include water, light, radiation, temperature, humidity, atmosphere, and soil, flow of energy and cycling of materials.

(iv) Energy flow is unidirectional.

71. (a) Food chain is a series of organisms interrelated in their feeding habits, the smallest being fed upon by a larger one, which in turn feeds a still larger one, etc. A food chain begins with a producer, usually a green plant or alga that creates its own food through photosynthesis. Limitation of the length of the food chain does not depend on one organisms ability to consume another.

72. (a) Fungi is not autotrophic. Autotrophs are producers which make their own food through the process of photosynthesis.

73. (a) 74. (b)

75. (d) Sedimentary nutrients cycle slowly because they settle on the bottom of the ocean and do not enter the atmosphere. The upwelling of the ocean floor is needed for them to be released and made available for biological use.

37 Biodiversity and Conservation

1. Alpha diversity is present:
(a) Within community
(b) Between community
(c) Ranges of communities
(d) All of the above

2. Genetic diversity in agricultural crops is threatened by:

Critical Thinking

(a) Introduction of high yielding varieties
(b) Intensive use of fertilisers
(c) Intensive intercropping
(d) Intensive use of biopesticides

3. *Antilopa cervicapra* or Black Buck is:
(a) Vulnerable
(b) Endangered
(c) Critically endangered
(d) Extinct in the wild

4. Which of the following species is endangered?
(a) *Panthera leo persica*
(b) *Python molurus*
(c) *Chelonia mydas*
(d) Siberian crane

5. Exotic species:
(a) are often endangered.
(b) usually increase biodiversity.
(c) often enhance the habitat for native species.
(d) usually reduce biodiversity.

6. What is genetic diversity?
(a) The total genetic information contained within all individuals of species.
(b) The total phenotypic information contained within all individuals of a species.
(c) The variety of life-forms on Earth.
(d) The variety of biotic communities in a region along with abiotic components.

7. Which of the following is the most serious threat to biodiversity?
(a) Competition from exotic species
(b) Commercial harvesting
(c) Habitat destruction
(d) Overexploitation

8. National Park associated with Rhinoceros is:
(a) Kaziranga (b) Ranthambore
(c) Corbett (d) Valley of Flowers

9. In which state "Periyar National park" is present?
(a) Karnataka (b) Maharashtra
(c) Madhya Pradesh (d) Kerala

10. In India, we find mangoes with different flavours, colours, fibre content, sugar content and even shelf-life. The large variation is on account of:
(a) Species diversity (b) Induced mutations
(c) Genetic diversity (d) Hybridisation

11. The proportionate number of individuals of different species in an area is called as:
(a) Species diversity (b) Species richness
(c) Diversity index (d) Species evenness

12. For conservation of tiger 'a tiger project' is being run in the country. How many tiger reserves are there under this project?
(a) 10 (b) 14
(c) 12 (d) More than 16

13. Biosphere reserve project was started in India during:
(a) 1984 (b) 1985
(c) 1986 (d) 1989

14. In a biosphere reserve, the area surrounding the core zone is called:
(a) Buffer zone (b) Transition zone
(c) Peripheral zone (d) Development zone

15. Which one of the following statements does not represent biodiversity of a geographical region?
(i) Genetic diversity present is in the dominant species of the region .
(ii) Species endemic to the region.
(iii) Endangered species found in the region.
(iv) The diversity in the organisms living in the region.
(a) (i) and (ii) (b) (iii) and (iv)
(c) (ii) only (d) (iv) only

16. Which one of the following shows concept of species-area relationship?
   (a) The number of species in an area increases with the size of the area.
   (b) Larger species require larger habitat areas than do smaller species.
   (c) Most species within any given area are endemic.
   (d) The larger the area, the greater the extinction rate.

17. The percentage of forest cover recommended by National Forest Policy (1988) is:

   (a) 33% for plains and 67% for hills
   (b) 37% for plains and 63% for hills
   (c) 20% for plains and 70% for hills
   (d) 23% for plains and 77% for hills

18. Which of the following does not constitute habitat destruction?
   (a) Fishing (b) Logging
   (c) Burning forests (d) Filling in wetlands

19. Why do migratory species present special preservation challenges?

Critical Thinking

   (a) Because they are endemic, they are especially susceptible to habitat destruction.
   (b) Their conservation may require international cooperation when they require habitats in different countries.
   (c) They are often prone to population number decline during their long migratory journeys.
   (d) They reside in biodiversity hot spots that are most susceptible to habitat degradation.

20. Biosphere Reserves differ from National Parks and Wildlife Sanctuaries because in the former:
   (a) human beings are not allowed to enter.
   (b) people are an integral part of the system.
   (c) plants are paid greater attention than the animals.
   (d) living organisms are brought from all over the world and preserved.

21. Earth Summit at Rio-de-Janeiro was related to:
   (a) Soil fertility
   (b) Survey of natural resources
   (c) Conservation of environment
   (d) Prevention of afforestation

22. Which of the following is not a concern regarding habitat fragmentation?
   (a) Habitats can change type.
   (b) Habitats can be too small to support some species.
   (c) It reduces dispersal ability of some animals.
   (d) It creates a large amount of vulnerable edge habitat.

23. Degree of biodiversity increases from:
   (a) Equator to poles
   (b) Polar to equator
   (c) Low latitude to high latitude
   (d) Low altitude to high altitude

24. The Red data Book, which lists endangered species of animals, is maintained by:
   (a) WWF (b) UNO
   (c) WHO (d) IUCN

25. Which animal is the symbol of WWF (World Wildlife Fund):
   (a) Dolphin
   (b) Tiger
   (c) Giant Panda
   (d) Great Indian Bustard

26. Which is preserved in National Park?
   (a) Flora (b) Fauna
   (c) Both (a) and (b) (d) None of these

27. New approach of conservation is the establishment of:
   (a) Sancturies (b) Reserve forests
   (c) National parks (d) Biosphere reserves

28. Approximately 50% of total world species are present on:

Critical Thinking

   (a) tropical rain forest
   (b) temperate rain forest
   (c) temperate deciduous forest
   (d) coral reefs

29. The country which hosted the First World Earth Summit on conservation of environment is?
   (a) Spain (b) Peru
   (c) India (d) Brazil

30. Blue whale is placed under:
   (a) Endangered
   (b) Critically endargered
   (c) Rare
   (d) Extinct

31. Which one of the following in India, is a hotspot of biodiversity?
   (a) Sunderbans (b) Eastern Ghats
   (c) Western Ghats (d) Gangetic Plain

32. Given below are pie diagram A, B, C, D, and E related to the percentage of various categories of threatened species of angiosperms, amphibians, reptiles, birds and mammals respectively. Identify the each shaded slice of the pie chart.

Critical Thinking

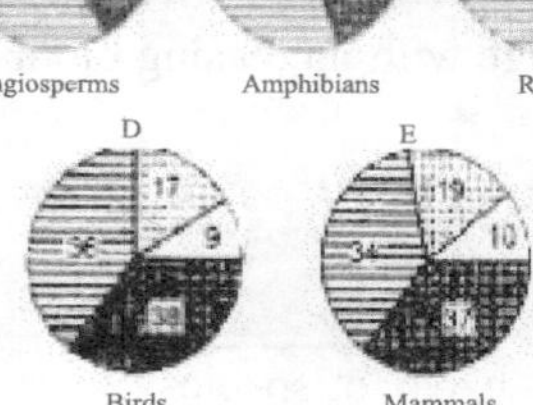

D E

Birds Mammals

Critically study and identify the following regions

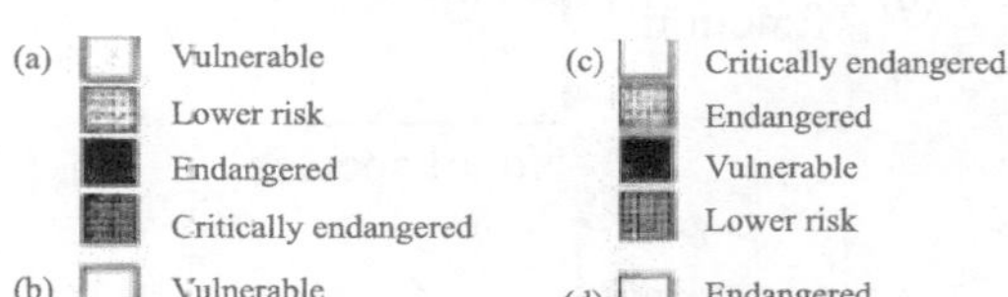

(a) Vulnerable
Lower risk
Endangered
Critically endangered

(b) Vulnerable
Lower risk
Endangered
Critically endangered

(c) Critically endangered
Endangered
Vulnerable
Lower risk

(d) Endangered
Critically endangered
Lower risk
Vulnerable

33. A number of natural reserves have been created to conserve specific wild life species. Identify the correct combination from the following.

(a) Gir Forest – Tiger
(b) Kaziranga – Elephants
(c) Rann of Kutch – Wild Ass
(d) Manas wild life Sancturay – Musk deer

34. Identify the **correct** statements and choose the option.

(i) Diversity in species provides stability to the ecosystem.
(ii) Communities with greater number of species tend to be more stable than those with less number of species.
(iii) Ecosystem with higher biodiversity is more productive than the ecosystem with lower biodiversity.
(iv) Biodiversity does not play any role in the maintenance and health of ecosystems.

(a) (i), (ii) and (iii) (b) (i), (ii) and (iv)
(c) (ii), (iii), and (iv) (d) (i), (ii), (iii), and (iv)

35. Diversity of habitat over the total landscape is called:

(a) β diversity
(b) γ diversity (gamma)
(c) Landscape diversity
(d) Ecosystem diversity

36. Those species whose populations have been seriously depleted and whose ultimate security is not assured are known as: **Critical Thinking**

(a) Threatened species
(b) Endangered species
(c) Vulnerable species
(d) Rare species

37. The greatest threat to global biodiversity is:

(a) natural disasters such as storms.
(b) pollution.
(c) overexploitation of natural resources.
(d) human alteration of habitats.

38. Which of the following is threatened by illegal trade in wildlife products? **Tricky**

(a) Nile perch
(b) Elephants
(c) Key deer
(d) Galapagos tortoises

39. The taxa believed likely to join the endangered category in near future is called:

(a) Extinct (b) Rare
(c) Vulnerable (d) Out of danger

40. Red data book provides data on:

(a) Red flowered plants
(b) Red coloured fishes
(c) Endangered plants and animals
(d) Red eyed birds

41. The most biodiversity rich zone in India:

(a) Gangetic planes (b) Trans himalayas
(c) Western Ghats (d) Central India

42. In a biosphere reserve, limited human activity is permitted in: **Tricky**

(a) Core zone (b) Buffer zone
(c) Manipulation zone (d) None of these

43. Which of the following represents the biodiversity of an area?

(a) Gene (b) Species
(c) Ecosystem (d) All of these

44. The Wildlife (Protection) Act 1972 was first amended in:

(a) 1991 (b) 1995
(c) 2001 (d) 2007

45. In the United States, the groups of organisms with the highest proportion of endangered or extinct species live in:

(a) Grasslands
(b) The deciduous forest biome
(c) Freshwater habitats
(d) Deserts

46. Hotspots are regions of high: Tricky
(a) Rarity
(b) Endemism
(c) Critically endangered population
(d) Diversity

47. Kaziranga wild life sanctuary is famous for
(a) Tiger
(b) Musk deer
(c) Elephant
(d) Rhinoceros

48. Which of the following is not done in a wild life sanctuary?
(a) Fauna is conserved.
(b) Flora is conserved.
(c) Soil and flora is utilised.
(d) Hunting is prohibited.

49. One of the *ex-situ* conservation methods for endangered species is:
(a) Wildlife Sanctuaries
(b) Biosphere Reserves
(c) Cryopreservation
(d) National parks

50. Match column-I with column-II and choose the correct option.

	Column-I		Column-II
A.	Narrowly utilitarian argument	I.	Conserving biodiversity for major ecosystem services.
B.	Broadly utilitarian argument	II.	Conserving biodiversity for philosophical or spiritual need to realise that every species has intrinsic value and moral duty to pass our biological legacy in good order to future generation.
C.	Ethical argument	III.	Conserving biodiversity for direct economic benefits like food, medicine, industrial products etc.

(a) A – I; B – II; C – III
(b) A – III; B – I; C – II
(c) A – II; B – I; C – III
(d) A – I; B – III; C – II

51. Which one of the following statement is **correct** for genetic diversity?
(a) The total genetic information contained within all individuals of species.
(b) The total phenotypic information contained within all individuals of a species.
(c) The variety of life-forms on earth.
(d) The variety of biotic communities in a region along with abiotic components.

52. If redundancy in the role of species is widespread in nature how would ecosystem function change with increasing biodiversity?

(a)
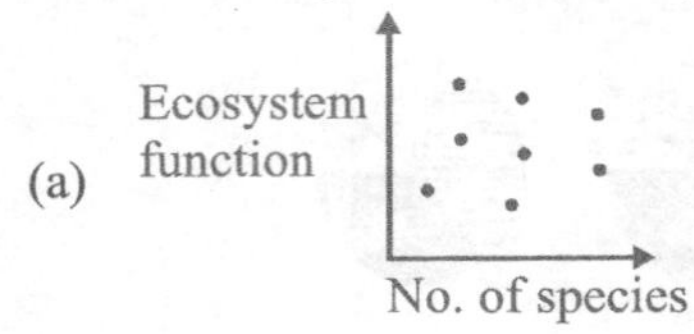

(b)
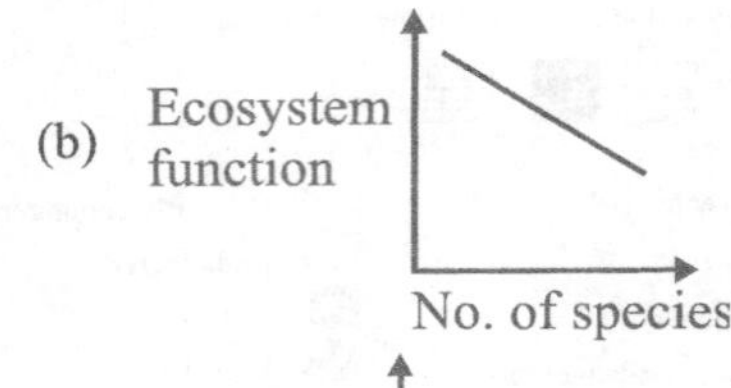

(c)
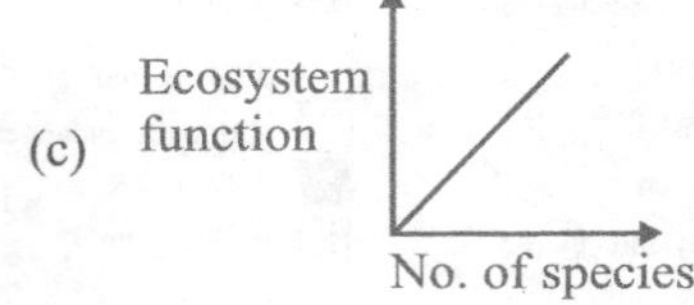

(d)
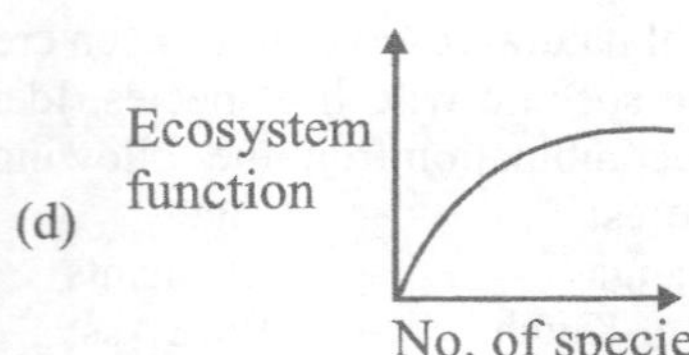

53. The largest brackish water lagoon in Asia is:

(a) Pulikat lake (b) Chilka lake
(c) Sukhna lake (d) Sultanpur lake

54. Gujarat forest 'Gir' is famous for:
(a) Asiatic lion (b) Tiger
(c) Birds (d) None of these

55. Which one of the following is an example of *ex situ* conservation?
(a) Zoo (b) Sanctuary
(c) National park (d) Biosphere reserve

56. In an effort to save an endangered population, wildlife managers might try to enhance ____ to help reduce ____.
(a) Population fragmentation . . . biodiversity hot spots
(b) Biological magnification . . . population fragmentation
(c) Movement corridors . population fragmentation
(d) Biological magnification . . . source habitats

57. Which one of these is an *in-situ* method of conservation?
(a) National park
(b) Botanical garden
(c) Tissue culture
(d) Genetic engineering

58. 'Project Tiger' in India was started in:
(a) 1970 (b) 1973
(c) 1981 (d) 1985

59. In cryopreservation, germplasm is maintained at
(a) – 196° F (b) 0° F
(c) – 100° F (d) None of these

60. First National Park of India is:
(a) Periyar
(b) Kaziranga
(c) Corbett National Park
(d) Palamau

61. Human activities such as clear-cutting and road building often create edges between areas of different habitat. Edges created by human activities:
(a) are usually more abrupt than natural boundaries between habitats.
(b) may have their own species not present in adjacent habitats.
(c) may be dominated by relatively few species.
(d) All of the above

62. Which of the following is the main factor of desertification?
(a) Tourism
(b) Irrigated agriculture
(c) Over-grazing
(d) All of these

63. Which of the following is **not** currently a major cause of the global reduction in biodiversity?
(a) Overexploitation **Tricky**
(b) Global warming
(c) Habitat destruction
(d) Introduction of foreign predators and disease

64. Which of the following is endangered because of the loss of pollinating agents?
(a) Furbish's lousewort
(b) Madagascar rosy periwinkle
(c) American chestnut **Toughnut**
(d) Hawaiian Lobelioids

65. An island ecosystem is said to be the most vulnerable.
(a) Due to its small size and lesser number of species.
(b) Due to its large size and large number of species.
(c) Due to its large size only.
(d) Due to its small size only.

66. Urbanisation and increase in human population affects biodiversity by:
(a) rendering the land as underproductive.
(b) leading to permanent loss of flora and fauna.
(c) reducing the dependence on nature.
(d) maintaining land fertility.

67. First biosphere reserve was established in 1986 at:
(a) Nilgiri
(b) Nanda Devi
(c) Little Rann of Kutch
(d) Sunderbans

68. The area where wild populations, traditional life styles and genetic resources are protected, is:
(a) Core Zone
(b) Buffer Zone
(c) Biosphere reserve
(d) Manipulation Zone

69. Which one of the following sets consists entirely of endangered wild life species of India?
(a) Himalayan musk deer, Black buck, Indian lion, Rhino, Sambar
(b) Himalayan musk deer, Indian lion, Rhino, Wild buffalo, Golden cat
(c) Egret, Black boar, Bison, Spotted deer
(d) Himalayan musk deer, Great Indian bustard, Snow leopard, Kashmir stag, Wild ass

70. Which one statement is the most important human activity leading to extinction of wildlife?
(a) Alteration and destruction of the natural habitats.
(b) Hunting for commercially valuable wildlife products.
(c) Pollution of air and water.
(d) Introduction of alien species.

71. Match column-I with column-II and choose the correct option.

	Column-I		Column-II
A.	Rhinoceros	I.	Bharatpur
B.	Tiger project in Karnataka	II.	Tropical evergreen forest
C.	Assemblage protection	III.	Kaziranga
D.	Silent valley	IV.	National park
		V.	Bandipur

(a) A – V; B – III; C – I; D – IV
(b) A – II; B – IV; C – III; D – V
(c) A – IV; B – III; C – II; D – V
(d) A – III; B – V; C – I; D – II

72. Which one of the following is the correct matching pair of a sanctuary and its main protected wild animals?
(a) North Eastern Himalayan region – Sambar
(b) Sundarbans – Rhino
(c) Gir – Lion
(d) Kaziranga – Musk deer

73. Diversity in the type of beaks of finches adapted to different feeding habits on the Galapagos Islands, as observed by Darwin, provides evidence for:
(a) intraspecific variations.
(b) intraspecific competition.
(c) interspecific competition.
(d) origin of species by natural selection

Answer KEYs

1	(a)	**9**	(d)	**17**	(a)	**25**	(c)	**33**	(c)	**41**	(c)	**49**	(c)	**57**	(a)	**65**	(a)	**73**	(d)
2	(a)	**10**	(c)	**18**	(a)	**26**	(c)	**34**	(a)	**42**	(b)	**50**	(b)	**58**	(b)	**66**	(b)		
3	(a)	**11**	(d)	**19**	(b)	**27**	(d)	**35**	(b)	**43**	(b)	**51**	(a)	**59**	(d)	**67**	(a)		
4	(b)	**12**	(d)	**20**	(b)	**28**	(a)	**36**	(b)	**44**	(a)	**52**	(d)	**60**	(c)	**68**	(c)		
5	(d)	**13**	(c)	**21**	(c)	**29**	(d)	**37**	(d)	**45**	(c)	**53**	(b)	**61**	(d)	**69**	(d)		
6	(a)	**14**	(a)	**22**	(a)	**30**	(a)	**38**	(b)	**46**	(b)	**54**	(a)	**62**	(c)	**70**	(a)		
7	(c)	**15**	(d)	**23**	(b)	**31**	(c)	**39**	(c)	**47**	(d)	**55**	(a)	**63**	(b)	**71**	(d)		
8	(a)	**16**	(a)	**24**	(d)	**32**	(b)	**40**	(c)	**48**	(c)	**56**	(c)	**64**	(d)	**72**	(c)		

Hints & Solutions

1. (a) Alpha diversity is present within community.
2. (a) 3. (a) 4. (b) 5. (d)
6. (a) 7. (c) 8. (a) 9. (d)
10. (c) Genetic diversity is the diversity in the number and types of genes as well as chromosomes present in different species and the variations in the genes and their alleles in the same species. Variations in the genes of a species increase with the increase in size and environmental parameters of the habitat. It helps in speciation and evolution of new species.
11. (d) 12. (d) 13. (c) 14. (a)
15. (d) Biodiversity is the variety of living forms present in various ecosystems. It includes variability among life forms from all sources including air, water and land. It is the diversity of biological organisation ranging from cellular macromolecules to biomes.
16. (a) According to the concept of species-area relationship, within a region, species richness gets increased when explored area is increased, but only up to a limit.
17. (a) 18. (a) 19. (b)
20. (b) Biosphere reserves are a special category of protected areas of land or coastal environments where people are an integral component of the system.
21. (c) 22. (a) 23. (b) 24. (d)
25. (c) 26. (c) 27. (d) 28. (a)
29. (d) 30. (a) 31. (c) 32. (b)
33. (c) Gir forest - Lion

    Kaziranga - Rhinoceros

    Manas - Elephant, Rhinoceros

    Wildlife - Sanctuary
34. (a) All are true except the (iv).

    It is species diversity and not biodiversity, which is important for maintaining higher levels of productivity and ecosystem health.
35. (b)
36. (b) Those species whose populations have been seriously depleted and whose ultimate security is not assured are known as endangered species.
37. (d) 38. (b)
39. (c) Vulnerable species are threatened species which are likely to join thê category of endangered species.
40. (c) The list of endangered species of plants and animals is given in Red Data Book.
41. (c) Largest region is Deccan, Peninsula and most biodiverstiy rich region is Western Ghats (4%) with a very large number of endemic amphibian species.
42. (b) A biosphere reserve is basically divided into three zones.
    (i) Core zone – It lies at centre where no human activity is allowed.
    (ii) Buffer zone – In this zone limited human activities are allowed.
    (iii) Manipulative zone – In this zone, multiple human activities are allowed.
43. (b) 44. (a)
45. (c) Habitat destruction and pollution have caused extinction or endangerment of a very high proportion of aquatic freshwater species.
46. (b) 47. (d) 48. (c) 49. (c)
50. (b)
51. (a) Genetic diversity is the diversity shown by a single species at genetic level.
52. (d) No matter how rapidly populations grow, they eventually reach a limit imposed by shortage of important environmental factors, such as space, light, water or nutrients, A population ultimately may stabilise at a certain size, called the carrying capacity of the particular place where it lives.

    Various experimental field studies support the hypothesis that species-rich communities are more stable. Species richness promotes ecosystem productivity and is fostered by spatial heterogeneity and stable climate.

53. (b) Migratory birds from Siberia and the Himalayan region make their winter abode in Chilka Lake, the largest brackish water lagoon in Asia. Nalbana Bird Sanctuary inside Chilka Lake is an oasis for the birds.

54. (a) 55. (a) 56. (c) 57. (a)

58. (b)

59. (d) The germ plasm is maintained at –196°C (not °F).

60. (c) 61. (d)

62. (c) Desertification is the process leading to the formation of desert. The main factor of desertification of land is over-grazing by herbivorous animals.

63. (b) Global warming is predicted to be a major cause of extinction in the future, but is not yet having an impact on populations.

64. (d) The declining population of the iiwi (*Vestiaria coccinea*), a Hawaiian honeycreeper, threatens the *Lobelia* plant, which has no other pollinator.

65. (a) Island ecosystem are the most vulnerable due to the small size and small number of the species.

66. (b) 67. (a) 68. (c) 69. (d)

70. (a) Habitat loss and its alteration is the most important human activity leading to extinction of wildlife. The transformation of the natural areas determines not only the loss of the vegetable species, but also a decrease in the animal species associated with them. Habitat alteration and destruction of the natural habitat cause the destruction of breeding grounds, shelter and sources of food.

71. (d)

72. (c) Nowadays Asiatic lions, *Panthera leo persica*, are very well protected in Gir National Park, Junagarh, Gujarat.

73. (d) Nature selects those set of characters that are best adapted to the environment. This has resulted in the great diversity seen in the population of finches.

38 Environmental Issues

1. Bhopal gas accident occurred in which year and due to which gas?
(a) 1986/Methyl isocynate
(b) 1984/Methyl isocynate
(c) 1985/CO
(d) 1985/Ethyl isocynate

2. Photochemical smog does not include:
(a) PAN (b) NO_x
(c) CO (d) O_3

3. DDT causes egg shell thinning in birds because it inhibits:
(a) calmodulin
(b) calcium ATPase
(c) magnesium ATPase
(d) carbonic anhydrase

4. The accompanying graph follows the progress of an algal bloom and two associated environmental factors over the course of a year in the North Atlantic Ocean.

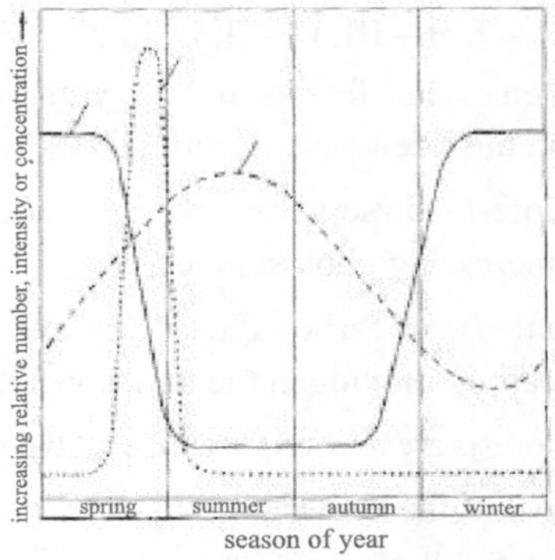

Which line in the table given the correct identity of W, X and Y ?

	Relative number of algae	Relative light intensity	Relative concentration of nutrate
(a)	W	Y	X
(b)	X	W	Y
(c)	W	X	Y
(d)	X	Y	W

5. Ozone layer in stratosphere is effected by:
(a) Excess CO_2
(b) Excess CO
(c) Chloro-fluoro-carbon
(d) Low rainfall

6. Why is the concentration of ozone less over the north and south poles?
(a) CFCs accumulate only in area where the air is cold
(b) CFC use is highest in these areas
(c) CFCs stick to frozen water vapour and are able to act catalysts
(d) UV rays are stronger in the atmosphere

7. The percentage of air pollution caused due to smoke is:
(a) 10-15% (b) 15-20%
(c) 20-25% (d) 2-5%

8. Increase in productivity most likely cause loss of species because:
(a) those species that increase in productivity outcompete other species.
(b) the causes of increased productivity in one species do not necessarily have the same effect in all species.
(c) there are inadequate resources available for productivity increases in all species.
(d) All of the above

9. The maximum biological magnification of DDT through food web is seen in:
(a) Algae
(b) Bacteria
(c) Higher plants
(d) Man

10. In case CO_2 of earth's atmosphere disappears, the temperature of earth's surface would:
(a) Increase
(b) Decrease
(c) Depend on oxygen concentration
(d) Remain the same

11. The below diagram shows electrostatic precipitator. Identify 1, 2 and 3.

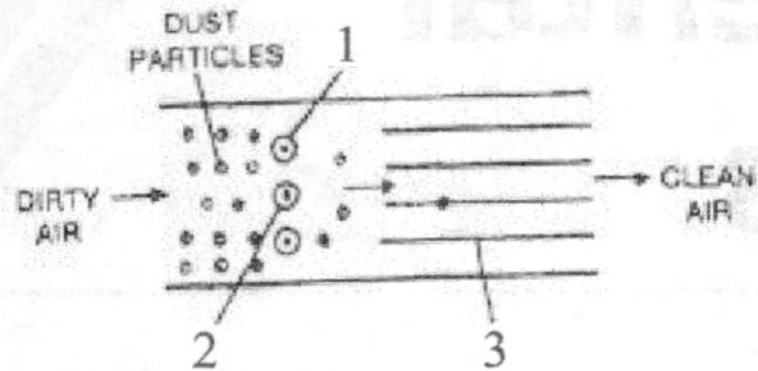

(a) 1 - Discharge corona, 2 - Negatively charged wire, 3 - Collection plate grounded
(b) 1 - Discharge corona, 2 - Positively charged wire, 3 - Collection plate grounded
(c) 1 - Discharge corona, 2 - Negatively charged wire, 3 - Collection plate burnt
(d) 1 - Uncharge corona, 2 - Positively charged wire, 3 - Collection plate never grounded

12. The diagram below shows the effect of polluting a river with untreated whey. What does graph X represent?

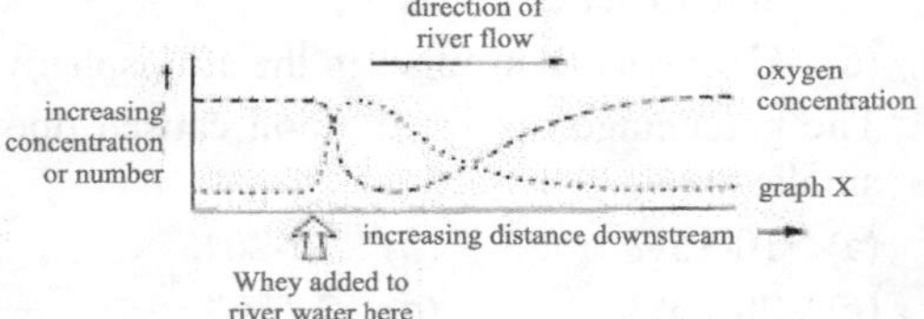

(a) Bacterial count
(b) Number of fish
(c) Mass of curds
(d) Concentration of rennet

13. Pollution of SO_2 destroys:
(a) Fungi (b) Lichen
(c) Algae (d) Fishes

14. Which organisms in a food chain consisting of algae, copepods (small crustaceans), small fish, large fish, and ospreys (fish-eating birds of prey) would have the highest concentrations of DDT in their tissues?
(a) Algae (b) Copepods
(c) Small fish (d) Ospreys

15. Major aerosol pollutant in jet emission is:
(a) Sulphur dioxide (b) Carbon monoxide
(c) Methane (d) Fluorocarbon

16. Pollution indicator plants:
(a) are resistant to pollution
(b) can purify the atmosphere
(c) are very sensitive to pollutants
(d) can indicate onset of rains

17. Noise pollution is created if noise is in excess to:
(a) 70-75 dB (b) 50-60 dB
(c) 80-99 dB (d) 40-65 dB

18. Match column-I with column-II and select the correct option. **Critical Thinking**

	Column-I		Column-II
A.	Environment (Protection) Act	I.	1987
B.	Air (Prevention and Control of Pollution) Act	II.	1986
C.	Water (Prevention and Control of Pollution) Act	III.	1980
D.	Concept of Joint Forest Management of Govt. of India	IV.	1974

(a) A – IV; B – III; C – II; D – I
(b) A – II; B – I; C – IV; D – III
(c) A – I; B – II; C – III; D – IV
(d) A – I; B – III; C – II; D – IV

19. Maintenance of forests in areas with increasing carbon dioxide levels is critical because:
(a) forests can serve as carbon dioxide sinks by increasing photosynthesis.
(b) the use of carbon dioxide by trees reduces carbon dioxide in the atmosphere.
(c) forests are needed to reduce global warming.
(d) All of the above

20. Effect of pollution is first marked on:
(a) Micro-organisms
(b) Green vegetation of an area
(c) Food crop
(d) Mammals

21. The Montreal Protocol refers to:
(a) persistent organic pollutants
(b) global warming and climate change
(c) substances that deplete the ozone layer
(d) biosafety of genetically modified organisms

22. Which of the following phenomenon is represented by the given figure?

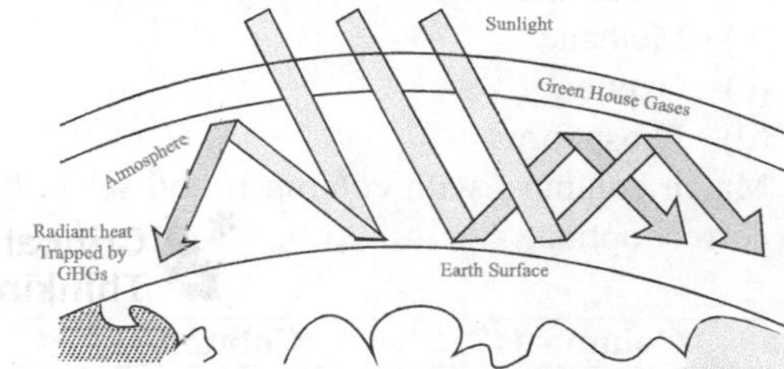

(a) Green house effect
(b) El Nino effect
(c) Ozone hole
(d) Eutrophication

23. Which is a non-degradable pollutant?
(a) Smoke
(b) Polythene
(c) Copper scraps
(d) Iron scraps

24. Which of the following is not used for disinfection of drinking water?
(a) Chlorine (b) Ozone
(c) Chloramine (d) Phenyl

25. Which of the followig is not a major contributor to the current global increase in methane level?
(a) Activity of aerobic bacteria in rice paddies.
(b) Burning of tropical biomass.
(c) Flatulence of herbivorous ruminants.
(d) Decomposition of waste in landfill dumps.

26. Which of the following statement (s) is/are correct about noise pollution?
(a) It is an undesired high level of sound.
(b) It is measured in dB.
(c) It can cause damage to heart, increase blood cholesterol and even raise blood pressure etc.
(d) All of the above

27. 'EcoSan' toilets, being used in Kerala and Srilanka are hygienic and cost effective solution to human waste disposal. Mark the correct statement regarding EcoSan.
(a) Recycled human waste from this can be used as a natural fertilizer.
(b) They are called composting toilets.
(c) They do not depend on water for the disposal of excreta.
(d) All of the above

28. Identify **correct** statement and choose the option.
(a) The major contribution in green house gases is of CH_4.
(b) Global warming is because of ozone depletion.
(c) When organic waste enters into a water body its BOD increases.
(d) All of the above

29. Consider the following statements. Choose the **correct** option. **Critical Thinking**
(i) By the end of twentieth century, the forest cover in India was reduced to 19.4 percent.
(ii) National Forest Policy was implemented in the year 1988.
(iii) The average annual production of dry grass or hay in India is about 250 million tons.
(iv) About 10% of the world's population lives in arid or semi arid regions.
(a) (i) and (ii) are true
(b) (i), (iii) and (iv) are false
(c) (i), (ii), and (iii) are true
(d) (iii) is false

30. Frequent occurrence of water blooms in a lake indicates:
(a) Nutrient deficiency.
(b) Oxygen deficiency.
(c) Excessive nutrient availability.
(d) Absence of herbivores in the lake.

31. In cities like Mumbai and Kolkata the major air pollutants are:
(a) Ozone
(b) CO and oxides of sulphur
(c) Hydrocarbons and hot air
(d) Algal spores and marsh gas

32. Thermal pollution of water bodies is due to:
(a) Discharge of heat (hot water) from power plants.
(b) Discharge of chemicals from industries.
(c) Discharge of waste from mining.
(d) Discharge of agricultural run off.

33. Gaseous pollutants can be controlled by:
(a) Arrestors
(b) Electrostatic precipitators
(c) Pyrolysis
(d) Adsorption

34. Which pollutant is absent in automobile exhausts?
(a) SO_2
(b) Fly ash
(c) Hydrocarbon gases
(d) Nitrogen oxides

35. It is said, the Taj Mahal may be destroyed due to:
(a) food in Yamuna river.
(b) decomposition of marble as a result of high temperature.
(c) air pollutants released from oil refinery of Mathura.
(d) All of the above

36. Mark the statement that describes the eco-friendly disposal of municipal solid- waste.
(a) It should be burnt to completion.
(b) It should be dumped in open waste land.
(c) It should be dumped in sanitary landfills.
(d) It should be sorted out into bio-degradable, non- biodegradable and recyclable wastes and treated separately.

37. Match Column- I with Column- II and select the correct answer using the codes given below.

	Column-I		Column-II
A.	Nitrous oxide	I.	Secondary pollutant
B.	Chlorofluoro-carbon	II.	Combustion of fossil fuels, wood, etc.
C.	Methane	III.	Denitrification
D.	Ozone (O_3)	IV.	Refrigerators
E.	Carbon dioxide	V.	Cattle, rice fields, toilets.

(a) A – III; B – IV; C – V; D – I; E – II
(b) A – V; B – I; C – III; D – IV; E – II
(c) A – IV; B – V; C – I; D – II; E – III
(d) A – I; B – III; C – IV; D – V; E – II

38. Which of the following does not cause pollution?
(a) Thermal power project
(b) Steel power plant
(c) Automobiles
(d) Hydro electric project

39. The enrichment of water by nutrients is:
(a) Mineralization (b) Biomagnification
(c) Eutrophication (d) All of these

40. Non-point source of water pollution is:
(a) Sewage outlet
(b) Effluents from a factory
(c) Agricultural runoff
(d) All of these

41. Photochemical smog always contains:
(a) Aluminum ions
(b) Methane
(c) Ozone
(d) Phosphorus

42. Match column-I with column-II and select the correct option. **Critical Thinking**

	Column-I		Column-II
A.	Catalytic converter	I.	Particulate matter
B.	Electrostatic precipitator	II.	Carbon monoxide and nitrogen oxides
C.	Earmuffs	III.	High noise level
D.	Land fills	IV.	Solid wastes

(a) A – I; B – II; C – III; D – IV
(b) A – II; B – I; C – III; D – IV
(c) A – IV; B – III; C – II; D – I
(d) A – III; B – II; C – IV; D – I

43. Lichens are important in the studies on atmosphere pollution because they:
(a) can grow in polluted atmosphere
(b) can readily multiply in polluted atmosphere
(c) efficiently purify the atmosphere
(d) are very sensitive to pollutants like sulphur dioxide

44. Carbon monoxide (CO) is harmful to man because:
(a) It forms carbolic acid.
(b) It generates excess CO_2.
(c) It is carcinogenic.
(d) It competes with O_2 for haemoglobin.

45. Identify the deleterious effect of deforestation.
(i) CO_2 concentration increases.
(ii) Habitat destruction leading to loss of biodiversity.
(iii) Disturbance in hydrologic cycle.
(iv) Desertification.
Which of the statement given above are **correct**?
(a) (i), (ii) and (iii)
(b) (ii), (iii) and (iv)
(c) (i), (iii) and (iv)
(d) (i), (ii), (iii) and (iv)

46. The greatest air pollution disaster during recent years had been witnessed in:
(a) Kolkata (b) Ahmedabad
(c) Agra (d) Bhopal

47. Which of the following is/are **correct** regarding Montreal Protocol?
    (i) Persistent organic pollutants.
    (ii) Global warming and climate change.
    (iii) To control the emission of ozone depleting substances.
    (iv) Biosafety of genetically modified organisms.
    (a) (i) and (iii) (b) (iii) only
    (c) (iii) and (iv) (d) (i) and (iii)
48. The sea water gets polluted mainly due to:
    (a) Industrial waste.
    (b) Oil spills.
    (c) Death of marine fauna.
    (d) None of these
49. The mine workers are most likely to suffer from: **Critical Thinking**
    (a) Beri-Beri
    (b) Osteomalacia
    (c) Scurvy
    (d) Xerophthalamia
50. 'Sonic booms' refer to:
    (a) Shock waves (b) Noise of traffic
    (c) Noise of siren (d) Nitrates
51. Which among the following is likely to have the highest levels of D.D.T. depositions in its body?
    (a) Eel (b) Crab
    (c) Sea gull (d) Phytoplankton
52. Match column-I with column-II and select the correct answer using the codes given below.

	Column-I		Column-II
A.	Carbon Monoxide	I.	Qualitative Pollutant
B.	DDT	II.	Freon
C.	Chlorofluoro-methane	III.	Lung Cancer
D.	Smoking	IV.	Quantitative Pollutant
E.	Acid rain	V.	Nitrogen and Sulphur oxides

    (a) A – II; B – I; C – IV; D – V; E – III
    (b) A – IV; B – I; C – II; D – III; E – V
    (c) A – III; B – II; C – IV; D – I; E – V
    (d) A – V; B – III; C – I; D – II; E – IV
53. Which of the following statements about eutrophication are **correct**?
    (i) It can be a naturally occurring process.
    (ii) It is commonly found in standing rather than running water.
    (iii) It can lead to oxygen depletion.
    (iv) It is commonly associated with high levels of phosphates and nitrates.
    (a) (iii) and (iv) (b) (i), (ii) and (iii)
    (c) (ii), (iii) and (iv) (d) All of the above
54. Main air pollutant is:
    (a) CO_2 (b) CO
    (c) N_2 (d) Sulphur
55. The scrubber is used mainly to remove the following gases from the exhaust after spraying water/lime: **Critical Thinking**
    (a) CO_2 (b) SO_2
    (c) O_2 & N_2 (d) CO & CO_2
56. Which one of the following is a most efficient device to eliminate particulate matters from the industrial emissions?
    (a) Cyclonic separators
    (b) Trajectory separators
    (c) Incineration
    (d) Electrostatic precipitator
57. Soil fertility can be destroyed by:
    (a) cutting down forests
    (b) acid rain
    (c) overgrazing and overirrigation
    (d) All of the above
58. Secondary air pollutant is:
    (a) PAN (b) CO
    (c) SO_2 (d) aerosols
59. The most harmful effect of ionizing radiations precisely involves:
    (a) Cell membranes
    (b) DNA
    (c) Mitochondria and ribosomes
    (d) Nucleus
60. Among the following environmental pollutants, which one is concerned with biological amplification? **Critical Thinking**
    (a) SO_2
    (b) O_3 and CO_2
    (c) Hg and fungicides
    (d) NO_2
61. The amount of freshwater of the earth frozen as polar or glacial ice is:
    (a) 0.5 % (b) 0.02 %
    (c) 0.01% (d) 1.97 %
62. Which one has the least green house effect?
    (a) Carbon dioxide (b) Nitrogen gas
    (c) Water vapours (d) Ozone

63. If global warming continues, the organism which may face more severe threat is:
(a) Cow (b) Dog
(c) Snow leopard (d) Dolphin

64. All organic substance found in the biosphere necessarily contain:
(a) Silica (b) Carbon
(c) Oxygen (d) Nitrogen

65. Which one of the following is an environment related disorder with the correct main cause?
(a) Black lung disease (pneumoconiosis) found mainly in workers in stone quarries and crushers.
(b) Blue baby disease (methaemoglobinaemia) due to heavy use of nitrogenous fertilizers in the area.
(c) Non-hodgkin's lymphoma found mainly in workers involved in manufacture of neem based pesticides.
(d) Skin cancer mainly in people exposed to benzene and methane.

66. Fishes die by sewage because:
(a) Of its bad smell.
(b) It replaces food material of fishes.
(c) It increases oxygen competition among fishes.
(d) CO_2 is mixed in large amount in water.

67. Water pollution can be stopped best by:
(a) Treating effluents to remove injurious chemicals.
(b) Rearing more fishes.
(c) Cultivating useful water plants.
(d) Spraying with DDT.

68. Industrial waste containing non-biodegradable substance left in a pool of water and a person eating a fish of that water, showing signs of toxicity is a typical example of: **Critical Thinking**
(a) biological multiplication.
(b) biological magnification.
(c) accidental care of toxicity.
(d) not fish but the person showing toxic effect.

69. The automobile emission can be reduced by adding the following in the petrol: **Critical Thinking**
(a) Calcium salts (b) Cobalt salts
(c) Silicon oil (d) Barium salts

70. Heavy metals are detoxified in plants by:
(a) ABA (b) Allelopathins
(c) Phytochelatins (d) Phytoalexins

71. In B.O.D. test, oxygen plays an important role to: **Critical Thinking**
(a) Destroy inorganic matter.
(b) Destroy pollution.
(c) Destroy waste organic matter.
(d) None of the above

72. Two lakes, A and B are identical in all aspects except that lake A has higher temperature. Which of the following is true? **Critical Thinking**
(a) A has higher rate of oxygen dissolution
(b) B has higher rate of oxygen dissolution
(c) Oxygen dissolution of both is the same
(d) Both have same BOD

73. Blue baby syndrome is due to:
(a) Mercury in drinking water.
(b) Excess of nitrates in drinking water.
(c) Cd-poisoning.
(d) As-poisoning.

74. Which of the following strategy is **incorrect** approach to reduce global warming?

(a) Reducing the green house gas emission by limiting the use of fossil fuels.
(b) Increase the vegetation cover particularly the forest for photosynthetic utilization of CO_2.
(c) Minimizing the use of nitrogen fertilizers in agriculture for reducing N_2O emission.
(d) Increasing the use of air conditioners, refrigeration unit and production of plastic foams and propellants in aerosal spray cans.

75. The antiknock agent added to unleaded petrol is:
(a) Tetraethyllead
(b) Tetramethyllead
(c) Dibromoethane
(d) Methyltertiary butyl ether

76. Deforestration has caused:
(a) more floods and decrease of temperature.
(b) more drought and decrease of temperature.
(c) shortage of food products.
(d) soil erosion and loss of wildlife.

77. Prolonged water logging in an agricultural fields is likely to create the problem of: **Critical Thinking**
(a) poor aeration and low salinity
(b) poor aeration and high salinity
(c) poor aeration and high acidity
(d) metal toxicity and proper aeration

78. Secondary treatment of sewage is:
(a) biological degradation of organic materials by microorganisms under controlled conditions.
(b) chlorination to kill pathogenic microbes.
(c) separation of heavy metals from sewage.
(d) sedimentation of heavy particles.

79. Agenda-21 was adopted in: **Critical Thinking**
(a) Kyoto protocol
(b) Earth summit
(c) Helsinki declaration
(d) Montreal protocol

80. So-called greenhouse gases have been given this name because they:
(a) trap the heat radiated from Earth's surface.
(b) are produced by plants.
(c) are necessary for plant growth.
(d) are similar to gases produced in a greenhouse.

Answer KEYs

1	(b)	**9**	(d)	**17**	(c)	**25**	(a)	**33**	(d)	**41**	(c)	**49**	(a)	**57**	(d)	**65**	(b)	**73**	(b)
2	(c)	**10**	(b)	**18**	(b)	**26**	(d)	**34**	(b)	**42**	(b)	**50**	(a)	**58**	(a)	**66**	(c)	**74**	(d)
3	(b)	**11**	(a)	**19**	(d)	**27**	(d)	**35**	(c)	**43**	(d)	**51**	(c)	**59**	(b)	**67**	(a)	**75**	(d)
4	(d)	**12**	(a)	**20**	(a)	**28**	(c)	**36**	(d)	**44**	(d)	**52**	(b)	**60**	(c)	**68**	(b)	**76**	(d)
5	(c)	**13**	(b)	**21**	(c)	**29**	(c)	**37**	(a)	**45**	(d)	**53**	(d)	**61**	(d)	**69**	(d)	**77**	(b)
6	(c)	**14**	(d)	**22**	(a)	**30**	(b)	**38**	(d)	**46**	(d)	**54**	(b)	**62**	(b)	**70**	(c)	**78**	(a)
7	(a)	**15**	(d)	**23**	(b)	**31**	(b)	**39**	(c)	**47**	(b)	**55**	(b)	**63**	(c)	**71**	(c)	**79**	(b)
8	(d)	**16**	(c)	**24**	(d)	**32**	(a)	**40**	(c)	**48**	(b)	**56**	(d)	**64**	(b)	**72**	(b)	**80**	(a)

Hints & Solutions

1. (b) 2. (c)

3. (b) DDT causes egg shell thinning in birds because it inhibits calcium ATPase.

4. (d) 5. (c) 6. (c) 7. (a)

8. (d) Resources are limited. An increase in a resource that benefits one species often causes another species (or several species) to suffer. An increase in resources may mean that one species outcompetes another, that resources are not equally available to all species, or that the increased resources do not have the same effect on all species.

9. (d)

10. (b) In case CO_2 of earth's atmosphere disappears, the temperature of earth's surface would decrease.

11. (a) 12. (a) 13. (b)

14. (d) The concentration of DDT increases as you move up the food chain. Thus ospreys would have the highest concentration.

15. (d)

16. (c) Pollution indicator plants such as lichens are very sensitive to pollutants and their absence in a polluted area is the indication of high pollution.

17. (c)

18. (b) A : Environment (protection) Act was enacted in 1986 with the objective of providing for the protection and improvement of the environment. The Act was last amended in 1991.

B : Air (Prevention and Control of Pollution) Act, 1981 is an Act of the Parliament of India to control and prevent air pollution. It was amended in 1987.

C : Water (Prevention and control of pollution) Act, 1974 is established for water pollution control. This act prevents and controls water pollution, maintains or restores the wholesomeness of water, establish boards for the prevention and control of water pollution and confer on and assign to the boards, the power and functions relating to the above mentioned.

D : Concept of Joint Forest Management of Govt. of India, 1980 to work closely with the local communities for protecting and managing forests.

19. (d) Natural forests are necessary to moderate human impact on the carbon cycle.

20. (a)

21. (c) The Montreal Protocol on substances that deplete the ozone layer was signed in Montreal, Canada, by over 150 countries at a convention in 1987 to cut use of CFCs (chlorofluorocarbons). The aim of the protocol was to protect the ozone layer in the stratosphere by decreasing and eventually eliminating the use of ozone depleting substances like CFCs. It is regarded as one of the most successful international treaties in modern history.

22. (a) The given figure shows the phenomenon of green house effect. Green house effect is a naturally occurring phenomenon that is responsible for heating of earth surface and atmosphere.

The sunlight or solar energy when enters atmosphere, about ¼ of it is reflected back due to clouds and gases in the outermost atmosphere, some of it is absorbed, and about half of solar radiation falls on the earth surface for heating it, though a small proportion of it is reflected back from earth.

23. (b)

24. (d) Chlorine, ozone and chloramine are used for disinfection of drinking water.

25. (a)

26. (d) All the statements regarding noise pollution are correct. Noise pollution is the unwanted sound which is released into the environment. It disturbs the human being and cause an adverse effect on the mental and psychological well - being. It is measured in the units of decibels and is denoted by dB. The noise which is more than 115 dB is tolerant. It affects the general health and hearing power of the human beings.
27. (d) All the statements regarding Eco San are correct. Eco san is a sustainable system for handling human excreta or faecal matter by using composting toilets. These are very useful in the rural areas where sewer system are not possible. These toilets are hygienic, efficient, practical and most effective for the disposal of human excreta.
28. (c) The major contribution in green houses is water vapour, carbon dioxide, methane, nitrous oxide, ozone, and any fluorocarbons. Global warming is because of green - house gases. Global warming is a gradual increase in the overall temperature of the earth's atmosphere generally attributed to the greenhouse effect caused by increased levels of carbon dioxide, CFCs, CH_4, and other pollutants.
29. (c)
30. (b) In polluted water nitrogen and phosphorus (from sewage) are accumulated, which results in excessive growth of algae on water surface. Excessive growth of algae called water bloom. Due to death and decomposition of organic matter, O_2 is not available to aquatic animals.
31. (b)
32. (a) Due to nuclear and other conventional sources, temperature of water bodies and atmosphere increases, which has adverse effect on different organisms.
33. (d)
34. (b) Fly ash is an air pollutant made by industries.
35. (c) Yellowing and blackening of Taj Mahal at Agra is due to SO_2 and other pollutants released by Mathura refinery.
36. (d) Eco friendly disposal of municipal wastes should be sorted out into biodegradable, non - biodegradable and recyclable wastes and treated separately.
37. (a) 38. (d) 39. (c)
40. (c) Choice (a) & (b) point source of water pollution.
41. (c) 42. (b)
43. (d) Lichens are important in the studies on atmospheric pollution because they are very sensitive to pollutants like sulphur dioxide.
44. (d) CO is highly toxic and impairs respiration. CO combine with haemoglobin of blood and reduces its O_2 carry capacity.
45. (d) Deforestation can be resulted into increase in carbon dioxide (CO_2) concentration in the atmosphere because trees that could hold a lot of carbon in their biomass are lost with deforestation. Deforestation also causes loss of biodiversity due to habitat destruction, disturbs hydrologic cycle, causes soil erosion and may lead to desertification in extreme cases.
46. (d) In Bhopal in 1986, there was the leakage of methyl isocyanate gas from factory of Union Carbide.
47. (b) Montreal protocol on substances that deplete the ozone layer is a protocol to the Vienna Convention for the Protection of the Ozone Layer. It is an international treaty designed to protect the ozone layer by phasing out the production of a number of substances believed to be responsible for ozone depletion. The treaty was opened for signature on September 16, 1987, and entered into force on January 1, 1989, followed by a first meeting in Helsinki, May 1989.
48. (b) 49. (a) 50. (a)
51. (c) DDT is a non-biodegradable pollutant. This often biologically magnified with their subsequent movement in food chain. Its concentration in 'sea gull' will be highest.
52. (b)

53. (d) All the statements regarding eutrophication are correct. Eutrophication is a natural process that occurs in an aging lake or pond as that body of water gradually builds up its concentration of plant nutrients.

54. (b) 55. (b) 56. (d) 57. (d)

58. (a) Secondary air pollutants are O_3, PAN and aldehydes produced from oxides of nitrogen and hydrocarbon by the reaction with UV rays.

59. (b) 60. (c) 61. (d) 62. (b)

63. (c) If global warming continues then all the polar ice and glaciers will melt first. Hence, in the given option, snow leopard is the organism, which will face more severe threat due to destruction of habitat and other influences of ecological imbalance.

64. (b)

65. (b) Blue baby syndrome is a pathological condition, called methemoglobinemia, in which blood's capacity for oxygen transport is reduced, resulting in bluish skin discoloration in infants. Blue baby syndrome begins when large amounts of nitrates in water are ingested by an infant and converted to nitrite by the digestive system. The nitrite then reacts with oxyhaemoglobin (the oxygen-carrying blood protein) to form metheglobin, which cannot carry oxygen.

66. (c)

67. (a) Water pollution can be stopped best by treating effluents to remove injurious chemicals.

68. (b) Every year millions of tonnes of pesticides are used for agricultural purpose and these are washed out from field into nearby water body, where these pollutants are absorbed by aquatic organism. These organisms can be further concentrated in the higher members like human beings. This process or this chain is called biological magnification.

69. (d) 70. (c) 71. (c) 72. (b)

73. (b) 74. (d) 75. (d)

76. (d) In deforestation, natural habitats are destroyed which leads to soil erosion and loss of wildlife.

77. (b) 78. (a) 79. (b)

80. (a) A blanket of gases around earth acts much like the glass of a greenhouse because it holds heat generated by earth inside the atmosphere.

Assertion & Reason

Directions: In the following questions, a statement of assertion is followed by a statement of reason. Mark the correct choice as:

(a) If both Assertion and Reason are true and Reason is the correct explanation of Assertion.

(b) If both Assertion and Reason are true but Reason is not the correct explanation of Assertion.

(c) If Assertion is true but Reason is false.

(d) If both Assertion and Reason are false.

Unit I: DIVERSITY IN THE LIVING WORLD (The Living World, Biological Classification, Plant Kingdom, Animal Kingdom)

1. **Assertion:** The science of classifiying organisms is called taxonomy.
   Reason: Systematics and taxonomy have same meaning.
2. **Assertion:** Both internal as well as external growth is shows by living organisms.
   Reason: Living organisms undergo the process known as accretion.
3. **Assertion:** One of the defining property of living organisms is consciousness.
   Reason: Human being is the only organism that has self consciousness.
4. **Assertion:** Whittaker's five kingdom classification for algae is not acceptable.
   Reason: Whittaker grouped algae in different kingdom.
5. **Assertion:** The scientific terms for the categories like dogs, cats, mammals, plants, animals etc. is taxa.
   Reason: The taxa can indicate categories at very different level.
6. **Assertion:** Flora contains the actual account of habitat and distribution of plants of a given area.
   Reason: Monographs contain detailed information on any taxon.
7. **Assertion:** Bacteria do not always move with the help of flagella.
   Reason: Rotary motion of flagellum is employed by flagellated bacteria for movement.
8. **Assertion:** Exotoxins are released by Gram +ve bacteria causing diseases to animals.
   Reason: Exotoxins are proteins that react with WBC of animals.
9. **Assertion:** Cyanobacteria are unicellular and colonial.
   Reason: Blue green algae are chemosynthetic.
10. **Assertion:** Chemosynthetic autotrophic bacteria oxidise various inorganic substances.
    Reason: Energy released during oxidation is used in ATP production.
11. **Assertion:** Malarial fever appear at merozoite stage of *Plasmodium*.
    Reason: The infective stage of *Plasmodium* is sporozoite.
12. **Assertion:** "Fungi imperfecti" does not show alternation of generation.
    Reason: The diploid phase is present in only zygote.
13. **Assertion:** TMV is a virus which causes mosaic disease.
    Reason: TMV has RNA as genetic material.
14. **Assertion :** Members of phaeophyceae vary in colour from olive green to various shades of brown
    Reason: Phaeophyceae possess chlorophyll a, c, carotenoids and xanthophylls.
15. **Assertion:** Isogamy is a primitive type of sexual reproduction.
    Reason: The gametes are of different sizes.
16. **Assertion:** The peristome is a fringe of teeth-like projections found at the mouth of the capsule.
    Reason: It may be of two types-nematodontous and orthodontous.

17. **Assertion:** Bryophytes are mostly land dwellers.
Reason: Water is necessary to complete their life-cycle.
18. **Assertion:** In *Funaria*, the antheridia are projected from surface.
Reason: Its antheridial cluster is surrounded by perigonial leaves.
19. **Assertion:** Bryophytes are known as the amphibians of plant kingdom.
Reason: They are found in swamps and the areas where land and water meet.
20. **Assertion:** *Adiantum caudatum* is called walking fern.
Reason: It can reproduce by its leaf tips.
21. **Assertion:** The male and female gametophytes do not have independent existence in gymnosperms.
Reason: They remain within the sporangia retained on the sporophyte.
22. **Assertion:** Angiosperm and Gymnosperms both form spermatophyta.
Reason: Angiosperm and Gymnosperms both form triploid endosperm.
23. **Assertion:** Platyhelminthes are generally hermaphrodites.
Reason: Fertilisation is internal in Platyhelminthes.
24. **Assertion:** *Obelia* is dimorphic in nature.
Reason: Polyp and gonangia form are exhibited by *Obelia*.
25. **Assertion:** *Plasmodium vivax* is responsible for malaria.
Reason: Malaria is caused by polluted water.
26. **Assertion:** Open circulatory system is found in most arthropods.
Reason: Arthropods contain haemolymph which directly bathes internal tissues and organ.
27. **Assertion:** Ambulacral system plays a major role in locomotion of echinoderm.
Reason: Hydraulic pressure of fluid and contraction of muscle of tube feet make possible movement of echinoderm.
28. **Assertion:** Respiration in *Amphioxus* is done by both water and blood.
Reason: *Amphioxus* is aquatic and possesses blood.
29. **Assertion:** Osteichthyes fish swim constantly to avoid sinking.
Reason: In fishes of class Osteichthyes, air bladder is absent.
30. **Assertion:** The fangs of snake is the maxillary teeth.
Reason: The poison apparatus in snake consists of poison gland, ducts and fangs.

Unit II: STRUCTURAL ORGANISATION IN PLANTS AND ANIMALS
(Morphology and Anatomy of Flowering Plants, Structural Organisation in Animals)

1. **Assertion:** Roots do not possess nodes, leaves or buds.
Reason: Root branches arise endogenously.
2. **Assertion:** Epiphytes are called space parasites.
Reason: Their roots possess velamen.
3. **Assertion:** A simple leaf has undivided lamina.
Reason: Leaves showing pinnate and palmate venations have various type of incisions.
4. **Assertion:** Onion leaves are centric and green.
Reason: Aerial leaves of onion store food.
5. **Assertion:** An incomplete flower can be perfect.
Reason: Perfect flowers (incomplete) are called neuter.
6. **Assertion:** Flower of racemose inflorescence are pollinated by insects.
Reason: In racemose head infloresence, the florets are arranged in a centripetal fashion.
7. **Assertion:** In cyathium, several male flowers surround a single female flower.
Reason: The involucre is nectariferous.
8. **Assertion:** Coconut tree is distributed in coastal areas over a large part of the world.
Reason: Coconut fruit can float and get dispersed over thousands of kilometers before losing viability.
9. **Assertion:** Seed coat is the outermost covering of a dicotyledonous seed.
Reason: The seed coat has two layers-outer testa and inner hilum.
10. **Assertion:** Higher plants have meristematic regions for indefinite growth.
Reason: Higher plants have root and shoot apices.
11. **Assertion:** Collenchyma forms the hypodermis of dicotyledon stems.
Reason: This is the reason for flexibility of dicotyledonous stems.
12. **Assertion:** Sclerenchyma cells do not possess plasmodesmata.
Reason: The cell walls of some permanent tissues are heavily lignified.

13. **Assertion:** In angiosperms, the conduction of water is more efficient because their xylem has vessels.
Reason: Conduction of water by vessel elements is an active process with energy supplied by xylem parenchyma which is rich in mitochondria.

14. **Assertion:** Phloem fibres or bast fibres are made up of collenchymatous cells.
Reason: These are generally found in primary pholem.

15. **Assertion:** Open vascular bundles are found in dicot stem & gymnosperm.
Reason: Cambium is absent in between xylem & phloem.

16. **Assertion:** All tissues lying inside vascular cambium are called as bark.
Reason: Bark is made up of phellogen, phellem and phelloderm lying inside secondary phloem.

17. **Assertion:** Sapwood is less durable than the heartwood.
Reason: Hollow tree trunks are due to the disappearance of sapwood.

18. **Assertion:** Secondary growth is usually seen in dicotyledonous stems.
Reason: The vascular cambium present between xylem and pholem possesses the ability to form secondary xylem and secondary pholem respectively.

19. **Assertion:** The squamous epithelium is made of a single thin layer of flattened cells with irregular boundaries.
Reason: They are found in walls of blood vessels and air sacs of wings.

20. **Assertion:** Presence of connective tissue inside the brain is essential for conduction of nerve impulse.
Reason: Connective tissue held together the nerve cells of brain.

21. **Assertion :** Mast cells in the human body release excessive amounts of inflammatory chemicals which cause allergic reactions.
Reason: Allergens in the environment on reaching human body stimulate mast cells in certain individuals.

22. **Assertion:** Erythropoietin hormone circulates to kidney where it increases stem cell mitosis and speed up development of RBCs.
Reason: The regulation of RBC production is accomplished by FSH.

23. **Assertion:** Smooth muscles are known as involuntary muscles.
Reason: Smooth muscles are controlled by autonomic nervous system.

24. **Assertion:** Earthworms are known as friends of farmers.
Reason: Earthworms make burrows in the soil and make the soil porous, which helps in respiration and penetration of developing plant roots.

25. **Assertion:** Blood in cockroach is colourless haemolymph with no respiratory pigment.
Reason: Respiration in cockroach occurs through diffusion in haemolymph.

Unit III : CELL : STRUCTURE AND FUNCTION (Cell: The Unit of Life, Biomolecules, Cell Cycle and Cell Division)

1. **Assertion:** Smaller cells are usually metabolically active cells.
Reason: In smaller cell nucleo-cytoplasmic ratio and surface volume ratio is higher.

2. **Assertion:** Rudolf Virchow modified the hypothesis of cell theory given by Schleiden and Schwann.
Reason: Cell theory says that all cells arise from pre-existing cells.

3. **Assertion:** Eukaryotic cells have more DNA than prokaryotic cells.
Reason: Eukaryotes are genetically more complex than prokaryotes.

4. **Assertion:** Cell wall is not found in animal cell.
Reason: Animal cells are covered by cell membrane.

5. **Assertion:** Mitochondria and chloroplasts are semi autonomous organelles.
Reason: They are formed by division of pre-existing organelles as well as contain DNA but lack protein synthesising machinery

6. **Assertion:** The arrangement of axonemal microtubules in cilia or flagella is called 9 + 2 array.
Reason: The axoneme usually has nine pairs or doublets of radially arranged peripheral microtubules, and pair of centrally located microtubules.

7. **Assertion:** DNA is associated with proteins.
Reason: DNA binds around histone proteins that form a pool and the entire structure is called a nucleosome.

8. **Assertion:** Human diet should compulsorily contain glycine, serine and tyrosine.
Reason: Essential amino acids can be synthesised in the human body.

9. **Assertion:** Unsaturated fats are more reactive compared with the saturated fats.
Reason: Unsaturated fats have only single bonds in their structure.

10. **Assertion :** In a DNA molecule, A–T rich parts melt before G–C rich parts.
Reason: In between A and T, there are three H–bond, whereas in between G and C, there are two H-bonds.

11. **Assertion :** Allosteric enzymes show feed back inhibition.
Reason : The inhibitor is competitive.

12. **Assertion :** Enzymes are defined as biological proteins.
Reason : Chemically all enzymes are globular proteins.

13. **Assertion:** Hydrolases are the enzymes which catalyse the hydrolysis of ester, ether, peptide, glycosidic, C – C or P – N etc. bonds.
Reason : Lyases are the enzymes catalysing the linking together of 2 compounds like joining of C – O, C – N, P – O etc. bonds.

14. **Assertion:** Enzymes lower the activation energy.
Reason: A substrate molecule can be acted upon by a particular enzyme.

15. **Assertion :** Cell growth is a continuous process in terms of cytoplasmic increase.
Reason : DNA synthesis occurs only during two specific stages in the cell cycle.

16. **Assertion :** Karyokinesis occurs in M-phase.
Reason : Cell division stops in M-phase.

17. **Assertion :** In animal cells, the cytokinesis is marked by the appearance of a furrow in plasma membrane.
Reason : In plant cells, the formation of the new cell wall starts with the formation of simple precursor called cell plate.

18. **Assertion :** Meiosis II is known as equational or homotypic division.
Reason : Meiosis II produces same number of chromosome in cell.

19. **Assertion :** Meiotic division results in the production of haploid cells.
Reason : Synapsis occurs during zygotene of meiosis.

20. **Assertion :** Meiosis is known as reductional division.
Reason : During meiosis, the chromosome number get reduce by half of its total number.

Unit IV: PLANT PHYSIOLOGY (Transport in Plants, Mineral Nutrition, Photosynthesis in Higher Plants, Respiration in Plants, Plant Growth and Development)

1. **Assertion:** A special type of diffusion of water through a semipermeable membrane is known as osmosis.
Reason: The net direction and rate of osmosis only depends on the pressure gradient.

2. **Assertion:** Turgor pressure is the positive pressure that develops in the plant cell due to entry of water.
Reason: During growth of cells, the turgor pressure is responsible for enlargement and extension.

3. **Assertion:** Pure water has maximum water potential.
Reason: The osmotic potential is zero in purewater.

4. **Assertion:** Light is very important factor in transpiration.
Reason: Light induces stomatal opening and darkness induces closing of stomata. Therefore, transpiration increases in light and decreases in dark.

5. **Assertion:** Guttation is the loss of water in its liquified form from the leaves.
Reason: Guttation takes place at night only.

6. **Assertion :** Cohesion force is also called as tensile strength.
Reason : Attraction of water molecule to polar surface or hydrophilic wall of the xylem tube causes adhesion force.

7. **Assertion :** Upward movement of water is called ascent of sap.
Reason : Upward movement of water occurs through xylem and phloem.

8. **Assertion:** Hydroponics is used for solution culture.
Reason: A balanced nutrient solution contains both essential and nonessential elements.

9. **Assertion:** Plants absorb sulphur in the form of sulphate ions.
Reason: Sulphur bacteria are required for the formation of sulphate.

10. **Assertion:** Magnesium is important in photosynthesis and carbohydrate metabolism.
Reason: Mg^{++} is involved in the synthesis of nucleic acids.

11. **Assertion:** The most effective symptom of manganese toxicity is the appearance of brown spots surrounded by chlorotic veins.
Reason: Excess of manganese may induce deficiencies of iron, magnesium and calcium.

12. **Assertion:** Nitrogen-fixing bacteria in root nodules of legumes survive in oxygen-depleted cells of nodules.
Reason: Leghaemoglobin completely removes oxygen from the nodule cells.

13. **Assertion:** Plants lack excretory organs.
Reason: Plant usually absorb essential nutrients and lead a passive life.

14. **Assertion:** *Rhoeo* leaves contain anthocyanin pigments in epidermal cells.
Reason: Anthocyanins are accessory photosynthetic pigments.

15. **Assertion:** Six molecules of CO_2 are fixed to form a hexose.
Reason: One molecule of CO_2 is fixed to produce 686 kcal in photosynthesis.

16. **Assertion:** Water splitting complex is associated with PS-II.
Reason: Water splitting complex and PS-II both are physically located on the outer side of the membrane of thylakoid.

17. **Assertion:** The atmospheric concentration of CO_2 at which photosynthesis just compensates for respiration is referred to as CO_2 compensation point.
Reason: The CO_2 compensation point is reached when the amount of CO_2 uptake is less than that generated through respiration because the level of CO_2 in the atmosphere is more than that required for achieving CO_2 compensation point.

18. **Assertion:** Mitochondria helps in photosynthesis
Reason: Mitochondria have enzymes for dark reaction.

19. **Assertion:** C_4 Photosynthetic pathway is more efficient than the C_3 pathway.
Reason: Photorespiration is suppressed in C_4 plants.

20. **Assertion:** CAM plants lack structural compartmentation of leaf, as found in C_4 plants.
Reason: Stomata of CAM plants are open during the day.

21. **Assertion:** The C_4 plants have a special type of leaf anatomy called Kranz anatomy.
Reason: Chloroplasts of bundle sheath cells have well developed grana and starch grains.

22. **Assertion:** Photorespiration decreases the rate of photosynthesis.
Reason: Rate of respiration in C_3 and C_4 is same.

23. **Assertion:** Cyclic pathway of photosynthesis first appeared in some eubacterial species.
Reason: Oxygen started accumulating in the atmosphere after the non-cyclic pathway of photosynthesis evolved.

24. **Assertion:** Stomata are absent in submerged hydrophytes.
Reason: Respiration occurs by means of air chambers in submerged plants.

25. **Assertion:** Substrate level phosphorylation is present in glycolysis.
Reason: Substrate level phosphorylation causes synthesis of ATP.

26. **Assertion:** The process of glycolysis is also known as EMP pathway.
Reason: It is the only process of respiration in aerobic organisms.

27. **Assertion:** Under aerobic conditions, pyruvate gives rise to lactate.
Reason: Under anaerobic condition, pyruvate gives rise to acetyl CoA.

28. **Assertion:** The inner membrane of mitochondria contains systems involving electron transport.
Reason: The mitochondrial matrix contains enzymes of Kreb's cycle.

29. **Assertion:** Respiratory pathway is also an amphibolic pathway.
Reason: In respiration, there is breakdown of many substances (catabolism) and synthesis of many substances (anabolism) by respiratory intermediates.

30. **Assertion:** Both hexokinase and glucokinase require divalent cations Mg^{++} or Mn^{++}.
Reason: The divalent cations act as catalysts.

31. **Assertion**: Auxins help to prevent fruit and leaf drop at early stages.
Reason: Auxins promote the abscission of older mature leaves and fruits.

32. **Assertion:** Gibberellins induce flowering in long day plants.
Reason: Genetically tall plant become dwarf by application of Gibberellin.

33. **Assertion:** Dark period plays more important part in flowering than light period.
Reason: Flowering occurs in short-day plant if the dark period is interrupted by light break.

34. **Assertion:** Vernalisation is acceleration of subsequent flowering by low temperature treatment.
Reason: Site of vernalisation is apical meristem.

35. **Assertion:** Flowering depends only on a combination of light and dark exposure in some plants.
Reason: The site of perception of light or dark duration are the shoot apices of plants.

Unit V: HUMAN PHYSIOLOGY (Digestion and Absorption, Breathing and Exchange of Gases, Body Fluids and Circulation, Excretory Products and their Elimination, Locomotion and Movement, Neural Control and Coordination, Chemical Coordination and Integration)

1. **Assertion:** Presence of HCl in stomach is necessary for the process of digestion.
Reason: HCl kills and inhibits the growth of bacteria in the stomach.

2. **Assertion:** Lipases of bile help in the emulsification of fats.
Reason: Lipases can break large fat droplets into smaller ones.

3. **Assertion:** Wall of small intestine consists of many tube like glands.
Reason: These glands secrete enzymes DNAse and RNAse into the intestinal juice.

4. **Assertion:** In human gut, small intestine is the longest part of digestive system.
Reason: Different type of food like proteins, fats and carbohydrates are digested completely in this longest part of human alimentary canal.

5. **Assertion:** In the condition of obstructive jaundice, large amounts of unabsorbed fats are eliminated out of the body.
Reason: Entry of bile into the small intestine is prevented during obstructive jaundice.

6. **Assertion:** Products of digestion are absorbed in the large intestine.
Reason: The mucosal lining of large intestine forms finger-like foldings called villi which aid in absorption.

7. **Assertion:** Scurvy is caused by deficiency of vitamin C.
Reason: Deficiency of ascorbic acid causes scurvy.

8. **Assertion:** Insects develop a complex system of air tubes called trachea for respiratory purpose.
Reason: Exchange through body surface is not possible in insects.

9. **Assertion:** Alveoli are the primary sites for exchange of gases.
Reason: All factors in our body are favourable for diffusion of O_2 from alveoli to tissues and that of CO_2 from tissues to alveoli.

10. **Assertion:** Histamine is involved in allergic and inflammatory reactions.
Reason: Histamine is a vasodilator.

11. **Assertion:** Oxyhaemoglobin dissociates near the organ tissue due to Bohr effect and oxygen is released.
Reason: Increased CO_2 concentration reduces the affinity of haemoglobin for oxygen.

12. **Assertion:** Severve Acute Respiratory Syndrome (SARS) originated in China.
Reason: China is the most populated country of the world.

13. **Assertion:** Type 'O' blood group individuals are called 'universal donors'.
Reason: RBCs of 'O' blood group consists both 'A' and 'B' surface antigens.

14. **Assertion:** Blood coagulates in uninjured blood vessels.
Reason: Uninjured blood vessels release an anticoagulant heparin.

15. **Assertion:** Prothrombinase enzyme act as antiheparin.
Reason: Heparin prevent coagulation of blood in blood vessels.

16. **Assertion:** Closed circulatory system is more effective than open type.
Reason: The closed circulatory system considerably enhances the speed, precision and efficiency of circulation.

17. **Assertion:** Blood pressure is arterial blood pressure.
Reason: It is measured by sphygmomanometer.

18. **Assertion:** Lub is a heart sound which is produced during each cardiac cycle.
Reason: It is associated with the closure of the tricuspid and bicuspid valves.

19. **Assertion:** Persons suffering from haemophilia fail to produce blood clotting factor VIII.
Reason: Prothrombin producing platelets in such persons are found in very low concentration.

20. **Assertion:** Both ammonia and urea are excreted by earthworms.
Reason: Excretion in earthworm depends on the environment.

21. **Assertion:** If human urine is allowed to stand for some time, it smells strongly of ammonia.
Reason: Main constituent of human urine is ammonia.

22. **Assertion:** Kidneys maintain the osmotic concentration of the blood.
Reason: Kidneys eliminate either hypotonic or hypertonic urine according to the need of the body.

23. **Assertion:** Ultrafiltration takes place in presence of effective filtration pressure.
Reason: In ultrafiltration process, blood is filtered in Bowman's capsule, filtered fluid contain protein & blood corpuscles also.

24. **Assertion:** For long periods, camel can go without water.
Reason: Camels stores water in the pouches of their rumen and fat in their hump.

25. **Assertion:** Mammals, living in deserts contain more concentrated urine.
Reason: They consists very long loop of Henle in their nephrones.

26. **Assertion:** During micturition, urine is prevented from flowing back into the ureters.
Reason: Urethral sphincters relax during micturition.

27. **Assertion:** Diabetes insipidus is featured by excessive urination and too much thirst of water.
Reason: The posterior lobe of pituitary gland secretes anti-diuretic hormone (ADH).

28. **Assertion:** Fatigue is inability of muscle to relax.
Reason: It is due to lactic acid accumulation by repeated contractions.

29. **Assertion:** On stimulation, a muscle cell releases calcium ions (Ca^{2+}) from sarcoplasmic reticulum.
Reason: By reacting with a protein complex, Ca^{2+} uncover active sites on the actin filaments.

30. **Assertion:** Red muscles depend on anaerobic process for energy.
Reason: Red muscles have few number of mitochondria in them.

31. **Assertion:** First seven pairs of ribs are called true ribs.
Reason: These ribs are not connected ventrally to the sternum.

32. **Assertion:** Ball and socket joints are the most mobile joints.
Reason: Synovial fluid is present here.

33. **Assertion:** The imbalance in concentration of Na^+, K^+ and proteins generates resting potential.
Reason: To maintain the unequal distribution of Na^+ & K^+, the neurons use electrical energy.

34. **Assertion:** Nerve fibre can become excited through touch, smell, pressure and chemical changes and there is a change in polarity.
Reason: It is called action potential.

35. **Assertion:** During physical exercise, the demand of oxygen is increased, which compels the lungs and heart to increase the rate of respiration and to increase the blood flow for proper supply of oxygen respectively.
Reason: Coordination is the process through which two or more organs interact and complement the functions of each other.

36. **Assertion:** A cerebellum is related with skillful voluntary movement and involuntary activity like body balance, equilibrium, *etc*.
Reason: It is a part of hind brain and is situated behind the pons.

37. **Assertion:** Cerebral cortex appears white.
Reason: It is because of the myelin sheath covering of the tracts.

38. **Assertion:** Tongue is a gustatoreceptor.
Reason: Receptors for gustatory sensations are located in the taste bud.

39. **Assertion:** Astigmatism is due to uneven curvature of lens.
Reason: It is treated with cylindrical lenses.

40. **Assertion:** Iris muscles show both dilation as well as constriction of pupil.
Reason: It is sympathetic and parasympathetic nervous system which have antagonistic functions.

41. **Assertion:** The Eustachian tube helps in equalising the pressures on either sides of the ear drum.
Reason: The Eustachian tube connects the middle ear cavity with the nasopharynx.

42. **Assertion:** Neurohypophysis is under the direct regulation of the hypothalamus.
Reason: Neurohypophysis stores and releases two hormones called oxytocin and vasopressin which are actually synthesised by the hypothalamus.

43. **Assertion:** The person with diabetes insipidus feels thirsty.
Reason: A person with diabetes insipidus suffers from excessive secretion of vasopressin.

44. **Assertion:** Melatonin influences the menstrual cycle, pigmentation and defense capability.
Reason: It plays an important role in the regulation of diurnal rhythm of our body.

45. **Assertion:** Hormone calcitonin has antagonistic effect to that of parathormone.
Reason: Calcitonin decreases blood calcium level while parathormone increases blood calcium level.

46. **Assertion:** Adrenal medullary hormones help in combating stress.
Reason: Both adrenaline and noradrenaline act on same organs and produce similar effects.

47. **Assertion:** Our body secretes adrenaline in intense cold.
Reason: Adrenaline raises metabolic rate.

48. **Assertion:** Females have less stature than males after puberty.
Reason: This happens because of the presence of hCG in the blood of females.

49. **Assertion:** Prolactin is also called the 'milk ejection hormone'.
Reason: It stimulates the smooth muscle contractions of the mammary glands.

Unit VI: REPRODUCTION (Reproduction in Organisms, Sexual Reproduction in Flowering Plants, Reproductive Health)

1. **Assertion:** The life span of plants is greater than animals.
Reason: Plants continue their growth at their tips.

2. **Assertion:** The main function of stem is reproduction.
Reason: Stem does not bear leaves to manufacture food for plants.

3. **Assertion:** Leaves of *Bryophyllum, Begonia* help in vegetative multiplication.
Reason: Leaves of these plants possess adventitious buds.

4. **Assertion:** Parthenogenesis does not play any role in organic evolution.
Reason: In parthenogenesis females develop new organisms without fertilisation.

5. **Assertion:** The zygote developed from sexual reproduction is diploid.
Reason: In sexual reproduction, haploid gametes fuse and form zygote.

6. **Assertion:** Endosperm in gymnosperms is haploid as it is formed before fertilisation.
Reason: Endosperms is formed post-fertilisation in angiosperms and is diploid.

7. **Assertion:** Insects visit flower to gather honey.
Reason: Attraction of flowers prevents the insects from damaging other parts of the plant.

8. **Assertion:** A typical microsporangium of angiosperms is generally surrounded by four wall layers.
Reason: The outer three wall layers perform the function of protection and help in dehiscence of anther to release the pollen.

9. **Assertion:** Hydrophily is a major mode of pollination in most of the aquatic plants in angiosperms.
Reason: Almost all the aquatic dicot and monocot plants require water for the transport of male gametes and for fertilisation.

10. **Assertion:** In angiosperms, the male gametophyte is the pollen grain.
Reason: Pollen grain contains stigma, style and ovary.

11. **Assertion:** If pollen mother cells has 42 chromosomes, the pollen has only 21 chromosomes.
Reason: Pollens are formed after meiosis in pollen mother cell.

12. **Assertion:** Nuclear endosperm is formed by subsequent nuclear division without wall formation.
Reason: Coconut is an example of such endosperm, where the endosperm remains nuclear throughout the development of the fruit.

13. **Assertion:** In apomixis, plants of new genetic variations are not produced.
Reason: In apomixis, reductional division takes place.

14. **Assertion:** Mango is a true fruit.
Reason: The thalamus also contributes to fruit formation in false fruits.

15. **Assertion:** The bulbourethral gland is a male accessory gland.
Reason: Its secretion helps in the lubrication of the penis, thereby facilitating reproduction.

16. **Assertion:** Vagina acts as copulation canal and fertilisation canal.
Reason: Both insemination and fusion of gametes occur in the vagina of female.

17. **Assertion:** Human male ejaculates about 200 to 300 million sperms during coitus.
Reason: Only few reach the isthmus ampullary junction for process of fertilisation.

18. **Assertion:** Primary spermatocytes of testes are haploid.
Reason: These are formed by meiosis-I in the spermatogonia.

19. **Assertion:** Graafian follicle ruptures at the mid of menstrual cycle releasing the ovum.
Reason: Both LH and FSH attain a peak level at the middle of cycle.

20. **Assertion:** Mammalian ova produces hyaluronidase.
Reason: The eggs of mammal are microlecithal and telolecithal.

21. **Assertion:** Vigrous contraction of the uterus at the end of pregnancy causes expulsion.
Reason: The stimulatory reflex between the uterine contraction and oxytocin results in weakening contractions.

22. **Assertion:** HIV infection can be avoided by use of condoms.
Reason: Condoms secrete anti-viral interferons.

23. **Assertion:** IVF in fertilisation is carried out outside the body of woman.
Reason: The zygote upto 8 blastomeres could be transferred into the fallopian tube.

24. **Assertion:** Cu-T and Cu-7 do not suppresses sperm-motility.
Reason: Hormones released by them do not affect sperm motility.

25. **Assertion:** Biologically, woman is superior to man.
Reason: A woman provides nutrition to developing foetus during gestation and also nourishes the baby on her milk after birth.

Unit VII: GENETICS AND EVOLUTION (Principles of Inheritance and Variation, Molecular basis of Inheritance, Evolution)

1. **Assertion:** The progeny produced have both the characters of parents.
Reason: The process by which characters pass from parent to progeny is known as inheritance.

2. **Assertion:** A pair of contrasting characters is termed as allele.
Reason: Only one gene of an allele is expressed in an individual.

3. **Assertion:** By means of dihybrid cross, the law of independent assortment can be studied.
Reason: The law of independent assortment is applicable only to linked genes.

4. **Assertion:** The F_1-generation resembles both the parents in codominance.
Reason: An example is different type of red blood cells that determine ABO blood grouping in humans.

5. **Assertion:** In humans, most sex-linked genes are present on the X-chromosome.
Reason: X-chromosome contains a large number of genes with major effects on phenotype.

6. **Assertion:** Haemophilia never occurs in women.
Reason: Gene for haemophilia is located on X chromosome.

7. **Assertion:** Sickle-cell anaemia is an autosome linked recessive trait.
Reason: A single allele of a gene controlls this disorder.

8. **Assertion:** Histones are basic in nature.
Reason: Histones are rich in the amino acids lysine and arginine.

9. **Assertion:** Synthesis of daughter or new strand occurs continuously along the parent 3' $\rightarrow$ 5' strand.
Reason: DNA polymerase can polymerise nucleotides in 3' $\rightarrow$ 5' direction on 5' $\rightarrow$ 3' strand.

10. **Assertion:** Central dogma is the flow of information from DNA to mRNA and then decoding the information present in mRNA in the form of protein.
Reason: In retroviruses, reverse of central dogma occurs.

11. **Assertion:** UAA, UAG and UGA terminate protein synthesis.
Reason: They are not recognised by tRNA.
12. **Assertion:** Polytene chromosomes have a high amount of DNA.
Reason: Polytene chromosomes are formed by repeated replication.
13. **Assertion:** The first cells used RNA as their hereditary molecule.
Reason: DNA evolved from RNA.
14. **Assertion:** Thorns and tendrils of *Bougainvillea* and cucurbita represent homology.
Reason: Homologous organs have similar functions but are different in their structural details and origin.
15. **Assertion:** Lichens act as pollution indicators.
Reason: Lichens will not grow in areas that are polluted.
16. **Assertion:** Homologous organs have common ancestry and similar function.
Reason: Analogous organs have unlike origin and dissimilar function.
17. **Assertion:** The chimpanzee is the closest relative of the present day humans.
Reason: The banding pattern in some autosomes of man and chimpanzee is remarkably similar.
18. **Assertion:** Ear muscles of external ear in man are poorly developed.
Reason: These muscles are useless.
19. **Assertion:** Industrial melanism of *Biston betularia* is an example adaptation.
Reason: The mutation saved their population.

Unit VIII: BIOLOGY IN HUMAN WELFARE (Human Health and Disease, Strategies for Enhancement in Food Production, Microbes in Human Welfare)

1. **Assertion:** There is no chance of malaria to a man on the bite of male *Anopheles* mosquito.
Reason: It carries a non-virulant strain of *Plasmodium*.
2. **Assertion:** Tapeworm, roundworm and pinworm are endoparasites of human intestine.
Reason: Improperly cooked food is the source of all intestinal infections.
3. **Assertion:** Virus-infected cells secrete proteins known as interferons.
Reason: Interferons protect the non-infected cells from bacterial infection.
4. **Assertion:** Immunisation is achieved by the successful delivery of vaccines.
Reason: Vaccine is a preparation of one or more microbial agents used to induce protective immunity.
5. **Assertion:** Organ transplantation patients are given immunosuppressive drugs.
Reason: Transplanted tissue has antigens which stimulate the specific immune response of the recipient.
6. **Assertion:** Many visitors of hills suffer from skin and respiratory allergy problem.
Reason: Conifer trees produce a large quantitiy of wind-borne pollen grains.
7. **Assertion:** Nicotine is an alkaloid found in tobacco.
Reason: It has several uses.
8. **Assertion:** Cattle breeds can be improved by superovulation and embryo transplantaion.
Reason: Superovulation in high milk-yielding cows is induced by hormonal injection.
9. **Assertion:** Mass selection is useful in homozygous plants.
Reason: Mass selection does not work with cross-pollinated plants.
10. **Assertion:** Use of fertilisers greatly enhances crop productivity.
Reason: Irrigation is very important in increasing crop productivity.
11. **Assertion:** Allopolyploidy involves numerical increase of the same genome.
Reason: Autopolyploidy develops through hybridisation between two species followed by doubling of chromosomes.
12. **Assertion:** Virus-free plants can be produced by meristem tissue culture.
Reason: The virus fails to grow during the growth of the host tissue in the artificial medium.
13. **Assertion:** Protoplast culture is an important technique of genetic engineering.
Reason: Protoplast culture technique results in the production of genetically modified crops.
14. **Assertion:** A major advantage of tissue culture is protoplast fusion.
Reason: A hybrid is formed by the fusion of naked protoplasts of two plants.
15. **Assertion:** Lichen is important for chemical industries.
Reason: Litmus and Orcein are formed from lichens.

16. **Assertion:** Beer and wine are called soft liquors while gin, rum, etc. are hard liquors.
Reason: Beer and wine are made without distillation.
17. **Assertion:** Secondary treatment of sewage is also called biological treatment while primary treatment is called physical treatment.
Reason: Primary sewage treatment depends only upon sedimentation properties of materials present in sewage and filteration.
18. **Assertion:** Biogas is used as fuel for cooking and lighting.
Reason: It is considered as ecofriendly and pollution free source of energy.
19. **Assertion:** Leguminous plants are best preferred in rotation of crops.
Reason: They have root nodules, which have nitrogen-fixing bacteria *Clostridium*.
20. **Assertion:** Biofertilisers are preferred to chemical fertilisers.
Reason: Chemical fertilisers are generally more expensive and hazardous to environment.

Unit IX: BIOTECHNOLOGY
(Biotechnology: Principles and Process, Biotechnology and its Application)

1. **Assertion:** Restriction enzymes recognise palindromic sequence.
Reason: Palindromic sequences read same in both directions of the two strands.
2. **Assertion:** Restriction enzymes cut the strand of DNA to produce sticky ends.
Reason: Stickiness of the ends facilitates the action of the enzyme DNA polymerase.
3. **Assertion:** Plasmids are extrachromosomal DNA.
Reason: Plasmids are found in bacteria and are useful in genetic engineering.
4. **Assertion:** *E.coli* having pBR322 with DNA insert at *Bam* H I site cannot grow in medium containing tetracycline.
Reason: Recognition site for *Bam* H I is present in tet^r region of pBR322.
5. **Assertion:** *Agrobacterium tumefaciens* is a pathogen of several monocot plants.
Reason: Retroviruses in plants have ability to transform normal cells into cancerous cell.
6. **Assertion:** A gene from *Bacillus thuringiensis* is incorporated in plant genome to increase yield.
Reason: It is *Bt* toxin producing gene which kills larvae of insects.
7. **Assertion:** "DNA finger printing" has become a powerful tool to establish paternity and identity of criminals in rape and assault cases.
Reason: Trace evidences such as hairs, saliva and dried semen are adequate for DNA analysis.
8. **Assertion:** Interferons are effective against viruses.
Reason: Proteins which can be synthesised only by genetic engineering are effective against viruses.
9. **Assertion:** *Cry* gene expressing crop is resistant to a group of insects.
Reason: *Cry* proteins produced from *Bacillus thuringiensis* are toxic to larvae of certain insects.
10. **Assertion:** ELISA test is a technique based on antigen-antibody interactions where a pathogen can be detected by the presence of antibodies (proteins, glycoproteins etc) on it.
Reason: The pathogen antibody to be identified is immobilised on the surface of specially constructed ELISA plates and is then tested.
11. **Assertion:** The term hybridoma is applied to fused cells.
Reason: They are formed by the fusion of lymphocyte cell and myeloma cell.
12. **Assertion:** In developing world, Green revolution was comparatively less effective where farmers were dependent on conventional breeding.
Reason: In developing world, inability to buy expensive agro-chemicals forced farmers to rely on conventional breeding.
13. **Assertion:** For gene transfer in plant cells Ti plasmid obtained from *Agrobacterium tumefaciens* is effectively used as a vector.
Reason: The part of Ti plasmid transferred into the DNA of plant cells is called as T-DNA.

Unit X: ECOLOGY
(Organisms and Population, Ecosystem, Biodiversity & Conservation, Environmental Issues)

1. **Assertion:** Indo-Gangetic plains have high population density.
Reason: These have favourable climate and fertile soil.
2. **Assertion:** Leaf butterfly and stick insect show mimicry to dodge their enemies.
Reason: Mimicry is a method to acquire body colour blending with the surroundings.

3. **Assertion:** Thick cuticle is mostly present in disease resistant plants.
Reason: Disease causing agents cannot grow on cuticle and cannot invade the cuticle.
4. **Assertion:** Aerenchyma is present in the leaves and petioles of hydrophytes.
Reason: It imparts buoyancy to the hydrophytes.
5. **Assertion:** In sigmoid growth curve, population finally stabilises itself.
Reason: Finally, the death rate increases than the birth rate.
6. **Assertion:** In mutualism both the population are benefitted and neither can survive under natural condition without other.
Reason: Both populations are benefitted by the association but their relationships are not obligatory.
7. **Assertion:** Net primary productivity is gross primary productivity minus respiration.
Reason: Secondary productivity is produced by heterotrophs.
8. **Assertion:** Food web consists of several food chains.
Reason: Food web decreases the stability of an ecosystem.
9. **Assertion:** The conversion of productivity at next trophic level is 10%.
Reason: Energy is lost in the respiration process.
10. **Assertion:** Pyramid of energy may be upright or inverted.
Reason: Only 20% of energy goes to next trophic level.
11. **Assertion:** Herbivores are also known as first order consumers.
Reason: Directly from plants herbivores obtain their food.
12. **Assertion:** Maximum contribution of oxygen is from phytoplankton.
Reason: 90% of the photosynthesis with oxygen evolution is contributed by them.
13. **Assertion:** Insectivorous habitat of plants is to cope up N_2 deficiency.
Reason: Insectivorous plants are partly autotrophic and partly heterotrophic.
14. **Assertion:** In tropical rain forests. O-horizon and A-Horizon of soil profile are shallow and nutrient-poor.
Reason: Excessive growth of micro-organisms in the soil depletes its organic content.
15. **Assertion:** Tropical rain forests are disappearing fast from developing countries such as India.
Reason: No value is attached to these forests because these are poor in biodiversity.
16. **Assertion:** A sanctuary is formed for the conservation of animals only.
Reason: Restricted human activities are allowed in sanctuaries.
17. **Assertion:** Red data book has a record of only those animals which are known to be in danger.
Reason: Endangered species no need to be noticed.
18. **Assertion:** Great Indian Bustard is a rare species.
Reason: Vulnerable species are rare species.
19. **Assertion**: Pollution is always caused by human activities.
Reason: Pollution is not different from contamination.
20. **Assertion:** Water pollutants are measured by BOD.
Reason: If BOD is more, the water is polluted.
21. **Assertion:** Agricultural output increased several times after introduction of DDT.
Reason: DDT was the first insecticide used on a wide scale.
22. **Assertion:** Deforestation is one main factor contributing to global warming.
Reason: Besides CO_2, two other gases methane and CFCs are also included under green house gases.
23. **Assertion:** Chlorofluorocarbons are responsible for ozone depletion.
Reason: Ozone level decreases by as much as 67% every year.
24. **Assertion:** Deforestation increases carbon dioxide concentration in the atmosphere.
Reason: Deforestation may lead to desertification generally.
25. **Assertion:** Emphysema disease is very common in Kanpur and Calcutta.
Reason: It is disease of respiratory tract occuring due to pollution.

Hints & Solutions

Unit I: DIVERSITY IN THE LIVING WORLD

1. (c) Taxonomy is the science of identification, nomenclature and classification of organisms. But taxonomy and systematics are different terms. Systematics is the branch of biology that deals with diversity of organisms at every level of classification.

2. (d) Living organisms show internal growth due to addition of materials and formation of cells inside the body. Such a method is called intussusception. While, non-living things grow due to addition of similar materials to their outer surface. The process is called accretion.

3. (b) All living organisms have an inherent ability to sense their surroundings or environment and respond to their environmental stimuli. Plants and animals both respond to the external stimuli. Responses of animals are rapid and can easily be observed because they escape by moving away from the site or environmental stress. The plants, on the other hand, cannot move. They face the stresses and adapt themselves by their slow physiological responses. Human beings are slightly different. They are the only organisms who are aware of themselves, *i.e.,* have self -consciousness.

4. (a) In Whittaker's classification, algae are grouped in three kingdoms - Monera (blue-green algae), Protista (dinoflagellates, diatoms, euglenoids) and Plantae (red algae, brown algae and green algae). Though plant kingdom of Whittaker is often called metaphyta or multicellular plants, the algae included in this kingdom also contain a number of unicellular and colonial forms. The most accepted classification of algae is given by Fritsch. He classified all algae into 11 classes.

5. (a) Taxon (taxa) may be defined as each unit or category of classification. It represents a rank for example, the basic level of classification is species, followed by genus, family, order, class, phylum or division in ascending order.

6. (b) Floras, manuals, monographs, etc. are some important taxonomic aids that help in the correct identification. Flora provide the index to the plant species found in a particular area. Manuals are useful in providing information for identification of names of species found in an area. Monographs have information of any taxon.

7. (b) Myxobacteria do not have flagella and move by gliding movement.

8. (a) Mostly pathogenic gram +ve bacteria releases exotoxins outside the cell thus, killing the WBC and causing disease.

9. (c) Cyanobacteria or Blue green algae are photosynthetic prokaryotes with chlorophyll a, carotenoids and phycobillins. They may be unicellular, colonial or filamentous and marine or terrestrial.

10. (b) Chemoautotrophic bacteria are able to manufacture their organic food from inorganic raw materials with the help of energy derived from exergonic chemical reactions involving oxidation of an inorganic substance present in the external medium. The chemical energy obtained from oxidation reaction is trapped in ATP molecules.

11. (b) When the mosquito bites man, sporozoites present in the salivary gland of female *Anopheles* mosquito are injected into the blood of the man. The erythrocytic schizont gives rise to merozoites. Malaria fever occurs when schizonts in red blood corpuscles burst and set free their contained merozoites and malarial pigment (haemozoin) in the blood plasma.

12. (c) Fungi imperfecti is that group of fungi where only the imperfect stage (asexual stages) is present. The only known method of reproduction is by conidia (asexual reproduction). As sexual reproduction is absent in this group therefore, diplophase is also absent.

13. (b) In TMV, the chromosome consists of single stranded, linear RNA molecule coiled into a regular spiral. TMV causes mosaic disease in tobacco and some other plants.

14. (a) Phaeophyceae possesses chlorophyll a, c, carotenoids and xanthophylls. Members of phaeophyceae show variations in colour from olive green to different shades of brown depending upon the amount of xanthophyll pigments.

15. (c) The fusing gametes in the lower forms of algae are similar in size, structure and behaviour. They are indistinguishable with respect to sex. Such gametes are called isogametes. Sexual reproduction involving the fusion of isogametes is termed isogamous. It is a primitive type of sexual reproduction.

16. (b) The peristome teeth are present at the mouth of capsule. The teeth may be solid cellular tissue or composed only of the thickened portions of the cell walls of adjacent cells. When the teeth of peristome are solid structures composed of bundles of dead cells, it is termed nematodontous peristome. It is found in *Polytrichum*, *Pogonatum* and *Tetraphis.* If peristome composed of thin, membranous, transversely barred teeth, each tooth is made up of the thickened portions of the cell walls of adjacent cells. Such a peristome is called orthodontous.

17. (b) Majority of the bryophytes are land dwellers and inhabit damp, shaded and humid localities. A few of them live in or float on water. The bryophytes cannot carry on their reproductive activities without sufficient moisture. Presence of water is necessary.

18. (b) In *Funaria,* the antheridia project from the surface of the receptacle and are aggregated to form a cluster. The leaves surrounding the antheridial cluster are known as the perigonial leaves. The antheridial cluster with the surrounding perigonial leaves is called the perigonium.

19. (a) Bryophytes are the dwellers of transitional habitat between the aquatic and terrestrial habitats. It is represented by the swamps and the areas where water and land meet. Bryophytes usually grow in amphibious situation and cannot complete their life cycle without external water.

20. (c) Leaf tips of *Adiantum caudatum*, develop adventitious buds for vegetative propagation. Leaf tip when reaches the ground, develops into new plant, therefore, it appears that the plant is walking.

21. (a) Gametophytes are retained within sporangia in gymnosperms.

22. (c) In angiosperms, endosperm is formed through triple fusion (fusion of sperms with two polar nuclei of the central cell) and is triploid. In gymnosperms, endosperm is formed before fertilisation and therefore, it is haploid.

23. (b) Platyhelminthes are generally hermaphrodites, *i.e.,* both the sexes are present in one organism. These show internal as well as cross fertilisation where male gametes of one organism fertilise female gametes of another organism.

24. (a) Hydroid colony of *Obelia* is dimorphic, exhibiting two types of individuals or zooids which differ both morphologically as well as physiologically. These two zooids are -
(i) Polyp – the nutritive zooid of the colony
(ii) Gonangium – the reproductive zooid.

25. (c) *Plasmodium vivax* is responsible for malaria. It is spread by the bite of female *Anopheles* mosquito. Thus, malaria is not caused by polluted water.

26. (a) Most arthropods, certain molluscs and tunicates contain open circulatory system. In them, a fluid composed of blood mixes with tissue fluid bathing internal tissues and organs directly. It oozes through spaces or cavities that surround the organs, this mixture of fluid is usually referred to as haemolymph.

27. (a) The water vascular system is a unique organ system that functions in locomotion, feeding, respiration and excretion. Ambulacral canal is connected to outside through external tube feet. Hydraulic pressure of fluid and contraction of muscle of tube feet make possible movement of Echinoderm.

28. (a) In *Amphioxus*, some exchange of O_2 and CO_2 occurs between the water current and blood through the gill-clefts, but this appears doubtful since the blood contains no respiratory pigment. The pharyngeal wall of *Branchiostoma* is richly vascular and the water current entering the pharyngeal cavity brings O_2. The blood flows so close to the surface that some exchange between blood and water can easily occur. It appears more probable that an exchange of gases occurs over the whole surface of the body and particularly in the walls of atrium.

29. (d) Members of class Chondrichthyes are marine with streamlined body having cartilaginous endoskeleton. Due to the absence of air bladder, these have to swim constantly to avoid sinking, air bladder is present in Osteichthyes thus, helping in regulation of buoyancy.

30. (b) The poison apparatus of snake consists of a pair of poison glands, their ducts and a pair of fangs. The poison glands are situated one on either side of the upper jaw. These glands are possibly the superior labial glands or parotid glands. The fangs are sharply pointed and are enlarged maxillary teeth.

Unit II: STRUCTURAL ORGANISATION IN PLANTS AND ANIMALS

1. (b) Root is typically a non green underground cylindrical structure forming the descending axis of the plant which gives rise to endogenous branches and does not possess nodes, leaves or buds. Presence of nodes and internodes is the characteristic feature of stems.

2. (b) Epiphytes live on the surface of other plants for shelter and space only, hence are also called space parasites. Epiphytic roots possess a covering of dead spongy tissue known as velamen. They are able to absorb water from moist atmosphere, dew and rain with the help of velamen.

3. (b) A leaf having a single or undivided lamina is called simple leaf. The lamina can have different types of incisions, which may reach upto half (-fid), more than half (-partite) or near the base or midrib (-sect). Depending upon the pinnate or palmate venation, the incisions are known as pinnatifid palmatifid, pinnatipartite, palmatipartite, pinnatisect and palmatisect etc.

4. (c) A leaf is called centric leaf when the leaf is more or less cylindrical and directed upwards or downwards, as in onion. A centric leaf is equally illuminated on all sides and, therefore, it is evenly green. Centric leaves of onion is aerial. It is underground scale leaves of bulb which store food.

5. (c) A flower is regarded as complete if it has all the four floral whorls, *i.e.*, calyx, corolla, androecium and gynoecium. If any of these whorl is missing, the flower is described as incomplete. An incomplete flower can either be perfect, having male as well as female sex organs or imperfect with either of the sexes missing. When both the essential organs are absent, the flower is spoken as neuter.

6. **(b)** In racemose head, the main axis or receptacle is suppressed, becoming almost flat. It bears a mass of small sessile flowers (florets) on its surface, with one or more whorls of bracts at the base forming an involucre.
The florets are arranged in a centripetal fashion, *i.e.,* younger towards the centre and older towards the periphery. The advantages of this kind of inflorescence are that the head as a whole becomes more showy and attractive and the florets being close together, one or a few insects can pollinate most of them within a short time.

7. **(b)** Cyathium is a special kind of inflorescence found in *Euphorbia, i.e., Poinsetia* and *Euphorbia.*
In cyathium, there is a cup – shaped involucre, often provided with nectar – secreting glands. The involucre encloses a single female flower in the centre, seated on a comparatively long stalk and a number of male flowers around this, seated on short slender stalks.

8. **(b)** Coconut tree requires the hot and wet climate hence it is grown in coastal areas. It is widely grown in coastal and deltic regions of tropical and subtropical countries.
Coconut fruit show hydrochory *i.e.,* the dispersal takes place through water medium. Being the habitat of coastal area the fruit is so adapted that it can float and dispersed over thousand of kilometers before losing viability.

9. **(c)** Seed is the ripened ovule which contains an embryo or miniature plant which has adequate reserve food for future development of the embryo. A seed may have one or two coverings called seed coat. The outer or the only seed coat (if one is present) is called testa while the inner one is named as tegmen. The hilum is a scar on the seed coat through which developing seed attaches to fruit. It is the place where funiculus or stalk of the seed is borne.

10. **(a)** The root apex and shoot apex are meristematic in nature. These meristematic tissues are embryonic in origin. They are primary in origin because it develops from embryonic tissues and primary in function because they form the primary structure of the plant cell, the root apex and shoot apex, that live till the death of the whole plant. Hence, plants have the feature of indefinite growth.

11. **(a)** Collenchyma cells are elongated cells with irregularly thick cell walls that provide support and structure. Their thick cell walls are composed of the compounds cellulose and pectin. These cells are often found under the epidermis, or the outer layer of cells in young stems and in leaf veins.

12. **(a)** Sclerenchyma is the supporting tissue in plants. Two types of sclerenchyma cells exist: fibers and sclereids. Their cell walls consist of cellulose, hemicellulose and lignin. Sclerenchyma provides the main structural support to a plant.

13. **(d)** Xylem is the water conducting tissue. It consists of living cells like parenchyma and dead cells like tracheary elements.

14. **(d)** Phloem is a complex tissue meant for transport of food from leaves to the other body parts of the plant. Phloem is also called bast. It consists of four types of cells: sieve tubes, companion cells, phloem parenchyma and fibres. Sclerenchyma fibres found in the phloem are called phloem or *bast* fibres. Phloem fibres provide mechanical strength. The textile fibres of flax (*Linum usitatissimum*), hemp (*Cannabis*) and jute (*Corchorus* species) are phloem fibres which are used for making ropes and coarse textiles.

15. **(c)** Open vascular bundles are the feature of dicot stem and gymnosperm. The cambium present between xylem and phloem is called fascicular cambium. In closed vascular bundles, the cambium will be absent (fascicular cambium absent) and they do not show secondary growth (closed for secondary growth). In open vascular bundles, a layer of cambium will be present between the xylem and phloem and thus, they shows secondary growth.

16. **(d)** Bark consists of all tissues outside the vascular cambium. Phellem, phellogen and phelloderm constitute periderm.

17. **(c)** Various types of plant products like oils, resins, gums and tannins are deposited in the cells of the heartwood. They are antiseptic. The heartwood is, therefore, stronger and more durable than the sapwood. It is, however, liable to be attacked by wood rotting fungi. Hollow tree trunks are due to their activity. Sapwood (outer light coloured wood) is less durable because it is susceptible to attack by pathogens and insects.

18. **(a)** Dicotyledonous stems usually display secondary growth in thickness, which is due to the activity of vascular cambium and cork cambium. Vascular cambium is present between phloem and xylem. Because of the presence of cambium, this vascular bundle possess the ability to form secondary xylem and phloem tissues, and exhibit secondary growth.

19. **(b)** Squamous epithelium is a single layer of flattened cells in contact with basal lamina of the epithelium. This type of epithelium is often permeable and occurs where small molecules need to pass quickly through membranes.

20. **(d)** Ordinary connective tissue is absent inside the central nervous system. *i.e.* brain and spinal cord and has no function in the conduction of nerve impulse. The neurons of nerve tissue inside the brain and spinal cord are held together by supporting cells called neuroglia cells. Neuroglia cells resemble neurons and have long radiating processes but no nissl granules.

21. **(a)** The symptoms of an allergic reaction develop in response to histamine. Mast cells release a large amount of histamine into the blood stream and it also act as initiator of the inflammatory response which aids the arrival of leucocytes at a site of infection. Histamine stimulates capillary dilation, increased capillary permeability, closure of bronchial tubes, mucus secretion, pain and swelling.

22. **(c)** Regulation of RBCs production is accomplished by hormone erythropoietin which is secreted by kidney. This hormone increases stem cell mitosis and speed up RBCs development in red bone marrow.

23. **(a)** Smooth muscles are found in the posterior part of oesophagus, stomach, intestine, blood vessels, iris of eye and dermis of skin. They are involuntary as their functioning cannot be directly controlled. Action of these muscles is controlled by autonomic nervous system.

24. **(a)** Earthworms are known as 'friends of farmers' because they make the soil loose and porous by their burrowing habit. Thus, the soil provides quick aeration and absorption of water, thereby the roots of the plant get penetrated easily. Earthworms continuously bring the lower soil on the surface, and deposit it there. They plough the land and share the work of the farmers. The process of increasing fertility of soil by earthworm is called vermicomposting.

25. **(c)** The colourless blood or haemolymph of cockroach has a clear plasma and numerous white corpuscles called haemocytes. Being devoid of any respiratory pigment it does not serve for gaseous exchange. Its plasma contains about 70% water. In all terrestrial insects, like cockroach every tissue of body is in direct contact with atmospheric air for gaseous exchange.

Unit III: CELL: STRUCTURE AND FUNCTION

1. **(a)** Metabolically active cells are usually smaller due to higher nucleocytoplasmic ratio and higher surface volume ratio. The former will allow the nucleus to have better control of metabolic activities, while the latter will allow quicker exchange of materials between the cells and its outside environment.

2. **(b)** Schleiden and Schwann together formulated the cell theory. The theory however, did not explain as to how new cells were formed. Rudolf Virchow (1855) first explained that cells divided and new cells are formed from pre-existing cells (*Omins cellua e cellua*). He modified the hypothesis of Schleiden and Schwann to give the cell theory a final shape. The present cell theory states that: (i) All living organism are composed of cells and products of cells. (ii) All cells arise from pre-existing cells.

3. **(a)** Eukaryotic cells have more DNA than prokaryotic cells because in eukaryotic cells complex chromosomes are composed of DNA and histone proteins. But in prokaryotic cells, histone protein is absent.

4. **(a)** Cell wall is characteristic feature of plant cells. Cell wall is absent in animals. Animals cells are covered by cell membrane.

5. **(c)** Mitochondria and chloroplasts both are double membrane organelles. They are semi autonomous as both their structure and functions are partially controlled by nucleus of the cell and partially themselves. Both possess their own DNA and arise from pre-existing cells. Synthesis of many amino acids occurs in mitochondria. The first formed amino acids are glutamic acid and aspartic acid.

   Also, plastid manufactures some of its own proteins, enzymes and other chemicals because of the presence of 70S ribosomes which can help translate the coded information contained in mRNAs transcribed over chloroplast DNA.

   DNA replication is the process of forming carbon copy of DNA. Whereas, transcription is the formation of RNA over DNA template.

6. **(a)** The study of a cilium or the flagella shows that they are covered with plasma membrane. Their core is called axoneme, it possesses a number of microtubules running parallel to the long axis. The axoneme usually has nine pairs or doublets of radially arranged peripheral microtubules, and a pair of centrally located microtubules. Such an arrangement of axonemal microtubules is referred to as the 9 + 2 array.

7. **(a)** A chain of DNA has 140 base pairs which makes $1\frac{3}{4}$ turns and twists around a histone octamer forming nucleosome. The core of nucleosome consists of 4 histones H_2A, H_2B, H_3 and H_4.

8. **(d)** Essential amino acids are those which are taken from food and not synthesised in the body whereas non-essential amino acids need not be supplied in the diet and are synthesised in the body. Glycine, serine and tyrosine are non-essential amino acids.

9. **(c)** Compounds having double bond in their structure are more unstable compounds in comparison to single bond holders. Unsaturated fats those have double bonds in their structures are more reactive than saturated fats.

10. **(c)** In a DNA molecule, A-T rich parts melt before G-C rich parts because there are two H-bond between A and T whereas in between G and C, there are three H-bond.

11. **(c)** Feed back inhibition is a type of reversible inhibition found in allosteric enzymes. The inhibitor is non-competitive and is usually a low molecular intermediate or product of metabolic pathway having a chain of reactions involving a number of enzymes.

12. **(a)** We know that all biological reactions are catalysed by special catalysts called enzyme, thus enzymes are defined as biological proteins. We also know that enzymes are small organic molecules which are weakly held to the protein and can be easily separated by dialysis. Therefore, chemically all enzymes are globular proteins.

13. (c) Enzymes are of various types depending on their action. Hydrolases catalyse the hydrolysis of ester, ether, peptide, glycosidic, C – C, C – halide, P – N bonds etc. which are formed by dehydration and condensation. Hydrolases break up large molecules into smaller ones with the help of hydrogen and hydroxyl groups of water molecules. The phenomenon is called hydrolysis. Lyases are the enzymes which cause cleavage, removal of groups without hydrolysis, addition of groups to double bonds or removal of a group producing double bond, e.g., histidine decarboxylase breaks histidine to histamine and CO_2.

14. (b) The external supply of energy which is needed for the initiation of the chemical reaction is activation energy. Activation energy required for such a large number of reactions cannot be provided by living systems. Enzymes lower the activation energy required for a reaction. Enzymes are usually specific for their substrates.

15. (c) DNA synthesis takes place only during one specific stage in cell cycle (interphase).

16. (c) M-phase represents the phase of actual division. It consists of karyokinesis (the division of nucleus) followed by cytokinesis (the division of cytoplasm). Cell divisions stop after M-phase.

17. (b) In an animal cell, cytokinesis is marked by the appearance of a furrow in the plasma membrane. The furrow slowly deepens and ultimately joins in the centre dividing the cell cytoplasm into two. Plants cells however, are enclosed by a relatively inextensible cell wall, therefore they undergo cytokinesis by a different mechanism. In plant cells, wall formation starts in the centre of the cell and grows outward to meet the existing lateral walls. The formation of the new cell wall begins with the formation of a simple precursor called the cell-plate that represents the middle lamella between the walls of two adjacent cells.

18. (a) Meiosis II is known as equational or homotypic division like mitosis. It ensures the maintenance of constant number of chromosomes from generation to generation of a species.

19. (a) Diploid cell produce haploid cell by meiosis. It occurs due to synapsis. A chromosome has two chromatids, hence a bivalent has 4 chromatids.

20. (a) After meiosis, the chromosomes are reduced by half, producing haploid cells. The sperm and the egg are haploid cells and when they fuse during fertilisation, they produce original diploid cell.

Unit IV: PLANT PHYSIOLOGY

1. (c) Osmosis is diffusion of water from its pure state (or dilute solution) into a solution (or stronger solution) when the two are separated by a semi-permeable membrane. The direction and rate of osmosis depend upon the sum of two forces, pressure gradient (gradient of Ψ_w) and concentration gradient (gradient of Ψ_s). The difference in the water potentials of solutions separated by a semipermeable membrane determine the net force or gradient.

2. (b) Due to entry of water into the plant cell, a positive pressure develops. This positive hydrostatic pressure is also known as turgor pressure. Turgor pressure keeps the cells and their organelles stretched, this is essential for proper functioning of a cell. It gives support to non-woody tissues like parenchyma. During growth, turgor pressure is essential for cell enlargement.

3. (b) Water potential is the potential energy of water relative to pure water in reference conditions. Pure water has maximum water potential, whereas, addition of solutes to water, lowers the water potential. Osmosis is a special type of diffusion in which water diffuses from its pure state i.e., higher water potential to concentrated solution i.e., lower chemical potential through a semipermeable membrane. Since pure water has no difference in chemical potential, it has zero osmotic potential.

4. (a) Light is an important factor in transpiration. The stomata opens well on days when light is brighter. It is also evident on cloudy days that the stomata does not open well. Hence, light induces stomatal opening and darkness induces closing.

5. (c) Guttation is the loss or excretion of water in the form of liquid droplets from the leaves and other parts of an uninjured or intact plant. All plants do not show guttation. It is restricted to about 345 genera of herbaceous and some woody plants. Guttation occurs through special structures called hydathodes that are usually found on the margins and tips of the leaves. Common examples are found in garden nasturtium, oat and other cereals, balsam, tomato, cucurbits. Guttation usually occurs during periods of active growth when conditions favour more water absorption and less transpiration. It takes place either at night or early in the morning.

6. (b) The force which is responsible to join water molecules to each other in water column is cohesion force. On the account of cohesion force, water column can bear a pull or tension up to 100 atm. Therefore, it is also known as tensile strength while the force between the wall of tracheary elements and water molecule is called as adhesion force which produces surface tension and which account for high capillarity through tracheids and vessels.

7. (c) Sap is water with dissolved ingredients. The upward movement of water from roots towards the tips of stem branches and their leaves is called ascent of sap. It occurs through the tracheary elements of xylem.

8. (c) Solution culture is being used for raising flowers and vegetables at home. This soilless production of plants is called hydroponics. A solution having all the essential elements in proper proportion is called normal or balanced nutrient solution

9. (a) Sulphur is a constituent of amino acids (cysteine, cystine and methionine). Sulphur is present in the soil in the form of oxides. Sulphur bacteria convert them into sulphate ions.The plants absorb sulphur in the form of sulphate ions.

10. (b) Magnesium is a constituent of the chlorophyll molecule, without which photosynthesis would not occur. Many of the enzymes involved in carbohydrate metabolism require magnesium as an activator. Magnesium is also an activator for those enzymes involved in the synthesis of nucleic acids (DNA, RNA) from nucleotide polyphosphate.

11. (b) The necessity of micronutrients is always in low amounts while their moderate decrease causes deficiency symptoms and a moderate increase causes toxicity. Toxicity levels for any element also varies for different plants. Many times, excess of an element may inhibit the uptake of another element. For example, the prominent symptom of manganese toxicity is the appearance of brown spots surrounded by chlorotic veins. Manganese competes with iron and magnesium for uptake and with Magnesium for binding with enzymes. Manganese also inhibits calcium translocation in shoot apex. Therefore, excess of manganese may induce deficiencies of iron, magnesium and calcium.

12. (c) *Rhizobium* species (e.g., *Leguminosarum*, *R. lupini* etc) live in symbiotic association in the root nodules of leguminous plants like, pea, gram etc. The bacteria cannot fix nitrogen in the soil. After entering the legume root, the bacterium induces a cell membrane. A pink-red pigment called leghaemoglobin lines the membrane and protects the bacteriods from the oxygen.

13. (b) Plants do lack excretory organs, but it is not due to absorption pattern or passive life. The carbon dioxide during respiration passes out through the stomata. The other waste materials come out in the form of alkaloids, gums and resins.

14. **(c)** Anthocyanin pigments only give colouration since the epidermal cells mainly have potential colouring pigments. It is responsible for blue, red, pink and purple colours, observed in different parts of plants such as petals, stamens and fruits etc. Anthocyanins are also important for attracting insects for pollination and seed dispersal. Hence, anthocyanin pigments are not accessory photosynthetic pigments.

15. **(c)** The overall equation of photosysthesis is $CO_2 + 2H_2O + \text{light energy} \rightarrow (CH_2O)_6 + O_2$
The standard free-energy change for the synthesis of hexose from CO_2 and H_2O is $\Delta G° = +\ 686$ kcal. As six molecules of CO_2 are involved in forming one molecule of hexose, the energy input per CO_2 molecule will be 114 kcal.

16. **(c)** Water splitting complex is associated with PS-II, which itself is physically located on the inner side of the membrane of the thylakoid.

17. **(c)** Compensation point is that value or point in the light intensity and atmospheric CO_2 concentration when the rate of photosynthesis is just equivalent to the rate of respiration in the photosynthetic organ. So that there is no net gaseous exchange.

18. **(d)** Mitochondria helps in cellular respiration by transferring energy from organic compounds in form of ATP. Chloroplast helps in photosynthesis. Dark reaction takes part in the stroma of the chloroplast.

19. **(a)** C_4 photosynthetic pathway is more efficient than C_3 pathway as C_4 plants can pick up CO_2 even when it is found in low concentration. PEP enzymes show high affinity for CO_2. C_4 plants contain two types of chloroplast (Kranz anatomy) : bundle sheath chloroplast and mesophyll chloroplast. Bundle sheath cells contain Calvin cycle enzymes. Due to high concentration of CO_2 in bundle sheath cells, RuBP carboxylase works only for Calvin cycle and not for photorespiration. Photorespiration is a wasteful process as it works to undo the act of photosynthesis in C_4 plants. No energy rich compound is produced in this process. When temperature increases, more and more photosynthetically fixed carbon is lost by photorespiration thus reducing the efficiency of C_3 plants.

20. **(c)** CAM plants do not exhibit the structural compartmentation (C_3 and C_4 cycles taking place in different cells) of conventional C_4 plants. CAM plants fix CO_2 at night because their stomata are open at night and closed during the day.

21. **(c)** C_4 plants show 'Kranz' type of anatomy. In 'Kranz' anatomy, the mesophyll is undifferentiated and its cells occur in concentric layers around vascular bundles. The vascular bundles are surrounded by large sized bundle sheath cells which are arranged in wreath like manner in one to several layers. Chloroplasts of mesophyll cells are smaller, have well developed grana and they do not produce starch. Chloroplasts of bundle sheath cells are larger and agranal. Starch is often present.

22. **(c)** Photorespiration is a highly wasteful process because, it may drop down the photosynthesis up to 50%. Rate of respiration in C_3 and C_4 plants are not same as they have different types of leaf anatomy (arrangement of chloroplast).

23. **(b)** Cyclic pathway of photosynthesis is appeared first in some eubacterial species. It is supposed to be the first evidence of production of ATP in the presence of light. During non-cyclic photophosphorylation photolysis of water takes place. Under the influence of light energy and the catalytic action of chlorophyll, water is split up into oxygen and hydrogen. Non-cyclic photophosphorylation is the only natural process which adds molecular oxygen to the atmosphere.

24. **(b)** Stomata are absent since gaseous exchange takes place through diffusion in submerged plants.

25. (b) One of the two phosphates of diphosphoglycerate is linked by high energy bond. It can synthesise ATP and form 3-phosphoglycerate. The enzyme is phosphoglycerate kinase. The direct synthesis of ATP from metabolites is called substrate level phosphorylation.
During formation of phosphoenol pyruvate the phosphate radical pick up energy. It helps in the production of ATP by substrate level phosphorylation.

26. (c) Glycolysis is also known as EMP pathway, as it was discovered by three German scientists-Gustav Embden, Otto Meyerhof and J. Parnas in 1930. In glycolysis, partial oxidation of glucose takes place that form two molecules of pyruvic acid. It occurs in the cytoplasm of the cell and is present in almost all living organisms. Aerobic respiration consists of diferent processes i.e., (i) Glycolysis, (ii) Oxidative decarboxylation of pyruvic acid, (iii) Kreb's cycle and (iv) Terminal oxidation and oxidative phosphorylation.

27. (d) Under aerobic conditions, the products are pyruvate and coenzyme NADH. Pyruvate directly enters the mitochondrial matrix and is converted into acetyl-CoA. Under anaerobic conditions, pyruvate is used as a hydrogen acceptor and converted into lactate. In anaerobic conditions, pyruvate remains in the cytosol.

28. (b) The inner membranes of mitochondria contain all systems involving electron transport. The mitochondrial matrix contains all the soluble enzymes of the citric acid or Kreb's cycle and those involved in the oxidation of fatty acids

29. (a) Amphibolic pathway is the pathway that functions for both catabolism and anabolism reactions. Respiratory pathway is mainly a catabolic process that serves to run the livng system by providing energy. The pathway produces a number of intermediates. Many of them are raw materials for building up both primary and secondary metabolites. Because the respiratory pathway is involved in both anabolism and catabolism, it would hence be better to consider the respiratory pathway as an amphibolic pathway rather than a catabolic one.

30. (c) Both kinases require a divalent cation (Mg^{2+}or Mn^{2+}), which first combines with ATP to form the true substrate, $MgATP^{2-}$ or $MnATP^{2-}$.

31. (b) Auxin delays abscission of young leaves and fruits. Its effect is through non-formation of abscission zone below a leaf or fruit. Abscission zone cuts off nutrients and water supply. However, auxin promotes the abscission of mature or older leaves and fruits.

32. (c) Application of GA can induce flowering in long day plants. It has no favourbale effect on size of flower and fruit in certain plants. The genetically dwarf plants can be made tall by application of GA.

33. (c) It has been demonstrated that flowering in plant is more of a response to the dark period than to the light period. In short day plants, plants can flower in complete darkness if supplied with exogenous nutrients. Flowering is prevented in them if dark period below the critical level is interrupted by a flash of light. Interruption of light by dark inhibits flowering under normal photoperiods.

34. (b) The physiological mechanism of flowering in plants is controlled by two factors– light period and low temperature. The cold treatment of plants to induce flowering is called vernalisation. Term vernalisation was first given by T.D.Lysenko(1928). As a result of vernalisation a flowering hormone called vernaline is formed. Site of vernalisation is apical meristem.

35. (d) To induce flowering, some plants require a periodic exposure to light. It is now also known that not only the duration of light period but the duration of dark period is also of equal importance. Hence, it shows that flowering in certain plants depends not only on a combination of light and dark exposures but also their relative durations. These response of plants to periods of day/night is termed photoperiodism. The shoot apices modify themselves into flowering apices prior to flowering, shoot apices cannot percieve photoperiods by themselves. The leaves are the site of perception of light/dark duration.

Unit V: HUMAN PHYSIOLOGY

1. (b) Presence of hydrochloric acid in stomach is necessary for digestion because acidic medium activates the action of gastric juice. HCl maintains a strong acidic pH of about 1-2 in the stomach. At this acidic pH, inactive pepsinogen is spontaneously hydrolysed to active pepsin and inactive pro-rennin is converted to active rennin. Pepsin and rennin digest proteins to peptones and proteoses. In addition, HCl helps to kill and inhibit the growth of bacteria and other harmful organisms that may enter in the stomach along with the food.

2. (d) It is not lipases but the bile salts which are responsible for the emulsification of fats. Bile salts are steroids secreted by the liver in the bile. In the intestinal lumen, they reduce the surface tension of fat dropletes, causing their breakdown into many smaller ones. A stable fine emulsion of fat is thereby formed. On the other hand, lipases are the enzymes which hydrolyse fats and oils. Lipases can digest fat in significant amounts only when large fat droplets are broken into tiny droplets to form a fine emulsion. Emulsificaion of fats by bile salts thus, increases the lipase action on fats.

3. (c) Numerous tube like glands are present in the wall of small intestine which secrete intestinal juice into the intestinal lumen. This juice contains a number of enzymes like enterokinase, aminopeptidases, dipeptidases, maltase, lactase, lipase etc. for digesting various types of food. On the other hand, the enzymes RNAse and DNAse are present in the pancreatic juice. These enzymes are secreted by the pancreas and are drained into the small intestine via hepato pancreatic duct.

4. (a) Small intestine of the human gut is a narrow long tube. Because of its small diameter, it is called small intestine. Its length is according to the height of individual. It is the major site of digestion and absorption of different type of food. Digestion and absorption of various food is carried out in a particular part of small intestine, like absorption of iron takes place in proximal part of duodenum. Food digestion starts from oral cavity in humans, which continues in stomach and is digested completely in small intestine. Undigested and unabsorbed food (faeces) is transferred to long intestine where the water and some vitamins are absorbed.

5. (a) In the condition of obstructive jaundice, the entry of bile into the small intestine is prevented due to an obstruction in the bile duct. As we know that bile salts helps in the digestion of fats by emulsification and also in their absorption by the formation of water soluble droplets called micelles from where fatty acids, glycerides, sterols and fat soluble vitamins are absorbed into the intestinal cells. Therefore, in the absence of bile, the fats remain unabsorbed and consequently are eliminated out of the body through the faeces.

6. (d) The innermost layer lining the lumen of the alimentary canal is the mucosa. This layer forms small finger-like foldings called villi in the small intestine. Villi help in increasing absorptive surface of the small-intestine. They are supplied with a network of capillaries and a large lymph vessel called the lacteal. Because of these modifications, about 90% of absorption of all nutrients occurs in the small intestine.

7. (a) Deficiency of ascorbic acid/vitamin C causes scurvy.

8. (a) As the integument of insect is thick and impermeable to minimise lose of body water, they can not carry out gas exchange through their body surface. To overcome this difficulty they have developed a complex system of air tubes called trachea to reach the air directly near the tissue cells. Each trachea communicates with the exterior through openings in the body wall.

9. (b) The solubility of the gases as well as the thickness of the membranes involved in diffusion are important factors that affect the rate of diffusion. A gradient of partial pressure is present for oxygen from alveoli to blood and blood to tissue. Similarly, a gradient of CO_2 is present in the opposite direction i.e., from tissues to blood and blood to alveoli. The pO_2 in alveoli is higher (104 mm Hg) than that in deoxygenated blood in the capillaries arising from pulmonary arteries (95 mm Hg), therefore, oxygen diffuses from the alveoli to the blood. Similarly, the partial pressure of carbon dioxide (pCO_2) is higher in deoxygenated blood (45 mm Hg) than that in alveoli (40 mm Hg), therefore carbon dioxide passes from the blood to the alveoli. It is further dependent on the solubility of the diffusing gases. As the solubility of CO_2 is 20-25 times higher than that of O_2, the amount of CO_2 that can diffuse through the diffusion membrane per unit difference in partial pressure is much higher compared to that of O_2. Therefore, all the factors in our body are favourable for diffusion of O_2 from alveoli to tissues and that of CO_2 from tissues to alveoli.

Alveolus is the primary site of exchange as it has an extensive network of blood capillaries and consists of squamous epithelium. Due to very intimate conact of blood capillaries with the alveoli, the exchange of gases takes place easily.

10. (a) Histamine is a derivative of the amino acid histidine produced by damaged cells of vertebrates. When released, it has the effect of dilating capillaries and lowering blood pressure. Histamine is involved in allergic and inflammatory reactions.

11. (a) Bohr's effect is the effect of CO_2 on oxyhaemoglobin. Body tissues obtain oxygen from oxyhaemoglobin because of its dissociation caused by low O_2 and high CO_2 concentration. The increased CO_2 concentration reduces the affinity of haemoglobin for oxygen.

12. (b) The world health organisation (WHO) reported China as the origin place of SARS. Several other countries have reported SARS cases after travel to China and its nearby areas/countries in Asia or close contact with a person affected with SARS. Microbiologists of Hongkong Central University initially detected a virus corona as causative agent of SARS. It is very known fact that the China is the most populated country. This fact cannot be correlated with SARS.

13. (c) If a blood transfusion is made between an incompatible donor and recipient, reaction of antigens on the cells and antibodies in the plasma will produce clots that will clog the capillaries. Type O blood group individuals are without A and B antigens on their RBCs, but consists antibodies for both these antigens in their plasma. Hence, these persons can donate blood to anyone. O blood group is the most important blood group for transfusion.

14. (d) When an injury is caused to a blood vessel, bleeding starts which is stopped by blood clotting. At the site of injury blood platelets release platelet factor - III and injured tissues release thromboplastin. The two combine to form prothrombinase enzyme which converts prothrombin to thrombin. The latter stimulates formation of fibrin thread or clot. Blood contains an anticoagutant heparin which prevents blood clotting in uninjured vessels.

15. (b) Prothrombinase enzyme is necessary for blood clotting. It acts as antiheparin. Coagulation of blood in vessels is prevented by heparin, a quick acting anticoagulant. It inhibits conversion of prothrombin to thrombin and is used in open-heart surgery.

16. (a) There are some reasons behind the effectiveness of closed circulatory system over the open type. These are -

(i) The blood flows more rapidly in closed blood cavities.

(ii) It takes much shorter time to circulate through the closed system and return to the heart.

(iii) This quickens the supply and removal of materials to and from the tissues by the blood.

(iv) In the closed system, the arteriolar diameter can be regulated to alter the blood flow, so the volume of blood flowing through a tissue or organ may be regulated according to its needs. No such regulation is possible in open type.

17. (b) Blood pressure is the arterial blood pressure taken in left brachial artery and the instrument used to measure blood pressure is sphygmomanometer.

18. (a) Lub and dub are two heart sounds, which occur due to the closure of cuspid valves and semilunar valves respectively. Lub is the first heart sound which is formed due to closure of atrioventricular valves at the beginning of ventricular systole. It is low pitched of long duration (0.15 sec).

19. (a) Haemophilia is caused by lack of activity of blood clotting factor VIII or IX and they show platelet function disorder.

20. (a) When sufficient water is available, earthworms excrete ammonia. Because a large volume of water is needed by the animal to dissolve ammonia and eliminate it from, the body. Whereas in drier environment the animal removes urea as it needs only considerable amount of water for the excretion of urea because urea is very soluble in water.

21. (c) Urea is the chief nitrogenous constituent of human urine, though it possesses small amount of ammonia. But when the urine is allowed to stand for sometime, bacterial degradation occurs and it leads to the production of ammonia from urine, and thus smells strongly.

22. (a) Kidneys play an essential role in maintaining the concentration and osmotic pressure (osmoconcentration) of blood. When water intake of an animal is very high, the urine excreted has to be hypotonic i.e., dilute and lower in osmotic pressure than their blood in order to remove the excess of water. Contary to this, when there is a threat of excessive water loss from the body; the urine needs to be hypertonic more concentrated and higher in osmotic pressure than their blood, to reduce the loss of water with urine. In this way, the osmotic concentration of the blood is maintained.

23. (c) Ultrafiltration takes place in renal corpuscle of uriniferous tubule. It takes place in presence of effective filtration pressure. During the process, blood is filtered and contains only blood plasma – proteins. The filtered blood entering into Bowman's capsule is called glomerular filtrate.

Glomerular filtrate = Blood – (Blood corpuscles + plasma proteins).

24. (c) For long periods, camels have the ability to withstand water deprivation. But they do not store any water in the pouches of their rumen. The fat of hump is not particularly useful as a source of water because respiration must be enhanced to oxidise fat to produce water and this enhances respiratory loss of moisture. They reduce urinary loss of water by secreting small volume of the urine much more hypertonic than the human urine. They lose far less water in the sweat, because they sweat only when their body temperature rises by as much as 60 °C. Moreover, compared to all other mammals. These factors are mainly responsible for the camel's ability to go without water for long periods.

25. (a) The functional unit of kidney is called as nephron. Nephron contains a tubular region which is present between proximal and distal convoluted tubule, the loop of Henle. The length of loop of Henle is proportional to the concentration of urine. The mammals (chordates) living in desert consists longer loop of Henle in their nephrons. So, these animals contain more concentrated urine.

26. (b) The act of releasing the urine is called micturition. Besides functioning as a temporary reservoir of urine, the urinary bladder also evacuates the urine by the process of micturition at suitable intervals. When enough urine has accumulated in the bladder to distend the bladder and raise its pressure sufficiently, a spontaneous nervous activity (reflex) is initiated; this causes the smooth muscles on the bladder wall to contract and the urethral sphincters, which guard the urethra, to relax. Urine consequently flows from the bladder through the urethra to the exterior. But it is prevented from flowing back into the ureters, because the terminal part of each ureter passes obliquely through the bladder wall and is consequently closed due to compression by the contracting bladder muscles.

27. (b) Antidiuretic hormone (ADH) or vasopressin is secreted from posterior pituitary gland. It is stimulated in response to a fall in the water content of blood plasma, that leads to an increase in the permeability to water of the distal and collecting tubules of the nephron. Deficiency or hyposecretion of ADH results in Diabetes insipidus. Diabetes insipidus is characterised by micturating dilute urine several times a day that results in excessive thirst (polydipsia) and dehydration.

28. (a) Due to the accumulation of lactic acid, muscles do not respond to a stimuli after a prolonged previous activity.

29. (b) A neural signal on reaching the neuromuscular junction releases a neurotransmitter (acetylcholine) which generates an action potential in the sarcolemma. This spreads through the muscle fibre and causes the release of calcium ions from sarcoplasmic reticulum into the sarcoplasm. Increase in Ca^{2+} level leads to the binding of calcium with a subunit of troponin on actin filaments and thereby, removing the mask of active sites for myosin.

30. (d) Muscle contains a red coloured oxygen storing pigment called myoglobin. Myoglobin content is high in some of the muscles which gives a reddish appearance to them. Such muscles are called the red fibres. These muscles also contain plenty of mitochondria which can utilise the large amount of oxygen stored in them for ATP production. These muscles, therefore, can also be called aerobic muscles.

31. (c) First seven pairs of ribs are called true ribs. Dorsally, they are attached to the thoracic vertebrae and ventrally connected to the sternum with the help of hyaline cartilage.

32. (b) Synovial fluid is a thick sticky fluid like egg white consistency, secreted by synovial membranes into the synovial cavity. Though the presence of synovial fluid is one of the reasons behind the mobility of the joints, but the most accurate reason is the arrangement of the bones at the joint, the spheroidal ball-like end of one bone articulates here with the cup-shaped depression in another. This allows the bone with the ball head to move freely in many planes. Shoulder joints and hip joints are the ball-and-socket joints.

33. (c) To maintain the unequal distribution of Na^+ and K^+ ion, the neurons use chemical energy in the form of ATP to actively transport Na^+ ion out of cell and more K^+ inside the cell.

34. (b) Action potential occurs in response to stimulation of nerve fibre as a result of various stimuli like touch, smell, pressure and chemical changes.

35. (a) Coordination between organs/organ system, is the key for better functioning of our body. In case of physical exercise, the increased demand of oxygen initiates a cascade, which coordinates lungs, heart and blood vessels of the body to fulfil the demand.

36. (b) Hind brain consists of cerebellum located dorsally to medulla oblongata and pons varoli. It contains centres for maintenance of posture and equilibrium of the body and for the muscle tone. All activities of the cerebellum are involuntary but may involve learning in their early stages.

37. (d) The layer of cells which covers the cerebral hemisphere is called cerebral cortex and is thrown into prominent folds. The cerebral cortex is referred to as the grey matter due to its greyish appearance. The neuron cell bodies are concentrated here giving the colour.

38. (a) Tongue is also called as gustatoreceptor, i.e., it is a receptor for taste. The receptor for gustatory sensation are located in taste buds on tongue.

39. (b) Astigmatism is a kind in defect of vision in which the image of an object is distorted. It is because all the light rays do not come to focus on retina due to abnormal curvature of the lens . It can be corrected by wearing cylindrical lenses.

40. (a) Autonomic nervous system includes sympathetic nervous system and parasympathetic nervous system which have antagonistic functions. Iris muscles, under sympathetic nervous syetem dilates pupil, while under parasympathetic nervous system constricts pupil.

41. (a) The Eustachian tube connects the middle ear cavity with the nasopharynx. It aerates the middle ear system and clears mucus from the middle ear into the nasopharynx. Opening and closing functions of the eustachian tube are physiologically and pathologically important. Normal opening of the eustachian tube equalises atmospheric pressure in the middle ear; closing of the eustachian tube protects the middle ear from unwanted pressure fluctuations and loud sounds.

42. (a) Neurohypophysis also known as posterior pituitary. It stores and releases two hormones called oxytocin and vasopressin which are actually synthesised by the hypothalamus and are transported axonally to neurohypophysis. Therefore, the posterior pituitary is under the direct neural regulation of the hypothalamus.

43. (c) Vasopressin or antidiuretic hormone is secreted by posterior pituitary gland. The deficiency of vasopressin results in a disorder known as diabetes insipidus. The main symptoms of diabetes insipidus are increase in thirst and increase in urination.

44. (b) Melatonin concentration in the blood follows a diurnal (day-night) cycle as it rises in the evening and through the night and drops to a low around noon. Because of this light mediated response, the pineal gland may act as a kind of 'biological clock' which may produce circadian rhythms (variations following a 24 hour cycle). In addition, melatonin also influences metabolism, pigmentation, the menstrual cycle as well as our defence capability.

45. (a) Calcitonin or thyrocalcitonin is secreted by parafollicular cells of thyroid stroma. It retards bone dissolution and stimulates excretion of calcium in urine. Thus, it lowers calcium level in extra cellular fluid (ECF). Parathormone is secreted by chief cells of parathyroid gland and is also known as Collip's hormone. It maintains blood calcium level by increasing its absorption from food in intestine and its reabsorption from nephrons in the kidney. Maintenance of proper calcium level is in fact, a combined function of parathormone and calcitonin. When calcium level falls below normal parathormone, it maintains by promoting its absorption, reabsorption and also by demineralisation of bones. When blood calcium level exceeds above normal then calcitonin hormone increases excretion of calcium in urine.

46. (b) The adrenal medulla secretes two hormones called adrenaline or epinephrine and noradrenaline or norepinephrine. Adrenaline and noradrenaline are rapidly secreted in response to stress of any kind and during emergency situations and are called emergency hormones or hormones of fight. These hormones increase alertness, pupillary dilation, piloerection, sweating, etc. Both the hormones increase the heart beat, the strength of heart contraction and the rate of respiration. They also stimulate the breakdown of glycogen resulting in an increased concentration of glucose in blood. In addition, they also stimulate the breakdown of lipids and proteins. Both the hormones act on the same organs and produce similar effects on them e.g., accelerate heart beat, raise blood pressure, slow peristalsis, etc.

47. (a) Adrenaline is an emergency hormone whose concentration increases under stress conditions. Adrenaline is secreted from adrenal medulla. It initiates many bodily responses, including the stimulation of heart action and an increase in blood pressure, metabolic rate, and blood glucose concentration.

48. (c) Males have more stature than females because of the action of male sex hormone-testosterone which is secreted by testis in males. Body starts secreting testosterone from the age of puberty. Its secretion is under the influence of Luteinizing Hormone (LH) of the anterior lobe of pituitary gland. Testosterone controls the development of secondary sexual characters in males like hoarseness of voice, development of facial hairs, bone growth, calcium retention, closing of epiphysial cartilage. The total quantity of bone matrix increases. The pelvic outlet is narrowed and lengthened. The strength of the pelvic bones increases to carry more loads. That is why males have more stature than females after puberty when this hormone is present in the blood. hCG (Human Chorionic Gonadotropin) is the hormone secreted by human placenta during pregnancy. hCG enlarges the corpus luteum in the mother's ovary and stimulates it to secrete progesterone.

49. (d) Oxytocin hormone is also known as 'milk ejection hormone' because oxytocin induces contractions of the mammary gland muscles and helps in the flow of stored milk from the mammary glands to the mouth of the suckling infant. Oxytocin contracts the smooth muscles of uterus and mammary glands. Uterine contractions, stimulated by oxytocin at the end of pregnancy, help in child-birth. That is why, it is also called 'birth hormone'.

Prolactin or lactogenic hormone or luteotrophic hormone (LTH) is secreted by the anterior pituitary which helps in the growth of mammary glands during pregnancy and initiates the secretion of milk after child-birth.

Unit VI: REPRODUCTION

1. (a) The plant at their tips possess meristematic tissues which help them to grow throughout their lives even though the rest of their bodies is mainly made up of sclerenchyma, xylem and cork.

2. (d) Stems have four main functions which are: Support for the elevation of leaves, flowers and fruits. The stems keep the leaves in the light and provide a place for the plant to keep its flowers and fruits. Transport of fluids between the roots and the shoots in the xylem and phloem.

3. (a) Vegetative propagation through leaves takes place in *Bryophyllum* and *Begonia*. The leaf in the plant is broad and has notches at its margins. Buds arise from notches and these buds are called epiphyllous buds.

4. (a) Parthenogenesis is the development of an egg (ovum) into a complete individual without fertilisation. Parthenogenesis occurs in its natural course in many invertebrates like insects, arachnids, some vertebrates and some birds (e.g., turkeys). Parthenogenesis eliminates variations in a population so it does not play any role in organic evolution.

5. (a) Gametes are haploid irrespective of whether the structures or cells producing them are haploid or diploid. It is due to meiosis which occurs in the life of all sexually reproducing organisms. The strucutre formed by the fusion of gametes is always diploid.
6. (c) The endosperm in angiosperms is of a special type formed post-fertilisation as a product of fusion of a male gamete with the diploid secondary nucleus present in the central cell, and is thus triploid in nature.
7. (d) Honey bee visit flowers to gather nectar and turn it into honey. Visiting of insects for nectar helps in pollination.
8. (b) A typical microsporangium consists of two parts, outer wall and central homogeneous sporogenous tissue. Microsporangial wall has four types of layers– epidermis, endothecium, 1–3 middle layers and tapetum. The outer three wall layers perform the function of protection in the young anther and mechanism of dehiscence in the ripe anther.
9. (d) Hydrophily is the mode of pollination which is accomplished through the agency of water. Pollination by water is quite rare in flowering plants and is limited to only about 30 genera, mostly monocotyledons e.g., *Vallisneria, Zostera, Ceratophyllum*, etc. In many aquatic plants with emergent flowers, pollination occurs by wind or insects, e.g., Lotus, Water Lily, Water Hyacinth, etc.
10. (c) Pollen grain does not contain the stigma, style and ovary. It is female gametophyte (gynoecium) which contains stigma, style and ovary.
11. (a) Pollen mother cells undergo meiosis and produce pollen grains. The pollen grains have haploid number of chromosomes.
12. (c) In nuclear type of endosperm the division of the primary endosperm nucleus and a few subsequent nuclear division are not accompanied by wall formation. So numerous nuclei are freely suspended in its sap. In coconut, the primary endosperm nucleus undergoes a number of free nuclear divisions. When the fruit is about 50 mm long before the embryo sac gets filled with a clear fluid in which float numerous nuclei of various sizes. At a later stage (about 100 mm long fruit) the suspension shows in addition to free nuclei, several cells each enclosing variable number of nuclei. Gradually these cells and free nuclei start setting at the periphery of the cavity and layers of cellular endosperm start appearing. In mature coconuts the liquid endosperm becomes milky and it does not contain free nuclei or cell.
13. (c) Apomixis is the type of asexual reproduction in which development of reproductive propagules takes place without meiosis and syngamy. There are two kinds of apomixis (asexual reproduction) in flowering plants: Agamospermy and vegetative propagation. Agamospermy is further divided into three types:
    (i) Adventive Embryony: Formation of embryo directly from the diploid sporophytic cells (nucellus integument) of ovule (other than zygote) is called adventives embryony, e.g., Citrus.
    (ii) Parthenogenesis: Formation of embryo from unfertilised egg.
    (iii) Apospory and apogamy: Formation of embryo from any other cell of embryo sac (other than egg) without fertilisation. During embryogenesis, an embryo develops from zygote inside the embryo sac and the embryo sac becomes an endosperm. Apomictic embryo, if develops, increases the number of embryos inside the seed. Occurrence of more than one embryo in a seed is called polyembryony.
14. (b) In false fruits the thalamus also contributes to fruit formation, e.g. apple, cashew nut, etc.
15. (a) Bulbourethral gland, also called Cowper's Gland, either of two pea-shaped glands in the male are located beneath the prostate gland at the beginning of the internal portion of the penis. These are responsible for adding fluids to semen during the process of ejaculation, thereby facilitating the process of reproduction.

16. (d) Vagina is the tubular female copulatory organ. Passageway for menstrual flow as well as birth canal. Vagina receives semen from male during mating but fertilisation (fusion of gametes) occurs in fallopian tube.
17. (b) The male releases large number of sperms inside female reproductive tract to increase chances of fertilisation.
18. (d) Primary spermatocytes of testes are diploid and formed by mitotic division in the spermatogonium.
19. (a) Graafian follicle is formed due to increase in FSH and ruptures due to rise in level of LH during middle of menstrual cycle.
20. (d) Hyaluronidase, a hydrolytic enzyme is an acrosomal content in mammalian sperm. It helps at the time of fertilisation during the penetration of the sperm into the ovum. Based on the amount of yolk, mammalian eggs are alecithal means egg without yolk. Microlecithal eggs contain very little yolk *e.g.*, sea urchin, starfish. On the basis of distribution of yolk telolecithal eggs are those eggs in which the yolk, is concentrated towards the vegetal pole and cytoplasm and nucleus lie near the animal pole, *e.g.*, birds and reptiles.
21. (c) Vigorous contraction of the uterus at the end of pregnancy causes parturition. Parturition is induced by a complex neuroendocrine mechanism. The signals for parturition originate from the fully developed foetus and the placenta which induce mild uterine contractions called foetal ejection reflex.
22. (c) The use of condoms have been shown to decrease the transmission of AIDS because condoms are contraceptive.
23. (b) In IVF, the fusion of ovum and sperm is done outside the body of woman, to form a zygote which is allowed to divide to form embryo. This embryo is implanted in the uterus where it develops into a foetus which in turn develops into a child. This is called as a test tube baby. In this method, ova from the wife/donor female and sperms from the husbands/donor male are induced to form zygote in the laboratory. The zygote is allowed to divide forming 8 blastomeres. The zygote or early embryo is transferred into the fallopian tube (ZIFT– zygote Intra Fallopian Transfer).
24. (d) Cu-7 and Cu-T are intrauterine contraceptive devices for females. They suppress the sperm motility. Their mode of action is different. Cu-T and Cu-7 discharge 50-75 micrograms of ionic copper into the uterus daily. These copper ions interfere with life-sustaining functions that regulate implantation in the uterus. No any hormone is released by them.
25. (a) Woman plays an important role in the continuity of the family and human race. She provide nutrition to the foetus in her womb about nine months through placenta. Even after birth, she nourishes the infant on her milk. Hence, biologically woman is superior to man.

Unit VII: GENETICS AND EVOLUTION

1. (b) According to Mendelian inheritance, genes come in different varieties called alleles. Somatic cells contain 2 alleles in which one allele provided by each parent of an organism.
2. (c) An allele is an alternative form of a gene. Diploid organisms, for example, humans, have paired homologous chromosomes in their somatic cells, and these contain two copies of each gene.
3. (c) The law of independent assortment states that two factors of each character separate or assort independent of the factors of other characters at the time of gamete formation and get randomly rearranged in the offspring producing both parental and traits of new combinations. The principle of law of independent assortment is applicable to only those factors or genes that are either located distantly on the same chromosome or occur on different chromosomes. Actually, a chromosome bears hundred of genes. All the genes of factors present on a chromosome are inherited together except when crossing over takes place.

4. (b) When two alleles for a single trait are equally expressed with neither being recessive or dominant, is known as codominance. As for example– A person with type AB blood, that means both A and B alleles are equally expressed.

5. (a) Most sex - linked genes in male heterogametic animals are present on the X-chromosome. Some animals however, may carry a few genes on the Y chromosome that produce visible effects on the phenotype of the organism. Y linkage is very rare in higher animal, particularly mammals. X linkage, on the other hand, is very common in all mammals that have been studied, the mammalian X chromosome contains a larger number of genes with major effects on phenotype.

6. (b) Haemophilia never occurs in women because it is a recessive gene disease if both the sex chromosome are X^hX^h, it will be lethal i.e. such female die before birth. Its gene is never found on chromosome Y.

7. (c) Sickle-cell anaemia is inherited from parents to their off spring. In an autosomal recessive trait, mutation is present in both alleles of father and mother.

8. (a) Histones contain a large proportion of the positively charged (basic) amino acids, lysine and arginine in their structure. DNA is negatively charged due to the phosphate groups on its backbone.

9. (c) DNA-polymerase can polymerise nucleotides only in 5' → 3' direction on 3' → 5' strand because it adds them at the 3' direction on 3' → 5' strand because it adds them at the 3' end. Replication over the two templates proceeds in opposite directions. One strand with polarity 3' → 5' forms its complementary strand continuously because 3' end of latter is always open for elongation. It is called leading strand. Replication is discontinuous on the other template with polarity 5' → 3' because only a short segment of DNA strand can be built in 5' → 3' direction due to exposure of a small stretch of template at one time. Short segments of replicated DNA are called Okazaki fragments. DNA strand built up of Okazaki fragments is called lagging strand. As one strand grows continuously while the other strand is formed discontinuously, DNA replication is semidiscontinuous.

10. (b) Biosynthesis of protein is under direct control of DNA in most cases or else under the control of genetic RNA where DNA is absent. Sequences of bases in a particular segment of a polynucleotide chain will determine the sequence of amino acids in a particular polypeptide. The relationship is popularly known as central dogma. Flow of information is one way, *i.e.*, from DNA, information is transferred to RNA (mRNA) and from RNA to protein. Temin (1970) reported that retroviruses operate a central dogma reverse inside host cells. Genomic RNA of these viruses first synthesises DNA through reverse transcription. DNA then transfers information to messenger RNA which takes part in translation of the coded information to form polypeptide.

$$\text{DNA} \xrightarrow{\text{trasnscription}} \text{RNA} \xrightarrow{\text{translation}} \text{proteins}$$

(i) One way flow of information (central dogma)

$$\text{DNA} \underset{}{\overset{\text{trasnscription}}{\rightleftarrows}} \text{RNA} \xrightarrow{\text{translation}} \text{proteins}$$

(ii) Reverse flow of transcription information

11. (a) Synthesis of polypeptide terminates when a nonsense codon of mRNA reaches the A - site. There are three nonsense codons - UAA, UAG and UGA. These codons are not recognised by any of the tRNAs. Therefore, no more aminoacyl tRNA reaches the A- site. The P-site of tRNA is hydrolysed and the completed polypeptide is released in the presence of release factor. Thus termination occurs.

12. (a) Polytene chromosomes show endomitosis and they have high DNA content.

13. **(b)** The RNA word is proposed as the first stage in the evolution of life in which RNA catalysed all molecules necessary for survival and replication. If the first cells used RNA as their hereditary molecule, DNA evolved from an RNA template. DNA probably did not evolve as a hereditary molecule until RNA based life became enclosed in a membrane. Once cells evolved, DNA probably replaced RNA as the genetic material for most organisms.

14. **(c)** The organs which have the same fundamental structure but are different in functions are called homologous organs. These organs follow the same basic plan of organisation during their development. For example, in plants, the homologous organs are thorns of *Bougainvillea* or a tendril of *Curcurbita*, both arising at the axillary position. This represents divergent evolution.

15. **(a)** Polluted areas show less growth of lichens.

16. **(d)** Organs which have the same essential structure, which they inherit from common ancestors though they may be very differently modified in adaptation to different functions are homologous. Analogous organs are structurally different organs which get modified to perform similar functions.

17. **(a)** It has been observed that the banding pattern of some human chromosomes is very similar to the banding pattern of the corresponding chromosomes in apes. It shows common origin of man and chimpanzee and their close relatedness.

18. **(c)** The outer pinna or external ear are useful that detects sound and focuses on ear drum.

19. **(d)** Industrial melanism in *Biston betularia* is the example of natural selection. The mutation caused due the industrialisation decreased its population size.

Unit VIII: BIOLOGY IN HUMAN WELFARE

1. **(c)** Male *Anopheles* mosquito do not have piercing and sucking type of mouth parts. So, they can not inject malarial parasite into man.

2. **(b)** Endoparasite lives inside its host. Bovine endoparasites may be divided into three categories: nematodes or roundworms, cestodes or tapeworms and trematodes or flukes.

3. **(b)** Virus-infected cells secrete proteins called interferons which protect non-infected cells from further viral infection. Interferons are a group of related low molecular weight regulatory cytokines produced by certain eukaryotic cells in response to a viral infection. Besides defending against viruses, they also help to regulate the immune response.

4. **(a)** A vaccine is a preparation of one or more microbial antigens used to induce immunity. It may consists of killed living, weakened microorganisms, inactivated bacterial toxins (toxoids), etc. Immunisation is achieved by the successful delivery of vaccines.

5. **(a)** Success of organ transplant depends on proper matching of histocompatibility of antigens that occurs in all cells of the body. As there are antigens which are likely to be attacked by recipient's T-cells and antibodies, the recipient of organ transplant is always given immunosuppressants to prevent immune rejection of the transplanted tissue.

6. **(a)** Gymnosperm like *Pinus* usually grow on the slopes of hills and form dense forests of evergreen trees. The pollination of these plants takes place through wind and during peak hours, the microspore of *pinus* are liberated as clouds of yellow dust (sulphur shower). On inhalation, these microspores (pollens) cause various allergies like dermatitis, asthma, allergic rhinitis, hay fever, ete., in human beings.

7. **(a)** Tobacco contains an alkaloid, nicotine which can be used as a raw material for preparation of the antipellagra vitamin, nicotinic acid (niacin). Nicotin is also used in physiological and pharmacological investigation and as insecticide and a fumigant.

8. (b) Superovulation is induced by hormone injection. Superovulation and embryo transplantation techniques have been used in India for cattle improvement.

9. (b) Mass selection is useful in self pollinated plants since they are usually homozygous and retains the selected traits in the progeny. However, cross pollinated plants are usually heterozygous. Their phenotypic expression is largely due to their hybrid vigour or heterozygous nature. Mass selection of such plants fails in subsequent generations.

10. (b) Irrigation relates to the supply of water to the crops. Fertilisers which are mainly NPK are required to increase the harvest of crops.

11. (d) Polyploidy is the phenomenon of having more than two sets of chromosomes or genomes which occur in nature due to failure of chromosomes disjunction at the time of division. It can be of three types -autopolyploidy, allopolyploidy and autoallopolyploidy. In autopolyploidy there is a numerical increase of the same genome, Allopolyploidy develops through hybridisation between two species followed by doubling of chromosomes. Autoallopolyploidy is the type in which genome is in more than diploid state.

12. (c) Virus-free plants can be produced from the virus infected plants by means of meristem tissue culture. By removing the meristem and growing *in vitro*, virus free plants can be obtained. Some plants developed by meristem culture are banana, sugarcane, potato, etc.

13. (b) Protoplast culture is an important technique of tissue culture and this technique results in the production of somatic hybrid plants.

14. (b) An important technique of tissue culture, somatic hybridisation results in the production of somatic hybrid plants. Two different plant varieties each with a desirable character can be made to undergo protoplast fusion, which further can be grown into a new plant.

15. (a) Litmus is an important and widely used dye in chemical laboratories as an acid-base indicator. It is obtained from *Rocella montagnei*. Orcein, a biological stain, is obtained from *Rocella tinctoria*.

16. (b) Beer, wine, gin, rum, vodka, etc. are fermentation products but beer and wine are formed without distillation. However, distillation of fermented broth is carried out in case of rum, gin, etc. Distillation increases the alcohol concentration of the beverage. Thus, gin with 40% alcohol, rum with 40% alcohol and brandy with 60-70% alcohol are more alcoholic than beer (3-6% alcohol) and wine (9-12% alcohol).

17. (a) Sewage treatment is divided into primary treatment and secondary treatment. Primary treatment involves filtration of the sewage and then sedimentation of the debris present on filtrate. No microbes are used.
The sediment of primary treatment, called primary sludge, is treated by aerobic microbes and then by anaerobic microbes to reduce its organic components during secondary treatment. Secondary treatment involves microbes, thus it is also called biological treatment.

18. (b) Gases present in bio-gas produce energy by combustion or oxidation which can be used for heating purpose or for generation of power through gas engines. Biogas is used as fuel for heating and cooking, lighting, power for irrigation and other purposes as an alternative of firewood, kerosene, dung cakes or even electricity and pollution free source of energy.

19. (c) In the rotation of crops, leguminous crops like pulses beans, peas, groundnut and Bengal gram are sown in between the seasons of cereal crops like wheat, maize and pearl millet. The leguminous plants are grown alternately with non-leguminous plants to restore the fertility of the soil.
When the cereal crops like rice, wheat, maize are grown in the soil, it uses up lot of nitrogenous salts from the soil. If another crop of cereal is grown in the same soil, the soil becomes nitrogen deficient.
So by rotation a leguminous crop is grown. These plants have the ability to fix atmospheric nitrogen to form nitrogen compounds through the help of certain bacteria (*Rhizobium*) present in their root.
Genus *Clostridium* includes about 100 sps., most of them are either free-living or pathogenic.

20. **(a)** Biofertilisers are micro-organisms that are used for increasing fertility of soil. For example, free-living bacteria and cyanobacteria that fix nitrogen increase soil productivity. Biofertilisers are preferred over chemical fertilisers because chemical fertilisers are more expensive, disturb soil fauna and flora and affect soil pH. Although chemical fertilisers initially show better response than the biofertilisers however they decrease soil fertility in the long run.

Unit IX: BIOTECHNOLOGY

1. **(b)** The palindrome in DNA is a sequence of base pairs that reads same on the two strands when orientation of reading strand is kept same. Restriction enzymes cut the strand of DNA a little away from the centre of the palindrome sites, but between the same two bases on opposite strands.

2. **(c)** Restriction enzyme, a type of endonuclease, functions by "inspecting" the length of a DNA sequence. Once it finds a recognition sequence, it binds and cut each of the two strands of the double helix at specific point leaving single stranded portions at the ends. This results in overhanging stretches called sticky ends. These are named so because they form hydrogen bonds with their complementary counter parts i.e., they can join similar complementary ends of DNA fragment from some other source with the help of DNA ligase. This stickness of the ends facilitates the action of the enzyme DNA ligase, not DNA polymerase.

3. **(b)** Plasmids are the extrachromosomal part in the bacteria and are useful in recombinant DNA technology.

4. **(a)** pBR322 has recognition sites for several commonly used restriction enzymes. Recognition site for *Bam* H I is present in tet^r region *i.e.*, region responsible for tetracycline resistance. When an insert is added at the *Bam* H I recognition site the gene for tetracycline resistance becomes non-functional and the recombinant bacteria with plasmid pBR322 that has DNA insert at *Bam* H I loses tetracycline resistance.

5. **(d)** *Agrobacterium tumefaciens* is pathogen of dicot plants. Retroviruses are vectors for cloning genes in animals.

6. **(a)** Strains of the *Bacillus thuringiensis* produce different *Bt* toxins, each harmful to different insects. Most notably, *Bt* toxins are insecticidal to the larvae of moths and butterflies, but are harmless to other forms of life. The gene coding for *Bt* toxin has been inserted into cotton as a transgene, causing it to produce this natural insecticide in its tissues.

7. **(a)** DNA finger printing has become a powerful tool due to its ability in tracing evidences of crime and to establish the paternity. These evidences can be collected from hairs, semen and saliva.

8. **(c)** Interferons are proteins that are effective against most viruses. They are naturally produced by virus infected cells. The proteins interact with adjacent cells and make them resistant to virus attack. Now interferons are also being manufactured through genetic engineering.

Interferons control the multiplication of virus particles by inhibiting their protein synthesis.

9. **(b)** A soil bacterium *Bacillus thuringiensis (Bt)* produces a *Cry* protein (crystal protein). This protein is toxic to the larvae of certain insects. There are several kinds of *Cry* proteins, which are toxic to a different group of insects. The gene encoding *Cry* protein, that is *cry* gene has been isolated and transferred into several crops. A crop expressing a cry gene is usually resistant to the groups of insects.

10. **(d)** ELISA (enzyme linked immuno-sorbant assay) is a technique based on the principle of antigen-antibody interactions. On their surface pathogens have specific antigens like protein, glycoproteins, etc. Thus, by using the antibodies specific to bind with those antigens their infection can be detected. In process antigens are immobilised on ELISA plates. Then antibody specific to the antigen is added and allowed to react with the adsorbed antigen. A complex procedure is followed.

11. (a) A hybridoma cell, is produced by the injection of a specific antigen into a mouse, procuring the antigen-specific plasma cells (antibody-producing cell) from the mouse's spleen and the subsequent fusion of this cell with a cancerous immune cell called a myeloma cell.

12. (a) The phase of enormous increase in the agricultural production, when high yielding improved varieties of crop plants were used was known as green revolution. But the increased yield was partly due to use of improved crop varieties but mainly due to use of better management practices and agrochemicals (chemicals and fertilisers). Generally these practices and agrochemicals were very expensive in developing world that the farmers could not afford. Hence, the farmers of developing world relied mainly on cheaper methods of conventional breeding thus green revolution could not become as productive or effective as it was in developed world.

13. (b) *Agrobacterium* is natural genetic engineer of plants. It is a pathogen of several dicot plants can transfer a piece of DNA called (T-DNA) to transform normal plant cells into a tumor cells. Ti plasmid of *Agrobacterium tumefaciens* and Ri plasmid of *A. rhizogens* are best known plant cloning vectors.

Unit X: ECOLOGY

1. (a) Soils of Indo-Gangetic plains and the Indus valley are highly fertile and suitable for agriculture, so these areas have high population density. Favourable climate increases population density of a geographical area because it favours increase in population growth. High soil fertility favours population density.

2. (a) Leaf butterfly is green in colour and stick insect also mimics in order to escape from the enemies and also to catch prey. The camouflaging mechanism helps the organism to get adapted to its surroundings. The Leaf butterfly is seen more near the leaves while the stick insect camouflages with its surroundings by living on the branches.

3. (a) Disease resistant plants possess thick cuticle. Infectious organisms can not grow or invade cuticle.

4. (a) Plants which remain permanently immersed in water are called hydrophytes. They may be submerged or partly submerged and show the presence of aerenchyma (large air space) in the leaves and petioles. Aerenchyma helps to transport oxygen produced during photosynthesis and permits its free diffusion to other parts, including roots located in anaerobic soils. These tissues also impart buoyancy to the plants. Presence of inflated petioles in *Eichhornia* (water hyacinth) keeps the plants floating on the surface of water.

5. (c) In sigmoid growth curve, finally, growth rate becomes stable because mortality and natality rates become equal to each other and finally the population shows zero growth rate as birth rate equals death rate.

6. (c) In Mutualism, two organisms of different species exist in a relationship in which each individual benefits from the activity of the other.

7. (b) Net primary productivity is the rate of organic matter built up or stored by producers in their bodies per unit time and area. Net productivity is equal to gross primary productivity minus loss due to respiration and other reasons. Rate of increase in energy containing organic matter or biomass by heterotrophs or consumers per unit time and area is known as secondary productivity.

8. (c) Food web is a network of food chains which become interconnected at various trophic levels so as to form a number of feeding connections amongst the different organisms of a biotic community. Food web is meant for increasing the stability of an ecosystem by providing alternate source of food and allowing the endangered population to grow in size.

9. (a) Ten percent law put forth by Lindemann states that while transferring organic food from one trophic level to the next, about ten percent of the organic matter is stored as flesh, the remaining is lost during transfer or broken down in respiration. The net productivity of the next higher trophic level shall be 10% of the first one.

10. (d) Energy flow in the ecosystem is in a unidirectional manner. There is a decline in the amount of energy passing from one trophic level to the next. Thus the pyramid of energy is always upright. According to Lindemann, only 10% of energy goes to the next trophic level.

11. (a) This group of organisms are consumers of first order and include the herbivores which directly feed upon green plants (producers).

12. (a) Phytoplanktons account for almost half of the photosynthetic activity on earth. Thus they are responsible for maximum (about 90%) of the oxygen present on the earth while the rest is because of the other plants.

13. (a) Insectivorous plants are those plants which capture and digest live prey (normally insects) to obtain nitrogen compounds that are lacking in its usual marshy habitat. These plants are partly autotrophic and partly heterotrophic.

14. (d) O-horizon occupies the topmost soil and is rich in mineral and decomposed organic matter (humus). A-horizon is dark coloured and has abundant minerals mixed with humus.

15. (c) Tropical rain forests are located in the equatorial regions where the annual rainfall exceeds 140 cm. They are also called jungles and cover one twelveth of earth's surface but contain more than half of the earth's flora and fauna (i.e., rich in biodiversity). Now-a-days these forests are disappearing due to excessive cutting of forests for domestic purposes like fuel, furnitures, accomodations, cloths, resin, gum, etc.

16. (a) A sanctuary is an area which is reserved for the protection of wild animals only. The activities like harvesting of timber, collection of minor forest products and private ownership rights are allowed, however, such activities should not have any adverse effect on animals.

17. (c) Red data book contains a record of species of an area which are known to be in danger. Endangered species need to be noticed because only after that the planning for their protection and improving their population can be given the proper direction.

18. (d) Rare species are those which have small population in the world. They are usually confined to limited areas or are thinly scattered over a more wide area. Great Indian Bustard are the species which are in danger of extinction, hence are endangered species. Black buck population is likely to be in danger of extinction, hence vulnerable.

19. (d) Pollution may be defined as an undesirable change in physical, chemical or biological characteristics of air, water and land causing harmful effects on living organisms. Pollution can be natural or man made. Natural pollution includes volcanic eruptions, soil erosion, UV- rays, *etc.* Pollution is different from contamination. Contamination is the presence of harmful organisms causing disease.

20. (a) Water pollutants are commonly measured by their main common denominator, called BOD (Biochemical Oxygen Demand), *i.e.*, the amount of free oxygen absorbed by extraneous substances from water. If water is polluted, it will consume more oxygen, thereby enhancing the BOD of water.

21. (a) DDT (dichloro diphenyl trichloroethane) is a organochlorine contact insecticide that kills by acting as a nerve poison. DDT was originally used during world war II to control typhus which was spread by body louse. Since then it has been used to control mosquito borne malaria and was used extensively as a general agricultural insecticide.

22. (b) Deforestation results in increase in green house gases which retains more and more UV radiations and leads to global warming. Global warming is the warming/heating up of the earth's atmosphere due to depletion of ozone in the stratosphere.

23. (b) Stratosphere zone of Earth's atmosphere contains a layer of ozone which protects us from harmful ultraviolet radiations of the sun. A group of chlorine containing compounds called chlorofluoro carbon (CFCs) used as coolants in air conditioners and refrigerators are primary chemicals responsible for ozone depletion. After their release into troposphere, CFCs go to stratosphere where they are broken down by UV radiations releasing chlorine.In presence of sunlight, chlorine breaks O_3 into O_2. Due to ozone depletion, its levels decreases by 67% every year causing higher levels of UV radiations reaching earth which may cause eye cataracts, skin cancer, etc.

24. (c) Deforestation is the conversion of forested areas to non-forested ones.One of the major effects of deforestation is enhanced carbon dioxide concentration in the atmosphere because trees that could hold a lot of carbon in their biomass are lost with deforestation. Deforestation also causes loss of biodiversity due to habitat destruction, disturbs hydrological cycle, causes soil erosion and may lead to desertification in extreme cases.

25. (a) Emphysema is a lung condition that causes shortness of breath. In people with emphysema, the air sacs in the lungs (alveoli) are damaged. Smoking is the most common cause, but emphysema can also be genetic. Air pollution in Kolkata and Kanpur are too much, all because of industrialisation.